T0362099

COLOR FOR SCIENCE, ART AND TECHNOLOGY

AZimuth

Volume 1   Color for Science, Art and Technology
            edited by **Kurt Nassau**

COLOR FOR SCIENCE, ART AND TECHNOLOGY

EDITED BY

KURT NASSAU

N·H

1998

ELSEVIER

Amsterdam - Lausanne - New York - Oxford - Shannon - Singapore - Tokyo

ELSEVIER SCIENCE B.V.
Sara Burgerhartstraat 25
P.O. Box 211, 1000 AE Amsterdam, The Netherlands

ISBN: 0 444 89846 8

This book is printed on acid-free paper.

Transferred to digital printing 2006

Printed and bound by Antony Rowe Ltd, Eastbourne

"Every gaudy color
Is a bit of Truth."

Natalia Crane

"It has been said that science demystifies the world. It is
closer to the truth to say that science, when at its best,
opens the world for us, bringing daily realities under a
kind of magic spell and providing the means to see the
limits of what we think we know, and the scope of what
we do not at all understand."

C. Emmeche, The Garden in the Machine

Dedication

This volume is dedicated to the proposition that there is much common ground between science, technology, and the arts, and that color is a major connecting bridge.

And to my wife Julia: ever helpful, ever patient.

CONTENTS

Cover Color Plate (top left-hand corner): Painting by Sanford Wurmfeld, *II-9* (*full saturation*), 1987, acrylic on canvas, 90″ × 90″

Preface ix

List of Contributors xiii

Biographical Notes xv

Section I. THE SCIENCE OF COLOR

1. Fundamentals of Color Science
   K. NASSAU 1
2. The Measurement of Color
   R.T. MARCUS 31
3. Color Vision
   J. KRAUSKOPF 97
4. The Fifteen Causes of Color
   K. NASSAU 123

Section II. COLOR IN ART, CULTURE AND LIFE

5. Color in Abstract Painting
   S. WURMFELD 169
6. Color in Anthropology and Folklore
   J.B. HUTCHINGS 195
7. The Philosophy of Color
   C.L. HARDIN 209
8. Color in Plants, Animals and Man
   J.B. HUTCHINGS 221

9. The Biological and Therapeutic Effects of Light
   G.C. BRAINARD 247
   Addendum:
   Double Blind Testing for Biological and Therapeutic Effects of Color
   K. NASSAU 270

Section III. COLORANTS, THE PRESERVATION AND THE REPRODUCTION
           OF COLOR

10. Colorants: Organic and Inorganic Pigments
    P.A. LEWIS 283
11. Colorants: Dyes
    J.R. ASPLAND 313
12. Color Preservation
    K. NASSAU 345
13. Color Imaging: Printing and Photography
    G.G. FIELD 353
14. Color Encoding in the *Photo CD* System
    E.J. GIORGIANNI and T.E. MADDEN 389
15. Color Displays
    H. LANG 423

Color Section 457

Index 473

PREFACE

The aim of this book is to assemble a series of chapters, written by experts in their fields, covering the basics of color – and then some more. This should supply what almost any reader might want to know about color in areas outside their own expertise. Thus the color measurement expert, as well as the general reader, can find here information on the perception, causes, and uses of color. For the artist there are details on the causes, measurement, perception, and reproduction of color. And there are few indeed who would not want to know more about color in anthropology, art, medicine, nature, and about the philosophical aspects of color.

It would be easy to decide on the topics to be included in a multivolume work covering all aspects of color. However there might be a problem in how to limit the number of volumes. To make the same decision for a one-volume work of reasonable size presents a real challenge. I was greatly helped in the early stages of the selection process by Dr. Eric Melse of Voorburg, the Netherlands; he subsequently resigned as co-editor because of the pressure of other obligations.

Each chapter easily could have been many times its final length; several of the contributors have indeed published book-length treatments on their specialties. Yet the publisher's limitation on length had to be taken seriously. The fact that most of the contributors were ultimately annoyed with me for length restrictions suggests that I was uniformly unfair! Contributors encompass both academia and industry (both in several instances) and were selected to be international in range. Some are specialists in advanced research in their field, others are educators or generalists with a broad overview. Diversity was the aim.

The attempt has been made to cover not just the fundamentals, but also to include work on the frontiers. Two examples are a new approach to testing for the biological and therapeutic effects of color in the addendum to Chapter 9, and the encoding of color in a photo compact disk system of Chapter 14 with its surprising complexity.

Within each chapter, authors were requested to indicate directions of future efforts, where applicable. One might reasonably expect that all would have been learned about color in the more than three hundred years since Newton established the fundamentals of color science. The situation is far from it: the measurement of color still has unresolved complexities (Chapter 2); many of the fine details of color vision remain unknown (Chapter 3); every few decades a new movement in art discovers original ways to use color (Chapter 5); the philosophical approach to color has not yet crystallized (Chapter 7); new pigments and dyes continue to be discovered (Chapters 10 and 11); the study of the biological and therapeutic effects of color is still in its infancy; and so on. Color remains vigorously developing toward maturity.

A most difficult decision involved the application of color in fields extending from the pure arts, such as painting, to the applied, decorative, and commercial areas such as fashion, interior decorating, packaging, and advertising. Since the concepts in the latter groups change periodically as styles change, often on a monthly basis, I decided not to include these topics. The use of color in painting, however, changes only slowly with time. There have been many discussions of the use of color in the paintings of earlier periods, but the twentieth century is mostly neglected; it is therefore a pleasure to have abstract painting covered.

The aim has been to avoid excess technicalities, yet some topics would be meaningless without them. Some chapters, such as those on color measurement, color perception, and color reproduction, require some mathematical details in view of the astonishing complexities involved. Others, such as those on the causes of color, colorants, and color preservation, need to be grounded in the chemistry and physics involved. To produce absolute uniformity and consistency in any such multi-author work might be desirable, but would have required unreasonable effort and ultimately would be of little value. The reader will find that some chapters have many references, while others have only a few or even none. Detailed referencing was felt to be unnecessary where a few extended treatments cited in the "Further Reading" section are available and where these treatments provide both great depth and more than adequate referencing.

The attempt has always been made to include the necessary fundamentals for those whose background lies in other fields. Some overlap was accepted so that chapters should be able to stand on their own to a significant extent, always excepting the basic concepts covered in Chapter 1, which are prerequisite for almost all the other chapters. Some parts of Chapter 1 may seem to be trivial or self-evident, but it is always essential to build from a solid foundation.

There exist many erroneous ideas on color. As one example, we are usually taught in school that there is just one set of three specific primary colors. Since erroneous elementary ideas could lead to the misinterpretation of advanced concepts, such essential basics are covered in Chapter 1. This chapter is part of the first section the science of color, which also contains chapters on the measurement of color, color perception, and on the fifteen physical and chemical causes of color.

The next section deals with color in art, culture, and life, covering the uses of color in abstract painting of the twentieth century; the views on color from anthropology and folklore and from philosophy*; color in plants, animals, and man; as well as the biological and therapeutic effects of light and color. This last is a subject of much controversy but of almost no well-controlled experimental studies; appended is the outline for a new approach to the necessary double-blind investigations.

The last section covers technological aspects: colorants, pigments, and their preservation; color printing and photography; the *Photo CD* system; and color displays as used in television and computer displays.

---

* It might perhaps be noted that some philosophers even maintain that "secondary qualities" such as colors do not exist, or at least that "... nothing exemplifies or has any color" and "... there are no colored things and, therefore, nobody knows that there are." (C. Landesman, Color and Consciousness; An Essay in Metaphysics, Temple University Press, Philadelphia, 1989, pp. 105 and 121)

*   *   *

I have also included in Chapter 1 a very brief discussion of three universal paradigms involved in the basics of modern science. Although not strictly necessary for understanding the nature of color, they are an essential part of the nature of light. These important paradigms are fundamental models of science which are at odds with our everyday experiences; each forced a radical change in our understanding of the working of the universe. Two of these three paradigms have been generally accepted and continued exposure has dulled us to their weirdness.

The first great paradigm, the constancy of the speed of light as well as the inability of matter to travel as fast as or faster than light, derives from Einstein's relativity theory. Here, there are the equivalences of space and time, of mass and energy, as well as of acceleration and gravity.

The second great paradigm, quantum theory, explains the equivalence of the particle and wave characteristics of light (as well as of small quantities of matter such as electrons and atoms). Quantum theory had its roots in the early disagreements between the views of Newton and Hooke on the nature of light some 300 years ago. It was not resolved to general satisfaction until about the middle of this century.

The third great paradigm, involving the non-locality part of quantum theory, was first taken seriously by Einstein, who never felt comfortable with quantum theory because of it. This problem was outlined in the Einstein, Podolski, and Rosen paper of 1934 and clarified as Bell's Paradox in 1964. Only in the last few years has a series of increasingly more elegant experiments demonstrated that quantum theory is correct and that it is very, very weird indeed, as is briefly indicated in Chapter 1. There is as yet no generally accepted interpretation of the non-locality problem and most physicists have preferred to ignore its complexities and implications.

I bring up these paradigms because I believe that there is need for one more paradigm, a global one that would resolve the dichotomy between the seemingly irreconcilable approaches and attitudes in the "Sciences" and in the "Arts"; the emphasis is meant to indicate what I consider to be the arbitrariness of such designations. The origins of this dichotomy can be traced back to the differing approaches to color of Newton and Goethe outlined in Chapter 1. It is only now being seriously faced by authors such as Zajonc, as also briefly discussed there. The nature of the paradigm required to resolve this issue has yet to be defined. This volume is dedicated to the proposition that there is much common ground between the sciences and the arts and that color is a major connecting bridge.

*   *   *

Many friends have looked over various parts of this book at various stages and I am appreciative of their invaluable advice. I apologize for the absence of specific names. There were so many and I did not keep track in the early stages, so I decided not to give a partial and seemingly biased listing.

Having written four books by myself, I had always assumed that the editor of a multi-author volume has a much easier time than does an individual book author. As I approach the end of this task, I realize that I was very much in error. I do wish to thank all the contributors for their efforts and their patience.

K. Nassau
Lebanon, New Jersey
October, 1996

LIST OF CONTRIBUTORS

J.R. Aspland, University of Clemson, Clemson, SC, USA

G.C. Brainard, Jefferson Medical College, Philadelphia, PA, USA

G.G. Field, California Polytechnic State University, San Luis Obispo, CA, USA

E.J. Giorgianni, Eastman Kodak Co., Rochester, NY, USA

C.L. Hardin, Syracuse, NY, USA (Syracuse University, retired)

J.B. Hutchings, Bedford, England (Unilever PLC, retired)

J. Krauskopf, New York University, New York, NY, USA

H. Lang, Broadcast Television Systems GmbH, Darmstadt, Germany

P.A. Lewis, Sun Chemical Corp., Cincinnati, OH, USA

T.E. Madden, Eastman Kodak Co., Rochester, NY, USA

R.T. Marcus, D&S Plastics International, Mansfield, TX, USA

K. Nassau, Lebanon, NJ, USA (AT&T Bell Telephone Laboratories, Murray Hill, NJ, USA, retired)

S. Wurmfeld, Hunter College of the City University of New York, NY, USA

BIOGRAPHICAL NOTES

Kurt **Nassau** is the editor and also wrote the Preface; Chapter 1: Fundamentals of Color Science; Chapter 4: The Fifteen Causes of Color; Chapter 9: Addendum: Double Blind Testing for Biological and Therapeutic Effects of Color; and Chapter 12: Color Preservation. Dr. Nassau is active as author, lecturer, and consultant. He holds degrees from the Universities of Bristol and Pittsburgh. He recently retired as Distinguished Research Scientist after 30 years at AT&T Bell Telephone Laboratories, where he worked in a wide range of areas covering the preparation, chemistry, and physics of laser crystals, semiconductors, superconductors, non-linear optical crystals, and glasses, among others. He has also performed medical research at the Walter Reed Army Medical Center, Washington, DC while in the US Army and taught graduate students as Visiting Professor at Princeton University. In addition to "The Physics and Chemistry of Color: The Fifteen Causes of Color," he has also written "Experimenting with Color" for teen-agers and two books on the history, science, and the state of the art of gemstone synthesis and of gemstone enhancement, as well as the "Colour" article used in the Encyclopaedia Britannica since 1988.

J. Richard **Aspland** (Chapter 11: Colorants: Dyes) is Professor of Textile Chemistry at Clemson University, Clemson, SC, USA. Dr. Aspland holds degrees from Leeds University and the University of Manchester. He has worked on dye applications research in the dyestuff manufacturing and textile processing industries and is writing a book on the coloration of textiles.

George C. **Brainard** (Chapter 9: The Biological and Therapeutic Effects of Light and Color) is Professor of Neurology and Associate Professor of Pharmacology at Jefferson Medical College, Philadelphia, PA, USA. Dr. Brainard holds degrees from Goddard College and the University of Texas. He has performed research on the effects of light on humans and animals.

Gary G. **Field** (Chapter 13: Color Printing and Photography) is Imaging Scientist and Professor of Graphic Communication at California Polytechnic State University, San Luis Obispo, CA, USA. He holds a degree from the University of Pittsburgh. He has also worked in the area of color printing in Australia and England. He has written two books: "Color and its Reproduction" (1988) and "Color Scanning and Imaging Systems" (1990). He teaches color technology, quality control, and printing management.

Edward J. **Giorgianni** (Chapter 14: Color Encoding in the *Photo CD* System) is a Research Scientist at Eastman Kodak Co., Rochester, NY, USA. He holds a degree from

Rhode Island University. He has worked on photographic imaging products and electronic and hybrid imaging systems.

C.L. **Hardin** (Chapter 7: The Philosophy of Color) has recently retired as Professor of Philosophy at Syracuse University, Syracuse, NY, USA and is now Professor Emeritus. Dr. Hardin holds degrees from Johns Hopkins University, the University of Illinois, and Princeton University. He has written the award-winning book "Color for Philosophers" (1988) and co-edited "Color Categories in Thought and Language" (1996). His interests lie in color categories in thought and language and the relationship between color perception and the mind–body problem.

John B. **Hutchings** of Colmworth, Bedford, England (Chapter 6: Color in Anthropology and Folklore; Chapter 8: Color in Plants, Animals, and Man) is active as lecturer, author, and consultant. He is retired from Unilever PLC, where he led a research team on the science and psychophysics of color and appearance of foods, cosmetics, and other substances. His book "Food Color and Appearance" was published in 1994 and he has been principal editor on two books on colour in folklore, for the second of which he was coordinator of the international survey. He is a Fellow of the Institute of Physics (London).

John **Krauskopf** (Chapter 3: Color Vision) is Professor in the Center for Neural Science at New York University and also holds an appointment in the Center for Visual Science at the University of Rochester. He has taught at Brown University, Rutgers University, and Bryn Mawr College and also was a research scientist at AT&T Bell Telephone Laboratories. He has done research on many aspects of vision with particular concentration on the psychophysics and electrophysiology of color vision.

Heinwig **Lang** (Chapter 15: Color Displays) is responsible for Optics and Colorimetry at Broadcast Television Systems Gmbh, Division of Philips, Griesheim, Germany (formerly Fernsehen GMBH, Division of Bosch). Dr. Lang holds degrees from the University of Tuebingen and the Technical University Berlin. He also lectures at the Technical University of Darmstadt. He has worked with colorimetry and color standards, particularly as applied to high definition television transmission. He is also interested in the physiology of color vision.

Peter Anthony **Lewis** (Chapter 10: Colorants: Organic and Inorganic Pigments) is Coating Industry Manager at Sun Chemical Corp., Cincinnati, OH, USA. Dr. Lewis holds degrees from the Royal Institute of Chemistry and Bristol University. He has also worked for Imperial Chemical Industries. His field is the chemistry, production, evaluation, and marketing of pigments.

Thomas E. **Madden** (Chapter 14: Color Encoding in the *Photo CD* System) is a Research Scientist at Eastman Kodak Co., Rochester, NY, USA. He holds degrees from Valparaiso University and the Rochester Institute of Technology. He has worked on color photography systems modeling and color digital imaging.

Robert T. **Marcus** (Chapter 2: The Measurement of Color) is Color Development Manager at D&S Plastics International, Mansfield, TX, USA. Dr. Marcus holds degrees from the Rensselaer Polytechnic Institute. He has also worked at Pantone, Inc., at the Munsell Color Laboratory of the Macbeth Division of Kollmorgen Instruments Corp., and at PPG Industries. His interests cover color matching systems, color measurements, and color tolerancing.

Sanford **Wurmfeld** (Chapter 5: Color in Abstract Painting) is an artist and Chairman of the Department of Art at Hunter College, University of New York, New York, NY, USA. He holds a degree from Dartmouth College. He also pursues his special interests in the history of color theory in relation to painting. He creates large scale abstract color paintings in two and three dimensions. He has been exhibiting internationally since the late 1960s and his art is included in major museums such as the Metropolitan and Guggenheim Museums of New York City.

chapter 1

FUNDAMENTALS OF COLOR SCIENCE

KURT NASSAU

Lebanon, NJ, USA
(AT&T Bell Telephone Laboratories
Murray Hill, NJ, USA, retired)

Color for Science, Art and Technology
K. Nassau (Editor)

CONTENTS

1.1. Introduction 3

1.2. Defining color 3

1.3. Early views on color 4

1.4. Newton, the spectrum, and "colored" light 4

1.5. Light and color 6

    1.5.1. Color descriptions 8

    1.5.2. Color appearance 9

    1.5.3. Interactions of light with matter 10

1.6. Color mixing 11

    1.6.1. Additive color mixing 12

    1.6.2. Subtractive color mixing 15

1.7. Color vision and color theories 17

    1.7.1. Opponent theory 18

    1.7.2. Trichromacy theory 18

    1.7.3. Modern theories 19

    1.7.4. Goethe's theory of color 19

    1.7.5. Color circles (wheels) 20

    1.7.6. Light sources and metamerism 22

1.8. What color do *You* see? 24

1.9. Color and "Science versus Art" 25

1.10. The nature of light 28

1.11. Further reading 29

References 30

1.1. Introduction

In this chapter we consider some basic concepts which are the essential underpinnings of all that follows. Some of these ideas are taught even as early as kindergarten, yet complexity is usually avoided at that level. Many may therefore continue to believe that there is just a single set of three primary colors, as one example. Again, it is often assumed that there is a unique color perception for each wavelength of the spectrum and that a given wavelength is perceived by everyone as the same color, whatever the circumstances, and that there is one unique, absolute white. So the aim here is to outline some color fundamentals and correct those misconceptions which may present difficulties in grasping more subtle advanced concepts later.

The reader should not be discouraged if some of this matter appears to be too abstruse or too naive. Merely skimming across such material should serve: this will acquaint the reader with its existence and will provide the setting for some of the items that follow. It is most desirable for such a reader to continue to the end, since the level in this, and indeed in all of the chapters of this book, varies considerably from section to section.

1.2. Defining color

Since color is a sensation unrelated to anything else, it is essentially impossible to give a meaningful definition except indirectly and circularly; for example: "Color is that aspect of perception which distinguishes red from green, etc." A useful functional definition might be: "Color is that part of perception that is carried to us from our surroundings by differences in the wavelengths of light, is perceived by the eye, and is interpreted by the brain". Again, we could say: "Our brain perceives color when a non-white distribution of light is received by the eye". Yet it is easy to find flaws in and exceptions to any such definition. The painful tribulations of the Committee on Colorimetry of the Optical Society of America (Committee on Colorimetry 1953) to reach consensus on an adequate definition of color makes interesting but ultimately frustrating reading.

The term 'color' describes at least three subtly different aspects of reality. First, it denotes a property of an object, as in "*green* grass". Second, it refers to a characteristic of light rays, as in "grass efficiently reflects *green* light (but see Section 1.4 below) while absorbing light of other colors more or less completely". And, third, it specifies a class of sensations, as in "the brain's interpretation of the eye's detection of sunlight selectively reflected from grass results in the perception of *green*". By careful wording one could always indicate which of these three (and other) types of meaning is intended in any given usage. In actual practice the distinction among such usages of color is

3

not usually made, nor is any effort made here to do so. The mere awareness that such different aspects exist enables one to identify the intended meaning and save the use of many additional words. At the same time, it will be obvious to the discerning that some philosophical discussions on color are meaningless just because of confusion among such different aspects.

Sometimes such differences are, indeed, critical: 'black' (in the strictest sense) as used for the 'color' of a paint or the surface of an object has an exact meaning, namely zero transparency and zero reflectivity for all visually-perceived light. As the characteristic of a light ray, 'black' has no meaning at all. And in perception, the ideal 'black' is merely the total absence of visual sensation from a given region. These distinctions are also important to those who are involved with precise color communication (see Chapters 2 and 13 to 15), particularly when such communication places them in contact with others in different professions: these are circumstances where misunderstanding and confusion can so easily result.

1.3. Early views on color

The Greek philosopher, Plato, about 428–348 BC, held a pessimistic view on the possibility of a science of color:

> "There will be no difficulty in seeing how and by what mixtures the colors are made ... For God only has the knowledge and also the power which are able to combine many things into one and again resolve the one into many. But no man either is or ever will be able to accomplish either the one or the other operation" (MacAdam 1970, p. 1).

The view of the Greek philosopher Aristotle, 348–322 BC, dominated in the pre-experimental stage of science. Aristotle (or at least the writing attributed to him) noted that sunlight always becomes darkened or less intense in its interactions with objects and therefore viewed color as a mixture of white and black. To some of the Greeks it was the eye that sent out rays, acting like feelers, which detected the color of objects. Others believed that luminous objects emitted particles which were detected by the eye, while yet others believed that these objects emitted waves.

1.4. Newton, the spectrum, and "colored" light

The beginning of the science of color was described by Sir Isaac Newton, English mathematician and astronomer (1642–1727) in a report in the Philosophical Transactions for 1671:

> "... in the beginning of the Year 1666 ... I procured me a Triangular glass-Prisme ... having darkened my chamber, and made a small hole in my window-shuts, to let in a convenient quantity of the Suns light, I placed my Prisme at its entrance, that it might be thereby refracted to the opposite wall. It was at first a very pleasing divertisement to view the vivid and intense colours produced thereby; but after a while applying myself to consider them more circumspectly, I became surprised to see them in an *oblong* form; which, according to the laws of Refraction ... I expected should have been *circular* ... Comparing the length of this coloured *Spectrum* with its breadth, I found it about five times greater; a disproportion so extravagant, that it excited me to a more than ordinary curiosity of examining from whence it might proceed" (Newton 1671).

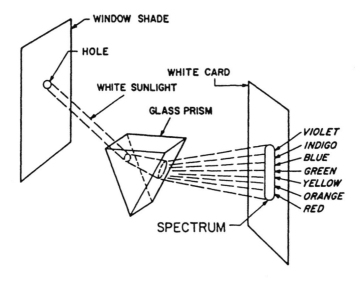

Fig. 1.1. The production of the spectrum by Newton in 1666.

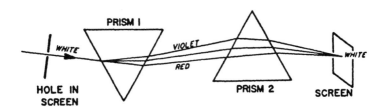

Fig. 1.2. Newton's recombination of the spectrum into white light by a second prism.

This production of a 'spectrum', a word coined by Newton, is illustrated in fig. 1.1. When Newton recombined these colors with a lens or with a second prism, as in fig. 1.2, he once again obtained white light, thus demonstrating irrefutably that ordinary white "Light is a heterogeneous mixture of differently refrangible Rays" (Newton 1671), that is a mixture of colors.

Newton chose to designate the spectrum as containing seven major colors: red, orange, yellow, green, blue, indigo, and violet (a modern mnemonic is "Roy G. Biv"), possibly by analogy with the seven notes, A through G, of the musical scale. Newton himself was careful to explain that there are many more colors in the spectrum in addition to these seven. All colors that we can possibly perceive are the colors in this pure spectrum and various combinations of these colors with each other and with white (namely by diluting a colored paint with white paint or a colored light beam with white light) and 'black' (namely by darkening a colored paint by adding black paint or by reducing the intensity of a colored light beam; but see below; this is frequently and incorrectly called "mixing with black", and we may also do so in what follows). These combinations of

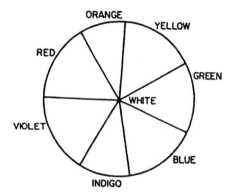

Fig. 1.3. Newton's color circle, simplified.

colors also include those colors that are not present in the Newton's spectrum, such as purple, magenta, pink, olive, and brown.

Newton also organized his spectral colors in the form of a color circle or wheel (Newton 1730) with white at the center of the circle, as shown in fig. 1.3, an approach frequently used by artists and in some versions of color theory, as discussed in Section 1.7.5 below.

It can be said realistically that color is in the eye of the beholder. It is based on the spectrum as perceived by the eye and interpreted by the brain. Newton knew that there is no 'colored' light: there is only the sequence of colors of the spectrum, as well as combinations of these. In Newton's own words:

> "And if at any time I speak of Light and Rays as coloured or endued with Colours, I would be understood to speak not philosophically and properly, but grossly, and accordingly to such Conceptions as vulgar People in seeing all these Experiments would be apt to frame. For the Rays to speak properly are not coloured. In them is nothing else than a certain Power and Disposition to stir up a Sensation of this or that Colour" (Newton 1730).

The eye in perceiving the range of colors functions quite differently from the ear in perceiving the range of tones. The ear is able to separate the tones of many instruments sounding together, recognizing the presence of a piccolo in a full orchestra. The eye always perceives only a single color at any point, whether this be a spectrally pure yellow, an identically visually-perceived equivalent mixture of green and red, and so on, as discussed in Section 1.6 below. Terms such as 'yellow light' are nevertheless widely used (for example in what follows) and need not produce any confusion if this range of possibilities is kept in mind. The eye of course has a much superior resolution of detail and of motion than does the ear.

To identify the actual spectral composition of light, it can be visually examined by using a spectroscope (using a prism or a diffraction grating, as in Sections 4.7.1 and 4.7.4 of Chapter 4, respectively), or measured in a spectrometer or spectrophotometer.

1.5. Light and color

Light in the strictest sense is that part of the electromagnetic spectrum that is perceived by the eye, with wavelengths between about 400 nm and about 700 nm as shown in

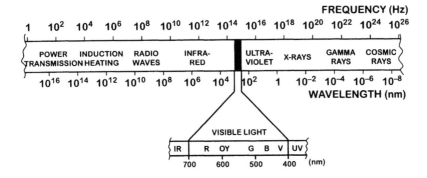

Fig. 1.4. The electromagnetic spectrum; Hz is one cycle per second.

TABLE 1.1
Various ways of specifying the spectrum.

Color	Wavelength* λ (nm)	Frequency ν (Hz)	Wavenumber $\bar{\nu}$ (cm^{-1})	Energy (eV)	Energy (kcal/mol)
Red (limit)**	700	4.28×10^{14}	14 286	1.77	40.8
Red	650	4.61×10^{14}	15 385	1.91	44.0
Orange	600	5.00×10^{14}	16 667	2.07	47.6
Yellow	580	5.17×10^{14}	17 241	2.14	49.3
Green, yellowish	550	5.45×10^{14}	18 182	2.25	51.9
Cyan (bluish green)	500	6.00×10^{14}	20 000	2.48	57.2
Blue	450	6.66×10^{14}	22 222	2.76	63.5
Violet (limit)**	400	7.50×10^{14}	25 000	3.10	71.4

* Typical values.
** Limits depend on the observer, eye adaptation, etc., ranging to 380 nm and 780 nm at high light intensities.

TABLE 1.2
Conversions for spectral light units.

(Energy in eV) × (Wavelength in nm)	= 1239.9
Wavenumber $\bar{\nu}$ in cm^{-1}	= $1 \times 10^7 / \lambda$(in nm)
Frequency ν in Hertz	= $\bar{\nu} \times 2.998 \times 10^{10}$
Energy in eV (electron Volts)	= $\bar{\nu} \times 1.240 \times 10^{-4}$
ergs	= $\bar{\nu} \times 1.986 \times 10^{-16}$
Joules	= $\bar{\nu} \times 1.986 \times 10^{-23}$
calories	= $\bar{\nu} \times 4.748 \times 10^{-24}$
cals/mol	= $\bar{\nu} \times 2.857$

fig. 1.4; vision extends to 380 nm and 780 nm under intense illumination. (Parts of the adjacent infrared and ultraviolet regions are sometimes also referred to inaccurately as light.) Various measurement units are used, based either on wavelength units, such as the nanometer (the accent is on the first syllable), frequency in Hertz or on energy units, such as the electron Volt. A summary of some of these is given in table 1.1 and some conversions are given in table 1.2.

The smallest possible quantity of spectrally pure, single color light is called a *photon*. The term *quantum* is a more general one. It includes the photon, but also applies to other regions of the electromagnetic spectrum, including infrared and ultraviolet (although 'photon' is also frequently used incorrectly for these regions adjacent to the visible), to radio waves, X-rays, and the other regions of fig. 1.4.

A photon or quantum may appear to act in two apparently very different ways. First a quantum can act as if it were a particle, called a corpuscle by Newton. This particle has no rest mass and no size, always moves at the velocity of light in vacuum, and carries energy as given in table 1.1. Or a quantum can act as if it were a wave spread out in space, with a wavelength as given in table 1.1, again carrying energy and moving at the velocity of light. Both approaches are merely different aspects of a single reality; details are found in physics and optics textbooks. The approach used at any given moment depends on the phenomenon being discussed. Both approaches are needed for the various causes of color covered in Chapter 4. Some additional aspects of these and other peculiar properties of quanta are discussed in Section 1.10.

1.5.1. Color descriptions

There are three attributes usually used to describe a specific color. The first of these attributes specifies one of the colors of the spectral sequence or one of the non-spectral colors such as purple, magenta, or pink. This attribute is variously designated in different descriptive systems as *hue, dominant wavelength, chromatic color*, or simply but quite imprecisely as *color*. It should be noted that these terms do not have precisely the same meaning and therefore are not strictly interchangeable. (The designation 'chromatic color' is redundant in the strict sense, since 'chromatic' already means 'colored', but it is validly used as a distinction from *achromatic colors*, itself an oxymoron in the strict sense, which refers to the sequence white, gray, black.)

A second attribute of color is variously given as *saturation, chroma, tone, intensity*, or *purity* (again, these terms are not strictly interchangeable). This attribute gives a measure of the absence of white, gray, or black which may also be present. Thus the addition of white, gray, or black paint to a saturated red paint gives an unsaturated red or pink, which transforms ultimately into pure white, gray, or black as the pure additive is reached; with a beam of saturated colored light, white light may also be added but the equivalent of adding black is merely a reduction of the intensity.

For a color having a given hue and saturation, there can be different levels variously designated as *brightness, value, lightness*, or *luminance* (once again these terms are not strictly interchangeable), completing the three dimensions normally required to describe a specific color.

One system for describing colors using these three dimensions is the Munsell System, illustrated in color plates 1 and 2. Details of this and other such systems are given in Chapter 2, Section 2.3.3 and following sections.

Color, in the colloquial sense, may at times imply merely the single dimension of the various hues of the spectrum (red to violet) joined by a few non-spectral hues such as purple, etc., as mentioned above. Color in the technical sense requires a multi-dimensional space defined by the above three parameters as the minimum dimensions, as discussed in Section 1.8 below and in Chapter 3. Additional variations in color term usage must be recognized. For example, when the brightness is reduced, an orange may become not a weak orange but a brown; a greenish yellow similarly gives olive; and a red or a reddish magenta when diluted with white gives pink. Of course such variations will differ considerably in different languages. An individual with normal color vision can distinguish about seven million different colors.

1.5.2. Color appearance

The appearance of color as detected by the eye and interpreted by the brain depends significantly on the exact viewing circumstances. In the everyday world one thinks of viewing a colored object in a normal environment by reflection from some type of illumination, i.e., in what is called the *object mode*. Viewing a light source directly represents the *illuminant mode*. Finally, there is the *aperture mode*, relevant when viewing a colored area through a hole in a screen. One's perception of a specific color differs significantly in these modes.

Consider the object mode. Here the eye/brain has the ability to compensate for a wide range of illuminants (direct 'white' sunlight, a blue cloudless northern sky, the yellow light from an incandescent lamp, or the reddish light from a candle) and infer something very close to the 'true' color. Various surface effects are significant as shown in table 1.3, adding additional color-description dimensions to the three basic ones. Even a non-metallic object with a deeply colored but glossy surface will reflect almost pure illuminant as glare at a glancing angle.

Consider just one characteristic of the aperture mode, seen when an orange sample is viewed through a hole in a gray screen. If the sample is illuminated as strongly as or stronger than the screen, then the sample is perceived as orange. When the screen is much brighter than the sample, the color of the latter is perceived as brown. The orange then is present at a low level of brightness compared to the environment of the screen, so that the perception is that of orange mixed with black, that is of brown.

The perceived color is strongly influenced by the color and brightness of adjacent areas also in the object mode, as well as by recently-viewed intense colors, as discussed in Chapter 2. These and additional color-appearance phenomena, also discussed in Section 1.8 below and in Chapter 2, can influence color perception, sometimes to a great

TABLE 1.3
Object mode perceptions (modified after Hunter and Harold 1987, p. 17).

Object	Dominant perception	Dominant attributes in approximate sequence of their importance
Opaque non-metal, matte	Diffuse reflection	Hue, saturation, brightness
Opaque non-metal, glossy	Diffuse and specular reflections	Hue, saturation, gloss, brightness
Transparent non-metal	Transmission	Hue, saturation, clarity
Translucent non-metal	Diffuse transmission	Translucency, hue, saturation
Matte metal	Diffuse reflection	Hue, saturation, brightness, gloss
Polished metal	Specular reflection	Reflectivity, gloss, hue

extent. Yet additional dimensions (Hunt 1987, Hunter and Harold 1987, Kaiser and Boynton 1996) may then be needed for a full specification of the perceived color; fortunately, this is normally not required.

1.5.3. Interactions of light with matter

When light is transmitted, reflected, or scattered toward our eye by a white (i.e., colorless) object, there is usually no change in color. A colored object, however, absorbs some of the spectral wavelengths. An orange object illuminated by white light may absorb all spectral colors except a narrow band centered about 600 nm, which remaining orange light is then reflected and scattered to our eye to be perceived as 'orange'. If all the spectral colors except the range from about 580 nm yellow to the 700 nm limit of red are absorbed by another object, this could give precisely the same perception of 'orange' to the eye; and so on. More details are given in the following Section 1.6.

The many phenomena of fig. 1.5 are possible in the most generalized interaction of light with a substance – be it solid, liquid, or gas. When a photon is absorbed, the energy it contained must appear elsewhere. Usually an electron in the absorbing substance is excited by the absorbed photon. Only some wavelengths may produce excitation, the substance thus selectively absorbing only those wavelengths. The absorbed energy may appear as heat or as *fluorescence*, an additional emitted light at a lower energy, i.e., longer wavelength. This effect is utilized in fluorescent brightening agents used in detergents, for example. If the fluorescence is delayed, then *phosphorescence* may occur.

Scattering of light may derive from irregularities of the surface, then giving a 'matte' or 'diffuse reflecting' surface as distinguished from a 'glossy' or 'specular reflecting' one. Random irregularities in the bulk of a substance (variations in composition, the presence of a second phase, foreign particles, etc.) may also produce scattering. Scattering (see Section 4.7.2 of Chapter 4) can lead to translucency and even opacity if there is pronounced multiple scattering, as well as to a change in color as in Rayleigh scattering. Interference effects can originate from regularly repeated patterns on the surface or within

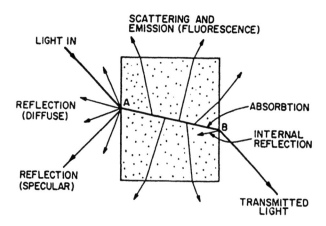

Fig. 1.5. The adventures of a beam of light passing through a block of a semi-transparent substance.

an object and can lead to intense colors as in a diffraction grating (see Sections 4.7.3 and 4.7.4 of Chapter 4).

1.6. Color mixing

Combine a red paint with a yellow paint and the resulting mixture is usually seen as having an orange color. Shine a red beam of light and a yellow beam of light onto a white paper and the combined light can also be seen as being orange. This agreement may seem reasonable, yet it is pure coincidence: yellow and blue paints usually mix to give a green paint, while yellow and blue light beams mix to give white. Complexity reigns!

The same perception of pink is given by: an unsaturated reddish-orange, namely a 620 nm reddish-orange beam mixed with some white light as at A in fig. 1.6; a mixture of beams of red with cyan (a blue-green) as at B; a mixture of red, green, and violet beams as at C; and so on. Again, two apparently equal-color yellow paints each mixed with an equal quantity of a red paint may give different shades of orange to red as

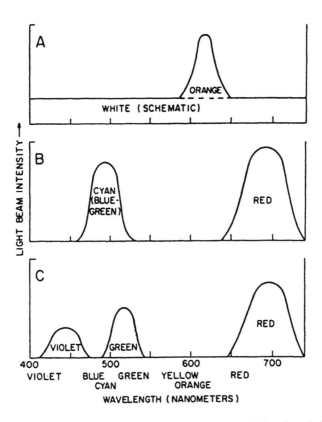

Fig. 1.6. Schematic representation of three different energy distributions of light, all producing the same pale pink color.

described below in Section 1.6.2. Even if the two yellow paints match under one type of illumination, their colors may not match under a different type of illumination, as described below in Section 1.8. More complexity!

An explanation of such phenomena involves the characteristics of the color vision process, as described in Chapter 2, interacting with the two different types of color mixing: additive and subtractive. Additive color mixing applies to the combination of light beams and to most color television receivers, as discussed in Section 1.6.1. When paints, pigments, dyes, or filters are combined, the subtractive color mixing of Section 1.6.2 applies.

1.6.1. Additive color mixing

The terms *additive mixing* or *light mixing* are used when two or more beams of light of different colors are combined. A typical example is stage illumination.

A diagram, which gives detailed and subtle light mixing information and predictions, is the color triangle of fig. 1.7 and color plate 4. This is the 1931 *CIE Chromaticity Diagram*. Here the saturated (pure, most intense) colors of the spectrum occur on the curved line extending from red to violet and the saturated non-spectral colors occur on the dashed straight line connecting the extreme red and violet ends of the spectrum. White occurs at the central achromatic point W, the position of which varies somewhat depending on which 'white' light (direct sunlight, diffuse daylight, incandescent lighting,

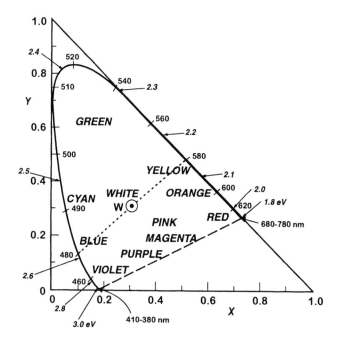

Fig. 1.7. The chromaticity diagram, showing wavelengths in nanometers (nm) and energies in electron Volts (eV); point W corresponds to "standard daylight D_{65}".

fluorescent lighting, etc.) is being used as the standard (also see Chapter 2, Section 2.2.8). A line such as that joining W and the 480 nm (or 2.58 eV) saturated blue point covers the unsaturated mixtures of this blue and white.

White is produced by mixtures of appropriate amounts of the two colored light beams that occur on either side of any straight line passing through W. An example is the line connecting the 480 nm (2.58 eV) blue with the 580 nm (2.14 eV) yellow in fig. 1.7. Such a pair of colors is called a complementary pair. Additional examples of complementary pairs taken from fig. 1.7 are: red with cyan; violet with yellowish green; and green with magenta, purple, or red, depending on the exact shade of the green; and so on. There is, in fact, and infinite number of complementary pairs. For any light mixtures anywhere on fig. 1.7, the "law of the lever" of fig. 1.8 applies. To obtain color C one mixes a quantity of color A corresponding to length a with a quantity of color B corresponding to length b. Note that as C approaches B, length a becomes smaller and length b larger, as would be expected. When B is at the central white point W, length b becomes length w and the percent saturation or purity is then given by $100a/(a+w)$.

The color perceived for any light beam or combination of beams can be specified by giving the resulting x, y coordinates, e.g., 0.20, 0.45 for the point G in fig. 1.9, together with a third number describing the luminous intensity of the beam (designated Y – see Section 2.2.8 of Chapter 2). Another way to specify the position of G gives the spectral wave length corresponding to the extension from W through G in fig. 1.9 together with the saturation or purity, i.e., 510 nm and 30% in this example. When the extrapolation corresponds to a non-spectral color, such as for the purple light beam P at 0.33, 0.15 in fig. 1.9, then an extrapolation in the opposite direction from P through W yields the "complementary dominant wavelength 540 nm", or "540c" with a purity of 70% in this example.

It will now be obvious from fig. 1.7 that the same visual sensation of yellow given by a saturated (pure) 580 nm beam of light can also be produced from a mixture of appropriate amounts of an orange to red beam in the 590 nm to 700 nm region combined with a yellowish green to green beam in the 545 nm to 570 nm region. The same yellow, but not quite as saturated, will result from a mixture of the same orange to red beam combined with a green beam in the 544 nm to 510 nm region. Because the connecting line now no longer passes through the saturated yellow 580 nm point on fig. 1.7, the resulting color corresponds to a mixture of this with white.

When equal amounts of three colored light beams are mixed, the resultant color is found at the center of the triangle, plotted on fig. 1.10 for the three light beams red R, green G, and blue B, resulting in a greenish-yellowish-white at point X. By varying the relative intensities of the three beams, any color within such a triangle can be obtained. It

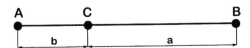

Fig. 1.8. The law of the lever; mixing amount a of color A with amount b of color B results in color C on the chromaticity diagram of fig. 1.7.

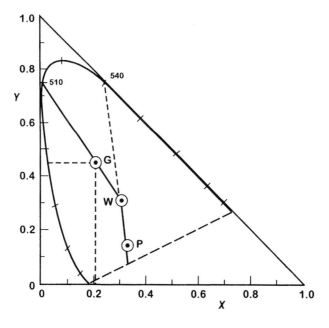

Fig. 1.9. Ways of describing color on the chromaticity diagram of fig. 1.7.

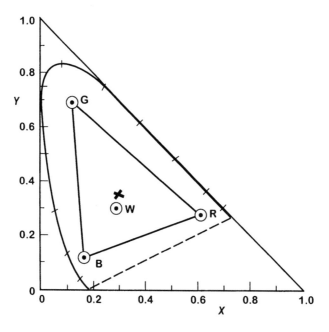

Fig. 1.10. The triangle outlining the gamut of colors available from the additive mixing of variable amounts of red R, green G, and blue B light on the chromaticity diagram of fig. 1.7.

is immediately obvious that the most complete coverage of the color triangle is achieved by mixing light beams as close as possible to the three corners, that is about 400 nm violet (or a nearby blue), about 520 nm green, and about 700 nm red. This is one of the sets of *additive primary colors*, relevant only to additive mixing. In appropriate amounts, a set of three additive primaries yields white.

Additive mixing applies to the colored spotlights used in stage lighting. The three cathodoluminescent phosphors used in most color television and computer monitor tubes also approximate this set of blue, green, and red additive primaries as in color plate 3 and can produce an almost complete gamut of colors on the screen.

Additive color mixing involving superimposed light beams can be considered to be *simultaneous additive mixing*; two other forms of additive mixing can be distinguished. *Temporal additive mixing* is demonstrated by rapidly spinning a disc on which colored sections have been painted. Due to the persistence of vision, only the single additive color is perceived. The pointillism technique of painting uses *spatial additive mixing* (also called *optical mixing*) from adjacent spots of color small enough so that they merge at a normal viewing distance. This permits the painter to achieve higher saturation than possible with the equivalent subtractive mixing of the paints themselves, discussed in the next section. It also permits a representation of high reflectivity, such as that of sparkling sunlight, by the use of white spots among those of other colors. This same technique is employed in quality color printing (see color plate 33), where great care is taken to ensure that the spots of the different colors are precisely aligned so as to avoid muddy colors. Television and computer monitor screens may use spatial (as in color plate 3) or temporal addition, the latter in most large projection sets.

1.6.2. Subtractive color mixing

Quite early it was recognized that many colors could be produced by mixing three paints, for example red (more precisely magenta, a bluish red), yellow, and blue (more precisely cyan, a greenish blue); these colors form yet another set of primary colors, the subtractive primaries, first used in color printing by the French printer Jaques C. Le Blon, about 1720 (Le Blon 1720).

When two or more paints, pigments, or dyes are mixed, then a chromaticity diagram cannot be used to predict the resulting subtractively mixed color. When such colorants are mixed with white, as one example, the resulting colors plotted on a chromaticity diagram follow curves, as shown in the dashed lines in fig. 1.11, rather than straight lines as in the additive mixing of the previous section. The solid lines on fig. 1.11 outline the area of colors accessible by mixing the specific three colorants cyan C, yellow Y, and magenta M. As can be seen, this area is bounded by curves and is not directly predictable, as is the case for additive mixing. In addition, different colorants, even if they plot at the same points on such a diagram, will yield differently-shaped areas.

It is necessary to use the individual absorption spectra, such as the curves of fig. 1.6, to understand and predict subtractive color mixing. It is important that chemical interactions, such as could occur between two dyes in solution or two pigments in a paint, are absent. Subtractive color mixing also applies when several color filters are superimposed in front of one light source.

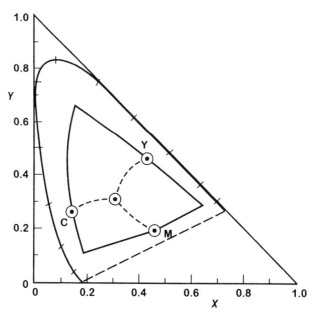

Fig. 1.11. The subtractive mixing of paints cyan C, yellow Y, and magenta M with white paint gives the curved dashed lines on the chromaticity diagram of fig. 1.7; also shown is the gamut triangle available from mixtures of variable amounts of the paints C, Y, and M, which also involves curves and is not simply predictable.

The subtleties of subtractive color mixing can be illustrated by considering two different yellow filters with light transmissions as shown in simplified schematic form at A and B in fig. 1.12. By suitably adjusting the colorant concentrations, these filters have the same visually perceived yellow color under a given illumination. Each of these filters is now combined with the same orange filter C.

The colorant in the yellow filter A, as seen in fig. 1.12, transmits all wavelengths from 540 nm to 700 nm; all other wavelengths are absorbed. This filter has a dominant wavelength of 580 nm (deduced from fig. 1.7) and the transmitted light is therefore perceived as yellow. The orange filter C transmits evenly from 575 nm to 700 nm, all other colors being absorbed. Filter C has a dominant wavelength of 600 nm (again deduced from fig. 1.7), perceived as orange. When filters A and B are combined in this idealized example, the light transmitted through both of them is the wavelength range 575 nm to 700 nm, as shown at C in fig. 1.12, resulting in the same light transmission as that of filter C by itself, which is perceived as orange.

Now consider the filter B in fig. 1.12, which transmits only from 575 nm to 585 nm. The dominant wavelength is 580 nm, with the perceived yellow exactly matching filter A. The combination of this filter with filter C results in light being transmitted through both filters only from 575 nm to 585 nm, as shown at E in fig. 1.12. This is exactly the same as with filter B by itself, perceived as yellow.

So, by combining two sets of visually equal color yellow and orange filters, the result can be either yellow or orange. Equivalent considerations apply to all occurrences of

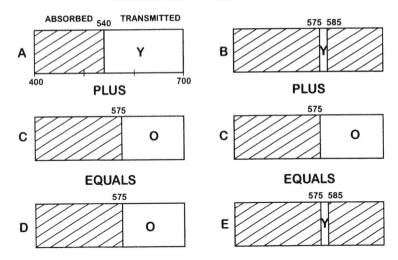

Fig. 1.12. Subtractive color mixing: the combinations of the absorptions of two equal-appearing yellow filters (A) or (B) combined with the absorption of the orange filter (C) results in orange (D) or yellow (E). Pigments can give similar results.

subtractive color mixing, including mixtures of paints, pigments, and dyes. Real colorants do not have the sharp cut-offs and uniform transmissions as used in the idealized representations of fig. 1.12, usually being even more complex than those of fig. 1.6, yet similar principles apply. These subtleties are also relevant to complementary subtractive color mixing, as when a little purple-producing manganese was added to "hide" the pale green color of an iron impurity in old glass.

The calculation of such subtractive mixing is performed by using the complicated Kubelka–Munk equation (see Section 2.8.3 of Chapter 2), which make allowance not only for absorption but also for the scattering of light which is particularly important in paints.

Artists use a wide range of paints and, while they do mix colors to obtain subtle differences, generally ignore the mixing of primary colors. Just one reason derives from subtractive color mixing considerations. The pigment for a deeply colored, saturated red paint absorbs most light except red, including some orange light. Consider mixing such a red paint with a yellow paint to obtain orange. The resulting orange paint would inevitably absorb some orange light and would therefore be unsaturated. A pure orange pigment, which ideally absorbs no orange light at all, could therefore provide a more saturated orange paint.

1.7. Color vision and color theories

In discussing color vision, it is appropriate to consider some important historical aspects involving opposition to Newton's description of the spectrum. Newton's insight that white light consists of a mixture of the full range of colors of the spectrum was not accepted immediately nor as widely as is usually assumed. This concept was viewed by many as being inconsistent with the then current knowledge about color vision. There were two major approaches to color vision which appeared at first to be inconsistent with Newton's explanation. It is now recognized that there is value in both of these views, the "truth" being in fact a combination using all of these as well as additional approaches.

1.7.1. Opponent theory

One major opposition to Newton arose from an approach involving the concept of primary colors. The Italian genius-of-all-trades, Leonardo da Vinci (about 1500), had concluded that any color could be produced by mixing yellow, green, blue, red, white, and black paints. Later the German philosopher Arthur Shopenhauer (1788–1860) noted the opposition of red and green and the opposition of yellow and blue, in addition to the opposition of white and black. This approach was formalized by the German physiologist Karl Ewald Konstantin Hering (1834–1918) in his 1878 "opponent" theory of color perception (also see Section 3.4.4 of Chapter 3).

A number of phenomena led to the conclusion that red–green, yellow–blue, and light–dark are three kinds of opposites which control vision. As one example, each of these pairs produces an after-image of the other. Again, while greenish and reddish blues and yellows, as well as bluish and yellowish reds and greens do occur, there are no reddish greens or greenish reds, no bluish yellows or yellowish blues. This postulated opponent approach was justified in the 1950s when two chromatic signals corresponding to opponent colors and a third achromatic light–dark signal were observed in the optical conduction system leading from the eye to the brain.

1.7.2. Trichromacy theory

The other major opposition arose from the concept that just three primary additive colors are adequate to produce essentially all colors. This was recognized early in the 18th century, e.g., by Le Blon as in Section 1.6.2 above, and was formalized in the "trichromacy" theory of the English scientist Thomas Young (1773–1829) in 1801. It was expanded by the German scientist Herman Ludwig Ferdinand von Helmholtz (1821–1894) and the Scottish scientist James Clerk Maxwell (1831–1879). This approach, usually called the Young–Helmholtz theory (also see Section 3.3.2 of Chapter 3), postulated three sets of receptors in the eye, sensitive to blue, green, and red light.

The English scientist John Dalton (1766–1844) was color blind and gave the first detailed description of this condition, sometimes called 'Daltonism'. He thought that a blue colorant in his eyes caused the problem and requested a posthumous examination of his eyes, which failed to demonstrate any blue. Already in 1807 Young, however, had given the correct explanation, namely that one of the three sets of the postulated color receptors in the eye is inactive or missing.

The Young–Helmholtz theory finally received full confirmation only in the 1960s, when three types of cones were indeed observed in the eye. It was found that these had quite broad and overlapping light sensitivities, with the maximum responses at short wavelengths in the blue near 440 nm, at medium wavelengths in the green near 540 nm, and at long wavelengths or 'red' near 580 nm (actually in the yellow region, but called

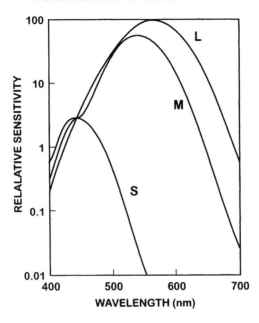

Fig. 1.13. The relative sensitivities of the three sets of cones of the human eye; after Smith and Pokorny (see Kaiser and Boynton 1996).

'red' for historical reasons). Hence the three sets of cones are usually designated as S, M, L (for short, medium and long wavelengths) or as b, g, r or β, γ, ρ (for the location of the maximum of each absorption). A set of response curves is shown in fig. 1.13.

1.7.3. Modern theories

The so-called "zone" theories combine the opponent and trichromatic approaches. The signals from the first step of the vision process, the trichromatic detecting cones in the eye, responding to the full range of Newton's spectral colors, are processed on their way to the brain and converted into three opponent signals, two chromatic and one achromatic. Yet modern investigations into the process of vision show that even this is too simplistic, as discussed in Section 1.8 below and in Chapter 3.

A recent development was the "Retinex" theory (Land 1959, 1977) of the American scientist and inventor Edwin Land (1909–1991). He showed that two photographs taken with, say, red and green filters sufficed to reproduce a scene in essentially full colors. While at first viewed as contradicting conventional color theories, these results can be understood on the basis of color constancy, our ability to see essentially true colors despite a wide variation in the color of the light illuminating a scene; for recent views and further references see Kaiser and Boynton (Kaiser and Boynton 1996, pp. 27 and 508).

1.7.4. Goethe's theory of color

Yet another opposition to Newton came from the German poet and dramatist, Johann Wolfgang von Goethe (1749–1832), who also studied color extensively. In his early

Fig. 1.14. Woodcut showing Goethe's eye being triumphant over Newton's prism and mirror (from Goethe 1791).

"Beitraege zur Optic" of 1791–1792 (Goethe 1791) and then in his masterful "Zur Far-benlehre" of 1810, translated into English in 1840 as "Theory of Colors" (Eastlake 1840), Goethe described many elegant and important experimental observations carried out over a twenty year period; he considered this his most important achievement. Goethe studied the complex relationships between physical stimuli and the resulting color perceptions. Some of these observations have been fully explained only recently. These observations are still of great value to artists in the understanding and use of subtle coloration effects. This was the evaluation of Helmholtz in 1856:

> "The experiments which Goethe uses to support his theory of colour are accurately observed and vividly described. There is no dispute as to their validity. The crucial experiments with as homogeneous light as possible, which are the basis of Newton's theory, he seems never to have repeated or to have seen. The reason of his exceedingly violent diatribe against Newton ... was more because the fundamental hypotheses in Newton's theory seemed absurd to him, than because he had anything cogent to urge against his experiments or conclusions. But Newton's assumption that white light was composed of light of many colours seemed so absurd to Goethe, because he looked at it from his standpoint which compelled him to seek all beauty and truth in direct terms of sensory perception" (Sloane 1991, p. 55).

Goethe's work is marred for us by his unjust criticism of Newton's discoveries (fig. 1.14). He also reverted to Aristotle's approach of Section 1.1, with color result-ing from a mixture of light and dark in some unspecified manner.

1.7.5. Color circles (wheels)

The sequence of the spectral colors to form a closed loop, completed with the colors of the magenta–purple line of fig. 1.7, was first used by Newton as shown in fig. 1.3. This approach has been widely adopted as the basis of color atlases and numerical descriptive and measurement systems, as in the chromaticity diagram of color plates 4, 5 and 6, in Munsell's scheme of color plates 1 and 2, and in other systems as described in Chapter 2. Color circles, also known as color wheels (as well as related two-dimensional geometrical displays, e.g., using three- or six-pointed figures, three-dimensional solids, as well as related color ordering systems), have been used extensively by artists, employing

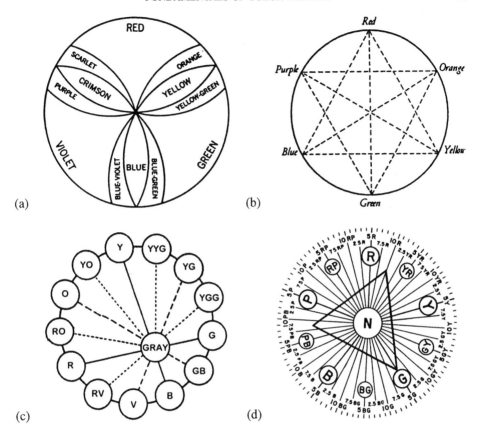

Fig. 1.15. Some color circles: (a) Michel Jacobs (1923), based on the additives red, green, and violet, after Jacobs (1923); (b) J.W. von Goethe 1840 (Eastlake 1967), apparently based on the subtractives red, blue, and yellow; (c) Faber Birren, 1934, based on the complementary opponents red/green and yellow/blue, modified after Birren (1987); (d) A.M. Munsell (1898), based on equal visual spacing, courtesy Munsell Color.

geometrical relationships within the circle to select color combinations for harmonious color design (Agoston 1987, Birren 1987, Geritsen 1988). Several color circles are shown in figs 1.3 and 1.15 (for early color wheels see Section 5.2.1 of Chapter 5).

An examination of the many arrangements used over the years shows that color circles can be grouped into four types. The first type can be viewed as being based on the approximately equal spacing of the three additive primary colors red, green, and blue. This includes the circles of Von Bezold (1876); Rood (1879); Jacobs (1923) (fig. 1.15); and Geritsen (1975). It should be noted that the precise "primary" colors used in these circles are not necessarily the same. Neither did the originators usually design their circles in terms of the approaches here used to classify them. These same remarks also apply to the other types that follow.

The second type can be viewed as being based on the approximately equal spacing of the three subtractive primary colors red or magenta, yellow, and blue or cyan. Names associated with this type of color circle include Le Blon (1720), Harris (1776); Goethe (fig. 1.15) (1793) (Eastlake 1840); Runge (1806); Hershel (1817); Hayter (1826); Brewster (1831); Chevreul (1839); Blane (1973); Van Gogh (1878); Holzel (1904); the CIE x, y chromaticity diagram (fig. 1.7 and color plate 4) (1931); Klee (1924); and Itten (1961).

The third type can be viewed as being based on an approximately equal spacing of the four opponent-type primary colors red, yellow, green, and blue. Such arrangements are employed by Hering (1878); Ostwald (1916); Plochere (1948); Birren (fig. 1.15) (1934) (Geritsen 1988); the Swedish NCS System, 1968; and the CIELAB chromaticity diagram (fig. 2.16 and color plate 6) (1976); among others.

Finally, a group of circles is based on attempts to achieve approximately equally-spaced colors as visually perceived. This includes the Munsell System (fig. 1.15 and color plates 1 and 2) (1898); the German DIN System (1953); CIELUV chromaticity diagram (fig. 2.19 and color plate 5) (1976); and the OSA Uniform Color Scales (1977), among others.

In some of these systems complementary colors are at opposite ends of a line passing through the central white, as in the Birren circle of fig. 1.15 and the CIE x, y chromaticity diagram of fig. 1.7; in some systems they are not. Interestingly, the color circle of Birren of fig. 1.15 is purposely distorted so that the artistically more useful 'warm' colors occupy a larger area than do the 'cold' colors.

In any of these systems black and white are or could be used to complete a three-dimensional color space; spheres as well as other geometries have been used. Three dimensional color spaces are necessary for a full description of all colors as in Section 1.5.1 above. This subject is covered in more detail in Chapter 3.

1.7.6. Light sources and metamerism

The energy distribution of different light sources varies widely (also see Section 2.2.3 of Chapter 2). The relative energy distributions of daylight from an overcast sky, of a tungsten incandescent light bulb, and of a fluorescent tube lamp are shown schematically in fig. 1.16. The fluorescent lamp consists of a tube containing mercury vapor. An electrical discharge excites this vapor to emit light mostly in narrow energy ranges, as shown by the tall spikes, e.g., at 430 nm, as described in Section 4.3.2 of Chapter 4. The overall color of light emitted from such a mercury vapor discharge is blue and much ultraviolet is also produced. A layer of fluorescent powder (see Section 4.6.3 of Chapter 4) inside the tube is used to convert this ultraviolet into the broad emission seen in the central region of the spectrum in fig. 1.16, resulting in a light much closer to the white of daylight.

Under most circumstances, the eye has the ability to compensate for different illuminating conditions and perceive essentially the same color in "color constancy". However, there are occasions when, for example, two fabrics will be color-matched in daylight but will show different colors in indoor lighting. This phenomenon is know as *metamerism* (this same word is also used with a different meaning when the color of a dye in solution changes with the concentration).

Again let us use the schematic absorption curves for the two yellow filters A and B of fig. 1.12. We combine these with the daylight and incandescent light curves of fig. 1.16. The results are given in the idealized schematic curves of fig. 1.17, where the absolute intensities are ignored for simplicity. The arrows point to the overall color perceived,

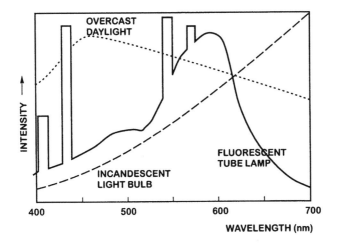

Fig. 1.16. Spectral energy distributions of three light sources.

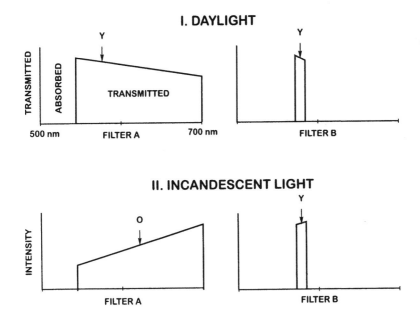

Fig. 1.17. Metamerism: two yellow filters, A and B, have the same color in daylight, with dominant wavelength Y; filter B remains Y under incandescent illumination but filter A now appears orange O.

making allowance for the color sensitivity variation of the eye. As seen, the relatively constant distribution of daylight in the long wavelength region of the spectrum gives a color match between the filters A and B as in the upper spectra of fig. 1.17, both having a dominant wavelength of about 580 nm, perceived as yellow Y. The lower spectra in fig. 1.17, involving the sharply rising intensity of incandescent light with wavelength of fig. 1.16, show a shift of the dominant wavelength from 580 nm yellow to 620 nm orange O for filter A, while the color of filter B remains essentially the same at 620 nm yellow Y.

The eye can compensate for this change in illumination if either of these filters is viewed by itself in a normally illuminated environment. When seen together, however, the eye cannot compensate for both simultaneously. Filter A is now seen as being more orange in color when next to the yellow of filter B. These are then called 'metameric colors'. This term implies that even though colors match under one illumination, the spectral absorptions are different and may no longer match under a different illumination. The designation 'non-metameric' indicates materials that have identical absorptions and that therefore will match under any illumination. Also see Section 2.2.7 of Chapter 2 and Section 3.4.3 of Chapter 3.

1.8. What color do *You* see?

Most of us have said to someone, at one time or another "take that brown book" and received the reply "do you mean the yellow one?" Describing colors in words can be difficult. Consider just one dictionary's definition of one color:

> "Puce: a dark red that is yellower and less strong than cranberry, paler and slightly yellower than average garnet, bluer, less strong, and slightly lighter than pomegranate, and bluer and paler than average wine".

We must also recognize that color names do not always translate readily from one language to another. In Russian our 'blue' is covered by both *siny* and *goluboy*, while in Japanese *asi* covers most of the range of our 'blue' and 'green'. Orange is merely a fruit but not a color term in early German; Goethe usually used 'red-yellow' for this color and 'red-blue' in place of our 'purple'. Homer and the early Greeks did not use 'blue' for sky or water: it was 'brazen sky' and 'wine-red ocean' (the exact reason for these early Greek usages are continuously under debate and no explanation proposed so far appears entirely satisfactory). And so on.

Consider that both you and I view a sample from which light of only a single wavelength is reflected to the eye. Even in these idealized conditions, the color that you perceive is not necessarily the same color that I perceive, because every person's perception depends on the following conditions (most of these are discussed in Chapter 3), in no particular order:

1) Are all three trichromatic detectors and transmission channels in the eye (see Section 1.7.2) working perfectly?
2) Are all three opponent transmission channels to the brain (see Section 1.7.1) working perfectly?
3) What colors at what intensities are immediately adjacent to the sample (see Section 3.3.5 of Chapter 3)?
4) What colors at what intensity are in the wider field of view?

5) What is the intensity and energy distribution of the light illuminating the wider field of view (see Section 1.7.6 above and Section 2.2.3 of Chapter 2)? Is this light direct or indirect, diffuse or focused?

6) Is there any periodic variation (flicker) in the illuminating light, at what intensity range and with what frequency?

7) What colors in what distribution and at what intensities were viewed immediately before looking at the sample?

8) Are there adequate boundaries to the sample (see Section 3.4.5 of Chapter 3)?

9) What is the mode of viewing (see Section 1.5.2)?

10) What is the sample size and flatness?

11) What is the sample texture (rough, matte, glossy, translucent, metallic, etc.; see table 1.3, and Sections 2.5 and 2.6 of Chapter 2)?

12) What is the angular relationship between the eye, the sample, and the source of light?

And so on.

A change in any one of these conditions can have an effect on the color that is perceived by the eye–brain system. This system obviously is extremely complex in view of all these aspects; new insights appear every few months in the scientific literature. Even given that all of these conditions are identical for both your and my viewing, do we see the same color? Perhaps – but there is no way we can be sure. Such questions arise with any of our senses, with hearing, touch, pain, etc., and there too we have no answer.

The best we can do is to have available a system by which every possible color stimulus can be described in some unique way, independently of what your or my eye–brain system perceives. This can involve color atlases, systems of comparison chips, such as those of Munsell (1929) (produced in both matte and glossy forms) and others, or systems based on instrument measurements that result in numerical or alphanumerical systems such as those of CIE, Munsell, and others (see Chapter 3).

1.9. Color and "Science versus Art"

The Newton–Goethe dichotomy survives as strongly as ever. Rare is the scientist who is adequately familiar with Goethe's work on color perception (other than that he was vehemently opposed to Newton) or with the artistic uses of color. And equally rare is the artist who is aquainted with enough of Newton's insights into color and of the subtleties to which this has led modern science and technology. Note that Newton's color work ultimately resulted in the particle/wave duality and quantum theory and its descendants as discussed in Section 1.10 below (also note that Newton's gravitational work was also groundbreaking and was only modified in small part by Einstein's relativity theory).

Consider first this quite reasonable quotation from Max Planck, 1858–1947, the Nobel Prize-winning German physicist, discussing the nature of light:

"The first problem of physical optics, the condition necessary for the possibility of a true physical theory of light, is the analysis of all the complex phenomena connected with light, into objective and subjective parts. The first deals with those phenomena which are outside, and independent of, the organ of sight, the eye. It is the so-called light rays which constitute the domain of physical research. The second part embraces the inner phenomena, from eye to brain, and this leads us into the realms of physiology and psychology. It is not at all self-evident, from first principles, that the objective light rays can be completely separated from the sight sense . . ." (Sloane 1991, p. 95).

Yet, next Planck proceeds to deride Goethe for ignoring the existence of this difference without any acknowledgment of the validity of most of Goethe's observations. In the words of P. Sloane, who compiled this useful collection of writings on color (Sloane 1991):

"... Planck argued that knowledge acquired through the reasoning of the physical sciences was pure knowledge and thus superior to, or significantly different from, knowledge acquired by other routes. The implication was that pure knowledge is revealed only to pure scientists and, as Goethe was not a scientist, his importance could be greatly diminished" (Sloane 1991, p. 94).

Faced with such attitudes, combined with most scientists' lack of knowledge of, or appreciation for, Goethe's work and the artistic uses of color, the antagonism of most artists for the scientific approach can readily be understood.

Yet contrast Sloan herself, who on the same page (but not in the section on Newton!), proceeds to cast doubt on Newton's objectivity because he possibly had a nervous breakdown and because of his (largely unpublished) interests in numerology and alchemy, as well as his unorthodox religious views. (She did not, but could also have referred to his most unpleasant personal disposition.) Surely Newton's objectivity is irrelevant; it is only the quality of his experimentation and adequacy of his interpretations that should concern us. Elsewhere, Sloan's statements such as:

"... brown light rays do not exist. Therefore the color of brown objects cannot be explained by saying that these objects are reflecting brown light rays" (Sloane 1991, p. 13)

show a surprising lack of understanding of Newton's dictum "the rays . . . are not colored", of the nature of brown as discussed above, as well as of the modern understanding of color perception, particularly as related to orange and brown, as discussed in Section 1.5.2 above and elsewhere in this book.

From the other extreme, consider this quotation from Naum Gabo (1890–1977), a Russian artist and color theorist:

"What kind of image comes to our consciousness when we say "red"? To a scientist in his laboratory, red is only one thing, the stripe produced by the longest waves on the visible band of the spectrum of light. To me, to you, as well as to the scientist as a human being, when he gets his experiences of color from life and not from his spectroscope, it may mean either nothing at all or an endless chain of images of a multitude of color experiences . . ." (Sloane 1991, p. 177).

Yet, despite his deep insights into the visual experiences of color, it is clear that Gabo does not understand either Newton's views or modern extensions of these. For example, he refers to

"... the seven visible colours of the spectrum, as science defines it" (Sloane 1991, p. 176).

Neither does he advance the artist's contact with scientists by implying that the latter in their work are less than "human" beings.

To me, the Goethe/Newton and the artist/scientist dichotomies do not make sense. After all, Newton's spectrum and related discoveries (once his views on the corpuscular nature of light are ignored) are beyond questioning; he never claimed to deal with the subtleties of the human color perception system. And Goethe (once his explanations of the origin of color in mixtures of black and white and his intentional avoidance of Newton's insights are ignored) provides much valuable information on many aspects of color perception which are not in the least incompatible with Newton's discoveries. The problem may be that only rarely are Newton and Goethe read in the original. Most of us rely on second or third-hand reports that focus on disagreements and limitations and ignore the positive and complementary aspects of these two approaches.

Zajonc uses an interesting analogy:

"To see ... light as vibration only is to reduce Michelangelo's *David* to marble dust ... in the process we lose the truth the work of art embodies" (Zajonc 1993, p. 106).

Absolutely true, as far as it goes. Yet let me suggest that this analogy can perhaps be taken a step further in this suggested continuation:

"Having seen the "truth" of the colors in a work of art, let us now also look at these colors as vibrations. To omit this is to ignore the fact that Michelangelo did construct his *David* out of marble. Had he used wood or bronze, the different techniques employed and the different nature of the medium surely would have resulted in a different *David*. These aspects, the nature and properties of the medium, are also part of the work of art".

Arthur Zajonc has written a truly multidisciplinary treatment of light, also including all aspects of color "Catching the Light: The Entwined History of Light and Mind" (Zajonc 1993) that aims to bridge the gaps between various disciplines. This immensely original treatment combines a wide variety of topics in a poetic scientist's approach:

"... light has gathered around it innumerable artistic and religious associations of extraordinary beauty. It has been treated scientifically by physicists, symbolically by religious thinkers, and practically by artists and technicians. Each gives voice to a part of our experience of light. When heard together, all speak of one thing whose nature and meaning has been the object of human attention and veneration for millenia. During the last three centuries, the artistic and religious dimensions of light have been kept severely apart from its scientific study. I feel the time has come to welcome them back, and to craft a fuller image of light than any one discipline can offer" (Zajonc 1993, p. 8).

The level of Zajonc's approach is inevitably deep; this has to be expected of any serious work as we approach the twenty-first century. Whether discussing Brunelleschi, Faust, Fermat's principle, Goethe, Hegel, Homer, Kandinsky, Land, Navajo gods, Newton, Schopenhauer, or the most recent experimental confirmation of John Bell's interpretation of the problematic Einstein, Podolsky, and Rosen thought experiment (see Section 1.10 below), not a source is left unquoted that might throw illumination on the confluence of a multiplicity of approaches. Does Zajonc succeed in his aim? I believe so, but the answer to this question undoubtedly depends on the background which readers bring to this work and on their ability to be flexible in overcoming previously formed attitudes.

More such treatments are needed. The specifications required for broadly-based authors such as Zajonc have been given (in a different context) by Alexander Pope in his Essay on Criticism (Pope 1733):

"But where's the man who counsel can bestow,
Still pleased to teach, and yet not proud to know?
Unbiass'd, or by favour, or by spite;
Not dully prepossess'd, nor blindly right;
Though learn'd, well-bred; and though well-bred; sincere;
Modestly bold, and humanly severe:
Who to a friend his faults can freely show,
and gladly praise the merit of a foe?
Bless'd with a taste exact, yet unconfined;
A knowledge both of books and human kind;
Generous converse; a soul exempt from pride;
And love to praise, with reason on his side?"

1.10. The nature of light

When Newton described his discovery of the spectrum and his many other experiments on color in his "Opticks" of 1704 and later editions (Newton 1730), he viewed light as consisting of 'corpuscles', being small solid bodies, the smallest indivisible units of light. However, he also had to attribute to light a type of vibration, his "fits of easy reflection" (which, indeed, correspond precisely with what we now call frequency, with its associated wavelength). Violent disagreement arose with the English physicist, Robert Hooke (1635–1703), who had proposed in 1665 that light consisted of a wave. Hooke's eminence and animosity prevented Newton from even attempting to publish his "Optics" until after Hooke's death. In turn, Newton's subsequent authority was such that the corpuscular view of light dominated for the next century, even though a careful reading shows that he himself was flexible on this aspect of light.

Hooke's wave theory had been taken up by the Dutch scientist Christian Huygens (1596–1687), but it was the interference experiments (see Section 4.7.3 of Chapter 4) by the English physician and scientist, Thomas Young (1773–1829) in 1802 which changed the attitude of most scientists. Light was now viewed as a wave (a wave carried by an intangible medium, the 'luminiferous ether'; many attempts have been made since that time to demonstrate the presence of such an ether, but all have failed). Many other experimenters added evidence on both sides of this controversy. After another century it was generally accepted that light had both corpuscular properties (such as the photoelectric effect) and wave properties (such as interference), but that the wave involved does not require any medium for its propagation as does sound, for example.

It was the attempt to explain black-body radiation (see Section 4.3.1 of Chapter 4) by the German physicist, Max Planck (1858–1947), that led to quantum theory. With further work by several physicists, including Albert Einstein (1879–1955), Niels Bohr (1885–1962), Werner Heisenberg (1901–1976), and others, there appeared quantum mechanics which ultimately gave a mathematical framework that could encompass both the corpuscular as well as the wave aspects of light. It was also found that not just light but also matter had both of these aspects.

To scientists in their work with light and color, quantum mechanics gives a more than adequate explanation for all observed phenomena for all practical purposes (FAPP), with agreement between theory and experiment being near perfect to many decimal places. Yet there are some extremely subtle effects which indicate that all is not quite well. Einstein finally felt that the story was not complete. In 1951 he said:

"All of the fifty years of conscious brooding have brought me no closer to the question, "What are light quanta?" Of course today every rascal thinks he knows the answer, but he is deluding himself" (Zajonc 1993, p. 279).

The problem is this. It seems that certain questions about a quantum or photon of light, such as "exactly where is the photon at this moment of time", cannot be answered in a way that is compatible with our everyday experience. As one example, under certain conditions a single, definitely indivisible photon acts as if it could follow two different paths and then interfere with itself. Einstein, Podolski, and Rosen in a 1935 paper (Einstein et al. 1935) had outlined the crux of the problem. The consequences of this 'EPR thought experiment', as it is now known, were elucidated by John Bell in the 1960s (Bell 1966, 1987), who also felt that quantum mechanics had to be superseded. The alternatives to quantum mechanics are 'local, realistic' theories, approaches more consistent with common sense and everyday experience, at least in some respects.

Starting in the late 1980s, experiments equivalent to the EPR thought experiment to resolve Bell's paradox finally became possible, ultimately using single photons. Every few months yet more precise and conclusive results appear in the scientific literature. And the results have been consistently uniform: quantum theory is correct, and it is very very weird even by super-sophisticated scientific standards. But there are significant consequences of these experiments that are not yet fully understood. Two major approaches to solve this dilemma seem equally discomforting to scientists. By necessity, both involve 'nonlocality', the concept that was so objectionable to Einstein and Bell: a photon seems to be able to know what another, distant photon is doing, and the knowledge appears to be communicated essentially instantaneously, faster than the speed of light, i.e., faster than any object or wave can possibly travel.

One way out, 'quantum realism', holds that certain properties of photons, such as polarization, do not exist until a measurement is made. This uncertainty only disappears when a measurement is made, involving an unknown mechanism that could only be described as very weird. The alternative, advocated by the late David Bohm and others, is just as weird: it proposes the 'quantum potential', a new equivalent of the old intangible ether. This pervades all space and instantaneously carries information between quanta. Even a near infinity of parallel and continuously branching universes has been proposed as a possible solution. Most scientists find the FAPP approach works and that these dilemmas can be ignored for all practical purposes. Interesting accounts have been given by Zajonc 1993, one of the 'Bell's paradox' experimenters and other, more technical authors (e.g., Bell 1987, Bohm and Hiley 1993).

1.11. Further reading

Reading both Newton (1671) and Goethe (Eastlake 1840) in the original is a fascinating experience. Other very useful sources are the collections of MacAdam (1970) and Sloane (1991).

For an original approach to light and color as a whole without disciplinary fragmentation, Arthur Zajonc's "Catching the Light: The Entwined History of Light and Mind" (Zajonc 1993) is highly recommended.

Other aspects of color are covered in the chapters that follow, where additional recommendations for further reading will be found.

References

Agoston, A.A., 1987, Color Theory and its Application in Art and Design, 2nd edn (Springer-Verlag, New York).

Bell, J.S., 1966, Rev. Mod. Phys. **38**, 447–452.

Bell, J.S., 1987, Speakable and Unspeakable in Quantum Mechanics (Cambridge Univ. Press, Cambridge).

Birren, F., 1987, Principles of Color (Schiffer, West Chester, PA).

Bohm, D., and B.J. Hiley, 1993, The Undivided Universe (Routledge, New York).

Committee on Colorimetry, 1953, The Science of Color (T.Y. Crowell, New York) pp. 1–15.

Eastlake, C.L. (transl.), 1840, Goethe's Theory of Colors (J. Murray, London; reprinted: F. Cass, London, 1967).

Einstein, A., B. Podolsky and N. Rosen, 1935, Physics Review **47**, 777–780.

Geritsen, F., 1988, Evolution in Color (Schiffer, West Chester, PA).

Goethe, J.W. von, 1791, Beitraege zur Optic.

Hunt, R.W.G., 1987, Measuring Colour (Wiley, New York).

Hunter, R.S., and R.W. Harold, 1987, The Measurement of Appearance, 2nd edn (Wiley, New York).

Jacobs, M., 1923, The Art of Colour (Doubleday Page, New York).

Kaiser, P.K., and R.M. Boynton, 1996, Human Color Vision, 2nd edn (Optical Soc. of America, Washington, DC).

Land, E.H., 1959, Experiments in color vision. Scientific American **200**(5), 84–99.

Land, E.H., 1977, The retinex theory of color vision. Scientific American **218**(6), 108–128.

Le Blon, J.C., 1720, Coloritto (about 1720; reprinted: Van Nostrand Reinhold, New York, 1980).

MacAdam, D.L., 1970, Sources of Color Science (MIT Press, Cambridge).

Munsell, 1929, Munsell Book of Color, various editions (Munsell Color Co., Baltimore, 1929 on).

Newton, I., 1671, New theory about light and colours. Philosophical Transactions of the Royal Society **6**, 3075–3085.

Newton, I., 1730, Opticks, 4th edn (W. Innys, London; reprinted: Dover, New York, 1952) pp. 124, 155.

Pope, A., 1733, Essay on Man and Other Poems (Dover Publications, Mineola, NY, 1994).

Sloane, P., 1991, Primary Sources. Selected Writings on Color from Aristotle to Albers (Design Press, McCraw-Hill, New York).

Zajonc, A., 1993, Catching the Light: The Entwined History of Light and Mind (Bantam, New York; paperback: Oxford Univ. Press, New York, 1994).

chapter 2

THE MEASUREMENT OF COLOR

ROBERT T. MARCUS

D&S Plastics International
Mansfield, TX, USA

Color for Science, Art and Technology
K. Nassau (Editor)

CONTENTS

2.1. Introduction 34

  2.1.1. Why measure color 34

  2.1.2. Materials that can be measured 35

2.2. The CIE system of colorimetry 35

  2.2.1. The CIE 35

  2.2.2. Color perception 36

  2.2.3. Standard light sources and illuminants 36

  2.2.4. Standard observers 38

  2.2.5. Objects 41

  2.2.6. Calculating tristimulus values 43

  2.2.7. Metamerism 48

  2.2.8. Chromaticity coordinates and the chromaticity diagram 49

2.3. Color collections and uniform color spaces 51

  2.3.1. Color collections 51

  2.3.2. Color order systems and uniform color spaces 51

  2.3.3. Munsell's color order system 52

  2.3.4. Other color order systems 54

  2.3.5. The 1976 CIE color spaces 55

  2.3.6. The 1976 CIE $L^*a^*b^*$ (CIELAB) color space 56

  2.3.7. The 1976 CIE $L^*u^*v^*$ (CIELUV) color space 59

  2.3.8. Color atlases based upon CIELAB space 60

2.4. Color differences and tolerances 60

  2.4.1. Describing color differences 60

  2.4.2. The CIELAB color difference equation 61

  2.4.3. The CIELUV color difference equation 62

  2.4.4. The CMC(l:c) color difference equation 63

  2.4.5. The CIE94 color difference equation 64

  2.4.6. Other color difference equations 64

  2.4.7. Product standards 65

  2.4.8. Color tolerances 65

  2.4.9. Visual evaluation of the color difference 67

2.5. Reflectance measurements 68

    2.5.1. Specular and diffuse reflectance 68

    2.5.2. Illuminating and viewing geometry 69

    2.5.3. Spectrophotometers 73

    2.5.4. Tristimulus colorimeters and spectrocolorimeters 76

    2.5.5. Measuring translucent samples 77

    2.5.6. Measuring fluorescent samples 78

    2.5.7. Measuring metallic and pearlescent samples with goniospectrophotometers 79

    2.5.8. Sample preparation for reflectance measurements 81

    2.5.9. Colorimetric accuracy, repeatability and reproducibility 82

2.6. Transmittance measurements 83

    2.6.1. Illuminating and viewing geometries 83

    2.6.2. Calibrating instruments for transmittance measurements 84

    2.6.3. Measuring transmittance 85

2.7. Radiometric measurements 86

    2.7.1. Radiometry and the calculation of tristimulus values 86

    2.7.2. Measuring lamps, light sources and displays 86

    2.7.3. Measuring retroreflection 87

2.8. Colorant mixing and color matching 87

    2.8.1. Introduction to color matching 87

    2.8.2. Transparent materials 88

    2.8.3. Translucent and opaque materials 88

    2.8.4. Computer color matching 89

2.9. Review and recommendations 89

2.10. Bibliography 90

References 91

2.1. Introduction

Life has never been more colorful. Many products are available in a wide variety of colors. Color measurement helps industry control and reproduce colors so that the consumer gets the "right" color all the time. This chapter shows how color measuring instruments and calculations simulate what people see, describes several of the multi-dimensional color spaces within which the results of color measurement can be plotted and color atlases which may be associated with those color spaces. It discusses how to calculate the color difference between two samples and provides some guidance on setting color tolerances. There is a section on the instruments and techniques for making reflectance measurements on opaque, translucent, fluorescent, metallic and pearlescent samples. There are also sections on measuring the transmittance of transparent materials and on measuring lamps, light sources and displays (television and computer monitors). Lastly, there is a discussion on colorant mixing and computer color matching. Many of the references given refer to standard test methods and practices that provide details on making color measurements and related calculations.

2.1.1. Why measure color

Earth, the blue planet, is truly a beautiful and colorful place – the clean white of winter snow; the black soot of a fireplace; the bright blue sky; the green grass; the brilliant yellows, oranges and reds of the American fall foliage. Throughout the centuries people have surrounded themselves with color. Not only have they decorated their bodies, they dyed their clothes and colored their home furnishing.

In the early days of the industrial revolution products were available in only a limited amount of colors. Cars were available in any color you could want, as long as it was black. Now consumers want a variety of colors. Materials available today allow manufacturers to respond to consumer demands.

For many centuries the only form of color control available was the good color eye of the manufacturer adjusting the color of his product in natural daylight. Dyers preferred to control the color of dyed fabric by visual examination with the light coming through a north sky window. North sky daylight was also a favorite of artists. Unfortunately, natural daylight is only available for a limited time each day. In addition, natural daylight is constantly changing throughout the day. Artificial lighting booths have eliminated many of the problems associated with natural daylight.

Master color matchers historically served long apprenticeships to learn how to judge color. It is much more difficult to find apprentices today. As consumers are exposed to more and more colored products, they also became more sophisticated. Consumers demand better reproducibility in the color of products.

Color measuring instruments provide the tools needed to satisfy today's consumer demands. The major use of color measuring instruments is for the quality control of colored products. Not only are the instruments available 24 hours a day, but they provide a quantitative evaluation of color. Instruments are the eyes of computer color matching programs, which are used to match new colors and adjust the color of products as they are being produced. Color measuring instruments are being used more and more in the manufacturing process, both to control and to improve the process. Other uses of color measuring instruments include the evaluation and control of raw materials and to help in solving product or process problems. They are also widely used to measure and describe animals, plants and minerals as well as in scientific research.

2.1.2. Materials that can be measured

Color is perceived when looking at a light source by itself. This color is measured with spectroradiometers and radiometric colorimeters. Much of the light reaching our eyes comes from light reflected from, or transmitted through, objects. Opaque materials scatter light (change its direction) back to the observer and do not transmit any light. Transparent materials only transmit light and do not scatter any light. Translucent and hazy materials both scatter and transmit light. Most color measurements are made on objects that reflect light, and this chapter will emphasize those materials. Spectrophotometers and colorimeters are used to measure these materials.

Some materials present special measurement problems. The safety vests that hunters wear are made from fluorescent cloth. Fluorescent materials both reflect and emit light. Although fluorescent materials can be measured with some spectrophotometers and colorimeters, a spectrofluorimeter can be used to separate reflection from emittance.

Cars are sometimes painted with paints containing metallic and/or mica flakes. These metallic and pearlescent paints change color when viewed at different angles and are measured using goniospectrophotometers.

The resins used in making paint and other coatings scatter light slightly. While they appear almost transparent, they have a hazy appearance. Hazy materials are measured with spectrophotometers and hazemeters.

Outdoor clothing often come with safety stripes. At night the stripes appear bright to a driver when the car's headlights shine on them. These warning stripes provide an example of retroreflection. Special instrumentation has been developed for the measurement of retroreflection.

2.2. The CIE system of colorimetry

2.2.1. The CIE

The Commission Internationale de l'Eclairage (International Commission on Illumination), commonly referred to as the CIE, is an international organization concerned with lighting and color. Scientists and engineers of the member countries of the CIE work together in committees that deal with the many aspects of light and color. Several of the

CIE committees deal with color, color measurement and color difference calculations. In 1931, the CIE issued its first major recommendations on colorimetry, the measurement of color. Since that time many additional recommendations have been made and adopted in the field of color measurement (CIE 1986a).

2.2.2. Color perception

Color is a perception. A light shines on an object. The light is reflected from (or transmitted through) the object and triggers light-sensitive cells in the eye. The cells send signals to the brain where the color of the object is perceived (also see Chapter 1, Sections 1.7 and 1.8, and Chapter 3, particularly Section 3.4). If the light is changed with the object and the observer staying the same, the perceived color of the object will change, but the eye–brain system attempts to compensate for the change in lighting. The more pronounced the change in lighting, the more pronounced the change in color will be. Some objects change color more than others for a similar change in lighting.

The CIE system of colorimetry attempts to simulate mathematically the perception of color and provide a standardized procedure for measuring and quantifying that perception. It should be emphasized, however, that since color is a perception, looking at the colored material being measured is a step that should not be ignored. The critical evaluation of colors should be done using the controlled lighting conditions of light booths. Most light booths contain a simulated daylight, a fluorescent lamp and an incandescent lamp.

2.2.3. Standard light sources and illuminants

Visible light, called the visible spectrum, is that portion of the electromagnetic spectrum having wavelengths from about 380 nm (nanometers or billionths of a meter) to 780 nm. Light of different wavelengths is perceived as having different colors.

A light source can be characterized by its spectral power distribution, the amount of power the source emits at each wavelength of the spectrum. A block of carbon at a temperature of absolute zero (0 K on the Kelvin scale, which is equal to $-273°C$) would look black. If the block of carbon is heated, its color will change from black at 0 K, to red at about 1000 K, to yellow at about 2500 K, to white at about 4500 K and to bluish-white above 6500 K. The carbon block simulates a full (blackbody) radiator, a theoretical material that changes color predictably with temperature. Light sources approximate this response and are often described by the temperature of the full radiator having the same (or closest) color of the light source. Thus an ordinary incandescent lamp would have a color temperature of about 2850 K and a 'cool white' fluorescent lamp about 4100 K (also see Chapter 4, Section 4.3.1).

The color temperature of daylight varies during the day, being lower (redder) at sunrise and sunset and higher (bluer) at noon. On a clear day, the color temperature of the sun plus skylight is about 5500 K. Overcast skylight can have a higher color temperature depending upon the altitude of the clouds. For the purposes of colorimetry, average daylight (diffuse skylight without direct sunlight) is considered to have a color temperature of about 6500 K. Indoor north sky daylight, preferred by many color matchers, has a color temperature of about 7500 K.

In 1931 the CIE recommended three light sources for use with colorimetry. Source A is an incandescent, tungsten filament light with a color temperature of 2854 K. Source B (no longer recommended) represents noon sunlight and is produced by placing a liquid filter in front of Source A. Source C approximates average daylight and is also produced by filtering Source A. Color measurement calculations require illuminants – mathematical descriptions of the spectral power distribution of lights. Illuminants A, B and C are radiometric measurements of the spectral power distribution of Sources A, B and C.

Light sources and illuminants are often confused. A light source can be turned on and off and be used to view an object. An illuminant is a mathematical description of a light source. Illuminants may describe light sources that do not actually exist in the laboratory. The current 'D' series of daylight simulating illuminants do not have comparable light sources.

Although Source C approximates average daylight, there are some significant differences. Because the Earth's atmosphere acts like a filter, the spectral power distribution of the light reaching the surface of the Earth from the sun changes considerably during the day. Clouds and other forms of overcast (such as smog) can also change the spectral character of sunlight and daylight. Color temperatures of daylight vary from about 2000 K in the late morning and late afternoon to in excess of 10 000 K. Real daylight contains significantly more ultraviolet energy than Source C.

Based on measurements of real daylight, the CIE recommended a series of 'D' illuminants to represent daylight. Illuminant D65 has a color temperature of 6500 K and represents average daylight. It is the most widely used of the daylight illuminants. Figure 2.1 compares the spectral power distributions of Illuminants C and D65. The graphic arts community prefers to evaluate color with sources having a relatively flat spectral power distribution with a color temperature of 5000 K and uses Illuminant D50 for color measurement.

Manufacturers of light booths and color measuring instruments have tried to duplicate D65 (and other daylight illuminants) with light sources. Procedures and techniques were

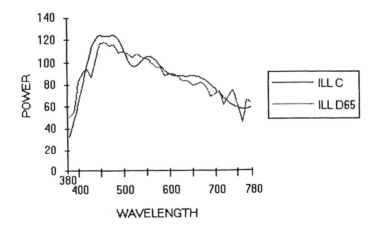

Fig. 2.1. The spectral power distribution of CIE Illuminant D65 has more energy in the ultraviolet wavelengths than CIE Illuminant C.

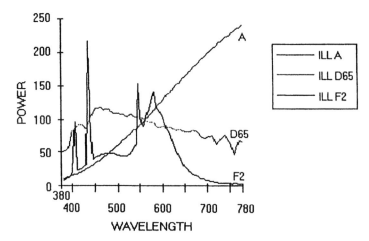

Fig. 2.2. Shapes of the spectral power distributions of CIE Illuminants D65 (daylight), F2 (a cool white fluorescent lamp) and A (an incandescent lamp) can be compared by adjusting them so that the power of each is 100 at 560 nanometers (compare with fig. 1.16 of Chapter 1).

developed to judge how close these new sources compared with the daylight illuminants (CIE 1981). A close simulation of daylight is important when measuring fluorescent materials.

Offices and stores often use fluorescent lights. Because of the commercial importance of these lamps, the CIE recommends a series of fluorescent (F) illuminants. F2 represents a cool white fluorescent lamp, the most common lamp used in offices and stores. Other F illuminants represent warm white and simulated daylight fluorescent lamps. Figure 2.2 compares the spectral power distributions of the three most common illuminants used for color measurement: A, D65 and F2.

2.2.4. Standard observers

As described in Chapters 1 and 3 rods react to all wavelengths of the visible spectrum at low levels of illumination resulting in monochromatic (black, gray and white) vision. Increasing the light level triggers the three sets of cones (L, M, S; or R, G, B), which produce color vision. Sharpest vision occurs in the part of the eye called the fovea. The central part of the fovea contains only cones. Rods begin to appear at an angle (from the line of sight) of about 1° away from the center of the fovea (see fig. 3.4) and predominate the retina at angles greater than 4° from the fovea. L, M, and S cones are distributed more or less randomly in the retina, but there are fewer S cones than L or M cones (Hunt 1991).

Since all color vision results from signals from the cones, all colors can be produced by shining combinations of red, green and blue lights on the cones. This is the principal behind color television and color computer monitors. These visual display units use different intensities of red, green and blue phosphors to produce the different colors seen on the screen. This is also the principal used to characterize the CIE Standard Observers.

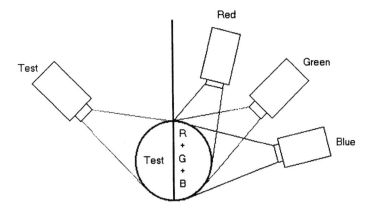

Fig. 2.3. The CIE standard observers were developed from experiments in which people matched a spectral color with mixtures of red, green and blue light.

If a single, or a very narrow band, wavelength of light from the visible spectrum is projected on one half of a screen, an observer should be able to match the color of that light by varying the intensities of red, green and blue lights projected on the other half of the screen. The amounts of red, green and blue light needed to match the spectral color are called color matching functions. An observer can be characterized by matching all of the wavelengths of the visible spectrum.

In the early standard observer experiments, subjects viewed a circular field that covered 2° of the visual field, approximately the size of dime at arm's length. Figure 2.3 illustrates the concept of this experiment. If red, green and blue lights existed that matched the sensitivity of the three types of cones, all of the spectral wavelengths could have been matched in this way. Unfortunately, when using available red, green and blue lights, some of the spectral colors could not be matched. For these colors, one of the lights had to be projected on the half of the screen containing the spectral color as shown in fig. 2.4. The light that was shifted was assigned a negative color matching function.

The results of this experiment for a number of individuals were averaged to create the first CIE Standard Observer recommended in 1931. To avoid the use of negative numbers in color calculations, the vision scientists created a special set of mathematical lights, X, Y and Z, to replace actual red, green and blue lights. The color matching functions for the X, Y, and Z lights are all positive numbers and are labeled $\bar{x}$, $\bar{y}$ and $\bar{z}$. Every color can be matched using the appropriate amounts of X, Y and Z light. The amounts of X, Y and Z light needed to match a color are called the color's tristimulus values. Because the observers in the experiment viewed a 2° visual field, the CIE 1931 Standard Observer is commonly called the CIE 2° Standard Observer.

Since most of the cones are located in the fovea, scientists believed that a 2° viewing field would result in a good characterization of a person with normal color vision viewing colors. For viewing signal lights, flares and other small objects, the 2° field is, in fact, a good approximation. Many individuals view visual fields larger than 2°. Another set of experiments were performed using a visual field of 10°. In 1964 the CIE recommended

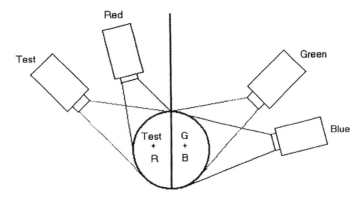

Fig. 2.4. Some spectral colors can not be matched with a combination of red, green and blue light. In this figure, red light had to be mixed with the spectral color so that it could be matched with a combination of green and blue light.

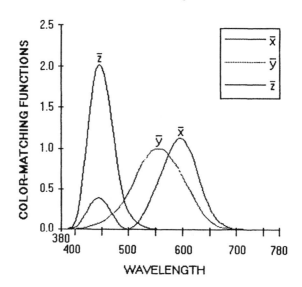

Fig. 2.5. The $\bar{x}$, $\bar{y}$ and $\bar{z}$ color matching functions for CIE 10° Standard Observer.

the CIE 1964 Supplemental Standard Observer based on the 10° field experiments – commonly called the CIE 10° Standard Observer. Figure 2.5 shows the color matching functions for the 10° Standard Observer. Most colorimetric calculations are made using the 10° Standard Observer because it more closely approximates industrial color matching and quality control viewing conditions.

It is important to remember several features about the Standard Observers. The determination of a person's color matching functions is a very difficult task. The experiments involved are tedious and time consuming. Very few people are capable of making the observations precisely and with sufficient reproducibility (ASTM 0001). Since a Standard

Observer is the average of the color matching functions of a number of individuals, any given person is likely to have slightly different color matching functions than the Standard Observer. Since the color matching functions are determined from a color matching experiment, even the color matching functions of a single individual will vary each time the experiment is performed. The experimental design and equipment will influence the determination of the color matching functions.

Researchers are attempting to better define the color matching functions with the hopes of improving the basis of mathematical colorimetry (CIE 1986b; Fairchild 1993; North and Fairchild 1993a, 1993b; Thornton 1993a, 1993b, 1993c; Thornton 1993d).

2.2.5. Objects

Objects can be characterized by the amount of light that they reflect or transmit at each wavelength of the visible spectrum, i.e., their reflectance or transmittance curve. Reflectance (or transmittance) is the ratio of the reflected (or transmitted) radiant flux (i.e., light) to that of the incident radiant flux, a number that goes from 0 to 1. This ratio is often changed to a percentage of the reflected (or transmitted) radiant flux, a number that goes from 0 to 100.

The reflectance curves for yellow, orange and red paint samples are shown in fig. 2.6. These paints absorb all the blue and some green wavelengths of the visible spectrum. The yellow paint reflects (scatters back to the observer) the green, yellow, orange and red wavelengths. The orange paint also absorbs all the green wavelengths, while the

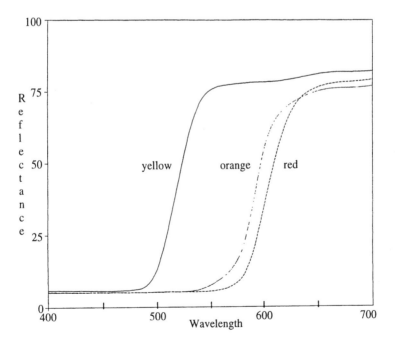

Fig. 2.6. The reflectance curves for yellow, orange and red paints.

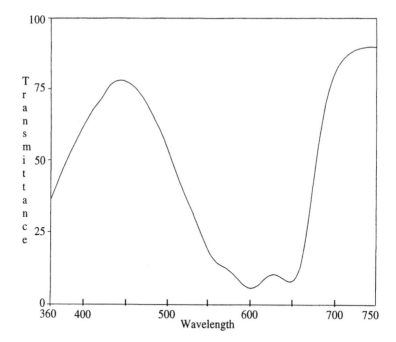

Fig. 2.7. The transmittance curve for a redish blue transparent plastic.

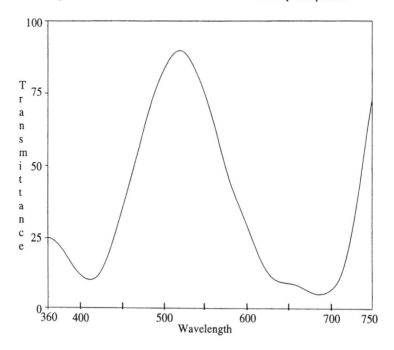

Fig. 2.8. The transmittance curve for a green transparent plastic.

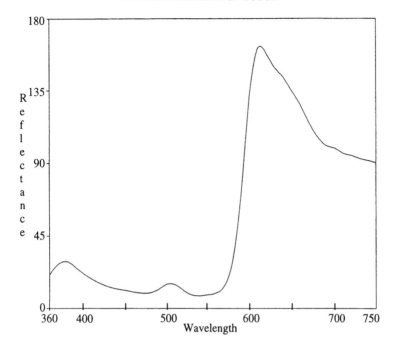

Fig. 2.9. Because fluorescent samples absorb light at some wavelengths and re-emit it at longer wavelengths, they may appear to have reflectances greater than 100%.

red paint absorbs all except the red wavelengths. Transparent yellow, orange and red materials would absorb the same wavelengths as the opaque samples, but transmit the other wavelengths rather than reflect them. The transmittance curve of a transparent red shade (i.e., reddish) blue plastic, shown in fig. 2.7, illustrates how this material absorbs much green, most of the yellow, orange and red wavelengths but transmits all the blue wavelengths. Figure 2.8 shows the transmittance curve of a transparent green plastic, which transmits the green wavelengths and absorbs almost all other wavelengths.

Fluorescent objects absorb light at some wavelengths and then emit light at other (longer) wavelengths, that is at lower energy. When measured on an instrument designed to measure reflectance, a fluorescent object may appear to reflect more than 100 percent of the light shining on it as shown in fig. 2.9. Fluorescent materials will be discussed in greater detail in Section 2.5.6.

2.2.6. Calculating tristimulus values

The CIE tristimulus values for a reflecting or transmitting sample are calculated by adding the product of the spectral power distribution of the illuminant, the reflectance or transmittance factor of the sample and the color matching functions of the observer at each wavelength of the visible spectrum. Figure 2.10 illustrates this concept for a red reflecting specimen. The areas under the three curves labeled X, Y and Z would be the numerical values of the tristimulus values.

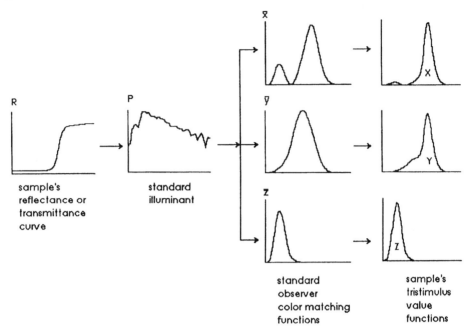

Fig. 2.10. CIE tristimulus values are found by combining a sample's reflectance or transmittance curve with a standard illuminant and with the color matching functions of a standard observer. The areas enclosed by the tristimulus value function curves would be the numerical value of the tristimulus values.

If the reflectance or transmittance factor of a sample is measured at intervals of 1 nm or 5 nm from 380 to 780 nm, eqs (2.1) through (2.4) are used to calculate the tristimulus values:

$$X = k \sum P(\lambda)\bar{x}(\lambda)R(\lambda), \tag{2.1}$$

$$Y = k \sum P(\lambda)\bar{y}(\lambda)R(\lambda), \tag{2.2}$$

$$Z = k \sum P(\lambda)\bar{z}(\lambda)R(\lambda), \tag{2.3}$$

$$k = \frac{100}{\sum P(\lambda)\bar{y}(\lambda)}, \tag{2.4}$$

where $P(\lambda)$ is the value of the spectral power distribution of the illuminant at the wavelength λ, $R(\lambda)$ is the reflectance factor of the sample at the wavelength λ, and $\bar{x}(\lambda)$, $\bar{y}(\lambda)$, $\bar{z}(\lambda)$ are the CIE color matching functions for the Standard Observer at the wavelength λ. The factor k normalizes the tristimulus value so that Y will have a value of 100 for the perfect white diffuser – a theoretical material that reflects or transmits 100 percent of the incident light. Table 2.1 shows the tristimulus calculation for the red specimen used to create fig. 2.10.

TABLE 2.1
CIE tristimulus value calculation for the red specimen of fig. 2.6 with 5 nm intervals.

Wvl	Refl*100 R	ILL D65	Color matching functions			D65*y	$R*D65*x$	$R*D65*y$	$R*D65*z$
			x	y	z				
380	5.28	49.98	0.0002	0.0000	0.0007	0.0000	0.053	0.000	0.185
385	5.19	52.31	0.0007	0.0001	0.0029	0.0052	0.190	0.027	0.787
390	5.10	54.65	0.0024	0.0003	0.0105	0.0164	0.669	0.084	2.927
395	5.04	68.70	0.0072	0.0008	0.0323	0.0550	2.493	0.277	11.184
400	4.98	82.75	0.0191	0.0020	0.0860	0.1655	7.871	0.824	35.440
405	4.95	87.12	0.0434	0.0045	0.1971	0.3920	18.716	1.941	84.998
410	4.92	91.49	0.0847	0.0088	0.3894	0.8051	38.126	3.961	175.281
415	4.89	92.46	0.1406	0.0145	0.6568	1.3407	63.569	6.556	296.959
420	4.87	93.43	0.2045	0.0214	0.9725	1.9994	93.048	9.737	442.491
425	4.86	90.06	0.2647	0.0295	1.2825	2.6568	115.857	12.912	561.339
430	4.84	86.68	0.3147	0.0387	1.5535	3.3545	132.026	16.236	651.742
435	4.83	95.77	0.3577	0.0496	1.7985	4.7502	165.461	22.943	831.931
440	4.82	104.86	0.3837	0.0621	1.9673	6.5118	193.932	31.387	994.323
445	4.82	110.94	0.3867	0.0747	2.0273	8.2872	206.780	39.944	1084.060
450	4.82	117.01	0.3707	0.0895	1.9948	10.4724	209.070	50.477	1125.044
455	4.82	117.41	0.3430	0.1063	1.9007	12.4807	194.109	60.157	1075.637
460	4.82	117.81	0.3023	0.1282	1.7454	15.1032	171.659	72.798	991.115
465	4.83	116.34	0.2541	0.1528	1.5549	17.7768	142.784	85.862	873.733
470	4.86	114.86	0.1956	0.1852	1.3176	21.2721	109.188	103.382	735.510
475	4.87	115.39	0.1323	0.2199	1.0302	25.3743	74.346	123.573	578.920
480	4.87	115.92	0.0805	0.2536	0.7721	29.3973	45.445	143.165	435.874
485	4.88	112.37	0.0411	0.2977	0.5701	33.4525	22.538	163.248	312.623
490	4.90	108.81	0.0162	0.3391	0.4153	36.8975	8.637	180.798	221.425
495	4.93	109.08	0.0051	0.3954	0.3024	43.1302	2.743	212.632	162.620
500	4.96	109.35	0.0038	0.4608	0.2185	50.3885	2.061	249.927	118.509
505	4.99	108.58	0.0154	0.5314	0.1592	57.6994	8.344	287.920	86.257
510	5.02	107.80	0.0375	0.6067	0.1120	65.4023	20.293	328.319	60.609
515	5.06	106.30	0.0714	0.6857	0.0822	72.8899	38.404	368.823	44.214
520	5.10	104.79	0.1177	0.7618	0.0607	79.8290	62.902	407.128	32.440
525	5.09	106.24	0.1730	0.8233	0.0431	87.4674	93.552	445.209	23.307
530	5.08	107.69	0.2365	0.8752	0.0305	94.2503	129.381	478.791	16.685
535	5.10	106.05	0.3042	0.9238	0.0206	97.9690	164.528	499.642	11.142
540	5.13	104.41	0.3768	0.9620	0.0137	100.4424	201.823	515.270	7.338
545	5.25	104.23	0.4516	0.9822	0.0079	102.3747	247.119	537.467	4.323
550	5.32	104.05	0.5298	0.9918	0.0040	103.1968	293.269	549.007	2.214
555	5.72	102.02	0.6161	0.9991	0.0011	101.9282	359.528	583.029	0.642
560	5.87	100.00	0.7052	0.9973	0.0000	99.7300	413.952	585.415	0.000
565	6.58	98.17	0.7938	0.9824	0.0000	96.4422	512.762	634.590	0.000
570	7.29	96.33	0.8787	0.9556	0.0000	92.0529	617.063	671.066	0.000
575	8.79	96.06	0.9512	0.9152	0.0000	87.9141	803.162	772.765	0.000
580	10.28	95.79	1.0142	0.8689	0.0000	83.2319	998.704	855.624	0.000
585	14.45	92.24	1.0743	0.8256	0.0000	76.1533	1431.900	1100.416	0.000
590	18.62	88.69	1.1185	0.7774	0.0000	68.9476	1847.100	1283.804	0.000
595	26.42	89.35	1.1343	0.7204	0.0000	64.3677	2677.659	1700.596	0.000
600	34.22	90.01	1.1240	0.6583	0.0000	59.2536	3462.080	2027.658	0.000
605	42.33	89.80	1.0891	0.5939	0.0000	53.3322	4139.924	2257.553	0.000
610	50.44	89.60	1.0305	0.5280	0.0000	47.3088	4657.266	2386.256	0.000
615	56.97	88.65	0.9507	0.4618	0.0000	40.9386	4801.406	2332.270	0.000

TABLE 2.1

(Continued)

Wvl	Refl*100 R	ILL D65	Color matching functions			D65*y	R*D65*x	R*D65*y	R*D65*z
			x	y	z				
620	63.49	87.70	0.8563	0.3981	0.0000	34.9134	4767.941	2216.650	0.000
625	67.26	85.49	0.7549	0.3396	0.0000	29.0324	4340.718	1952.719	0.000
630	71.02	83.29	0.6475	0.2835	0.0000	23.6127	3830.128	1676.975	0.000
635	72.84	83.49	0.5351	0.2283	0.0000	19.0608	3254.163	1388.386	0.000
640	74.67	83.70	0.4316	0.1798	0.0000	15.0493	2697.448	1123.728	0.000
645	75.63	81.86	0.3437	0.1402	0.0000	11.4768	2127.871	867.988	0.000
650	76.60	80.03	0.2683	0.1076	0.0000	8.6112	1644.759	659.620	0.000
655	77.15	80.12	0.2043	0.0812	0.0000	6.5057	1262.831	501.918	0.000
660	77.70	80.21	0.1526	0.0603	0.0000	4.8367	951.052	375.809	0.000
665	77.98	81.25	0.1122	0.0441	0.0000	3.5831	710.885	279.412	0.000
670	78.25	82.28	0.0813	0.0318	0.0000	2.6165	523.443	204.741	0.000
675	78.38	80.28	0.0579	0.0226	0.0000	1.8143	364.327	142.207	0.000
680	78.51	78.28	0.0409	0.0159	0.0000	1.2447	251.362	97.718	0.000
685	78.66	74.00	0.0286	0.0111	0.0000	0.8214	166.476	64.611	0.000
690	78.81	69.72	0.0199	0.0077	0.0000	0.5368	109.343	42.309	0.000
695	79.06	70.67	0.0138	0.0054	0.0000	0.3816	77.103	30.171	0.000
700	79.28	71.61	0.0096	0.0037	0.0000	0.2650	54.502	21.006	0.000
705	79.45	72.98	0.0066	0.0026	0.0000	0.1897	38.269	15.075	0.000
710	79.84	74.35	0.0046	0.0018	0.0000	0.1338	27.306	10.685	0.000
715	80.00	67.98	0.0031	0.0012	0.0000	0.0816	16.859	6.526	0.000
720	80.33	61.60	0.0022	0.0008	0.0000	0.0493	10.886	3.959	0.000
725	80.50	65.74	0.0015	0.0006	0.0000	0.0394	7.938	3.175	0.000
730	80.71	69.89	0.0010	0.0004	0.0000	0.0280	5.641	2.256	0.000
735	80.86	72.49	0.0007	0.0003	0.0000	0.0217	4.103	1.758	0.000
740	81.02	75.09	0.0005	0.0002	0.0000	0.0150	3.042	1.217	0.000
745	81.01	69.34	0.0004	0.0001	0.0000	0.0069	2.247	0.562	0.000
750	81.01	63.59	0.0003	0.0001	0.0000	0.0064	1.545	0.515	0.000
755	81.00	55.01	0.0002	0.0001	0.0000	0.0055	0.891	0.446	0.000
760	81.00	46.42	0.0001	0.0000	0.0000	0.0000	0.376	0.000	0.000
765	81.00	56.61	0.0001	0.0000	0.0000	0.0000	0.459	0.000	0.000
770	81.00	66.81	0.0001	0.0000	0.0000	0.0000	0.541	0.000	0.000
775	81.00	65.09	0.0000	0.0000	0.0000	0.0000	0.000	0.000	0.000
780	81.00	63.38	0.0000	0.0000	0.0000	0.0000	0.000	0.000	0.000

Sums   2324.1474   57257.950   34893.609   12093.827

Tristimulus Value = (Sum of $r * D65 * [x, y$ or $z]$)/(Sum of D65 $* y$) $X = 24.636$ $Y = 15.014$ $Z = 5.204$

Most commercially available spectrophotometers measure with wavelength intervals greater than 5 nm. Ten and 20 nm wavelength increments are most common. Simply summing the product of the spectral power distribution, the observer and the sample's reflectance or transmittance factor for measurements made at wider wavelength increments will not produce the adequate tristimulus values. To overcome this problem, tables of weighting factors were developed (Foster Jr. et al. 1970; Stearns 1981, 1985; Fairman 1985; Billmeyer Jr. and Fairman 1987; ASTM 0002). Since there are several ways to compute weighting factors, the use of the tables in ASTM E 308 Standard Test Method

TABLE 2.2

CIE tristimulus value calculation as in table 2.1 but with 20 nm weighting factors.

Wvl	Refl	Weighting Functions			$W(X)*$Refl	$W(Y)*$Refl	$W(Z)*$Refl
		$W(X)$	$W(Y)$	$W(Z)$			
380	0.052	− 0.044	− 0.004	− 0.207	− 0.002	0.000	−0.011
400	0.050	0.378	0.035	1.667	0.019	0.002	0.083
420	0.049	3.138	0.320	14.979	0.153	0.016	0.731
440	0.048	6.701	1.104	34.461	0.323	0.053	1.663
460	0.048	6.054	2.605	35.120	0.292	0.126	1.696
480	0.049	1.739	4.961	15.986	0.085	0.242	0.779
500	0.050	0.071	8.687	4.038	0.004	0.431	0.200
520	0.051	2.183	13.844	1.031	0.111	0.702	0.052
540	0.052	6.801	17.327	0.229	0.354	0.901	0.012
560	0.062	12.171	17.153	0.002	0.758	1.068	0.000
580	0.122	16.465	14.150	− 0.003	2.013	1.730	0.000
600	0.345	17.230	10.118	0.000	5.952	3.495	0.000
620	0.615	12.872	6.012	0.000	7.922	3.700	0.000
640	0.740	6.248	2.593	0.000	4.623	1.918	0.000
660	0.775	2.126	0.832	0.000	1.648	0.645	0.000
680	0.785	0.544	0.210	0.000	0.427	0.165	0.000
700	0.793	0.105	0.041	0.000	0.083	0.033	0.000
720	0.803	0.023	0.009	0.000	0.018	0.007	0.000
740	0.809	0.005	0.002	0.000	0.004	0.002	0.000
760	0.810	0.001	0.000	0.000	0.001	0.000	0.000
780	0.810	0.000	0.000	0.000	0.000	0.000	0.000
Sums		94.811	99.999	107.303	$X = 24.787$	$Y = 15.235$	$Z = 5.207$

for Computing the Colors of Objects by Using the CIE System is recommended (ASTM 0002).

Weighting factors are derived from the combination of the illuminant and observer data along with other adjustments based on the wavelength interval and range selected. The tristimulus values are computed by summing the product of the appropriate weighting factor, $W_x(\lambda)$, $W_y(\lambda)$ or $W_z(\lambda)$ and the reflectance at the same wavelength as shown in eqs (2.5) through (2.7):

$$X = \sum W_x(\lambda)R(\lambda), \tag{2.5}$$

$$Y = \sum W_y(\lambda)R(\lambda), \tag{2.6}$$

$$Z = \sum W_z(\lambda)R(\lambda). \tag{2.7}$$

For many industrial applications, the tristimulus values calculated using weighting factors designed for 10 and 20 nm wavelength intervals are sufficiently accurate (Strocka 1973). The spectrophotometer used to measure the sample's reflectance determines the wavelength increment which should be used for the weighting factors. That is, if the instrument measures the reflectance at 20 nm wavelength increments, then the tables of

20 nm weighting factors are used. Table 2.2 demonstrates the use of 20 nm weighting factors for the red specimen used to create fig. 2.10 and table 2.1.

A set of the CIE tristimulus values X, Y and Z represents the perception of a single color on a neutral gray surround of equal lightness. The gloss and texture of the material are not part of the equations, but may influence the measurement of the reflectance or transmittance factor. If two materials have the same tristimulus values, gloss and texture, they should match for a person having normal color vision who is viewing the specimens in a light source simulating the standard illuminant. This set of observing and viewing conditions is sometimes referred to as basic colorimetry and has wide application in industry, particularly in color matching and quality control.

Human color perception is influenced not only by the gloss and texture of the specimen but also by the background and adjacent colors, as well as by recently-viewed intense colors (see Section 1.8 of Chapter 1). Colors in a complex scene (surrounded by other colors and objects) may not appear to be the same even if they do have the same tristimulus values. Determining the calculations needed to describe complex scene colors is referred to as advanced colorimetry and is a much more difficult problem.

2.2.7. Metamerism

Metamerism is one of the fascinating phenomenon of color perception (ASTM 0003). Metamerism is probably the largest single cause of industrial color matching and acceptance problems. There are two types of metamerism related directly to the tristimulus values of two colors – illuminant metamerism and observer metamerism.

Illuminant metamerism is the most common form of metamerism and occurs when two samples match to a single observer when viewed under one light source and set of viewing and illuminating conditions, but no longer match when the light is changed. Such samples are called 'metameric'. The formal, mathematical definition of metamerism requires the tristimulus values of the two samples to be the same when calculated for the reference illuminant but different when calculated for a second illuminant. Because of the sensitivity of the human eye, two samples may still match visually and appear to be metameric when their tristimulus values differ by very slight amounts. Illuminant metamerism occurs most often when different colorants are used to make each sample but can occur when the same colorants are used in proportions which result in varying spectral curves (also see Chapter 1, Section 1.7.6, and Chapter 3, Section 3.4).

Figure 2.11 shows the spectral reflectance curves of two metameric samples. These specimens match under Illuminant D65 but cease to match under Illuminants A or F2. The spectral reflectance (or transmittance) curves of two metamers will cross at least three times. The severity (amount of mismatch) of a pair of metamers decreases as the number of reflectance curve cross-overs increases.

Observer metamerism occurs when two samples have differing spectral curves match to one observer but do not match to a different observer when viewed under identical conditions. The more different the spectral curves of the two specimens, the more likely observer metamerism will occur. Tristimulus values are calculated using a Standard Observer. Any particular person with normal color vision will most likely have different color matching functions than the Standard Observer. If the second observer has normal

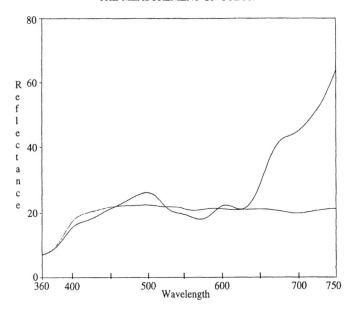

Fig. 2.11. The reflectance curves of a pair of metameric samples that cross at least three times.

color vision but has color matching functions significantly different than the first observer, observer metamerism can occur. Mathematically, the tristimulus values calculated for the two samples using the first observer's actual color matching functions would be equal (or very nearly so). If the second observer's actual color matching functions were used to calculate tristimulus values, they would be different for each sample.

Metamerism requires that the two samples have different spectral curves. If two samples differ in gloss or texture, or are made with metallic or pearlescent pigments, they may appear to match when viewed under one set of illuminating and viewing angles and not match if either the illuminating or viewing angle is changed. This phenomenon, goniochromatism, has been called geometric metamerism, a term to be avoided. Goniochromatic sample pairs may or may not be metameric. They are only metameric if each sample has a different spectral curve under the matching conditions.

Industrial color matches may not have to be perfect visually to be acceptable. If two samples differ slightly in color when viewed under one light source but differ by a noticeably different amount in a different light source, most industrial color matchers would say that the two samples were metameric. By this they mean that the two specimens would be metameric if they were made to match. Since the definition of metamerism requires a match, the use of the term metamerism to describe this condition is not accurate. The term paramerism has been introduced to describe this phenomenon and two parameric samples would be called paramers instead of metamers.

2.2.8. Chromaticity coordinates and the chromaticity diagram

Visualizing a color, or the relationship among colors, from their tristimulus values is impossible for most people. Making a graph or plot can be a useful tool, but is not easily done since color is three dimensional. A color plot can be more easily generated by separating colors into two components, lightness and chromaticity.

Lightness is described by the terms light and dark. Chromaticity is a combination of a color's hue and chroma. Hue is described with common color names such as blue, red, green, etc. A color's chroma is its colorfulness. A high chroma color may be described as highly saturated, such as the bright red of a tomato or the bright yellow of a plastic toy. A dull red brick (of equal lightness as the tomato) would have a low chroma as would the yellow of a lemon whose lightness was equal to the toy.

The Y tristimulus value was designed so that it can be used to represent the lightness. Chromaticity coordinates, x, y and z, are used to represent a color's chromaticity. They are calculated from the color's tristimulus values with eqs (2.8) and (2.9).

$$x = \frac{X}{X + Y + Z},$$ (2.8)

$$y = \frac{Y}{X + Y + Z}.$$ (2.9)

It is not necessary to calculate z, which is $1 - x - y$.

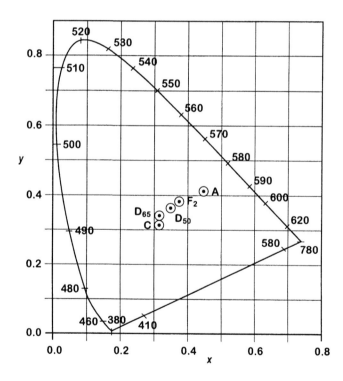

Fig. 2.12. When the chromaticities of spectral colors are plotted on a CIE x, y chromaticity diagram, they lie on a horseshoe shaped curve. Purple colors lie on the line connecting the ends of the horseshoe (also see color plate 4).

A chromaticity diagram based on illuminant C and the 1931 Standard Observer is shown in fig. 2.12 and in color plate 4. This is often called the color triangle. The chromaticity of the wavelengths of the visible spectrum form a horseshoe shape on the diagram. The non-spectral purples formed by mixing the 380 nm blue and 780 red wavelengths of the spectrum fall on the straight line connecting the ends of the horseshoe. Lightness is ignored on a chromaticity diagram, but the range of chromaticities possible for real colors will vary with lightness. Color plate 4 contains an artist's representation of where colors lie on the 1931 CIE x, y chromaticity diagram. The CIE does not associate specific colors with regions on the x, y chromaticity diagram.

Chromaticity diagrams can be very useful for predicting the color of a mixture of two or more lights or for the quality control of materials (ASTM 0004). A mixture of two lights will fall on the line connecting the chromaticities of each at a position predicted by the relative luminances of each light in the mixture (see Section 1.6.1 of Chapter 1). Industrial color tolerances can be set by plotting the chromaticities of accepted batches relative to the standard color. Since color space is three dimensional and a chromaticity diagram is two dimensional, lightness (Y) limits must also be given when using the chromaticity diagram for quality control applications.

2.3. Color collections and uniform color spaces

2.3.1. Color collections

Communicating color and appearance is a difficult task. Color names are not very precise and can cause a great deal of confusion. What color is *grass green*? Is it a bluish green or a yellowish green? Other names, such as *firewater*, *murmur*, and *butterfly*, are even more difficult to interpret. Even more confusing, two different colors may have the same name.

Color collections are groups of samples identified by a name and/or number. Color cards make up the largest category of color collections. Product manufacturers issue color cards to help the consumer pick the color of the product which they are purchasing. Several companies such as ICI-Glidden, TOYO Ink, and DIC publish color collections based upon the paints and inks that they sell. Color cards come in a wide variety of materials. Governments and other organizations may issue color cards as "standards" to specify products they wish to purchase. The 595a series of paint-on-paper standards issued by the United States government is an example of this category of color cards. Some companies produce color standards and color specifiers that are not related to a particular product either manufactured or purchased. Pantone, Inc., for example, publishes color specifiers consisting of ink on paper, paint on paper, dyed textiles and plastic chips (see color plate 7).

2.3.2. Color order systems and uniform color spaces

Color order systems are three-dimensional arrangements of color according to appearance. Each color has a notation relating to its position in the arrangement. Most color order systems have an accompanying atlas containing real samples at selected points inside the system. The atlas is used as a teaching tool and as an aid for assigning the proper notation to colors not included in the atlas. Color order systems are often based upon perceptually uniform color spaces.

A common problem centers on evaluating and quantifying the color difference between two colors when they do not match. To solve this problem, it is useful to develop a color difference ruler, that is a color difference equation. Color difference equations are easiest to use when they are associated with a uniform color space.

2.3.3. Munsell's color order system

Albert H. Munsell, an artist and teacher, was faced with two problems in the early 1900s. Munsell wanted to be able to specify transitory colors so that they could be recorded on canvas when he returned to his studio at a later time. He also wanted to teach his students the artistic concepts of color. Munsell created a color system to accomplish these tasks. The first edition of Munsell's *A Color Notation* was published in 1905. A set of charts exemplifying this notation appeared as an *Atlas of the Munsell Color System* in 1915.

Charts in the Atlas were based on careful psychophysical observation and experimental measurements and designed to represent a uniform color space. Munsell developed a ten-step 'Value' (lightness) scale ranging from white to black in perceptually equal steps. A ten-step 'Hue' circle was developed at equal 'Chroma' (saturation) with each hue being perceptually equally distant from the two adjacent hues. Munsell developed his Chroma scale from the highest saturation contemporary materials available at that time. Neutral gray samples have a Chroma of 0.0 and Munsell assigned his maximum chroma red a Chroma of 10.0. The remaining maximum-chroma hue samples were assigned Chroma values relative to that of the red. Munsell color space is a cylinder with Value being the central axis, Chroma radiating outward from the Value axis and circles of constant Hue forming the outside boundaries of the cylinder (see fig. 2.13 and color plates 1 and 2). One Value step of lightness difference is visually the equivalent to that of two steps of Chroma difference.

Munsell color notations are written in the form of Hue Value/Chroma, such as 5PB 4/8. There are ten Munsell Hues designated by the first letter or letters of the Hue name: Red, Yellow-Red, Yellow, Green-Yellow, Green, Blue-Green, Blue, Purple-Blue, Purple and Red-Purple. There are ten Hue units associated with each name so that 5PB would be the Hue in the middle of the purple-blue Hue range. A value of 4 would be a little less than the middle of the lightness scale and a chroma of 8 indicates a fairly pure color. Since each of the Munsell scales is continuous, they can be sub-divided into as small of an increment as desired by using decimals, such as 5.36PB 4.27/8.41. Neutral colors, i.e., usually colors having a Chroma of less than 0.2, are designated by the letter N followed by the Value, such as N 3.92/4.2.

In 1929 the *Munsell Book of Color* superseded the Atlas. Chips presented in the *Munsell Book of Color* were designed to have the perceptual characteristics of an ideal color space. In such a space, equal spatial distances would have equal perceptual steps (Nickerson and Newhall 1943; Judd 1970; ASTM 0005).

Munsell color space can be considered to be a perceptually uniform color space. Figure 2.14 shows a Munsell hue circle having a Munsell Value of 5 and a Munsell Chroma of 8. Munsell color space can be used to illustrate the concept of a color difference ruler. If the distance between Munsell colors 5B 5/8 and 5PB 5/8 is considered to be one

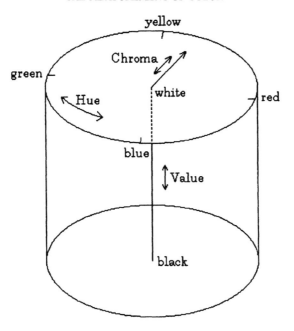

Fig. 2.13. Munsell color space is a cylinder with Value (lightness–darkness) making up the center axis, Chroma radiating outward from the Value axis with circles of constant Hue forming concentric circles around the Value axis (also see color plates 1 and 2).

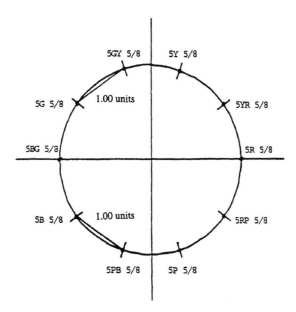

Fig. 2.14. A Munsell hue circle is a perfect circle in a visually uniform color space. Measured distances between pairs of colors that look equally different visually, would be the same; compare with color plate 6.

unit, the distance between colors 5G 5/8 and 5GY 5/8 (or any two adjacent colors on the circle) is also one unit. This ruler could be used to measure the distance between any two pairs of colors in Munsell color space. If each pair of colors measured the same distance, they would appear to have equal visual color differences. Cooperative efforts between the Munsell Color Company[1] and the Optical Society of America's Committee on Uniform Color Scales improved the correlation of the Munsell system with psychophysical scales and instrumental color measurements and has led to the currently accepted set of Munsell notations (Newhall et al. 1943). Computer programs have been developed that allow the calculation of Munsell notations from a sample's X, Y and Z tristimulus values (ASTM 0005).

2.3.4. Other color order systems

NCS, the Natural Color System, is widely used in Europe and is a Swedish Standard (Hård and Sivik 1981; Swedish Standard 0001, 1979). Hering's six elementary colors, White, Black, Yellow, Red, Blue and Green, form the basis of this system. An elementary color is one in which no trace of another color can be found. Thus the blue contains no red or green. The two achromatic colors, White and Black, form the central axis of the system and the remaining colors are arranged to form a hue circle. Each of the four elementary colors are placed 90° apart, with Yellow opposite Blue and Red opposite Green. Colors are described by their resemblances, in percentages, to the elementary color pairs White–Black, Yellow–Blue and Red–Green. The Chromaticness of a color is equal to the sum of its resemblances to the two elementary chromatic pairs. Thus a color notated as 20 40 Y30R would have a resemblance to Black (the first number) of 20%, a Chromaticness (the second number) of 40 and a Hue (the third number) of Y30R. Y30R indicates a 70% resemblance to Yellow and a 30% resemblance to Red.

In an effort to achieve the best possible uniform visual spacing of colors, the Optical Society of America's Committee on Uniform Color Scales developed the OSA-UCS (Uniform Color Scales) color order system (MacAdam 1974; Nickerson 1981; ASTM 0006). A 558-chip set of OSA-UCS colors was produced to illustrate the best uniform visual spacing that could be achieved on a regular rhombohedral lattice, allowing for the closest uniform spacing in three dimensions. The three dimensions of the OSA-USC color space are lightness, L, a yellow-blue dimension, j (from jeune, the French word for yellow), and a green-red dimension, g. A color in the OSA-USC system is designated as $L : j : g$ in which L ranges from -7 to $+5$, j ranges from -6 to $+15$ and g ranges from -10 to $+6$. The lightness scale is related to the recommended viewing background of the sample. Positive values of L are used for samples that are lighter than the preferred background, and negative values for samples that are darker. The unit length of the L axis is shorter than the j and g axes by $\sqrt{2}$.

The DIN (Deutsche Industrie Normung – German Standardization Institute) color system is a national standard in Germany (Hunt 1991; Billmeyer Jr. 1987; Richter and Witt 1986). Extensive visual scaling studies were carried out to develop a system demonstrating the equality of visual spacing. The three dimensions of the DIN system are hue, T, saturation, S, and relative brightness, D. Colors in the DIN system are designated

---

[1] The Munsell Color Company is now part of GretagMacbeth, LLC, New Windsor, New York.

in the form of $T : S : D$, for example 16 : 6 : 4. There are 24 equally spaced hues in the hue circle with CIE lines of constant dominant and complementary dominant wavelength used as lines of constant hue. Colors with constant lightness were scaled to provide constant saturation around the hue circle. Only eight hues were used to set up the saturation scale and the saturation determined by interpolation and extrapolation on a CIE u, v chromaticity diagram (see section 2.3.7). Darkness degree, D, is used as a lightness scale in the DIN system. A perfect white has a D of 0 and a perfect black has a D of 10.

A Hungarian color order system, Coloroid, was developed for architects for use in the design of colored environment (Hunt 1991; Nemcsics 1980). Coloroid attempts to achieve equality of aesthetic spacing. The three dimensions of the Coloroid system are hue, A, saturation, T, and lightness, V. Colors are designated in the form A–T–V, for example 13–22–56. Hue is based on a 48-step aesthetically evenly spaced hue circle. Saturation is the percent of a pure spectral color or purple in a mixture with a pure white and a pure black. Lightness is defined as the square root of the luminance factor, Y.

The COLORCURVE® system provides a systematic array of color samples whose arrangement has a simple relationship to the CIE 10° Standard Observer, illuminant D65 tristimulus values (Stanziola 1992; ASTM 0007). The system was based on aim points displayed on a CIE a^*, b^* chromaticity diagram at constant levels of L^* and there are atlases available exemplifying the system. Colors in the atlases are displayed on grids similar to, but not identical with, CIELAB a^*, b^* (see section 2.3.6). The three dimensions of the COLORCURVE system are lightness, L^*, a red–green axis, R/G, and a yellow–blue axis, Y/B. There are 18 lightness levels ranging from $L^* = 30$ to $L^* = 95$. COLORCURVE coordinates are based on the aim points of the colors in the atlases. Thus color L65 R3 Y4 would be a color whose lightness, L^*, is 65 and which lies three steps from the achromatic axis along the R axis and four steps from the achromatic axis along the Y axis. Computer software was developed that allows a COLORCURVE notation to be determined directly from the CIE Tristimulus Values.

2.3.5. The 1976 CIE color spaces (CIE 1986a)

A weakness of the CIE X, Y, and Z color space is its lack of visual uniformity. Figure 2.15 shows a Munsell hue circle plotted on a CIE x, y chromaticity diagram. The Munsell hue circle no longer looks like a circle. Using the same color difference ruler that was used to examine Munsell color space, the distance between the colors 5B 5/8 and 5PB 5/8 is set equal to one unit. The distance between the colors 5G 5/8 and 5GY 5/8 on the x, y chromaticity diagram is measured to be 3.86 units, which is considerably greater. Since the visual color differences between those sets of colors would appear equal, the chromaticity diagram is not a uniform color space.

Creating a uniform color space would have two major advantages. It would allow plots showing the perceptually relative positions of two or more colors in color space, and it would facilitate the creation of a good color difference ruler between two samples.

Throughout the years, a number of attempts have been made at developing color difference equations and uniform color spaces. More complete discussions of color difference equations and uniform color spaces can be found in books by Wyszecki and Stiles (1982)

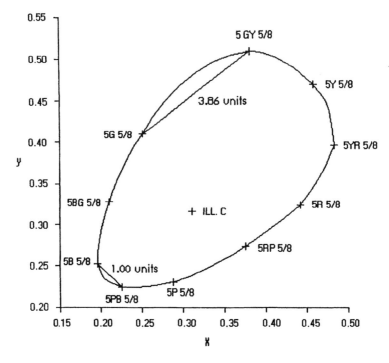

Fig. 2.15. Plotting a Munsell hue circle on a CIE x, y chromaticity diagram illustrates the lack of visual uniformity of x, y space. Although the observed color difference between Munsell colors 5G 5/8 and 5GY 5/8 appears to be equal to the color difference between 5B 5/8 and 5PB 5/8, the measured distance is 3.86 times greater.

and by Hunter and Harold (1987). In an attempt to unify the practice of color-difference evaluation and to assist those people concerned with setting and describing color tolerances, the CIE recommended two uniform color spaces and associated color difference equations in 1976. Although it would have been preferable to recommend only one color space for all needs, the experimental evidence available was considered insufficient to warrant the selection of a single color-difference formula which would be satisfactory for most industrial applications on merit alone (CIE 1978). The 1976 CIE $L^*a^*b^*$ (CIELAB) color space is widely used in the paint, plastic and textile industries while the 1976 CIE $L^*u^*v^*$ (CIELUV) color space is widely used in the television and video display industries.

2.3.6. *The 1976 CIE $L^*a^*b^*$ (CIELAB) color space*

L^* correlates with perceived lightness in CIELAB color space. A perfect white would have an L^* of 100, and a perfect black would have an L^* of 0. The coordinates a^* and b^* have their history in the opponent color theory of Ewald Hering (1964). Hering proposed that three pairs of opposing color sensations produce all colors: red and green; yellow and blue; and black and white. The CIELAB coordinate a^* correlates with red

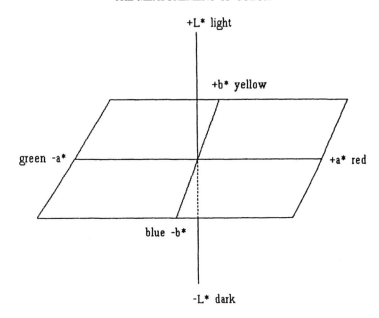

Fig. 2.16. CIELAB $L^*a^*b^*$ color space has coordinates representing lightness–darkness, redness–greenness and yellowness–blueness.

$(+a^*)$ and green $(-a^*)$ while the CIELAB coordinate b^* correlates with yellow $(+b^*)$ and blue $(-b^*)$. Figure 2.16 is a diagram of the CIELAB coordinate system.

CIELAB L^*, a^* and b^* coordinates are calculated from the tristimulus values according to eqs (2.10) through (2.12).

$$L^* = 116\, f(Y/Y_n) - 16, \tag{2.10}$$

$$a^* = 500\, [f(X/X_n) - f(Y/Y_n)], \tag{2.11}$$

$$b^* = 200\, [f(Y/Y_n) - f(Z/Z_n)], \tag{2.12}$$

in which X, Y and Z are tristimulus values and the subscript n refers to the tristimulus values of the perfect diffuser for the given illuminant and standard observer; $f(X/X_n) = (X/X_n)^{1/3}$ for values of (X/X_n) greater than 0.008856 and $f(X/X_n) = 7.787(X/X_n) + 16/116$ for values of (X/X_n) equal to or less than 0.008856; and the same with Y and Z replacing X in turn.

Recognizing the usefulness of the concepts of chroma and hue, the CIE also defined the chroma, C^*_{ab}, and hue angle, h_{ab}, for CIELAB space as

$$C^*_{ab} = \left[a^{*^2} + b^{*^2}\right]^{1/2}, \tag{2.13}$$

$$h_{ab} = \arctan[b^*/a^*]. \tag{2.14}$$

The hue angle has a value of $0°$ (or $360°$) along the $+a^*$ axis, a value of $90°$ along the $+b^*$ axis, $180°$ along $-a^*$ and $270°$ along $-b^*$. CIELAB chroma and hue are illustrated

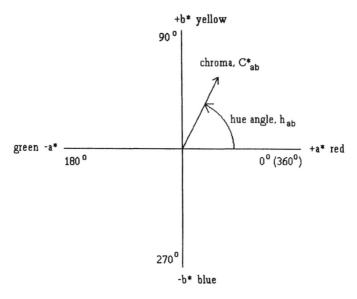

Fig. 2.17. The CIE defined hue and chroma for CIELAB color space (also see color plate 6).

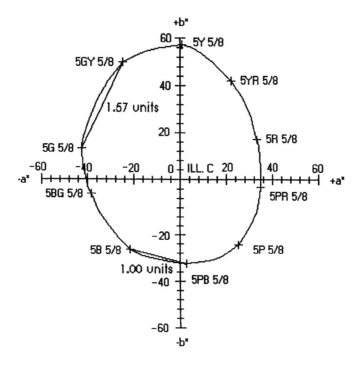

Fig. 2.18. CIELAB color space is more uniform than CIE x, y space, but it is still not perfect. The measured distance between Munsell colors 5GY 5/8 and 5G 5/8 is still 1.57 times greater than the distance between 5B 5/8 and 5PB 5/8.

in fig. 2.17. CIELAB chroma is not the same as Munsell Chroma. Color plate 6 contains an artist's representation of where colors lie on a CIELAB a^*, b^* diagram. The CIE does not associate specific colors with regions on the CIELAB a^*, b^* diagram. Surrounding the hue circle are approximations of the Munsell hues for several lightness ranges.

Although CIELAB provides a more uniform color space than using the chromaticity diagram, it is still not perfect. Figure 2.18 shows the Munsell hue circle in CIELAB space. Still using the distance between 5B 5/8 and 5PB 5/8 to represent 1 unit, the distance between 5G 5/8 and 5GY 5/8 is only 1.57 units.

2.3.7. The 1976 CIE L*u*v* (CIELUV) color space

The main advantage of CIELUV color space is that it has a chromaticity diagram associated with it as shown in fig. 2.19. Mixtures of colored lights or phosphors can be represented very simply on a chromaticity diagram which makes this color space popular in the television and video display industries. Color plate 5 contains an artist's representation of where colors lie on the CIELUV u', v' chromaticity diagram. The CIE does not associate specific colors with regions on the u', v' chromaticity diagram.

CIELUV L^*, u^* and v^* coordinates are calculated from eqs (2.15) through (2.22).

$$L^* = 116\,(Y/Y_n)^{1/3} - 16, \quad \text{when } Y/Y_n \text{ is greater than } 0.008856, \tag{2.15}$$

$$L^* = 903.3\,(Y/Y_n), \quad \text{when } Y/Y_n \text{ is less than or equal to } 0.008856, \tag{2.16}$$

$$u^* = 13\,L^*(u' - u'_n), \tag{2.17}$$

$$v^* = 13\,L^*(v' - v'_n), \tag{2.18}$$

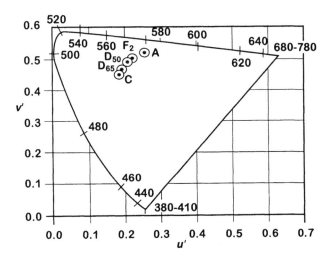

Fig. 2.19. Having a chromaticity diagram is an advantage of CIELUV color space (also see color plate 5).

where the chromaticity coordinates u', v', u'_n, v'_n are given by

$$u' = \frac{4X}{X + 15Y + 3Z}, \tag{2.19}$$

$$v' = \frac{9Y}{X + 15Y + 3Z}, \tag{2.20}$$

$$u'_n = \frac{4X_n}{X_n + 15Y_n + 3Z_n}, \tag{2.21}$$

$$v'_n = \frac{9Y_n}{X_n + 15Y_n + 3Z_n}. \tag{2.22}$$

The CIE also defined chroma, C^*_{uv}, and hue angle, h_{uv}, for CIELUV space with eqs (2.23) and (2.24).

$$C^*_{uv} = \left[u^{*2} + v^{*2} \right]^{1/2}, \tag{2.23}$$

$$h_{uv} = \arctan[v^*/u^*]. \tag{2.24}$$

In addition, the CIE defined a correlate to perceived saturation, S^*_{uv} with the eq. (2.25) or its equivalent form in eq. (2.26).

$$S^*_{uv} = C^*_{uv}/L^*, \tag{2.25}$$

$$S^*_{uv} = 13 \left[(u' - u'_n)^2 + (v' - v'_n)^2 \right]^{1/2}. \tag{2.26}$$

The perceived saturation, S^*_{uv}, remains constant for a series of colors of constant chromaticity as the lightness increases or decreases (Hunt 1991).

2.3.8. Color atlases based upon CIELAB space

Although the CIE does not publish any atlases showing colors in CIELAB color space, at least two have been created. The RAL Design System[2] arranges colors according to their CIE $L^*a^*b^*$ notation, and the Eurocolor Atlas[3] arranges colors based on their CIE L^*C^*h notations.

2.4. Color differences and tolerances

2.4.1. Describing color differences

Much of the time spent on color in industry involves describing color differences. Color measurement and color difference equations allow a quantitative description of color and color differences.

---

[2] RAL – Deutsches Institut für Gütesicherung und Kennzeichnung, Siegburger Strabe 39, 53757 Sankt Augustin, Germany.
[3] Eurocolor Limited, Wiltshire Road, Dairycoates, Hull HU4 6PA, England.

It is important to distinguish between perceptibility and acceptability. A color difference between two samples is perceptible if it can be seen. Even when a color difference is perceived between a standard and a batch, the color of the batch may still be acceptable for use. The amount of difference allowed is described by a color tolerance and varies with the use of the product. For example, the plastic used to line the interior of a refrigerator does not have to be a perfect match to the exterior color. However, if two bolts of cloth are used to make the two legs of a single pair of pants, the customer would not tolerate any visual difference between the legs.

Verbal descriptions of color differences are quite varied. Terms such as stronger, duller, slightly blue and too gray are used to describe color differences. Unfortunately one person's slightly may be another person's strongly. Instrumental color differences should agree with perceived color differences and both should be described in very specific terms, preferably the terms hue, chroma and lightness. Each describes a dimension of color independent of the other two, and each has instrumental correlates. Thus a batch can be equal in lightness to, or lighter or darker than the standard. The batch's chroma can be equal to, higher or lower than the standard. Hue differences from red or green standards would be described as yellower or bluer. Similarly, hue differences from yellow or blue standards would be redder or greener. Any of the hue difference terms could be used to describe the hue difference of a gray batch relative to its standard.

Instrumental measurements will correlate best with visual evaluation when standardized visual evaluation procedures (ASTM 0008, 0009) are followed and when the standard and the trial have the same gloss and texture (ASTM 0010, 0011).

2.4.2. The CIELAB color difference equation

The first step in calculating a CIELAB color difference is to subtract the L^*, a^* and b^* values of the trial or batch from the standard:

$$\Delta L^* = \Delta L^*_{\text{trial}} - \Delta L^*_{\text{standard}}, \tag{2.27}$$

$$\Delta a^* = \Delta a^*_{\text{trial}} - \Delta a^*_{\text{standard}}, \tag{2.28}$$

$$\Delta b^* = \Delta b^*_{\text{trial}} - \Delta b^*_{\text{standard}}. \tag{2.29}$$

A positive value of ΔL^* indicates that the trial is lighter than the standard whereas a negative value indicates that the trial is darker. A positive value of Δa^* indicates that the trial is redder than the the standard; a negative value indicates that the trial is greener. A positive value of Δb^* indicates that the trial is yellower than the standard; a negative value indicates that the trial is bluer. The description of a color difference in terms of Δa^* redness–greenness or of Δb^* yellowness–blueness can be incorrect if the trial and standard differ greatly in chroma.

The total CIELAB color difference, ΔE_{ab}, is given by:

$$\Delta E^*_{ab} = \left[(\Delta L^*)^2 + (\Delta a^*)^2 + (\Delta b^*)^2 \right]^{1/2}. \tag{2.30}$$

The chroma difference, ΔC^*_{ab}, between trial and standard is given by:

$$\Delta C^*_{ab} = C^*_{ab,\text{trial}} - C^*_{ab,\text{standard}}. \tag{2.31}$$

A negative value of ΔC^*_{ab} indicates that the trial has a lower chroma and a positive value indicates that the trial has a higher chroma.

CIELAB hue differences are *not* calculated by subtracting hue angles. The hue difference, ΔH^*_{ab}, is that part of the color difference that is left after accounting for lightness and chroma differences and is given by:

$$\Delta H^*_{ab} = \left[(\Delta E^*_{ab})^2 - (\Delta L^*)^2 - (\Delta C^*_{ab})^2 \right]^{1/2}. \tag{2.32}$$

The sign of ΔH^*_{ab} is positive if the hue angle of the trial or batch is greater than that of the standard and is negative if the hue angle is less than that of the standard.

In many computer programs and typed reports the English letter D is often substituted for the Greek letter Δ. Most of the time the superscript * and the subscript $_{ab}$ will be dropped from the CIELAB color difference notations. For example, the total CIELAB color difference, ΔE^*_{ab} will often be typed as DE.

2.4.3. The CIELUV color difference equation

A CIELUV color difference is found in a similar manner to a CIELAB color difference. The first step in calculating a CIELUV color difference is to subtract the L^*, u^* and v^* values of the trial from the standard:

$$\Delta L^* = \Delta L^*_{\text{trial}} - \Delta L^*_{\text{standard}}, \tag{2.33}$$

$$\Delta u^* = \Delta u^*_{\text{trial}} - \Delta u^*_{\text{standard}}, \tag{2.34}$$

$$\Delta v^* = \Delta v^*_{\text{trial}} - \Delta v^*_{\text{standard}}. \tag{2.35}$$

A positive value of ΔL^* indicates that the trial is lighter than the standard whereas a negative value indicates that the trial is darker. Δu^* and Δv^* represent differences in the chromaticity coordinates, but are not associated with color names as are Δa^* and Δb^*. The total CIELUV color difference, ΔE^*_{uv}, is given by:

$$\Delta E^*_{uv} = \left[(\Delta L^*)^2 + (\Delta u^*)^2 + (\Delta v^*)^2 \right]^{1/2}. \tag{2.36}$$

The chroma difference, ΔC^*_{uv}, between a trial and standard is given by:

$$\Delta C^*_{uv} = C^*_{uv,\text{trial}} - C^*_{uv,\text{standard}}. \tag{2.37}$$

A negative value of ΔC^*_{uv} indicates that the trial has a lower chroma and a positive value indicates that the trial has a higher chroma.

CIELUV hue differences are *not* calculated by subtracting hue angles. The hue difference is that part of the color difference that is left after accounting for lightness and chroma differences and is given by:

$$\Delta H^*_{uv} = \left[(\Delta E^*_{uv})^2 - (\Delta L^*)^2 - (\Delta C^*_{uv})^2 \right]^{1/2}. \tag{2.38}$$

The sign of ΔH^*_{uv} is positive if the hue angle of the trial or batch is greater than that of the standard and is negative if the hue angle is less than that of the standard.

2.4.4. The CMC(l:c) color difference equation

When the CIE recommended the CIELAB and CIELUV color spaces and color difference equations, they realized that those spaces were not perfect. In their recommendation, the CIE suggested guidelines for future work. It is interesting to note that the CIE recommended that (CIE 1978): "Future work should be directed to the perceptibility of color differences instead of the acceptability of color differences. Acceptability judgments may vary significantly from one application to another." However, many practical industrial problems involve the acceptability of products. Research done on the acceptability of textiles (Clark et al. 1984) led the Color Measurement Committee (CMC) of the Society of Dyers and Colorists of Great Britain to endorse a new color-difference equation which has become known as the CMC($l : c$) color difference equation. The CMC($l : c$) color difference equation became a British Standard (British Standard 1988) and has been adopted by the American Association of Textile Chemists and Colorists as a standard (AATCC 0001).

The British research showed that acceptability data plotted in CIELAB space gave ellipsoids rather then the spheres one would want in a perfectly uniform (for acceptability) color space. In the CMC($l : c$) equation, the CIELAB differences ΔL^*, ΔC_{ab}^* and ΔH_{ab}^* are modified to make the acceptability data plots more spherical. The two constants l and c allow the user to change the weighting of the lightness difference relative to the chroma difference. The textile industries prefers an $l : c$ weighting of 2 : 1. Different weightings may be more appropriate to different industries. One approach to determining the weightings would be to start with a weighting of 2 : 1 and adjust it downward (or upward) if the acceptance rates indicate a change. A second approach would be to try various ratios of $l : c$ on existing data until the best fit to past acceptances is found.

CMC($l : c$) uses the CIELAB L^*, C_{ab}^* and h_{ab}^* coordinates to calculate the CMC($l : c$) color differences with eqs (2.39) through (2.45).

$$S_L = \frac{0.040975 \, L^*}{1 + 0.01765 \, L^*} \tag{2.39}$$

unless $L^* < 16$, in which case $S_L = 0.511$,

$$S_C = \frac{0.638 \, C^*}{1 + 0.0131 \, C_{ab}^*} + 0.638, \tag{2.40}$$

$$S_H = (FT + 1 - F)S_C, \tag{2.41}$$

$$F = \left(\frac{(C_{ab}^*)^4}{(C_{ab}^*) + 1900} \right)^{1/2}, \tag{2.42}$$

$$T = 0.36 + \mathrm{abs}[0.4 \cos(35 + h_{ab})], \tag{2.43}$$

unless h_{ab} is between $164°$ and $345°$, in which case

$$T = 0.56 + \mathrm{abs}[0.2 \cos(168 + h_{ab})], \tag{2.44}$$

in which the notation "abs" is used to indicate the absolute, i.e., positive value, of the
term inside the square brackets

$$\Delta E = \left[\left(\frac{\Delta L^*}{lS_L} \right)^2 + \left(\frac{\Delta C_{ab}^*}{cS_C} \right)^2 + \left(\frac{\Delta H_{ab}^*}{S_H} \right)^2 \right]^{1/2}.$$
(2.45)

Recently the CMC($l : c$) equation has found acceptance in the plastics, (Liebeknecht
1992), colorant (West 1992) and paint (Adebayo and Rigg 1986) industries.

2.4.5. The CIE94 color difference equation

The Industrial Color-Difference Technical Committee (TC 1-29) of the CIE recommended
a new color difference equation, CIE94, based on the CMC($l : c$) equation (CIE 1995).
The Committee retained some of the features of the CMC($l : c$) formula but felt that others
did not contribute to its success. Although they kept the same form of the CMC($l : c$)
equation, they changed the weighting factors S_L, S_C and S_H.

A perceived color-difference, ΔV, is related to the measured color difference through
an overall sensitivity factor, k_E.

$$\Delta V = k_E^{-1} \Delta E_{94}^*,$$
(2.46)

$$\Delta E_{94}^* = \left[\left(\frac{\Delta L^*}{k_L S_L} \right)^2 + \left(\frac{\Delta C_{ab}^*}{k_C S_C} \right)^2 + \left(\frac{\Delta H_{ab}^*}{k_H S_H} \right)^2 \right]^{1/2},$$
(2.47)

$$S_L = 1,$$
(2.48)

$$S_C = 1 + 0.045 \, C_{ab}^*,$$
(2.49)

$$S_H = 1 + 0.015 \, C_{ab}^*.$$
(2.50)

The overall sensitivity factor, k_E, is not intended to be used as a commercial color
difference tolerance factor, but to account for variation in the illuminating and viewing conditions. A person in the textile industry who is using CMC(2 : 1) and would
like to compare the results with CIE94 would set $k_L = 2$ and $k_C = k_H = 1$, i.e.,
CIE94(2 : 1 : 1).

2.4.6. Other color difference equations

A number of other color difference equations have been developed in the past but do not
enjoy the popularity that they once did (ASTM 0012). The most widely used of these are
the FMCI (Chickering 1967), FMCII (Chickering 1971) and the Hunter LAB equation.

There are also a number of single dimensional scales that have been developed over the
years for specific applications. These include whiteness scales (ASTM 0013), yellowness
scales (ASTM 0014) and a special scale for near-colorless diamonds developed by the
Gemological Institute of America.

2.4.7. Product standards

Perhaps the greatest industrial use of color measurement is in evaluating the color of a product batch to the color standard for that color. High quality color standards thus form the keystone of a successful color control program. Although the exact requirements for a product sample will be dependent upon the product being evaluated, there are some general guidelines which are applicable for all products (ASTM 0015; SAE 0001).

The product standard should be the only standard by which the product should be accepted or rejected. It should be made out of the same materials as the final product and in a similar manner. Although the search for permanence of a product standard has led some to substitute more permanent materials than the product, this should be avoided. The probability of having problems with metamerism usually outweighs any gain in permanence (Fairman 1981). Product standards should have the same non-spectral attributes, such as gloss, texture and patterning, as the product itself. Product standards should be stored in the dark and kept away from chemical fumes. They should replaced when they get dirty, change color because of aging, suffer surface defects (scratching, etc.) or are otherwise damaged.

Since it is often difficult to reproduce product standards exactly at different times, it is recommended that a number of standards be produced at the same time and that three levels of product standards be set up. A master standard serves as the ultimate reference for judging color. The master standard is the standard that the customer officially approved and should not be used for routine evaluations. Duplicate master standards should be made at the same time as the master standard and are intended to be identical in every aspect. Duplicate master standards become the working standards for routine evaluation. Thus the number of duplicate master standards created depends upon both the expected life of the standard itself and the expected life of that particular product color. The criteria for replacement of a working standard should be based on agreement between the supplier and the customer. One of the duplicate master standards serves as a master standard at other locations where the product in that color is produced.

Extreme care is needed in the storing of product samples. The plasticizer used in the manufacture of some plastic bags can leach out of the bag and stain the sample. Temperature and/or humidity may have effects. If the standards are frozen or refrigerated, they should have an outer wrapping. When they are removed from the freezer or refrigerator, they should reach room temperature before they are unwrapped to prevent condensation from damaging the standard.

Developing permanent, or even semi-permanent, physical product standards may be impossible for some products. Often this is because of the stability of the product, such as short shelf life of a liquid or color changes due to the rapid aging of the product. In those instances a numerical standard may be acceptable. A numerical standard consists of either the reflectances or the color coordinates, such as $L^*a^*b^*$, of the master standard for the product. Tolerance limits set up for numerical standards must account for the repeatability and reproducibility in sample preparation and of the color measuring instrument.

2.4.8. Color tolerances

Color difference equations provide the ruler for instrumentally determining the color difference between a standard and a trial. Tolerance limits specify how far a batch can differ from the standard and still be acceptable for use.

The major problem with the setting of color tolerances is that not only do they vary from industry to industry, but from customer to customer within any given industry. Finishes of high priced consumer products, such as automobiles and major appliances, typically have very tight color tolerances. Products which are small, made from difficult to color material, have large texture and/or gloss variations or are viewed at a distance, typically have liberal color tolerances (ASTM 0004). Examples of products with liberal tolerances are color-coded electronic wires, colored masonry and the porcelain insulators on electric poles.

A color tolerance is an agreement between the supplier and the customer. Both should be able to live with the tolerance. If a tolerance is too loose, the product might not be acceptable to the purchaser. If a tolerance is too tight, the supplier might not be able to supply the product at all or have to charge a higher price for the product. The tolerance should agree with a visual evaluation of acceptability and may vary with the color of a product. Tolerance limits should be based on the process capability of the supplier to make the color. Sometimes a temporary standard is used until the process capability can be determined. It can be useful to set a warning tolerance to let the supplier know when the coloring process might be drifting out of limits.

One CIELAB unit of color difference was sized to signify the limit of a commercially acceptable color match in the textile industry. In the CIELAB color difference equation the lightness, chroma and hue components of the color difference are weighted equally. Inspectors in the textile industry are the least tolerant of hue differences, more tolerant of chroma differences and the most tolerant of lightness differences. That is why textile acceptability data plots as ellipsoids in CIELAB color space.

The CMC(l : c) and CIE94(k_L : k_C : k_H) color difference equations attempt to compensate for tolerance preferences in lightness, chroma and hue. Fixing the ratios of l to c or of k_L to k_C to k_H will define the shape of the acceptability ellipsoid in color space. The CMC(l : c) equation will be used to discuss the setting of color tolerances. The same discussion would apply to CIE94(k_L : k_C : k_H).

Although the textile industry seems to have settled on a CMC l : c ratio of 2 : 1, different industries may prefer different ratios. In addition, the size of the ellipsoid needed to define the acceptability of a product will vary from customer to customer. It is because of the customer variability in acceptability limits that a single total color difference tolerance calculated from a color difference equation is rarely adequate.

Some computer programs multiply the CMC(l : c) color difference by a commercial factor, CF. The commercial factor allows the user to change the size of the acceptability ellipsoid and still have a 1.0 unit of color difference be acceptable. For example, if a customer only accepts a product that has a CMC(2 : 1) color difference of 0.5 units or less, a commercial factor of 2.0 would be used for that customer. Since the calculated CMC(2 : 1) color difference will be multiplied by the commercial factor, the technician making the measurement would accept a batch that had a computed color difference of 1.0 units or less.

Using a commercial factor to adjust the size of the acceptability ellipsoid can simplify a laboratory's procedure. Even though they may have a different commercial factor for each customer, it is more likely that different commercial factors would be set up for different classes of customers. In the paint industry, office furniture paint may have a commercial factor of 2.0 while wall paint may have a commercial factor of 0.5. In both cases, the technician making the measurement would consider the paint acceptable as long as the color difference was less than 1.0.

Ellipsoids generated with CMC($l : c$) equation still assume that the lightness, chroma and hue differences can be evenly distributed around the standard's color. That assumption is not always valid. It may be a particularly bad assumption when the product the customer receives must still be processed, such as with paints and colored plastic resin, or when a hue may be associated with poor quality or spoilage, such as with green butter. Color tolerances for those customers may have to be set individually for lightness, chroma and hue. Sometimes an overall color difference tolerance is also specified.

There are many ways to set up instrumental tolerances (ASTM 0004; Ehr 1992, 1993; Vance 1983; Allen and Yuhas 1984). Perhaps the best technique is to create colors showing the standard and the acceptable limits of lightness, chroma and hue for non-gray colors and the standard and the lightness, red, green, yellow and blue limits for gray colors. Not only do these samples provide a visual reference, but they can be measured to accurately determine the instrumental acceptability limits (Ehr 1992, 1993).

Another method of determining tolerances is to produce a large number of variations from the standard and to statistically evaluate the acceptability judgments of a number of observers (Vance 1983). Historical data can be used to determine the tolerances if a sufficiently large number of batches of the same color have been produced (Allen and Yuhas 1984).

The least preferred, but probably most common, technique is to attempt to estimate an initial instrumental tolerance based on either past experience or industry experience. The initial tolerance is then adjusted as necessary to agree with either the visual assessments made before the product is shipped or on the actual acceptance and rejection of product. The main disadvantage of this method is that it may result in unnecessarily tight tolerances for the product.

One danger in setting instrumental tolerances is what this author calls the "zero syndrome". Simply put, the zero syndrome is the desire to have an instrumental color difference of zero. Color measuring instruments are like other measuring tools – they have tolerance limits on their ability to reproduce measurements. Sampling plans, samples and sample preparation procedures can also cause variation in the measurements. When no color difference is seen between a sample and a batch, the color difference does not have to be 0. How large the color difference can be before seeing a visual difference depends on the color and the direction of the color difference. Two dark blues may have an CIELAB color difference of 0.3 and have a visually perceptible color difference. If the color difference is in hue, 0.3 may even be unacceptable. On the other hand, two bright yellows may have a CIELAB color difference of 1.5 or even 2.0 and not appear to have a visual color difference, especially if the difference is in lightness. Visual and instrumental tolerances should account for the variability in the instrument, sampling plan, sample preparation and other appropriate factors.

2.4.9. Visual evaluation of the color difference

Color measurement does not eliminate the need for a visual evaluation of the standard and batch. It can not be emphasized enough that the supplier should look at the standard and batch. The visual examination should be done following standardized procedures (ASTM 0008). Visual evaluation complements, and adds validity to, instrumental measurements. It can also pick up defects, such as color mottle, that the instrument would not detect. Discrepancies between the visual and instrumental evaluation usually cause a more thorough examination of the standard and the batch. Unless the visual examination does not at all agree with the instrumental measurement, the final acceptance of the product will often be the result of the measurement.

2.5. Reflectance measurements

2.5.1. Specular and diffuse reflectance

If a beam of light shines on a mirror at 45° to its normal, it will be entirely reflected in the opposite direction at the same angle as shown in fig. 2.20 at (a). This mirror-like effect is called specular or regular reflection. For all practical purposes, the light does not enter the material from which the mirror is made but is reflected only by its front surface.

If a beam of light shines on the flat surface of a compressed pellet of barium sulfate powder at 45° to its normal, the beam will penetrate the surface and enter into the body of the pellet. Particles of barium sulfate will scatter, i.e., redirect, the light in many different directions. After being scattered many times the light will find its way out of the pellet.

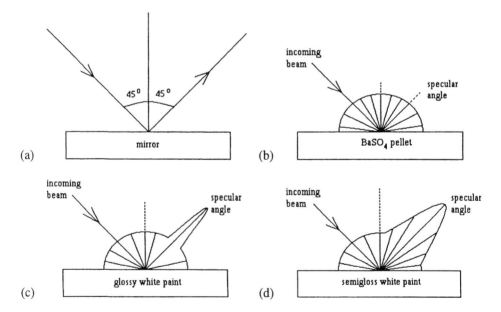

Fig. 2.20. (a) Specular reflection is sometimes call mirror-like reflection. (b) Barium sulfate diffuses light very well. (c) A glossy white paint has a narrowly defined specular reflection. (d) The specular reflection of semi-gloss paints is spread over a much wider area than glossy paints.

The light leaving the pellet will exit at many angles from the normal as shown in fig. 2.20 at (b). This is called diffuse reflection.

Light reflected from a highly glossy white paint film will have both a specular and a diffuse component as shown in fig. 2.20 at (c). A paint film is made up of pigment particles held together in a resin binder. Specular reflection occurs at the resin/air interface and constitutes about 4 percent of the light incident on the film. The remaining light enters the film, is scattered by the titanium dioxide pigments in the paint and is diffusely reflected. High gloss materials have very smooth surfaces and the specular component of the reflection becomes a narrow beam.

As the gloss of a material decreases, its surface becomes rougher. The specular reflection is a wider beam of less intensity at the specular (mirror) angle as shown in fig. 2.20 at (d). As the gloss decreases further, the specular peak widens until it disappears for a very matte material. A very matte (flat) white paint will exhibit only diffuse reflectance.

2.5.2. Illuminating and viewing geometry

Instruments made for measuring reflectance consist of an illuminator, a sample holder and a receiver. The illuminator contains the light source with associated optics, and may also contain a diffraction grating or filters, diffusers and electronics. The receiver contains components needed to gather and analyze the light reflected from the sample being measured. There are a number of ways these components can be arranged, and the CIE has recommended four illuminating and viewing geometries for measurement of diffusely reflecting specimens (CIE 1986a).

Figure 2.21 illustrates the two bidirectional geometries recommended by the CIE. In the normal/45°, 0/45, geometry of fig. 2.21 at (a), the illuminator shines a beam of light along the sample's normal ($\pm 10°$) and the receiver views the reflected beam at an angle of 45° ($\pm 2°$) from the sample's normal. In the 45°/normal, 45/0, geometry of fig. 2.21 at (b), the positions of the illuminator and the receiver are reversed. The illuminator

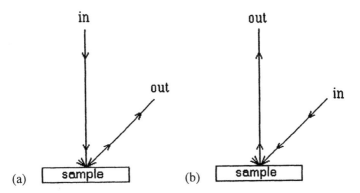

Fig. 2.21. (a) A color measuring instrument with 0°/45° illuminating/viewing geometry attempts to duplicate a common viewing condition used to visually evaluate samples. (b) A color measuring instrument with a 45°/0° illuminating/viewing geometry will produce measurements equivalent to those obtained with a 0°/45° geometry.

sample
normal

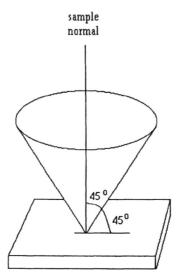

Fig. 2.22. When a 45° line is rotated around the 360° azimuthal angle, it will trace a cone shape.

now illuminates the specimen at a 45° ($\pm2°$) from the sample's normal and the receiver views the reflected beam along the sample's normal ($\pm10°$). Either of these illuminating and viewing geometries produce equivalent results for samples that reflect light diffusely (Billmeyer Jr. and Marcus 1969; ASTM 0016).

Bidirectional illuminating and viewing geometries will not measure any of the specular reflection from a sample. Depending upon the design of a bidirectional instrument, the measurement can also be greatly influenced by texture of the sample. If the reflected light from the sample is polarized in a particular direction, a bidirectional measurement could be affected. If a line drawn from the center of the sample at a 45° angle is rotated in space around the azimuthal angle, it will trace a cone as in fig. 2.22. Instruments that illuminate the sample with a single beam of light from, or have only a single 45° receiver at, only one position on this cone are most prone to these problems. To overcome these problems, instrument manufacturers may illuminate or view the samples from several positions along this cone. Texture and polarization problems are minimized or eliminated by using annular or circumferential illumination or viewing. Circumferential describes the bidirectional geometry in which the illuminator provides radiation (or the receiver possesses responsivity) in many beams distributed at uniform intervals around the 45° cone. Annular describes the bidirectional geometry in which the illuminator provides radiation (or the receiver possesses responsivity) continuously and uniformly around the 45° cone.

Figure 2.23 illustrates the two diffuse geometries recommended by the CIE. In the normal/diffuse, 0/d, geometry of fig. 2.23 at (a), the illuminator shines a beam of light along the sample's normal ($\pm10°$) and the light reflected from the sample in all forward directions is collected by means of an integrating sphere. An integrating sphere is a hollow metal sphere coated with an efficient white diffusing material such as barium

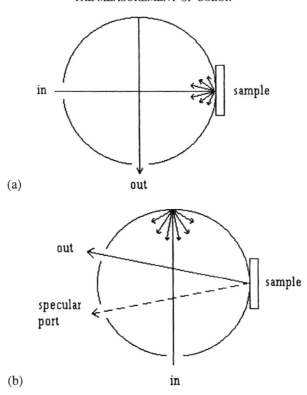

(a)   out

(b)   in

Fig. 2.23. (a) The 0°/diffuse illuminating/viewing geometry of a color measuring instrument with an integrating sphere minimizes sample texture and gloss. (b) A color measuring instrument with a diffuse/0° illuminat-
ing/viewing geometry will produce measurements equivalent to that from a 0°/diffuse geometry. By placing
the receiver (or illuminator) at an angle slightly different from the sample's normal, the specular reflection
from the sample can be either included or excluded.

sulfate, BaSO$_4$, or polytetrafluoroethylene, PTFE. The integrating sphere has ports (holes)
for the sample, the illuminator and the receiver. Light reflected from the sample is in
turn reflected many times by the wall of the integrating sphere. The receiver views a
point on the wall of the sphere. The sphere also contains baffles to prevent light from
passing directly between the specimen being measured and the spot on the sphere wall
being viewed. This geometry is rarely used.

When the illuminator is positioned directly on the sample's normal, 0°, any spec-
ularly reflected light will be directed back to the illuminator and not included in the
measurement. The specular component of the reflection can be included by positioning
the illuminator at a slight angle, usually either 6° or 8° from the sample's normal. A
specular inclusion port is then placed at the specular angle. With the specular component
included, SCI, a diffusing white plug is placed in the specular port to "complete" the
sphere wall. With the specular component excluded, SCE, a black plug or gloss trap is
used in the specular port. When the specular component of the reflection is included in
the measurement, the geometry is sometimes referred to as normal/total or 0/t.

In the diffuse/normal, d/0, geometry of fig. 2.23 at (b), the illuminator shines a beam of light directly on the wall of the integrating sphere. The light is reflected many times by the sphere wall and eventually illuminates the sample from all directions. The receiver views the specimen along the normal ($\pm 10°$). As with the normal/diffuse geometry, the specular component of the reflection can be included or excluded by positioning the receiver at a slight angle from the sample normal. The port then eliminates that area of the integrating sphere wall which illuminates the sample from the specular angle. When the specular component is included in the measurement, the geometry is sometimes referred to as total/normal, t/0.

Both the 0/d (0/t) and the d/0 (t/0) illuminating and viewing geometries produce equivalent results for samples that reflect light diffusely (British Standard 1988; AATCC 0001).

Which illuminating and viewing geometry is best for color measurement? There is no simple answer to that question (Hunt 1991; ASTM 0016; Rich 1988; Mabon 1992). Plastic chips serve to illustrate the differences in illuminating and viewing geometry nicely. A single plastic chip can be molded with one half having a glossy surface and the other half having a matte surface. The glossy side of the chip will be perceived to be darker and to have a higher chroma than the matte side when illuminated at an angle of 45° and viewed along its normal.

If the glossy plastic is used as the standard in a 45/0 or 0/45 measurement, the matte plastic will have a color difference indicating that it is lighter and has a lower chroma. This measurement will agree with the visual assessment of the chip. However, since the only difference between the glossy and matte plastic is the surface texture, people can argue that, despite the appearance difference, each side is the same material and should measure the same.

When the two halves are measured using d/0 or 0/d with the specular component included (SCI, i.e., t/0 or 0/t), the measured color difference between the two sides is practically zero. If the two halves are measured using d/0 or 0/d with the specular component excluded (SCE), the measured color difference will be very similar to that found with the 45/0 or 0/45 geometry.

All of the CIE recommended illuminating and viewing geometries will yield similar results for non-directional, matte diffusing specimens.

Because visual evaluations usually are made with the sample illuminated at 45° and viewed along its normal, 45/0 and 0/45 instruments are often said to measure total appearance and account for differences in gloss and surface texture. Instruments that measure d/0 or 0/d with the specular component included (SCI) are said to measure color only and minimize differences in gloss or texture. Excluding the specular component (SCE) in the d/0 or 0/d for glossy samples simulates a 45/0 or 0/45 instrument well. However, as the gloss decreases, the specular reflection spreads out and the gloss trap does not fully eliminate the specular component. Since the shape of the specular reflection varies from material to material and is ill-defined, it is impossible to design a universal gloss trap which just traps the specular component without affecting the measurement of the diffuse reflectance (Budde 1980). Specular component trapping will also depend upon the design of the instrument and the integrating sphere.

Deciding upon which geometry to use is often a matter of compromise, since few people wish to purchase multiple instruments. If quality control of diffusely reflecting materials is the primary use of the instrument, then a 45/0 or 0/45 instrument is often preferred. Many prefer the d/0 or 0/d instruments for computer color matching applications. For some applications, such as the measurement of metallized plastic films, such as used in snack food packages, a d/0 or 0/d instrument is required (Mabon 1992; ASTM 0017). The d/0 or 0/d geometry is also preferred for some transmittance applications.

2.5.3. Spectrophotometers

In addition to the illuminant and the standard observer's color matching functions, the spectral reflectance of the object is needed to calculate tristimulus values. A spectrophotometer measures the reflection (or transmission) characteristics of an object at different wavelengths of the visible spectrum. Spectrophotometers have been made in all four of the CIE recommended illuminating and viewing geometries. Some spectrophotometers can characterize the reflectance of a specimen continuously at all wavelengths from one end of the visible spectrum to the other. Other spectrophotometers, referred to as abridged spectrophotometers, measure at only selected wavelengths throughout the visible spectrum.

The first spectrophotometers used filters or a monochromator to illuminate the specimen with a "single" wavelength of light and viewed the reflected light with a receiver that responded to all visible wavelengths. In actuality, the "single" wavelength was a narrow band of wavelengths centered on the "single" wavelength. Thus a beam of light from a monochromator (or filter) at 560 nm might actually illuminate the sample with light from 540 through 580 nm. The range of wavelengths illuminating the sample is described by the instruments spectral bandwidth or bandpass. Some modern spectrophotometers have bandwidths as low as 1 nm while others may have bandwidths of 20 nm or even higher. Bandwidth definitely influences color measurements. The amount of that influence depends upon the sample (Strocka 1973, Schmelzer 1986). For best color measurement results, the spectral bandpass should be equal to the wavelength measurement increment used in the calculations (ASTM 0018).

Most of the current reflectance spectrophotometers illuminate the sample with polychromatic light containing all of the wavelengths of the visible spectrum and analyze the light reflected from the specimen with a polychromator, which has a diffraction grating or filters to separate the light into the various wavelengths of the visible spectrum. Although the instruments treat the analyzed light as if it contained only a single wavelength, polychromators work in the same manner as monochromators and have a definite spectral bandwidth.

Although a variety of sources are used to provide polychromatic illumination, they are usually filtered to simulate CIE illuminant D65. A good simulation to D65 is required to accurately measure fluorescent specimens.

Most spectrophotometers measure reflectance factor rather then reflectance. Reflectance is the ratio of the amount of light reflected from the sample compared to the amount of light illuminating the sample. Reflectance factor is the ratio of the amount of light reflected from the sample compared to the amount of light that would be reflected from a perfect reflecting diffuser under the same geometric and spectral conditions of measurement. This distinction becomes very important when measuring metallic and pearlescent materials.

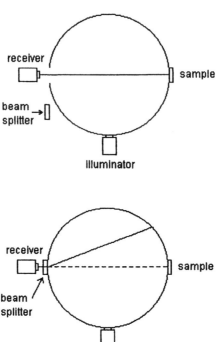

Fig. 2.24. Double beam in time color measuring instruments alternately view (or illuminate) the sample or a reference material. In this figure the wall of the integrating sphere serves as the reference material.

Any color measuring instrument requires stability. One means of increasing the stability of a spectrophotometer is to use a double beam optical design. Double beam instruments compensate dynamically for fluctuations in source output, detector response and atmospheric absorption (Zwinkels 1989). In a double beam instrument, light reflected from a reference standard is constantly monitored. Most double beam spectrophotometers use either the d/0 or 0/d geometry. Older instruments used an external reference standard placed at a reference port in the integrating sphere, while newer instruments monitor the sphere wall. Some of the newer 45/0 instruments are also double beam and monitor a reference standard built into the instrument. Figure 2.24 diagrams a double beam-in-time instrument which uses a beam splitter to alternately view the sample and the sphere wall. The double beam-in-space design of fig. 2.25 uses two independent polychromators: one viewing the sample and the other viewing the sphere wall. Several of the newer instruments increase stability even more by monitoring the output of the instrument's light source directly.

Single beam spectrophotometers have optical designs that use the CIE recommendations of figs 2.21 and 2.23. Although single beam instruments offer a simpler design with fewer optical components and higher sensitivity (Zwinkels 1989), they generally did not possess the stability of a double beam instrument. As a result, they required more frequent

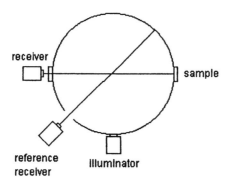

Fig. 2.25. Double beam in space color measuring instruments have two receivers. On receiver views light reflected from the sample while the second receiver views the reference material.

calibration to standard values. Improved electronic and optical components have allowed single beam instruments to achieve the stability thought only possible in double beam instruments (Stanziola et al. 1979). When a single beam instrument uses an integrating sphere, a correction for the reduction of sphere efficiency caused by sample absorption is necessary (CIE 1986a). The corrected reflectance is given by eq. (2.57).

$$\rho_s(\lambda) = R(\lambda) \frac{1 - \rho_w(\lambda)[1 - \sum_i f_i]}{1 - \rho_w(\lambda)[1 - \sum_i f_i] - f_s[\rho_r(\lambda) - R(\lambda)]} \tag{2.51}$$

in which $R(\lambda)$ is the uncorrected reflectance factor of the specimen, $\rho_w(\lambda)$ is the diffuse/diffuse spectral reflectance of the sphere wall, f_i is the fractional area of the ith port, f_s is the fractional area of the sample port, and $\rho_r(\lambda)$ is the reflectance of the reference standard. With the exception of the diffuse/diffuse spectral reflectance of the sphere wall, $\rho_w(\lambda)$, all of the quantities in the correction equation are accurately known. This reflectance will vary with wavelength and time. A gray calibration standard of known reflectance has been used by at least one instrument manufacturer to determine the reflectance of the sphere wall (Stanziola et al. 1979).

Before making reflectance factor measurements, the instrument must be calibrated, i.e., the zero point and high end of the photometric scale are determined at each wavelength, λ, to be measured. The zero point of the photometric scale is usually set by measuring the reflectance factor of a light trap or a black calibration standard, $R_z(\lambda)$. The high end of the scale is set by measuring the reflectance factor of a white calibration standard, $R_w(\lambda)$. A number of materials have been used as the white calibration standard including optical grade barium sulfate ($BaSO_4$), optical grade polytetrafluoroethylene (PTFE), opal glasses and ceramic plaques (ASTM 0019; Carter et al. 2000). The reflectance factor of the white calibration standard relative to the perfect reflecting diffuser, R_{prd}, should be known or determined. National standardizing laboratories, such as the National Institute of Standards and Technology in the United States, provide calibration services for white calibration standards. Instrument manufacturers supply white calibration standards having known reflectance factors. Since none of the white calibration standards are perfect diffusers, they must be calibrated at the same illuminating and viewing geometry as the color measuring instrument. The reflectance factor of the sample, $R(\lambda)$, is calculated by adjusting the initial reflectance measurement of the sample, $R_i(\lambda)$, for the photometric scale with eq. (2.52).

$$R(\lambda) = R_{prd}(\lambda)\left[R_i(\lambda) - R_z(\lambda)\right]/R_w(\lambda). \tag{2.52}$$

Commercially purchased reflectance spectrophotometers are supplied with the correction equations either built in to the instrument's computer or in computer software supplied with the instrument. The user of the instrument calibrates the instrument by measuring a light trap (when required), and a white calibration standard. For single beam instruments with an integrating sphere, a gray (or other sphere correction) calibration standard may also be measured at this time. All the necessary corrections are then done by either the instrument or the computer. Instruments must be calibrated at regular intervals. The interval will depend on the instrument's stability but is commonly four to eight hours. An instrument will usually have to be calibrated when the measurement conditions are changed, such as when changing from specular included to specular excluded measurements. When corrections are built-in to an instrument, usually only the set of calibration standards supplied with a given instrument can be used. If a calibration standard is damaged or lost, a new computer chip may have to be put into the instrument and/or the computer software changed. To ensure accurate reflectance factor measurements, the calibration standards should be cleaned or replaced regularly.

Many spectrophotometers allow the user to measure samples of different size. It is good practice to use the largest size possible to make the measurement. Doing this will compensate for lack of uniformity in the samples and average surface gloss and texture differences. Reflectance measurements and reports are best made using standardized procedures (SAE 0001; ASTM 0016, 0017, 0018, 0020, 0021, 0022; AATCC 0002; American National Standard 1899).

2.5.4. Tristimulus colorimeters and spectrocolorimeters

Early spectrophotometers were expensive and required skilled technicians to run and maintain them. Colorimetric calculations had to be done by hand and with mechanical calculating machines. Industrial color measurement became a common practice with the development of the colorimeter. Colorimeters were developed in the 1940s and, because of their lower cost and simplicity of operation, were in common use by the 1950s. Colorimeters have been made in all four of the CIE recommended geometries. They use three or four broad response filters to modify their light source in an attempt to duplicate a CIE illuminant and standard observer combination. Sample measurements do not result in spectral reflectance readings but give a direct conversion to either CIE tristimulus values or the coordinates of a uniform color space, such as $L^*a^*b^*$.

The main problem with colorimeters is their inability to measure metamerism. Colorimeters generally allow for only one illuminant/observer combination and the determination of metamerism requires a minimum of two such combinations. Colorimeters using polychromatic illumination have been modified by filtering their light sources to have two illuminant/observer combinations. Another problem with colorimeters is their lack of accuracy caused by the difficulty of finding filters to accurately duplicate the CIE illuminant and observer functions. Colorimeters can determine the color difference between two materials more accurately and are often referred to as color difference meters. Measurements with filter colorimeters should be made in accordance with standardized practices (AATCC 0002; American National Standard 1899; ASTM 0023).

As optics and computer technology progressed, spectrophotometers became much less expensive, faster and easier to operate. When the difference in cost between a spectrophotometer and a colorimeter narrowed, it appeared that spectrophotometers would entirely replace colorimeters. About that time, portable colorimeters were developed. Portability allowed color measurement to be taken directly to the shop floor or even to a remote location. The recent development of relatively inexpensive portable spectrophotometers make them likely candidates to replace the portable colorimeters. This is especially true for spectrocolorimeters – a spectrophotometer that only outputs tristimulus values and/or uniform color space coordinates. Spectrocolorimeters are less expensive than full function spectrophotometers, but can be used to determine of metamerism.

Colorimeters have been developed using video camera technology. These instruments are coupled with advanced image technology to make color measurements on small areas in a complex scene. This technology is particularly useful for materials whose appearance is naturally or intentionally non-uniform. Woven fabrics, patterned fabrics and marbled materials are examples of materials where conventional color measurement technology has not been particularly successful. Video based instruments may also be used for on-line applications, such as controlling the production of colored paper made as a continuous sheet.

2.5.5. Measuring translucent samples

Translucent samples can range from nearly opaque to nearly transparent. Reflectance measurements are common on the nearly opaque specimens whereas transmittance measurements are used on the more transparent specimens. Materials that are in the middle of the range should be measured according to the way they will be viewed in use. If they will be viewed by reflected light, measure them with a reflectance instrument. If they will be viewed by transmitted light, measure them with a transmittance instrument. Plastics, papers and high chroma paints are among the more common translucent materials whose reflectance is measured.

Light hitting a translucent sample will not only be scattered back towards the observer but also sideways within the sample. Some of the light will pass entirely through the sample. People looking at a colored translucent sample will usually ignore slight color differences at the edges by concentrating on the center of the sample. If they are holding the sample, some light might even find its way from the back of the sample to the observer. Overall, the sample's color appears uniform except at the edges.

When a translucent sample is measured, the area viewed by the instrument is limited to the size of the sample port. Some of the light reaching the sample will be lost by the sideways scattering and not detected. The light that passes through the sample may be lost if the sample holder is black, or it may be reflected back through the sample if the sample holder is white. The difference between two such readings is a rough measure of the translucency. The smaller the sample port, the more edge losses will affect the measurement. Edge losses result in measurements erroneously low in lightness and chroma (Spooner 1993).

Best agreement with visual evaluations will happen if the area illuminated by the instrument is very much larger than the area viewed (ASTM 0024). This is equivalent to illuminating a small area and viewing an area very much larger. This may be possible in instruments that have illuminating and viewing areas that can be varied in size. To obtain agreement between instruments, the sample port size should be the same, the sample holders should be the same color, the sample positioning must be the same and the sample should be backed with a standardized color.

2.5.6. Measuring fluorescent samples

Fluorescent samples absorb light at some wavelengths and emit light at longer wavelengths. The amount of light emitted depends on the intensity and spectral characteristics of the light source illuminating the sample.

When monochromatic light illuminates the sample at an excitation wavelength, a receiver sensitive to all wavelengths will not be able to distinguish the light emitted at the longer wavelength from the light reflected at the illuminating wavelength. As a result, the material will appear to reflect an erroneously high amount of light at the excitation wavelength. Monochromatic light illuminating the sample at a wavelength where light is only emitted, will show the true reflectance of the sample since no extra light will be emitted at any wavelength. Measurements of fluorescent samples with monochromatic light will not give accurate reflectance curves.

Illuminating the sample with a sufficiently broad band polychromatic light source results in the measurement of a combination of reflectance and emittance – the total spectral reflectance factor. The intrument's light source is usually filtered to simulate CIE Illuminant D65. The accuracy of the calculated tristimulus values for D65 depends upon the quality of the simulation. Tristimulus values calculated for any illuminant other than the one being simulated will be incorrect. Some instruments include filters for adjusting the amount of ultraviolet light given off by the source to better simulate D65. Good fluorescent standards are needed to accurately set those filters.

Color coordinates calculated for fluorescent samples illuminated with monochromatic light will indicate that the sample has a lower chroma than would be found when illuminated with polychromatic illumination. This is because monochromatic illumination will produce higher reflectance factor values in the area of excitation but lower reflectance factor values in the area of emittance.

Only the two bidirectional geometries, 45/0 and 0/45, are recommended for measuring fluorescent samples (ASTM 0025). The presence of light emitted by the fluorescent sample lowers the efficiency of the integrating sphere used in the d/0 or 0/d geometries. This results in lower total spectral reflectance factors than when using the bidirectional geometries (McKinnon 1987). According to Gundlach and Terstiege, fluorescent calibration standards are needed to calibrate a single polychromator instrument before fluorescent samples can be measured properly (Gundlach and Terstiege 1994).

The reflectance of the sample must be separated from the emittance to completely analyze a fluorescent sample. An instrument designed to separate these components is called

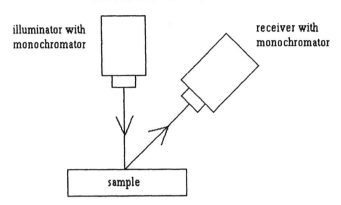

Fig. 2.26. Spectrofluorimeters use a monochromator to illuminate the sample and a second monochromator to view the sample.

a spectrofluorimeter or a fluorescence spectrometer. Figure 2.26 contains a diagram of a spectrofluorimeter. A monochromator is used to isolate a single (narrow band) wavelength with which to illuminate the sample. Light reflected from the sample passes through a second monochromator to the detector. The wavelength setting of each monochromator can be adjusted independently of each other. The true reflectance of the sample can be determined by scanning the sample with both monochromators synchronized to be at the same wavelength. Either the excitation spectrum or the emission spectrum can be studied independently by setting the wavelength of one monochromator and scanning with the other. A matrix approach in which both monochromators scan the sample at varying wavelengths can be used with the true reflectance curve to determine the color of a fluorescent material under any illuminant.

It is often necessary to know whether or not a sample is fluorescent before making color measurements. Industrial light booths usually have ultraviolet lamps which are helpful in determining the presence of fluorescence. There is also a spectrophotometric method for determining the presence of fluorescence (ASTM 0026).

2.5.7. Measuring metallic and pearlescent samples with goniospectrophotometers

Today automobiles are available in a wide variety of colors. Some of these colors, such as Metallic Red or Pearl White, contain effect pigments. Effect pigments cause pronounced changes in color or other appearance attributes with a change in illuminating or viewing angle, a phenomenon called goniochromatism. In addition to paints and printing inks, effect pigments are used in cosmetics and plastics.

Metallic paints contain little flakes of metal. When the paint dries, the flakes lie parallel to the surface on which the paint was applied. When viewed near the specular angle, metallic paint films look lighter than when viewed at angles far away from the specular angle. Although there may be a hue change as the illuminating and viewing angles change, the dominating change is in lightness. The amount of lightness change, called "flop" in the coatings industry, increases as the metal flake becomes larger. The presence of metal flakes give a sparkle appearance to the paint.

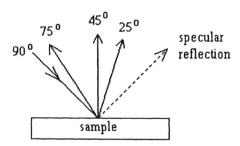

Fig. 2.27. Although a goniospectrophotometer may have variable illuminating and viewing geometries, many commercial instruments provide only one illuminating (or viewing) angle and three viewing (or illuminating) angles. The instrument in the figure illuminates at 45° from the sample's normal, an aspecular angle of 90°, and views at aspecular angles of 25°, 45° and 75°. Measurements at the aspecular viewing angle of 45° are equivalent to those of a 45°/0 instrument.

To create a more dramatic effect, pearlescent pigments are sometimes added to the metallic flake. Pearlescent pigments have color because of the interference of light waves passing through the pigment (see Section 4.7.3). Pearlescent pigments can change not only lightness but also hue as the illuminating and viewing angles change. Other industries, such as plastics and cosmetics, use pearlescent pigments without metallic flakes. A pearlescent pigment will cause a sparkle effect in those materials.

The four CIE recommended geometries should not be used with metallic or pearlescent materials. A goniospectrophotometer is needed to measure and characterize these materials properly. Figure 2.27 contains a diagram of a goniospectrophotometer, which is a spectrophotometer having the capability of measuring with a variety of illuminating and viewing angles using bidirectional geometry. In fig. 2.27, the illuminating angle is fixed at 45° from the sample normal and the receivers are in the same plane as the illuminator. The position of the receivers is described by the aspecular angle, the viewing angle measured from the the specular direction. Positive values of the aspecular angle are in the direction toward the illuminator axis. The three aspecular angles of the goniospectrophotometer in fig. 2.27 are 25°, 45° and 75°. Measurements on this instrument at the aspecular angle of 45° are equivalent to the 45/0 CIE recommended geometry. Typically only one receiver is active at a time.

An alternative instrument to the one in the figure has several illuminators, one at each of the three aspecular angles, and the receiver is fixed at 45° from the sample normal. Only one illuminator can be active at a time. The illuminating and viewing geometries of this alternative instrument would produce equivalent results as the one in the figure.

Goniospectrophotometry is still in its infancy. Researchers are trying to determine the minimum number of angles needed to fully characterize a material. They are also trying to determine which are the best angles.

Calibrating a goniospectrophotometer usually requires measurements of a light trap (or black calibration standard) and a white calibration standard. If a perfect white diffuser existed as a calibration standard, its reflectance at each illuminating and viewing angle could be mathematically predicted. Unfortunately, the materials used to make the white calibration standards do not reflect in the same way as the perfect white diffuser (Erb

1987; Grum et al. 1987). A complete calibration therefore requires that the reflectance of the calibration standard relative to that of the perfect reflecting diffuser be known at each of the illuminating and viewing geometries. Since these reflectances are not readily available for white calibration standards, many users have been forced to use standards calibrated only for the 45/0 or 0/45 geometry. Using these standards at other geometries produces erroneous reflectance factor values. The amount of error depends on how much the calibration standard deviates from the perfect reflecting diffuser at other geometries.

2.5.8. Sample preparation for reflectance measurements

The ideal sample for reflectance measurements is flat, of uniform gloss and surface texture, completely opaque, non-directional and totally uniform. Most samples are not ideal. Exactly how to prepare a sample for measurement depends on the material being measured. Whenever possible, it is probably best to prepare samples according to standard test methods and practices (SAE 0001; AATCC 0002; ASTM 0027, 0028, 0029). Books and articles also give advice on sample preparation (Hunter 1987; Connelly Sr. 1983; Wilson and Sterns 1983; Sterns and Prescott 1983). It is possible, however, to provide some general suggestions and guidelines.

Samples should always be prepared in ways that will make them easy to reproduce. Preparation uniformity will help make the measurements more repeatable. Sample must be clean and free from smudges and fingerprints. It is important that the sample preparation procedure is documented so that it may be reproduced by others.

The samples may have to be conditioned before measurement. For example, some pigments used in plastics are thermochromic and change color with temperature. Before measuring thermochromic materials, one must make sure they are at the proper temperature. Heat given off by the light source of some instruments will affect thermochromic samples, and one may have to wait for a specified period of time. Some textile dyes are hydrochromic and change color with moisture content. Before measuring hydrochromic materials, one must condition them to make sure that they have the proper moisture content.

The texture of some materials results in them having different colors in different directions. High pile carpets are a good example of directional samples. Directional samples should be measured in several orientations. Some textile materials are spun in special sample holders during measurement. Brushing the surface or stretching the material may also reduce directional affects.

Opacity can sometimes be achieved in translucent materials by making samples with multiple layers. Translucent materials should be backed with an opaque backing of a standardized color. It is extremely important to document the number of thickness used and the backing to obtain reproducible measurements.

Flexible materials can be backed with a rigid material to hold them flat for measurement. A cover glass may be used on the front of the sample if there is the possibility that it might pillow into the sample port.

Powders and other granular materials are sometimes pressed into pellets before measurement. At other times they are put into special holders and measured through a transparent window in the holder.

Threads and yarns are often wrapped on cards before measurement. The tension of the wrapping and the number of wraps should be controlled. It should be assumed that these samples will be directional and they must be spun or measured in various orientations.

Opaque liquids can be measured in special sample holders having either a transparent side or a transparent bottom. Sometimes a measurement of the top surface of the liquid is made. Whenever liquids are measured, extreme care must be taken to prevent the liquid from getting inside the instrument.

2.5.9. Colorimetric accuracy, repeatability and reproducibility

The concepts of colorimetric accuracy, repeatability and reproducibility have caused a great deal of confusion (Rich 1990). An instrument is accurate if it measures the "correct" values. Accuracy is determined by comparing the instrumental values to a known standard. No such standards exist for color measurement. The repeatability of an instrument is how close the same operator, using the same measurement procedures, can duplicate the measurement of a test sample on the same instrument over a short period of time. The reproducibility of an instrument is how close the measurements of a single test sample are when the same measurement procedures are used but when the operator, the instrument and/or the laboratory are changed. Measurements made by the same operator on the same instrument using the same procedure over a long period of time can be used to determine reproducibility.

Most of the modern color measuring instruments have extremely good repeatability, of the order of 0.1 CIELAB color difference units or lower (Billmeyer Jr. and Alessi 1981). Repeatability is usually measured by taking many readings of a stable standard over a limited time without moving the position of the standard. This provides a good measure of the instrument's short term stability. Everyone who uses color measuring instruments should set up a program to regularly monitor their instrument's short term repeatability. An increase in the repeatability can mean that a lamp is deteriorating or some other component of the instrument is beginning to fail.

Reproducibility provides a measure of a single instrument's long term stability and is of the order of 0.2 CIELAB units (Stanziola et al. 1979; Marcus 1978). Long term instrument stability should also be monitored as part of a good quality assurance program. A set of stable ceramic standards was designed for this purpose (Clarke 1971), but any set of stable standards could also be used. Some ceramic tiles are thermochromic and their color will vary with the temperature in the room (Compton 1984). Very few laboratories have precise temperature and humidity control, so thermochromic and hydrochromic effects increase the variability of some standards. Some standards, such as dark blue ceramic tiles, are very sensitive to dirt. All standards should be inspected before use and cleaned according to the manufacturer's instructions when necessary. Care is required in selecting the cleaning materials – some detergents contain fluorescent brighteners which might be left on the standard.

When we think of reproducibility as a measure of agreement between two instruments, the numbers can become much greater. Many instrument manufacturers use a set of stable ceramic tiles to maintain agreement between their instruments (Clarke 1971). Within the same model of the same manufacturer's instrument, one can often expect inter-instrument agreement to be about the same as the reproducibility of a single instrument or slightly greater.

Comparing similar instruments from different manufacturers or instruments with different illuminating and viewing geometries is much more difficult. Part of the problem is because there are no fundamental standards for color measurement as there are for length, mass or time (Rich 1990). Calibrating services provided by national standardizing laboratories help keep different instruments close. Differences in instrument design between the standardizing laboratory and instrument manufacture cause some "errors" in accuracy. Two different manufacturers may use two different sets of weighting factors to calculate tristimulus values which will appear as differences in accuracy.

A very matte sample will measure almost the same for each of the CIE recommended geometries, whereas a gloss sample can produce different results depending upon the illuminating and viewing geometry. Standards for accuracy must be calibrated for the illuminating and viewing geometry of the instrument. Measurements (CTS 0001) of two sets of matte samples provide a good illustration of inter-instrument agreement. Both sets have similar color differences but one set is parameric. These sets of samples were measured on several hundred different instruments with varying illuminating and viewing geometries and tristimulus calculation methods. The standard deviation of the differences in L^*, a^*, b^* and the total color difference ΔE^* for the non-parameric set was 0.05 or less CIELAB units (CTS 0001). The standard deviation for the parameric set was 0.18 or less CIELAB units. However, the standard deviation of the absolute values of L^*, a^* and b^* was between 0.19 and 0.28, almost 10 times greater. When the standard deviation of the reflectance values for one of the samples was calculated, it was of the order of 0.3 for most of the visible spectrum. The large standard deviation in the reflectance value derives partly from the varying bandwidths of the different instruments. It is generally safe to say that the reproducibility of color difference measurements is much better than that of either reflectance values or tristimulus values.

More variation in a color measurement comes from the sample rather than from the instrument. Careful sampling and sample preparation can help reduce the variability. Multiple measurements and sample averaging can also reduce the variability of a color measurement considerably (SAE 0001; ASTM 0030).

2.6. Transmittance measurements

2.6.1. Illuminating and viewing geometries

Four illuminating and viewing geometries are recommended by the CIE (1986a). Figure 2.28 at (a) illustrates the normal/normal, 0/0, geometry common in the transmittance spectrophotometers used in analytical chemistry for a sample that transmits light regularly with no scattering. The illuminator shines a beam on the sample along the sample's normal ($\pm 5°$), and the receiver views the transmitted light at the same angle. Although the intensity of the beam will be decreased because of absorption, the direction is unchanged. Hazy and translucent materials will scatter light and spread the incident beam of light as it passes through the sample as shown in fig. 2.28 at (b). This effect is called diffuse transmittance. A 0/0 illuminating and viewing geometry will not capture all of

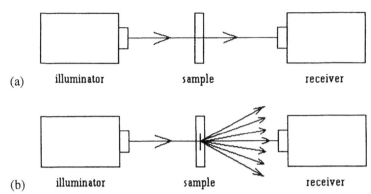

Fig. 2.28. (a) Spectrophotometers designed to be used in analytical chemistry usually measure transmittance with the 0°/0° illuminating/viewing geometry. (b) Measurements of the transmittance of hazy or translucent materials with a 0°/0° instrument will not capture all of the transmitted light.

the light passing through such a sample, producing transmittance factors that are too low. Two of the four geometries recommended for reflectance measurements, normal/diffuse (0/d) and diffuse/normal (d/0), are also recommended for transmittance measurement (see fig. 2.23). Both regular and diffuse transmittance can be measured using these geometries so they can be used to measure translucent and hazy materials. Regular transmittance is measured by keeping the sample as far away as possible from the integrating sphere. Diffuse transmittance is measured by placing the sample in contact with the integrating sphere (ASTM 0031). Sphere instruments designed only for transmittance measurements do not have a sample port in the sphere wall. The sample is placed in front of either the illuminator port or the receiver port. Sphere instruments designed for reflectance measurements generally can also make transmittance measurements. When a reflectance instrument is used to measure transmittance, a white material as close to the reflectance of the integrating sphere itself is placed in the reflectance sample port. Some of the older sphere instruments will only measure regular transmittance because their design does not allow a sample to be placed in contact with the sphere.

The fourth illuminating and viewing geometry recommended by the CIE is the diffuse/diffuse, d/d, shown in fig. 2.29. An integrating sphere is used to illuminate the sample, and a second integrating sphere is used to view the sample. This illuminating and viewing geometry is not used very much for color measurement.

Although not recommended by the CIE for transmittance measurements, the two bidirectional geometries used for reflectance measurements, 45/0 and 0/45, can be used to measure regular transmittance.

Fluorescent materials should be illuminated with polychromatic light simulating the source that will be used to visually evaluate the sample.

2.6.2. Calibrating instruments for transmittance measurements

There are three common ways to calibrate a color measuring instrument for measuring transmittance. Each technique will produce different results so the method of calibration should be documented.

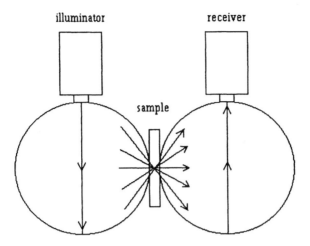

Fig. 2.29. One of the CIE recommended geometries for measuring transmittance is diffuse/diffuse. This geometry is not often available in commercial instruments.

The easiest method is to calibrate the instrument relative to air. A measurement is made with no sample in the sample compartment and the instrument set so that it measures a transmittance factor of 1.0. Measurements of the sample will then be relative to the empty sample compartment, i.e., air. Air is usually used as the standard for solid transmitting samples such as glass or plastic.

Liquids are put into a holder before measurement. An empty holder can be measured and the instrument set so that it measures a transmittance factor equal to 1.0. Transmittance measurements of liquids placed into the holder would still be relative to air. It is possible to fill the sample holder with water or a solvent of similar optical properties of the sample (ASTM 0032). The transmittance measurements of the liquids would then be relative to water or the solvent.

Instruments can be calibrated against a "colorless" blank. For example if glass or plastic filters are to be measured, a clear piece of glass or plastic can be measured and the instrument set so that it measures a transmittance factor equal to 1.0. The transmittance measurements would then be relative to the clear glass or plastic.

2.6.3. Measuring transmittance

The ideal samples for transmittance measurements are flat and have parallel sides. Liquid sample holders may be made from optical glass to minimize the effect of the sample holder on the measurement. If a second liquid sample holder or a clear piece of the material is used as the calibration standard, its path length or thickness shall be equal to the material being measured. Transmittance measurements change with sample thickness and the concentration of the colorant. When two different samples must be compared, they must be prepared the same way. When a numerical standard is used for the reference material, it is especially important to find out how the original physical standard was prepared so that the sample being measured can be prepared the same way.

Hazy materials that scatter light should be measured in contact with the integrating sphere. That produces a measurement of the diffuse transmittance. A haze index can be calculated from diffuse transmittance measurements by making four measurements (ASTM 0033). Two measurements are made in the usual way with a white material in the reflectance sample port, and two measurements are made using a light trap in the reflectance sample port. Special instruments called hazemeters are available just for the measurement of a haze index.

It is extremely difficult to measure the transmittance of objects that are not flat. Optical lenses are especially difficult to measure. This is because the curvature of the object deflects the beam of light passing through the object. A diffuse transmittance measurement may work if the object's curvature is not too great.

2.7. Radiometric measurements

2.7.1. Radiometry and the calculation of tristimulus values

Spectral radiometers and radiometric colorimeters are used to measure lamps, displays and other self-luminous objects. These instruments are similar to reflectance and transmittance spectrophotometers and colorimeters except that they do not have a light source. The object itself is the light source. Special standard calibration lamps are used to calibrate radiometric instruments. These lamps are available from many standardizing laboratories.

Calculating tristimulus values from 1 or 5 nm interval data is straightforward. In radiometric measurements the light source is both the object and the source of light. To calculate the tristimulus values, the measurement is first multiplied by the value of the appropriate standard observer color matching function at each wavelength and summed. The sum is multiplied by the maximum spectral luminous efficacy function, K_m (683 lumens/watt), and divide by the sum of the $\bar{y}$ color matching function to properly normalize the values (ASTM 0034).

2.7.2. Measuring lamps, light sources and displays

If the light source has enough intensity, it is convenient to let the source illuminate a stable white reflecting surface whose spectral power distribution is then measured (Hunt 1991). The instrument must be calibrated in a similar manner, by letting the calibration source illuminate the same white surface. It is also possible to measure the light source directly.

Radiation from a light source decreases by the square of the distance from the source. It is important that the calibration source and light source either be at the same distance from the instrument, or that the appropriate distance adjustment be made to the data.

For self-luminous colors that are not light sources, such as television and computer displays, the light must usually be imaged directly on the optics of the radiometer or radiometric colorimeter. Displays are not continuous colors but have images that repeat (refresh) at intervals too short to notice visually. One method for measuring displays is to synchronize the radiometer integration time interval with the display refresh cycle and integrate for an integral number of cycles (ASTM 0035). A second method has the instrument integrate for a large number of refresh cycles.

Radiometric colorimeters are often used to measure displays because they are less expensive than radiometers. As with other types of colorimeters, these instruments have difficulty matching the standard observer color matching functions. These errors can be minimized by following the calibration procedures outlined in ASTM E1455, Standard Practice for Obtaining Colorimetric Data from a Visual Display Unit using Tristimulus Colorimeters (ASTM 0036).

2.7.3. Measuring retroreflection

Retroreflectors come in a variety of shapes and sizes. Common retroreflectors are highway signs and the high visibility materials worn at night for individual safety (ASTM 0037). When a beam of light shines on a retroreflector, it is preferentially returned in directions very close to that of the incident beam. This will happen over a wide variation of directions of the incident beam.

Retroreflection is of most interest to automobile safety. Only a few specially built instruments are available to measure retroreflection. Usually a projector will be used to illuminate the retroreflector and a teleradiometer or radiometric telecolorimeter will be used to measure the retroreflection. A teleradiometer or radiometric telecolorimeter is similar to a standard radiometer or radiometric colorimeter, but has optics to view a limited area at a distance. The illuminating and viewing geometries for measuring retroreflection must be carefully described (ASTM 0037). For most measurements they are set to duplicate illumination from an automobile headlight and viewing by the driver of the automobile (ASTM 0039, 0040, 0041). Incident light from the projector typically simulates CIE Source A.

Two methods have been used in making measurements of retroreflectors. In the first method, the receiver is placed in the sample's position and the illuminating source measured. Then the sample is placed into position and the receiver moved to the viewing position. In the second method, a white calibration standard is placed in the sample position and measured (ASTM 0042). Then the white calibration standard is removed and the sample illuminated and measured. The CIE tristimulus values can be calculated from the reflectance factors.

2.8. Colorant mixing and color matching

2.8.1. Introduction to color matching

Many common materials are colored with pigments and dyes (see Chapters 10 and 11). Dyes are soluble in the material being colored while pigments are not. Dyes color materials by absorbing some wavelengths of light while transmitting other wavelengths. In addition to absorbing and transmitting wavelengths of light, pigments may also scatter (bend) light in different directions. White pigments scatter most wavelengths of light.

In order to match the color of a material, an experienced color matcher must know which dyes or pigments to use and determine how much of each is needed to match the color. Years of practice and experience are needed to visually color match materials efficiently. Color measuring instruments and computer color matching programs give the color matcher tools that greatly speed up the color matching process and provide less costly and less metameric color matches to the standard.

At the heart of computer color matching programs is a set of mathematical equations simulating how the light interacts with the dyes or pigments.

2.8.2. Transparent materials

Clear liquids and plastics are examples of transparent materials that can be colored with dyes. The dyes will dissolve in the material and color by absorbing some wavelengths of light. At any given wavelength of light, λ, the absorption, $A(\lambda)$, of a dye is given by the Beer–Lambert law:

$$A(\lambda) = \log[1/T(\lambda)] = a(\lambda)bc \tag{2.53}$$

in which $T(\lambda)$ is the transmittance at that wavelength, $a(\lambda)$ is the absorptivity or specific absorber at that wavelength, b is the length or thickness of the absorbing path, and c is the concentration of the dye. The absorptivity is a property of the dye in the material.

When more then one dye is present in the material, the total absorption can be calculated, in the absence of dye interactions, from:

$$A(\lambda) = \sum_i a_i(\lambda)bc_i, \tag{2.54}$$

in which the subscript i is used to indicate one of the specific dyes.

2.8.3. Translucent and opaque materials

Translucent and opaque materials are usually colored by pigments. Because there is a significant amount of scattering in these materials, different equations are used to model how light interacts in the material. The most commonly used equations are those derived by Kubelka and Munk (Kubelka 1948, 1954).

For completely opaque materials the reflectance at a given wavelength, λ, can be calculated from:

$$R_\lambda = 1 + (K_\lambda/S_\lambda) - \left[(K_\lambda/S_\lambda)^2 + 2(K_\lambda/S_\lambda)\right]^{1/2}, \tag{2.55}$$

in which K_λ is the Kubelka–Munk absorption coefficient at that wavelength and S_λ is the Kubelka–Munk scattering coefficient at that wavelength. When more than one pigment is present in the material, the absorption and scattering coefficients at a given wavelength are calculated from:

$$K_\lambda = \sum_i c_i K_{\lambda i}, \tag{2.56}$$

$$S_\lambda = \sum_i c_i S_{\lambda i}, \tag{2.57}$$

in which the subscript i is used to indicate one of the specific pigments.

For translucent materials, the thickness of the material and the background over which the material is viewed effect the reflectance. For these materials, the reflectance is calculated from:

$$R = \frac{1 - R_g\{a - b[\text{ctgh}(bSX)]\}}{a + b[\text{ctgh}(bSX)] - R_g}, \tag{2.58}$$

$$a = \frac{S + K}{S} = \frac{1}{2}\left(\frac{1}{R_\infty} + R_\infty\right), \tag{2.59}$$

$$b = (a^2 - 1)^{1/2} = \frac{1}{2}\left(\frac{1}{R_\infty} - R_\infty\right) \tag{2.60}$$

in which R_g is the reflectance of the background, ctgh is a hyberbolic cotangent, X is the thickness of the material and R_∞ is the reflectance of the material at complete opacity (reflectivity).

2.8.4. Computer color matching

Having a model to predict the reflectance of a colored material is only part of a complete computer color matching program. Techniques are required to predict the initial pigment or dye concentrations and to iterate to a good match prediction. More complete explanations of computer color matching can be found in books by Kuehni (1975), Park (1993), McDonald (1993), and Pierce and Marcus (0001).

2.9. Review and recommendations

The CIE system of basic colorimetry shows how to combine a standard illuminant with a standard observer and the reflectance, transmittance or emittance of an object to simulate how people see color. Once two or more colors are measured, a color difference equation can be used to calculate the difference in color between them. Color tolerances are set so that the manufacturer knows how large the color difference can be and still be acceptable to the consumer. Tolerances must be developed that include the variation in sample preparation and measurement repeatability and reproducibility.

Spectrophotometers and colorimeters are used to measure the reflectance or transmittance of materials using illuminating and viewing geometries recommended by the CIE. Special techniques are used to measure translucent, fluorescent, metallic and pearlescent samples. Sample preparation techniques depend on the type of material being measured.

Spectroradiometers and radiometric colorimeters are used to measure lamps and self-luminous displays such as televisions and computer monitors. The equations used to calculate the tristimulus values must be modified because the light source and object are combined. Radiometric measurements are used to measure retroreflecting materials such as highway signs and high visibility materials worn for safety.

Step-by-step instructions on color measurement and calculations are given in standard test methods. Many of the references provided in this chapter are to standard guides, practices and test methods. It is recommended that the reader use these or similar standards to develop sampling plans, prepare samples, take measurements and make the calculations.

Instrument manufacturers may provide help and advice on preparing samples and making measurements for a specific application. Many also conduct courses on color measurement.

Computer color matching programs use mathematical models of how light interacts with pigments and dyes to predict the reflectance of materials. These programs are relatively inexpensive and are a great aid to color matching many materials.

Some colleges and universities have courses on color science and measurement. At least one school, the Rochester Institute of Technology in Rochester, New York, has a course of study leading to a degree in color science.

Lastly, many professional organizations have groups or divisions devoted to color, including the American Association of Textile Chemists and Colorists and the Society of Plastics Engineers. The Inter-Society Color Council, ISCC, is an organization of groups with an interest in color. Individuals can also belong to the ISCC. Contact the ISCC office at Suite 301, 11491 Sunset Hills Road, Reston, VA 22090 (703-318-0263) for further information on the Council.

2.10. Bibliography

Recommended books

1. Billmeyer Jr., F.W., and M. Saltzman, 1981, Principles of Color Technology, 2nd edn (Wiley, New York).

2. CIE, 1986a, Publication 15.2, Colorimetry, 2nd edn (Commission International de l'Eclairage (CIE), Central Bureau of the CIE, Vienna, 1986). Available from USNC/CIE Publications, c/o TLA – Lighting Consultants, Inc., 72 Loring Avenue, Salem, MA 01970.

3. Field, G.G., 1988, Color and Its Reproduction (Graphics Arts Technical Foundation, 4615 Forbes Avenue, Pittsburgh, PA 15213).

4. Hunt, R.W.G., 1991, Measuring Colour, 2nd edn (Ellis Horwood, Chichester, West Sussex, England).

5. Hunter, R.S., and R.W. Harold, 1987, The Measurement of Appearance, 2nd edn (Wiley, New York).

6. McDonald, R., ed., 1997, Colour Physics for Industry, 2nd edn (Society of Dyers and Colourists, P.O. Box 244, Perkin House, 82 Grattan Road, Bradford, West Yorkshire BDI 2JB, England).

7. Wyszecki, G., and W.S. Stiles, 1982, Color Science, Concepts and Methods, Quantitative Data and Formulae, 2nd edn (Wiley, New York).

Recommended journal

Color Research and Application (Wiley, 605 Third Ave., New York, NY 10158).

References

AATCC, 0001, AATCC Test Method 173 CMC: Calculation of small color differences for acceptability. AATCC Technical Manual/1993 (American Association of Textile Chemists and Colorists, P.O. Box 12215, Research Triangle Park, NC 27709-2215).

AATCC, 0002, AATCC Test Method 153 color measurement of textiles: Instrumental. AATCC Technical Manual/1993 (American Association of Textile Chemists and Colorists, P.O. Box 12215, Research Triangle Park, NC 27709-2215).

Adebayo, O., and B. Rigg, 1986, Relative tolerances in the CMC colour-difference formula for paint samples. J. Oil Chemists and Colourists Association 11, 302–309.

Allen, E. and B. Yuhas, Setting up acceptability tolerances: A case study. Color Research and Application 9, 37–48.

American National Standard, 1899, American National Standard CGATS.5 Graphic technology – spectral measurement and colorimetric computation for graphic arts images (NPES The Association for Suppliers of Printing and Publishing Technologies, 1899 Preston White Drive, Reston, VA 22091-4367).

ASTM, 0001, ASTM E 1499 Standard guide to the selection, evaluation and training of observers. Annual Book of ASTM Standards (American Society for Testing and Materials, 100 Bar Harbor Drive, West Conshohocken, PA 19428).

ASTM, 0002, ASTM E 308 Standard test method for computing the colors of objects by using the CIE system. Annual Book of ASTM Standards (American Society for Testing and Materials, 100 Bar Harbor Drive, West Conshohocken, PA 19428).

ASTM, 0003, ASTM D 4086 Standard practice for visual evaluation of metamerism. Annual Book of ASTM Standards (American Society for Testing and Materials, 100 Bar Harbor Drive, West Conshohocken, PA 19428).

ASTM, 0004, ASTM D 3134 Standard practice for establishing color and gloss tolerances. Annual Book of ASTM Standards (American Society for Testing and Materials, 100 Bar Harbor Drive, West Conshohocken, PA 19428).

ASTM, 0005, ASTM D 1535 Standard test method for specifying color by the Munsell system. Annual Book of ASTM Standards (American Society for Testing and Materials, 100 Bar Harbor Drive, West Conshohocken, PA 19428).

ASTM, 0006, ASTM E 1360 Standard practice for specifying color by using the Optical Society of America uniform color scales system. Annual Book of ASTM Standards (American Society for Testing and Materials, 100 Bar Harbor Drive, West Conshohocken, PA 19428).

ASTM, 0007, ASTM E 1541 Standard practice for specifying and matching color using the Colorcurve system. Annual Book of ASTM Standards (American Society for Testing and Materials, 100 Bar Harbor Drive, West Conshohocken, PA 19428).

ASTM, 0008, ASTM D 1729 Standard practice for visual evaluation of color differences of opaque materials. Annual Book of ASTM Standards (American Society for Testing and Materials, 100 Bar Harbor Drive, West Conshohocken, PA 19428).

ASTM, 0009, ASTM D 2616 Standard test method for evaluation of visual color difference with a gray scale. Annual Book of ASTM Standards (American Society for Testing and Materials, 100 Bar Harbor Drive, West Conshohocken, PA 19428).

ASTM, 0010, ASTM D 4449 Standard test method for visual evaluation of gloss differences between surfaces of similar appearance. Annual Book of ASTM Standards (American Society for Testing and Materials, 100 Bar Harbor Drive, West Conshohocken, PA 19428).

ASTM, 0011, ASTM D 3928 Standard test method for evaluation of gloss or sheen uniformity. Annual Book of ASTM Standards (American Society for Testing and Materials, 100 Bar Harbor Drive, West Conshohocken, PA 19428).

ASTM, 0012, ASTM D 2244 Standard test method for calculation of color differences from instrumentally measured color coordinates. Annual Book of ASTM Standards (American Society for Testing and Materials, 100 Bar Harbor Drive, West Conshohocken, PA 19428).

ASTM, 0013, ASTM E 313 Standard test method for indexes of whiteness and yellowness of near-white, opaque materials. Annual Book of ASTM Standards (American Society for Testing and Materials, 100 Bar Harbor Drive, West Conshohocken, PA 19428).

ASTM, 0014, ASTM D 1925 Standard test method for yellowness index of plastics. Annual Book of ASTM Standards (American Society for Testing and Materials, 100 Bar Harbor Drive, West Conshohocken, PA 19428).

ASTM, 0015, ASTM D 5531 Standard guide for preparation, maintenance, and distribution of physical product standards for color and geometric appearance of coatings. Annual Book of ASTM Standards (American Society for Testing and Materials, 100 Bar Harbor Drive, West Conshohocken, PA 19428).

ASTM, 0016, ASTM E 179 Standard guide for selection of geometric conditions for measurement of reflection and transmission properties of materials. Annual Book of ASTM Standards (American Society for Testing and Materials, 100 Bar Harbor Drive, West Conshohocken, PA 19428).

ASTM, 0017, ASTM E 429 Standard test method for measurement and calculation of reflecting characteristics of metallic surfaces using integrating sphere instruments. Annual Book of ASTM Standards (American Society for Testing and Materials, 100 Bar Harbor Drive, West Conshohocken, PA 19428).

ASTM, 0018, ASTM E 1164 Standard practice for obtaining spectrophotometric data for object-color evaluation. Annual Book of ASTM Standards (American Society for Testing and Materials, 100 Bar Harbor Drive, West Conshohocken, PA 19428).

ASTM, 0019, ASTM E 259 Standard practice for preparation of pressed powder white reflectance factor transfer standards for hemispherical geometry. Annual Book of ASTM Standards (American Society for Testing and Materials, 100 Bar Harbor Drive, West Conshohocken, PA 19428).

ASTM, 0020, ASTM E 805 Standard practice for identification of instrumental methods of color or color-difference measurement of materials. Annual Book of ASTM Standards (American Society for Testing and Materials, 100 Bar Harbor Drive, West Conshohocken, PA 19428).

ASTM, 0021, ASTM E 1331 Standard test method for reflectance factor and color by spectrophotometry using hemispherical geometry. Annual Book of ASTM Standards (American Society for Testing and Materials, 100 Bar Harbor Drive, West Conshohocken, PA 19428).

ASTM, 0022, ASTM E 1349 Standard test method for reflectance factor and color by spectrophotometry using bidirectional geometry. Annual Book of ASTM Standards (American Society for Testing and Materials, 100 Bar Harbor Drive, West Conshohocken, PA 19428).

ASTM, 0023, ASTM E 1347 Standard test method for color and color-difference measurement by tristimulus (filter) colorimetry. Annual Book of ASTM Standards (American Society for Testing and Materials, 100 Bar Harbor Drive, West Conshohocken, PA 19428).

ASTM, 0024, Draft document "Standard guide for reflectance and transmittance measurement of translucent materials" (American Society for Testing and Materials, 100 Bar Harbor Drive, West Conshohocken, PA 19428).

ASTM, 0025, ASTM E 991 Standard practice for color measurement of fluorescent specimens. Annual Book of ASTM Standards (American Society for Testing and Materials, 100 Bar Harbor Drive, West Conshohocken, PA 19428).

ASTM, 0026, ASTM E 1247 Standard test method for identifying fluorescence in object-color specimens by spectrophotometry. Annual Book of ASTM Standards (American Society for Testing and Materials, 100 Bar Harbor Drive, West Conshohocken, PA 19428).

ASTM, 0027, ASTM D 3925 Standard practice for sampling liquid paints and related pigmented coatings. Annual Book of ASTM Standards (American Society for Testing and Materials, 100 Bar Harbor Drive, West Conshohocken, PA 19428).

ASTM, 0028, ASTM D 3964 Standard practice for selection of coating specimens for appearance measurements. Annual Book of ASTM Standards, American Society for Testing and Materials, 100 Bar Harbor Drive, West Conshohocken, PA 19428.

ASTM, 0029, ASTM D 823 Standard practices for producing films of uniform thickness of paint, varnish, and related products on test panels. Annual Book of ASTM Standards (American Society for Testing and Materials, 100 Bar Harbor Drive, West Conshohocken, PA 19428).

ASTM, 0030, ASTM E 1345 Standard practice for reducing the effect of variability of color measurement by use of multiple measurements. Annual Book of ASTM Standards (American Society for Testing and Materials, 100 Bar Harbor Drive, West Conshohocken, PA 19428).

ASTM, 0031, ASTM E 1348 Standard test method for transmittance and color by spectrophotometry using hemispherical geometry. Annual Book of ASTM Standards (American Society for Testing and Materials, 100 Bar Harbor Drive, West Conshohocken, PA 19428).

ASTM, 0032, ASTM E 450 Standard test method for measurement of color of low-colored clear liquids using the Hunterlab color difference meter. Annual Book of ASTM Standards (American Society for Testing and Materials, 100 Bar Harbor Drive, West Conshohocken, PA 19428).

ASTM, 0033, ASTM D 1003 Standard test method for haze and luminous transmittance of transparent plastics. Annual Book of ASTM Standards (American Society for Testing and Materials, 100 Bar Harbor Drive, West Conshohocken, PA 19428).

ASTM, 0034, ASTM E 1341 Standard practice for obtaining spectroradiometric data from radiant sources for colorimetry. Annual Book of ASTM Standards (American Society for Testing and Materials, 100 Bar Harbor Drive, West Conshohocken, PA 19428).

ASTM, 0035, ASTM E 1336 Standard test method for obtaining colorimetric data from a video display unit by spectroradiometry. Annual Book of ASTM Standards (American Society for Testing and Materials, 100 Bar Harbor Drive, West Conshohocken, PA 19428).

ASTM, 0036, ASTM E 1455 Standard practice for obtaining colorimetric data from a visual display unit using tristimulus colorimeters. Annual Book of ASTM Standards (American Society for Testing and Materials, 100 Bar Harbor Drive, West Conshohocken, PA 19428).

ASTM, 0037, ASTM F 923 Standard guide to properties of high visibility materials used to improve individual safety. Annual Book of ASTM Standards (American Society for Testing and Materials, 100 Bar Harbor Drive, West Conshohocken, PA 19428).

ASTM, 0038, ASTM E 808 Standard practice for describing retroreflection. Annual Book of ASTM Standards (American Society for Testing and Materials, 100 Bar Harbor Drive, West Conshohocken, PA 19428).

ASTM, 0039, ASTM E 809 Standard practice for measuring photometric characteristics of retroreflectors. Annual Book of ASTM Standards (American Society for Testing and Materials, 100 Bar Harbor Drive, West Conshohocken, PA 19428).

ASTM, 0040, ASTM E 810 Standard test method for coefficient of retroreflection of retroreflective sheeting. Annual Book of ASTM Standards (American Society for Testing and Materials, 100 Bar Harbor Drive, West Conshohocken, PA 19428).

ASTM, 0041, ASTM D 4061 Standard test method for retroreflection of horizontal coatings. Annual Book of ASTM Standards (American Society for Testing and Materials, 100 Bar Harbor Drive, West Conshohocken, PA 19428).

ASTM, 0042, ASTM E 811 Standard practice for measuring colorimetric characteristics of retroreflectors under nighttime conditions. Annual Book of ASTM Standards (American Society for Testing and Materials, 100 Bar Harbor Drive, West Conshohocken, PA 19428).

Billmeyer Jr., F.W., 1987, Survey of color order systems. Color Research and Application 12, 173.

Billmeyer Jr., F.W., and P.J. Alessi, 1981, Assessment of color-measuring instruments. Color Research and Application 6, 195–202.

Billmeyer Jr., F.W., and H.S. Fairman, 1987, CIE method for calculating tristimulus values. Color Research and Application 12, 27–36.

Billmeyer Jr., F.W., and R.T. Marcus, 1969, Effect of illuminating and viewing geometry on the color coordinates of samples with various surface textures. Applied Optics 8, 763–768.

Billmeyer Jr., F.W., and M. Saltzman, 1981, Principles of Color Technology, 2nd edn (Wiley, New York).

British Standard, 1988, BS6923:1988 Standard method for calculation of small colour differences (British Standards Institution, 2 Park Street, London W1A 2BS).

Budde, W., 1980, The gloss trap in diffuse reflectance measurements. Color Research and Application 5, 73–75.

Carter, E.C., F.W. Billmeyer Jr. and D.C. Rich, 2000, Guide to material standards and their use in color measurement, ISCC Technical Report 89-1. Available from Dr. D.C. Rich, Secretary of the Inter-Society Color Council, Datacolor International, 5 Princess Road, Lawrenceville, NJ 08648.

Chickering, K.D., 1967, Optimization of the MacAdam-modified 1965 Friele color-difference formula. J. Opt. Soc. Am. 57, 537–541.

Chickering, K.D., 1971, FMC color-difference formulas: clarification concerning usage. J. Opt. Soc. Am. 61, 118–122.

CIE, 1978, Supplement 2 to CIE Publication 15 (1971), Recommendations on uniform color spaces – color-difference equations; Psychometric Color Terms (Commission International de l'Eclairage (CIE), Central Bureau of the CIE, Vienna). These recommendations have since been incorporated into CIE Publication 15.2 (see CIE, 1986a).

CIE, 1981, Publication 51, A method for assessing the quality of daylight simulators for colorimetry (Commission International de l'Eclairage (CIE), Central Bureau of the CIE, Vienna). Available from USNC/CIE Publications, c/o TLA – Lighting Consultants, Inc., 72 Loring Avenue, Salem, MA 01970.

CIE, 1986a, Publication 15.2, Colorimetry, 2nd edn (Commission International de l'Eclairage (CIE), Central Bureau of the CIE, Vienna). Available from USNC/CIE Publications, c/o TLA – Lighting Consultants, Inc., 72 Loring Avenue, Salem, MA 01970.

CIE, 1986b, Publication x007, Proceedings of the CIE Symposium on Advanced Colorimetry (Commission International de l'Eclairage (CIE), Central Bureau of the CIE, Vienna). Available from USNC/CIE Publications, c/o TLA – Lighting Consultants, Inc., 72 Loring Avenue, Salem, MA 01970.

CIE, 1995, Technical Report 116, Industrial colour-difference evaluation, Commission International de l'Eclairage (CIE) (Central Bureau of the CIE, Vienna). Available from USNC/CIE Publications, c/o TLA – Lighting Consultants, Inc., 72 Loring Avenue, Salem, MA 01970.

Clark, F.J.J., R. McDonald and B. Rigg, 1984, Modification to the JPC 79 colour-difference formula. J. Soc. Dyers and Colorists 100, 128–132.

Clarke, F.J.J., 1971, Ceramic colour standards. Die Farbe 20, 299–306.

Compton, J.A., 1984, The thermochromic properties of the ceramic colour standards. Color Research and Application 9, 15–22.

Connelly Sr., R.L., 1983, Preparation and mounting textile samples for color measurement, in: Color Technology in the Textile Industry, eds G. Celikiz and R.G. Kuehni (American Association of Textile Chemists and Colorists, P.O. Box 12215, Research Triangle Park, NC 27709).

CTS, 0001, Report No. 87 of the color and appearance collaborative reference program, Collaborative Testing Services, Herndon, VA 22070.

Ehr, S.M., 1992, "Color Tolerancing: The Practical Way", presented at the Regional Technical Conference of the Society of Plastics Engineers, Color Tolerances: Measuring Up to Today's Standards (Cherry Hill, NJ, September 14–16).

Ehr, S.M., 1993, Color tolerancing: A practical case. Plastics Compounding, November/December, pp. 48–51.

Erb, W., 1987, "Gonioreflectometry – instrumentation and data", presented at the 1987 Inter-Society Color Council Williamsburg Conference, Appearance, Williamsburg, Virginia, February 8–11.

Fairchild, M.D., 1993, The CIE Standard Observer: Mandatory retirement at age 65? Color Research and Application 18, 129–134.

Fairman, H.S., 1981, A standards program for color control. Color Research and Application 6, 5–6.

Fairman, H.S., 1985, The calculation of weight factors for tristimulus integration. Color Research and Application 10, 199–203.

Field, G.G., 1988, Color and Its Reproduction (Graphics Arts Technical Foundation, 4615 Forbes Avenue, Pittsburgh, PA 15213).

Foster Jr., W.H., R. Gans, E.I. Stearns and R.E. Stearns, 1970, Weights for calculation of tristimulus values from sixteen reflectance values, Color Engineering 8(3), 35–47.

Grum, F., M.D. Fairchild and R.S. Berns, 1987, "Goniospectrophotometric characteristics of common transfer standards with respect to the cie normal/45 geometry", presented at the 1987 Inter-Society Color Council Williamsburg Conference, Appearance, Williamsburg, Virginia, February 8–11.

Gundlach, D., and H. Terstiege, 1994, Problems in measurement of fluorescent materials. Color Research and Application 19, 427–436.

Hård, A., and L. Sivik, 1981, NCS – Natural Color System: A Swedish standard for color notation. Color Research and Application 6, 129–138.

Hering, E., 1964, Outlines of a Theory of the Light Sense, 1874, translated by L.M. Hurvich and D. Jameson (Harvard University Press, Cambridge, MA).

Hunt, R.W.G., 1991, Measuring Colour, 2nd edn (Ellis Horwood, Chichester, West Sussex, England).

Hunter, R.S., and R.W. Harold, 1987, The Measurement of Appearance, 2nd edn (Wiley, New York).

Judd, D.B., 1970, Ideal Color Space. Color Eng. 8(2), 36–53.

Kubelka, P., 1948, New contributions to the optics of intensely light-scattering materials. Part I. J. Opt. Soc. Am. 38, 448–457.

Kubelka, P., 1954, New contributions to the optics of intensely light-scattering materials. Part II: Non-homogeneous Layers. J. Opt. Soc. Am. 44, 330–335.

Kuehni, R.G., 1975, Computer Colorant Formulation (D.C. Heath, Lexington, MA).

Liebeknecht, A.L., 1992, "CMC after one year in production: How well is it working?", presented at the Regional Technical Conference of the Society of Plastics Engineers, Color Tolerances: Measuring Up to Today's Standards, Cherry Hill, NJ, September 14–16.

Mabon, T.J., 1992, "Color Measurement of Plastics: Which Geometry Is Best?", presented at the Regional Technical Conference of the Society of Plastics Engineers, Color Tolerances: Measuring Up to Today's Standards, Cherry Hill, NJ, September 14–16.

MacAdam, D.L., 1974, Uniform color scales. J. Opt. Soc. Am. **64**, 1691–1702.

Marcus, R.T., 1978, Long-term repeatability of color-measuring instrumentation: Storing numerical standards. Color Research and Application **3**, 29–33.

McDonald, R. (ed.), 1993, Colour Physics for Industry (The Society of Dyers and Colourists, P.O. Box 244, Perkin House, 82 Grattan Road, Bradford BD1 2JB).

McDonald, R. (ed.), 1997, Colour Physics for Industry, 2nd edn (Society of Dyers and Colourists, P.O. Box 244, Perkin House, 82 Grattan Road, Bradford, West Yorkshire BD1 2JB, England).

McKinnon, R.A., 1987, Methods of measuring the colour of opaque fluorescent materials. Rev. Prog. Coloration **17**, 56–60.

Munsell, A.H., 1905, A Color Notation, 1st edn (Geo. H. Ellis Co., Boston); 11th edn (edited and rearranged) (Munsell Color Company, Baltimore, 1946). Now available from the Munsell Color Group, Macbeth Division of Kollmorgen Instruments Corporation, New Windsor, New York.

Munsell, A.H., 1915, Atlas of the Munsell Color System (Wadsworth-Howland, Malden, MA) (preliminary Charts A and B were published in 1910).

Munsell, A.H., 1929, Munsell Book of Color (Munsell Color Company, Baltimore). Now available from the Munsell Color Group, Macbeth Division of Kollmorgen Instruments Corporation, New Windsor, New York.

Nemcsics, A., 1980, The Coloroid color system. Color Research and Application **5**, 113–120.

Newhall, S.M., D. Nickerson and D.B. Judd, 1943, Final report of the O.S.A. subcommittee on the spacing of the Munsell Colors. J. Opt. Soc. Am. **33**, 385–418.

Nickerson, D., 1981, Uniform color scales samples: A unique set. Color Research and Application **6**, 7–33.

Nickerson, D., and S.M. Newhall, 1943, A psychological color solid. J. Opt. Soc. Am. **33**, 419–422.

North, A.D., and M.D. Fairchild, 1993a, Measuring color-matching functions. Part I. Color Research and Application **18**, 155–162.

North, A.D., and M.D. Fairchild, 1993b, Measuring color-matching functions. Part II. New data for assessing observer metamerism. Color Research and Application **18**, 163–170.

Park, J., 1993, Instrumental Colour Formulation: A Practical Guide (The Society of Dyers and Colourists, P.O. Box 244, Perkin House, 82 Grattan Road, Bradford BD1 2JB).

Pierce, P.E., and R.T. Marcus, 0001, Color and Appearance (Federation Series on Coatings Technology, Federation of Societies for Coatings Technology, 492 Norristown Road, Blue Bell, PA 19422).

Rich, D.C., 1988, The effect of measuring geometry on computer color matching. Color Research and Application **13**, 113–118.

Rich, D.C., 1990, Colorimetric repeatability and reproducibility of chroma-sensor spectrocolorimeters. Die Farbe **37**, 247–261.

Richter, M., and K. Witt, 1986, The story of the DIN color system. Color Research and Application **11**, 138.

SAE, 0001, SAE J1545 Recommended practice for instrumental color difference measurement for exterior finishes, textiles, and colored trim (Society of Automotive Engineers, 3001 W. Big Beaver, Troy, MI 48084).

Schmelzer, H., 1986, Influence of the design of instruments on the accuracy of color-difference and color-matching calculations. J. Coatings Technology **58**(739), 53–59.

Spooner, D.L., 1993, "Lateral diffusion errors caused by layered structure of graphics arts products", presented at Annual TAGA meeting, Minneapolis, MN, April 26. TAGA Proceedings.

Stanziola, R., 1992, The COLORCURVE® system. Color Research and Application **17**, 263–272.

Stanziola, R.H., H. Hemmendinger and B. Momiroff, 1979, The spectro sensor: A new generation spectrophotometer. Color Research and Application **4**, 157–163.

Stearns, E.I., 1981, The determination of weights for use in calculating tristimulus values. Color Research and Application **6**, 210–212.

Stearns, E.I., 1985, Calculation of tristimulus values and weights with the revised CIE recommendations. Textile Chemist and Colorist **17**(8), 162/53-168/59.

Sterns, E.I., and W.B. Prescott, 1983, Measurement of translucent cloth samples, in: Color Technology in the Textile Industry, eds G. Celikiz and R.G. Kuehni (American Association of Textile Chemists and Colorists, P.O. Box 12215, Research Triangle Park, NC 27709).

Strocka, D., 1973, "Are intervals of 20 nm sufficient for industrial colour measurement?", presented at the 2nd AIC Congress "Colour 73", York, England, July, 1973. Long abstract in: R.W.G. Hunt (ed.), Colour 73 (Halsted Press, New York) pp. 453–456.

Swedish Standard, 0001, Swedish Standard SS 01 91 00 D, Colour Notation System, Swedish Standards Commission.

Swedish Standard, 1979, SS 01 91 02 Colour Atlas (Swedish Standrds Institution, SIS, Stockholm).

Thornton, W.A., 1993a, Toward a more accurate and extensible colorimetry. Part I. Introduction. The visual colorimeter–spectroradiometer. Experimental results. Color Research and Application **17**, 79–122.

Thornton, W.A., 1993b, Toward a more accurate and extensible colorimetry. Part II. Discussion. Color Research and Application **17**, 162–186.

Thornton, W.A., 1993c, Toward a more accurate and extensible colorimetry. Part III. Discussion (continued). Color Research and Application **17**, 240–262.

Thornton, W.A., 1993d, Response. Color Research and Application **18**, 134–136.

Vance, L.C., 1983, Statistical determination of numerical color tolerances. Modern Paint and Coatings, November, pp. 49–51.

West, B., 1992, "The Application of CMC Color Difference to Lot Acceptance of Colorants", presented at the Regional Technical Conference of the Society of Plastics Engineers, Color Tolerances: Measuring Up to Today's Standards, Cherry Hill, NJ, September 14–16.

Wilson, Ch., and E.I. Sterns, 1983, The spectrophotometric reflectance measurement of small samples, in: Color Technology in the Textile Industry, eds G. Celikiz and R.G. Kuehni (American Association of Textile Chemists and Colorists, P.O. Box 12215, Research Triangle Park, NC 27709).

Wyszecki, G., and W.S. Stiles, 1982, Color Science, Concepts and Methods, Quantitative Data and Formulae, 2nd edn (Wiley, New York).

Zwinkels, J.C., 1989, Errors in colorimetry caused by the measuring instrument. Textile Chemist and Colorist **21**(2), 23–29.

chapter 3

COLOR VISION

JOHN KRAUSKOPF

New York University
New York, NY, USA

Color for Science, Art and Technology
K. Nassau (Editor)
© 1998 Elsevier Science B.V. All rights reserved.

CONTENTS

3.1. Introduction 99

3.2. Anatomy of the visual system 99

    3.2.1. The eye 99

    3.2.2. The midbrain nuclei 103

    3.2.3. The cerebral cortex 104

3.3. Neurophysiology of color vision 105

    3.3.1. Microelectrode recording 105

    3.3.2. Electrical recording from the retina 106

    3.3.3. Spectral sensitivity of monkey cone receptors 106

    3.3.4. The lateral geniculate nucleus 107

    3.3.5. Receptive fields 107

    3.3.6. Color in the visual cortex 109

3.4. The psychophysics of color vision 109

    3.4.1. Color matching 110

    3.4.2. Color deficiency (color blindness) 112

    3.4.3. Color appearance of complex arrays 114

    3.4.4. Adaptation 114

    3.4.5. Spatial interaction 117

3.5. In conclusion 120

3.6. Further reading 120

References 121

3.1. Introduction

In this chapter we discuss the physiological basis of color vision and the methods of studying the psychophysics of color vision. The first topic includes an overview of the anatomy of the visual system (the eye, the lateral geniculate nucleus and the visual cortex) and an account of some of the results of electrophysiological investigations of the visual system. We then go on to the psychophysics of color vision. The mathematics of color matching is presented and color appearance of the complex arrays of everyday vision is considered. The changes in visual sensitivity produced by viewing both steady and time varying stimuli are discussed. The role of eye movements in determining color is considered. Experiments in which the effects of eye movements are eliminated bring home the point that color vision is not the result of the simple mapping of the light distribution on the retina into the visual image. The fundamentals of color vision are covered in Section 1.7 of Chapter 1.

3.2. Anatomy of the visual system

The visual system consists of the eyes, the optic nerve, the thalamic relay nuclei (the superior colliculi and the lateral geniculate nuclei), the optic radiations and a large portion of the cerebral cortex (fig. 3.1). In this section we consider the anatomical relations of these units.

3.2.1. The eye

A cross section view of the eye is presented in fig. 3.2. The cornea and the lens provide the dioptric power requisite for the formation of images of external objects on the retina, which is a membrane about a half millimeter thick that lines the back of the eyeball. By altering the thickness of the lens (assisted by spectacles, if necessary) one is able to vary the dioptric power so as to focus objects located at different distances on the retina. By means of head and eye movements one brings the images of objects of interest to the neighborhood of the fovea. Anatomically, the fovea appears as a depression in the retina. The fovea can be seen by looking at the retina through the pupil with an ophthalmoscope. In this case the fovea is seen as a yellow spot (called the macula lutea), because of the presence of a pigment layer overlaying the fovea which absorbs strongly in the short-wavelength end of the spectrum. Although we are generally not aware of this pigment it does have a small effect on some aspects of color vision and can be demonstrated in various ways.

One way to reveal the existence of the macular pigment is by intermittent observation of a uniform field of short-wavelength light. One sees a shaded spot, Maxwell's spot,

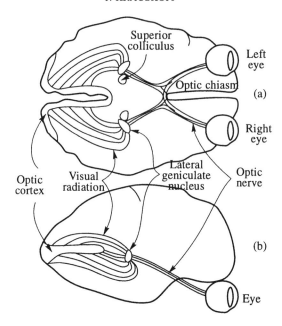

Fig. 3.1. Simplified schematic diagram of the visual pathways of the eye and brain seen from (a) above and (b) the side.

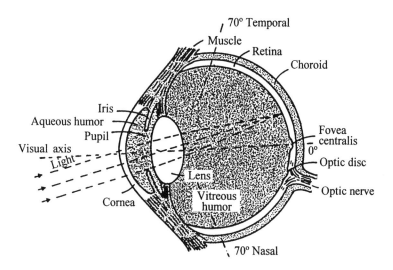

Fig. 3.2. Simplified schematic diagram of the right eye, seen from above.

named for its discoverer James Clerk Maxwell. Another method is to look at a uniform white field of light through a polarizing filter, while slowly rotating the filter. What is seen is known as Haidinger's brushes which appear as a faint cross made up of a yellowish

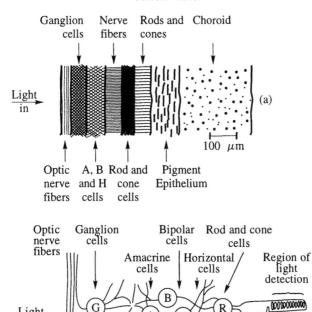

Fig. 3.3. Simplified schematic diagrams of the retina (a) as layers and (b) showing cells and nerve connections. There are several types of bipolar cells, including horizontal ones; the crossing of nerve fibers does not imply connection, but branching does.

line crossing the center of vision in one direction and a perpendicular bluish line which rotates as the polarizer is rotated. The phenomenon is attributed to there being a set of radiating dichroic fibers in the upper layers of the retina.

A cross section through the retina (fig. 3.3) reveals a complex network of receptors and neural elements. People are often dismayed by the fact that the receptors lie at the back of the retina and that the light passes through all the neural layers before reaching the receptors. The pictures one sees of the retina are histological sections of stained tissue or drawings made from such tissue which suggest that the neural elements within the retina would act to distort the images. In reality the neural apparatus is quite transparent and even if it were slightly absorbing this would be of little significance visually.

Light is absorbed by the photopigments which reside in the tips of the rods and cones. These lie at the back of the retina in contact with a strongly absorbing layer known as the choroid. Absorption of light in the photopigments results in a change in the currents flowing through the cell membranes of the receptors. The bases (pedicles) of the receptors make contact with bipolar cells which convey signals toward the inner layer of the retina and with horizontal bipolar cells which make intra-retinal connections. The bipolar cells,

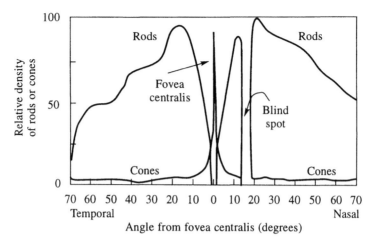

Fig. 3.4. Distribution of rods and cones in a horizontal section across the eye of fig. 3.2. After Cornsweet (1970).

in turn, make contact with the ganglion cells and to a second set of transverse elements, the amacrine cells. The ganglion cells have long axons (nerve fibers) which carry the information out of the eye. They traverse the retina, avoiding the fovea, and leave the eye through the optic disk (fig. 3.2), the blind spot, to form the optic nerve.

The approximately 120 000 000 rods and 5 000 000 cones in each retina are not uniformly distributed (fig. 3.4). The fovea centralis is devoid of rods which dominate the peripheral retina while the density of cones is greatest in the central retina, with the exception that none of the class of cones maximally sensitive to short-wavelength light are found within 20 minutes of the fovea centralis.

The retina is well supplied with blood. The retinal artery and vein enter the eye in the optic disk and smaller vessels branch out from there and, like the ganglion cell fibers, avoid traversing the fovea. It is possible to visualize one's own retinal circulation with the aid of a penlight. One holds the penlight in contact with the bone at the outer corner of the eye socket so that the light is directed toward the nose and enters the cornea at a glancing angle. One then slowly vibrates the light either up and down or in the horizontal plane while staring at a blank wall.

Another way to get this result is to image a small spot on an oscilloscope face in the plane of the observer's pupil with a positive lens. A support, a chin rest, is arranged so that the observer holds his or her head steadily with the image of the small spot entering the center of the pupil. (This optical arrangement is known as the Maxwellian view and has many uses in visual science.) Now a sinusoidally varying signal is applied to either the x- or y-input to the oscilloscope at a frequency in the neighborhood of 1 Hz. This causes the image to traverse the pupil and causes the shadows of the retinal blood vessels to move relative to the receptors. This phenomenon, like Haidinger's brushes and Maxwell's spot discussed above, serve to demonstrate that the visual system is relatively insensitive to steady light signals compared to those varying in time.

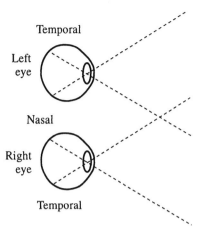

Fig. 3.5. The overlapping visual fields of the eyes, seen from above.

Except for the very peripheral parts of the visual field, both retinas form images of the same visual field as in fig. 3.5. The outer (temporal) half of the right retina receives an image of the left half of the visual world, and the inner (nasal) half of the right eye receives an image of the right half of the visual world, and so on. In general there is right-left reversal in the representation of various functions in the cerebral cortex. For example, the movement of the left arm is controlled by centers in the right motor cortex. The optic nerves from the two eyes meet at the optic chiasm (fig. 3.1). Fibers from the nasal halves of each retina cross over and join the fibers from the temporal halves and continue on to serve as inputs to the visual thalamic nuclei described in Section 3.2.2.

The optic nerve does not represent all parts of the retina equally. There are only about 1 000 000 nerve fibers in the optic nerve compared to the approximately 125 000 000 receptors in the retina. Convergence of signals is greatest for the rods. In the peripheral retina signals from 1000 or more rods may be carried on a single nerve fiber. Signals from peripheral cones also converge but not so drastically. On the other hand, it is believed that cones from the central retina may each provide input to two optic nerve fibers, one carrying information about stimulus increments and the other carrying information about stimulus decrements.

3.2.2. The midbrain nuclei

The smaller visual thalamic nucleus is the superior colliculus and is involved particularly in the control of eye movements. The other nucleus is the lateral geniculate nucleus (LGN). It serves as a relay station between the retina and the primary visual cortex. While we know a good deal about its role in the processing of signals being carried from the retina to the cortex, there is clearly much that we don't know because 90% of the fibers entering the LGN conduct signals down from the cortex and we know almost nothing about their function.

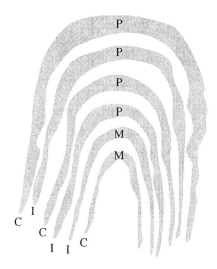

Fig. 3.6. Schematic cross-section of a lateral geniculate nucleus, serving as a relay station between the eye and the visual cortex. P stands for parvo-cellular, M magno-cellular, C contralateral, I ipsilateral.

The structure of the LGN (fig. 3.6) reflects the origin of the signals from the same half of the visual field. An histological stain reveals a pattern of six cellular layers intercalated with fiber layers. Furthermore, it is known that the cells in the upper layer of the LGN on one side derive their inputs from the nasal retina of the eye on the other (contralateral) while the cells in the second layer derive their inputs from the temporal retina of the eye on the same (ipsilateral) side. This pattern, contralateral–ipsilateral, is repeated in the third and fourth layers but, curiously, not in the fifth and sixth layers which proceed ipsilateral–contralateral.

The cells found in the bottom two layers are larger than those from the upper four and, therefore, are called magno cells while the smaller upper layer cells are called parvo cells. The magno cells are known to derive their inputs from a distinctive class of ganglion cells which derive their inputs from relatively larger areas of the retina than those ganglion cells which provide input for the parvo cells in the same general region of the retina. There are 8–10 times as many parvo cells than magno cells.

3.2.3. The cerebral cortex

The areas of the cerebral cortex that have been found to have visual inputs are depicted in fig. 3.7. Area V1 is known also as the primary visual cortex because all inputs to the cerebrum go to it. Neuro-anatomists have been working out the extremely complex pathways connecting the many retinotopic projections. Our knowledge of the functions of these areas is still rather limited, although great progress has been made recently. Areas that are believed to be particularly involved in the perception of color include V1, V2 and V4.

There is considerable support for the notion that not only do the signals from different sense organs, e.g., eyes and ears, travel in separate streams in the nervous system, but

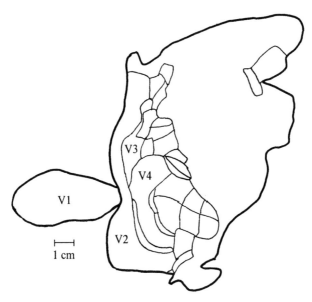

Fig. 3.7. A map of the unfolded cerebral surface of the macaque, showing areas known to be involved with vision. After Felleman and Van Essen (1987).

that within a modality there exist parallel channels devoted to different aspects of the information conveyed by that modality. Much of the evidence in favor of this view comes from the anatomy and physiology of the visual system. The magno-cellular stream seems to be more involved in the perception of movement while the parvo-cellular stream seems more involved in color vision. We will see that the distinctness of these streams seems to be maintained in the cerebellar representations of the visual inputs.

That representation is an evolving story. The output from the LGN consists of a large bundle of nerve fibers called the visual radiation which terminates in a part of the cerebral cortex known by several names: Area 17 (from an early system of classification due to the anatomist Brodmann), the striate cortex (because of its striped appearance) and more recently V1 (because it is now just one, albeit the largest representation of the visual input). The representation is said to be retinotopic, that is, the topological relations of space on the retina are maintained in V1 and in the other cortical visual areas. The amount of cortex devoted to different portions of the visual field is not equal, many more cells conveying information about the central visual field than about the peripheral field.

3.3. Neurophysiology of color vision

3.3.1. Microelectrode recording

Much of our knowledge of the way signals are conducted in the visual system comes from recording of the electrical activity in various structures in the visual systems of animals using microelectrodes, i.e., probes with very thin tips on the order of one micrometer in diameter. Some microelectrodes have a central conducting core of tungsten metal insulated with glass except at the tip. Others are hollow glass tubes that are heated and drawn to a fine tip and filled with a conducting electrolyte solution. The outputs of the electrodes are fed into high gain amplifiers, displayed on oscilloscopes, and preserved on paper recordings, tape, or computer files.

Two kinds of signals are found in the visual system. Within the retina, graded potentials are generated by the rod and cone receptors and by the other intra-retinal elements – the horizontal, bipolar and amacrine cells. These potentials vary according to the strength of the stimuli presented. On the other hand, the ganglion cells whose axons carry the signals from the eye to the central nervous system generate electrical spikes of uniform size and shape. It is the variation in spike frequency that conveys information about the strength of stimuli. The discrete nature of these signals has encouraged some to draw analogies between the nervous system and digital computers which are particularly apt for analysis of experimental results.

3.3.2. Electrical recording from the retina

The Young–Helmholtz theory postulated three receptive processes with different spectral sensitivities. Hering's theory was viewed in his time as in conflict with that of Young and Helmholtz (see Chapter 1, Section 1.7) but in the modern view is concerned with a second level of processing of the visual input. In this section we will consider the electrophysiological evidence that has contributed to our understanding of the nature of the mechanisms performing these functions. We will also consider evidence relevant to theories that postulate additional processing beyond the second level.

The occurrence of color vision follows an almost random pattern in the phylogenetic tree. We and most other primates have color vision but many mammals do not. However, turtles and some fish do. The fact that fish have larger cells in their retinas which generate larger signals contributed to the fact that the first physiological manifestation of opponent (see Chapter 1, Section 1.7.1) processing was obtained from gold fish retinas. The potentials generated in response to brief pulses of monochromatic light varying in wavelength were recorded. At some locations the responses were all of the same electrical sign and reflected the overall sensitivity of the eye. However, at other locations the responses were positive in electrical sign to some wavelengths and negative to others, a pattern consistent with the expectations of the Hering theory.

3.3.3. Spectral sensitivity of monkey cone receptors

More relevant for the understanding of human vision is the physiology of primates. The visual system of the macaque monkey appears to be very similar to that of humans in anatomy and in histological features. Therefore, much work has been done on this species.

Generally microelectrodes are inserted into tissue so that the tip lies either near or within the cell whose activity is recorded. The former is called extracellular and the latter intracellular recording. Recently, another technique has been developed in which a single cell is drawn partially into the electrode by the application of suction. This method

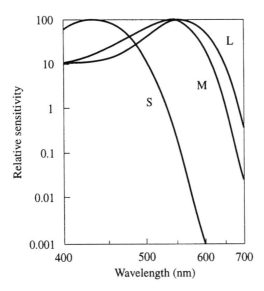

Fig. 3.8. Spectral sensitivities of the three types of cones in the macaque. After Baylor (1987). The S stands for short, M medium, and L long wavelength receptors (these are sometimes called blue-, green-, and red-sensitive receptors, respectively).

has been applied to individual rod and cone receptors in primates to achieve very secure measurements of their spectral sensitivity.

Some measurements made with this method are illustrated in fig. 3.8. These measurements yield very precise estimates of the spectral sensitivity of the receptors. However, if one wishes to relate these findings to human or monkey psychophysical results, then other factors have to be considered. The preretinal medium serves as a spectrally selective filter in front of the receptors. In addition, the cone receptors have been shown to be selective with respect to the angle of incidence of the light from the pupil and, further, this effect is spectrally selective. These factors are relatively minor but should not be neglected.

3.3.4. The lateral geniculate nucleus

The lateral geniculate nucleus (LGN) is the relay station between the eye and the visual cortex and receives input from the ganglion cells of the retina (fig. 3.3). Although its function is incompletely understood, it is clear that a primary function is to carry signals efficiently in a linear fashion to processes in the central visual system which perform various non-linear operations on the information in order to perform such functions as detection and recognition of shape, color, motion, etc.

3.3.5. Receptive fields

It is useful to describe neural elements in terms of their receptive fields (fig. 3.9). A diagram of a neuron's receptive field represents the contributions of regions in the visual

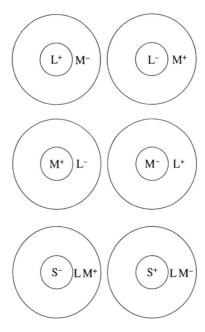

Fig. 3.9. Receptive fields of the six types of retinal ganglion cells.

field to the activity of that cell. A single cone receptor has a simple receptive field whose shape is determined by the light gathering properties of the receptor itself and the effects of the optics on the retinal image. All neurons in the visual system including the ganglion cells of the retina, those of the LGN, and the visual cortex have more complex receptive fields, with the complexity generally increasing as one goes deeper into the visual system.

The receptive fields of cells in the lateral geniculate are radially symmetrical as illustrated in fig. 3.9. All of the cells receive input from more than one class of cone receptors, that is, cones differing in spectral sensitivity. One class of cells provides the input to the center of the receptive field. Its effect may be to increase or decrease the frequency of firing of the cell. This is symbolized in the diagrams in fig. 3.9 by the '+' or '−' in the center of the receptive field. The region encircling the center, known as the "surround" provides an antagonistic influence on the response. That is, if stimulating the center is excitatory, causing an increase in firing, stimulating the surround is inhibitory.

Furthermore, the inputs serving the center and surround are chromatically opponent. The most common types of LGN cells are those which derive opposed inputs from the cone receptors maximally responsive to lights from the long and middle wavelength region of the spectrum. They occur in four types: L^+M^-, L^-M^+, M^+L^-, M^-L^+ as seen in the upper two rows of fig. 3.9. These cells are also known as "Red on-center", "Red off-center", "Green on-center" and "Green off-center" cells, respectively. The largest proportion of the cells in the top four layers of the LGN are of this sort. These cells receive inputs of approximately equal strength from their centers and their surrounds.

The remaining cells in these layers get one of their inputs from cones maximally responsive to lights in the short wavelength portion of the spectrum, the 'S' cones, and opposed inputs from cones maximally responsive to the long and middle wavelengths (the L and M cones). These come as $S^+(L+M)^-$ and $S^-(L+M)^+$ varieties as seen in the lower row of fig. 3.9, and are referred to as Blue on-center and Blue off-center cells. However, since the eye exhibits considerable chromatic aberration, it is actually difficult to determine the center-surround organization of the latter types of cells.

The cells in the lower two, magno-cellular, layers of the lateral geniculate nuclei also exhibit spatial and spectral opponency but tend to have less balanced input from the opponent inputs. The receptive fields of these cells tend to be larger than those of the parvo cellular neurons. The magnos are more sensitive to stimuli of low contrast and high temporal frequency. All of these features tend to support the idea that the magno layers are involved in the perception of motion and of flicker while the more numerous parvo cellular neurons are involved in color vision and in the perception of spatial detail.

3.3.6. Color in the visual cortex

Although there has been a certain amount of evidence and a good deal of speculation, little is known with certainty about the representation of color in the visual cortex. One difficulty is that we do not know quite what to expect of cortical cells nor just where among the many known areas that respond to visual stimuli are the cells of particular importance in color vision.

New techniques for determining regions in the brain responsive to different classes of stimuli are being explored. When a particular class of stimulus is presented, activity immediately increases in that part of the brain specialized to handle such stimuli. This activity may be detected by measuring variations in the electrical potentials at various locations on the skull or by measuring the variations in the magnetic fields near the head. These methods have been used for several years with modest success. New ideas about how to evaluate these kinds of measurements may significantly improve their utility.

New methods such as magnetic resonance imaging (MRI), positron emission tomography (PET) and optical methods have been introduced more recently and offer the promise of yielding a finer grain picture of the regions of increased activity when certain classes of stimuli are presented. These methods are expensive and are in their infancy but we have reason to believe that they will provide information to guide the study of specialized brain areas by more traditional methods.

3.4. The psychophysics of color vision

Psychophysics is the study of the relationships between physical stimuli and experience. Since we are unable to know whether our experiences are the same as those of other people, we have to ask somewhat different questions than we might naively pose. Thus, we cannot ask what a light of a certain wavelength looks like, only whether it is judged by an observer to look the same as another light.

The best established facts about color vision concern matching, that is, the rules which allow us to predict which lights differing in spectral distribution will appear identical to an observer. For example, a monochromatic light of say 575 nm may be seen as identical in all respects to an appropriate mixture of monochromatic lights of say 550 nm and 650 nm. In the color matching experiment the lights to be matched are presented in identical environments, regions of the retina that are similar to one another and have the same light history and are surrounded by lights of the same spectral content. Both the observer and the experimenter might call the three lights "yellow", "green" and "red", respectively, but we have no way of knowing what they look like to another person.

The primary function of color vision is to contribute to the identification of objects in the environment. In real life situations, the context in which we see things varies markedly from time-to-time. The spectral distributions of the lights that arrive from objects depend on their spectral reflectivity and on the spectral distribution of the light which illuminates them. Therefore, the same object would look very different from one time to the next if our sensations were directly correlated with the spectral distributions arriving from them. First we will consider the facts of color matching and then go on to the more difficult question of what factors control the appearance of objects in complex everyday situations.

3.4.1. Color matching

We have color vision under daylight illumination because light is absorbed by three different photopigments housed in different cone receptors. This hypothesis was put forth more than 200 years ago by several authors, clarified by Thomas Young, championed by Herman Helmholtz, but best enunciated by James Clerk Maxwell. Estimates of the sensitivities of human cone photoreceptors are plotted in fig. 1.13 of Chapter 1. These curves were derived from measurements of color matches made by people with normal color vision and by people with certain forms of defective color vision caused by not possessing one of the three sets of receptors that normal observers have.

A typical color matching experiment is illustrated in fig. 3.10. The subject of the experiment is confronted with a bipartite field, one half of which is illuminated with the color to be matched, in this case a monochromatic light of wavelength (λ). The other half is illuminated by a mixture of three primaries which may be an arbitrary set of lights meeting the restriction that no one of them can be matched with a mixture of the other two. In the illustration the primaries are three monochromatic lights of λ_1, λ_2 and λ_3. The subject is required to set the intensity of the three primaries so that the two halves of the field look identical, that is, the same hue, saturation and brightness. This can be done for lights of all wavelengths if the subject is allowed sometimes to add one of the primary lights to the light to be matched.

The essential rule that governs color matching is that two colored patches viewed in the same environment (that is, surrounded by lights of the same spectral composition and detected by regions of the retina with the same light histories) will appear the same color if the stimuli evoke the same responses in each of the three photoreceptors. So it is useful to consider the process by which responses are generated.

When a photon is absorbed by a photopigment in a receptor, a process is initiated that leads, through a complex biochemical process, to the movement of a large number of electronic charges. For present purposes the important fact is that the response depends

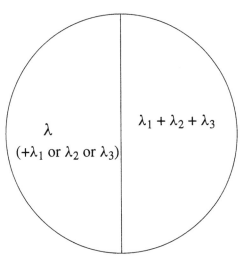

Fig. 3.10. The observer's view in a color matching experiment.

solely on the number of photons absorbed and not their wavelengths. This has come to be called the principle of univariance and was first enunciated by Maxwell who wrote:

"In order fully to understand Young's theory, the function which he attributes to each system of nerves must be carefully borne in mind. Each nerve acts, not, as some have thought, by conveying to the mind the knowledge of the length of an undulation of light, or of its periodic time, but simply by being more or less affected by the rays which fall on it. The sensation of each elementary nerve is capable only of increase and diminution, and no other change. We must also observe that the nerves corresponding to the red sensation are affected chiefly by the red rays but also to some degrees by those of every part of the spectrum; just as red glass transmits red rays freely, but also transmits those of other colors in some quantity."

We can represent the effect of a monochromatic stimulus on the response of a photopigment by:

$$R_\lambda = E_\lambda S_\lambda,$$

where R_λ is the magnitude of the response generated by a stimulus whose energy is E_λ and S_λ is the sensitivity of the photopigment for that wavelength.

If only one photoreceptor, with a single photopigment, is involved in processing the stimulus, the principal of univariance leads to the prediction that lights of two wavelengths will be identical in their effects and look exactly the same if $R_{\lambda_1} = R_{\lambda_2}$. This, in fact, happens in dim illuminations when only the rod receptors, which contain the single photopigment rhodopsin, are active. In these conditions we can make perfect matches between two monochromatic lights by merely adjusting the intensity of one of them.

The principal of univariance also predicts that the response of a single photopigment to a stimulus that consists of two or more monochromatic lights is determined by the total number of photons absorbed at each wavelength. Thus we can write:

$$R = \sum_{\lambda=400}^{700} E_\lambda S_\lambda.$$

Therefore two lights of different spectral composition will look identical in dim illumination if the responses they evoke in the rods are equal.

Turning now to the cone receptors, the same rules apply to each of them. Thus we can write:

$$R_L = \sum_{\lambda=400}^{700} E_\lambda L_\lambda, \tag{3.1}$$

$$R_M = \sum_{\lambda=400}^{700} E_\lambda M_\lambda, \tag{3.2}$$

$$R_S = \sum_{\lambda=400}^{700} E_\lambda S_\lambda, \tag{3.3}$$

where R_L, R_M, and R_S, are the responses generated in the long-, middle-, and short-wavelength sensitive cones, respectively, and L_λ, M_λ and S_λ are their spectral sensitivities, respectively.

3.4.2. Color deficiency (color blindness)

Defective color vision is a topic of particular interest, especially to men, since they are much more likely to have some form or reduced color vision than women. About 10% of males have reduced ability to make color discriminations while only about 1% of females do. This is because the genes which are responsible for the type of photopigments are determined by the X chromosomes. A person is male if they inherit one X and one Y chromosome from their parents and female if they inherit 2 X chromosomes.

The color deficient traits are recessive and therefore females must inherit genes for color deficiency from both parents and thus must have the same deficiency coded on both of their X chromosomes. If they only inherit one such gene, they may be carriers of color deficiency but do not manifest the trait. On the other hand, since males have but one X chromosome they must be color deficient if they have such a gene.

Color deficiency comes in several varieties. One distinction is concerned with the number of types of cone photoreceptors individuals possess. People who have three cone receptors are called trichromats, people with two cone receptors dichromats and people with one type of cone receptor are called monochromats. People who are classified as normal are trichromatic because they have three functioning sets of color receptors, having pigments with three distinct photopigments whose spectral response curves are different from one another. From the discussion of color matching it is clear that such people require three primaries in order to match any arbitrary light.

The largest fraction, about 90%, of the people who are classified as color deficient, are also trichromats. However, they are much more variable in their color matches and will accept a much wider range of matches than the normal trichromats. Such people can be identified by the fact that they may fail to distinguish subtle difference in color such as are found in the "Ishihara Pseudoisochromatic Test", the familiar patterns of pastel dots that form a pattern of numbers which can be read by people with normal trichromatic vision but may be incorrectly perceived by such marginally color deficient people. Since these people still require three primaries to match any arbitrary color, they are called "anomalous trichromats".

The second major class of color deficients are called "dichromats" because they need only two primaries to match any arbitrary color and thus must have only two functioning types of photoreceptor. There are three subtypes of dichromats according to which class of photoreceptors they are missing. The more common dichromats are called "protanopes" and "deuteranopes". The former have no long wavelength sensitive cone receptors while the latter have no middle wavelength sensitive receptors. Members of the third, quite rare, class are called "tritanopes" and have no short wavelength sensitive cone receptors.

Dichromats accept color matches that normal trichromats make but, in addition, will accept many mixtures of primaries that normals will say appear not to match. This fact was used by early students of color vision to establish the spectral sensitivities of the cone receptors. Although methods of establishing these spectral sensitivities based on "physical" methods have since been developed, those based on the difference between normal trichromats and dichromats are the most generally useful measures.

People often ask what does the world look like to the color-blind but this question is as meaningless as asking what the world looks like to anyone else. We can only know about the matches or discriminations others make, not what things look like. However, it is meaningful to ask what kinds of confusions color-blind people make.

Since lights longer in wavelength than about 500 nm stimulate the short wavelength sensitive receptors very little (fig. 1.13 of Chapter 1), both protanopes and deuteranopes confuse lights in the green, yellow and red portion of the spectrum. The matches they make are subtly different but the easiest way to distinguish one from the other is in terms of the brightness matches they make between normally red and green lights. For the protanope, missing the long wavelength sensitive receptor, the red end of the spectrum is much dimmer than it is to either the normal trichromat or the deuteranope. On the other hand, the deuteranope judges the brightness of spectral lights much as the normal trichromat.

Although some people are embarrassed to be color deficient, it is not a significant handicap except in certain technical situations. In fact it is so little a handicap that it wasn't commented on much until about 200 years ago when the physicist Dalton described his own color-blindness. It was such a novelty that in much of Europe color-blindness is known as *Daltonism*

3.4.3. Color appearance of complex arrays

As noted above, color vision serves to helps us to identify objects. The persisting attribute of an object is not the amount or spectral distribution of the light that arrives from it but the reflectance spectrum of the object. For example, a lump of coal in sunlight appears black, while a piece of chalk in the shade looks white even though the coal may reflect much more light than the chalk. This is referred to as "lightness constancy". Similarly, when we take a collection of objects of different spectral reflectance from one illuminant to another, the color of the objects tend to persist. This is referred to as "color constancy". These terms are somewhat misleading because neither lightness nor color are ever completely constant. Sometimes objects appear very different in different illuminants. This phenomenon is called metamerism and is described in Chapter 1, Section 1.7.6 and in Chapter 2, Section 2.2.7.

In principle, lightness and color constancy could be achieved by deduction if one knew the nature of the illuminant and used this information to interpret the spectral distribution of the lights coming from objects. This idea has its root in the thinking of the "empiricist" philosophers and was embodied, in the nineteenth century, by the great physicist-physiologist Helmholtz in the notion of "unconscious inference". Another view that was supported by other philosophers and early students of perception held that our perceptions were determined by built-in mechanisms. The second view is more in keeping with modern thinking but some theories of perception contain some aspects of the former view.

Here we consider some of the processes which serve to determine our perceptions of color in everyday life. These include adaptation and color induction.

3.4.4. Adaptation

The visual system operates over a 10^{10} range of light intensities but the dynamic range over which neurons operate is about 10^2. Several processes cooperate to make this possible. One of these is directly visible, namely, variation in pupil size over a range from about 2.5 to 7 mm in diameter. This factor can account for about one decade of the range. The rest of the adjustment for intensity levels is accomplished by: 1) the existence of two receptoral systems, 2) photopigment bleaching, 3) receptoral adaptation and 4) post-receptoral adaptation.

We are all familiar with the fact that it takes some time to adjust when we go into a dim environment from a bright one. If we formalize this observation with an experiment we get the results shown in fig. 3.11. In this experiment the amount of light needed to just see a test spot falling on the near peripheral part of the retina is plotted as a function of the time after the observer went from a bright environment to a dark one. Thresholds for seeing are high initially and drop along two upward concave curves. In the first phase which lasts about 10 minutes the test flashes appear colored if monochromatic or spectrally limited test lights are used. In the second phase which ends after about 30 minutes all manner of test stimuli appear neutral in hue.

If the experiment is repeated with a variety of monochromatic lights it is found that the spectral sensitivity curves are different in the two phases. In the first phase the peak sensitivity is at about 550 nm while in the second it is about 505 nm. These facts

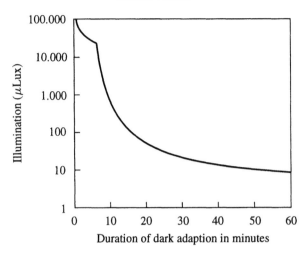

Fig. 3.11. The dark adaptation curve.

demonstrate that there are two classes of photoreceptors: cones which operate under higher light levels and mediate color vision, and rods which operate at lower levels but yield only monochrome sensations.

In the last century it was observed that fresh frog retinas were initially bluish-red and bleached on exposure to light. Subsequent research determined that the colored substance was a photopigment, rhodopsin, which was housed in the rod photoreceptors. In living tissue rhodopsin, also known as visual purple, regenerates in the dark. It is now known that similar pigments reside in human cone receptors and in the cone receptors of other species.

Early in this century it was thought that the recovery of visual sensitivity revealed in the dark adaptation curves was due to the regeneration of photopigments that had been bleached in the light. However, direct measurements of photopigment bleaching and regeneration in human eyes have been made by recording the amount of light reflected from the retina before and after exposure to light of varying intensity. These experiments demonstrated that insignificant bleaching occurs until the lights are exceedingly bright. So while photopigment bleaching has some influence on visual sensitivity, other processes are responsible for the changes in visual sensitivity we experience at ordinary light levels.

These other processes are not completely understood. There are at least two different additional processes. To understand these processes it is useful to consider the wiring diagram of the early stages of the visual pathway depicted in fig. 3.12. This diagram is generally accepted view of how signals from the three types of cones feed into the three opponent signals. While these mechanisms were originally postulated to explain facts of perception, the similarity to the classes of physiological mechanisms residing in the ganglion cells and in the lateral geniculate nucleus discussed earlier should be obvious (see fig. 3.9). The circles with the arrows imply the location of gain control mechanisms which determine the sensitivity of that portion of the pathway.

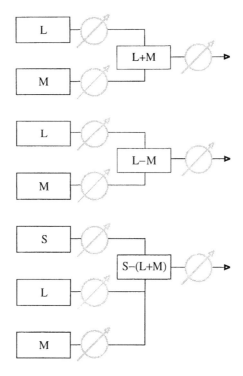

Fig. 3.12. The first and second stages of the visual pathway.

The sensitivity of the first stage mechanisms, the portion of the pathway in which signals are determined by the excitation of individual classes of cones, varies with the level of excitation of the cones. The system thus protects itself from signal overload by adjusting its gain according to the level of the input. There are several ways to demonstrate the existence of such a process. For example, in classical experiments performed by W.S. Stiles (see Kaiser and Boynton (1996) for a detailed overview) the amount of light necessary to detect a flash of monochromatic red light was measured on steady backgrounds of light of different wavelengths. More light was needed to detect the test flashes as the background intensity was raised. It was found that the amount of light necessary to raise the detection threshold to the same level varied with the background wavelength inversely with the sensitivity of the long wavelength receptors.

While such an experiment requires complex equipment, we are all familiar with effects that demonstrate the selectivity of the adaptation of the visual system to steady lights. We have all stared at a pattern of colored patches and then switched our gaze to a uniform white field and seen the "after image" of the pattern in "complementary" colors.

While the first stage sensitivity is controlled by the steady (DC) level of excitation of the individual cone classes, the second stage sensitivity is altered according to the history of varying (AC) activation of these mechanisms. In fig. 3.12 the locus of the sensitivity control is indicated by the circles with the arrows through them. DC levels of light are believed to control the gain in the first stage cone pathways, AC adaptation the gain in the second stage interactive pathways.

The effects of AC adaptation are not easily demonstrated without special equipment. The original experiments were done with a computer-controlled laser colormixer. This device used acousto-optic modulators to control the intensity of three laser sources rapidly so that one could modulate the target precisely in any direction in color space. The experiment involved measuring the threshold changes in saturation before and after viewing a target modulated in color. The test field is initially white and is changed briefly in some direction, say red. If the change is large, one would see a saturated red. If the change is small, one would see a very weak pink. In the first stage of the experiment thresholds are measured in all color directions, red, yellow, green, blue and intermediate hues. Then the observer is exposed to a light varying between maximally saturated complementary colors, say, red and green, after which the thresholds are measured again. After red–green modulation it is harder to see changes in the red and green directions but not in the yellow–blue directions. The reverse is true if the adapting light varies between yellow and blue. When the adapting light varies in color directions between these, it is harder to see all color changes.

While it is difficult to set up this experiment it is possible to show the effects of transient changes of light on sensitivity with a simple observation. One looks at the top of a light bulb at the label which gives the wattage with the light turned off but with sufficient light to read the label. If one keeps looking at the bulb and turns it on, the label at first can no longer be read. As one continues to look at the bulb, the label again becomes visible. Turning off the bulb results in a similar temporary loss of ability to see the label.

3.4.5. Spatial interaction

It is often said that the eye is like a camera because both have lenses and a photosensitive surface but this is, in fact, in many ways a poor analogy. The image on film in a particular place on the film is determined by the light falling on the local region. The lens and the film both show "spreading effects" so that the image is effected by the regions in the object space near each image point. In vision what we see in a particular place may be influenced by light coming from objects far removed from the object of interest. In the case of film, effects of light from remote regions are always additive. In the case of vision, interaction are almost always subtractive. Sometime, as we will see, the percept in one region of visual space may have no dependence on the stimuli coming from that region of object space.

One important example of the influence of remote stimuli on the appearance of local stimuli is the phenomenon of simultaneous contrast which is illustrated in color plate 8. In this figure all the central discs have the same spectral composition, of a neutral gray, but the color they appear to be depends on the color of the surrounding patch of light. This phenomenon plays an important part in the appearance of objects in everyday life. For example, it was mentioned earlier that a lump of coal in sunlight appears black, while a piece of chalk in the shade looks white. One could take this figure in to the sunlight or into a shaded place and the colors would appear much the same, indicating that it is the relations between stimuli and not their absolute intensities that are determinant in perception.

No simple rules determine color appearance. For example, comparing the appearance of the large and small discs in color plate 8, one can see that the color appearance depends on the dimensions of the patches of light as well as their spectral composition. The color appearance of the central discs is more influenced when the surrounding area is larger and when the discs themselves are smaller. This suggests a theory in which the amount of stimulation in the central and surrounding regions determine the amount of the effect, which does very easily lead to an account of color constancy.

Studies of the properties of single cells in the visual system demonstrate that the responses are strong to stimuli that vary in space and time. Many cells fail to fire in the presence of steady stimuli. When we look at large uniform colored objects they appear quite definite in color even though the neurons that receive input directly from these patches may not be experiencing time varying stimulation. So the question arises: how is the color of these objects determined? Some understanding of this question is suggested by considering the role of eye movements.

It turns out that our eyes are constantly in motion even when we stare at an object as fixedly as possible. This has been demonstrated in experiments in which observers wear contact lenses with mirrors embedded in them which serve to deflect the beam from a light source as the eye moves while viewing an object as in fig. 3.13. The beam falling on the mirror projects the image of a rectangle of light which falls partially on a straight

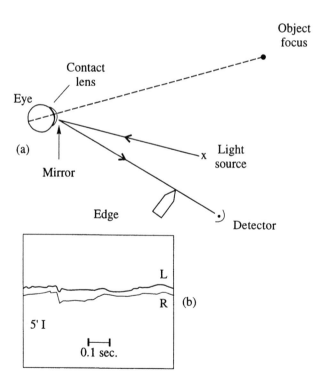

Fig. 3.13. Apparatus for recording eye movements (a) and typical results (b).

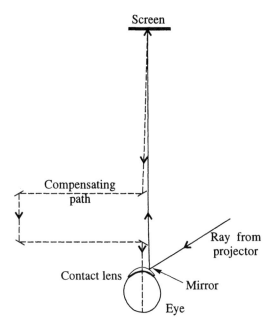

Fig. 3.14. Apparatus for stabilizing an image.

edge, behind which is located a sensitive photocell detector. As the eye rotates, more or less light reaches the photocell. The electrical signal is plotted as a function of time on the tracing in the figure.

It can be seen that the eyes are always in motion, albeit keeping the object of interest close to the central fovea. It might be thought that such motions would "blur" the image and be detrimental to vision, but it has been demonstrated that these little movements, called "saccades", are actually beneficial to vision. This has been revealed in experiments using "stabilized images" which do not move on the retina. A more complex optical arrangement, in which the light reflected from the contact lens mirror provides the target viewed by the observer, is shown in fig. 3.14. The beam reflected off the mirror moves at twice the angle that the eye rotates. By making the return path to the viewing eye twice the distance from the mirror to the screen, the image can be made to move in direct correspondence with the observer's eye movements and thus remain fixed on the retina. Targets viewed in this way initially appear sharp and clear but tend to disappear in time, thus demonstrating the importance of time-varying inputs for vision.

In one experiment, a target consisting of a disc of one color surrounded by an annulus of another color, was stabilized on the retina. By viewing the stabilized image through an aperture, the outer border of the annulus was destabilized so that only the border between the disc and the annulus was immobile on the retina.

The result was quite dramatic. The disc in the field disappeared in time even though it was a degree or so in diameter and fell on the central fovea where we image objects of greatest interest. When the disc disappeared the whole field appeared to be a disc of one color, that of the annulus. The explanation of this effect appears to be that the visual system performs what might be called an inferential process. Areas of uniform stimulation generate no signals in the visual system because they provide no time varying inputs. At borders, when the eyes move, time-varying inputs provide signals from which may be inferred a transition from the surrounding color to the surrounded color. In the absence of such signals, the disc and annulus together appear in the annulus color because the outer border of the annulus is not stabilized and provides information concerning the transition from its surround.

It takes specialized equipment to produce a stabilized image but we can simulate to some degree the effect by blurring the borders between stimulus regions. Eye movements result in abrupt variations in the inputs of receptors lying under borders when the borders are hard edges, that is, when the spectral distribution changes rapidly in space. When the normal eye movements move a graded image over the retina, the time-varying changes are smoother and result in less neural activity. This allows us to simulate the conditions of the two-color stabilized image in color plate 9. If you stare fixedly at the center of this pattern you may see the disc disappear and the whole field take on the color of the annulus.

In a simple experiment described by Kaiser and Boynton (1996, p. 357), a table tennis ball is cut in half. The translucent halves are placed over the open eyes and are illuminated with uniformly colored light. At first the color is perceived, but it fades after a few seconds. The color returns on closing and reopening the eyes, only to fade once again.

3.5. In conclusion

Color vision seems, to those who have it, so simple and direct that it needs for them no explanation. It is only when confronted with phenomena like those just discussed that one realizes that the eye is not simply like a camera. We tend to see something that approximates reality, but a "reality" that is constructed by our visual systems. For example, objects tend to appear *almost* the same color under very different illuminations. The same mechanisms which allows us to maintain a reasonably constant picture of the world are responsible for optical illusions. Both reality and illusions are manifestations of the lawful processes of our visual systems. The illusions are the more instructive since they often make it clear where the holes in our understanding are. As Kurt Koffka (1935) has pointed out, the problem in understanding perception is not why do things look as they are, but why do things look as they do.

3.6. Further reading

A comprehensive recent text covering the whole area of vision is Human Color Vision by Kaiser and Boynton (1996); this should be consulted for further details and for additional references to the literature. Useful general treatments are Foundations of Vision by Wandell (1995); Color Science by Wyszecki and Stiles (1982); Sources of Color Science by MacAdam (1970); Visual Perception by Cornsweet (1970); and Light, Colour and Vision by Le Grand (1957).

References

Baylor, D.A., 1987, Photoreceptor signals and vision. Investigative Opthalmology and Visual Science **28**, 34–49.

Cornsweet, T.N., 1970, Visual Perception (Academic Press, New York).

Felleman, D.J., and D.C. Van Essen, 1987, Receptive field properties of neurons in area V3 of macaque monkey extrastriate cortex. J. Neurophysiology **57**, 889–920.

Kaiser, P.K., and R.M. Boynton, 1996, Human Color Vision, 2nd edn (Optical Society of America, Washington, DC).

Koffka, K., 1935, Principles of Gestalt Psychology (Harwart, Brace, San Diego, CA).

Le Grand, Y., 1957, Light, Colour and Vision (Wiley, New York).

MacAdam, D.L., 1970, Sources of Color Science (MIT Press, Cambridge, MA).

Wandell, B.A., 1995, Foundations of Vision (Sinauer Associates, Sunderland, MA).

Wyszecki, G., and W.S. Stiles, 1982, Color Science (Wiley, New York).

chapter 4

THE FIFTEEN CAUSES OF COLOR

KURT NASSAU

Lebanon, NJ, USA
(AT&T Bell Telephone Laboratories
Murray Hill, NJ, USA, retired)

Color for Science, Art and Technology
K. Nassau (Editor)
© 1998 Elsevier Science B.V. All rights reserved.

CONTENTS

4.1. Introduction 125

4.2. Some fundamentals 127

4.3. Color from simple excitations and vibrations 129

    4.3.1. Mechanism 1: Color from incandescence 129

    4.3.2. Mechanism 2: Color from gas excitation 130

    4.3.3. Mechanism 3: Color from vibrations and rotations 132

4.4. Color from ligand field effects 134

    4.4.1. Mechanism 4: Color in transition metal compounds 137

    4.4.2. Mechanism 5: Color from transition metal impurities 137

4.5. Color from molecular orbitals 138

    4.5.1. Mechanism 6: Color in organic compounds 138

    4.5.2. Mechanism 7: Color from charge transfer 140

4.6. Color from band theory 142

    4.6.1. Mechanism 8: Color in metals and alloys 142

    4.6.2. Mechanism 9: Color in semiconductors 145

    4.6.3. Mechanism 10: Color in doped semiconductors 147

    4.6.4. Mechanism 11: Color from color centers 149

4.7. Color from geometrical and physical optics 153

    4.7.1. Mechanism 12: Color from dispersion 153

    4.7.2. Mechanism 13: Color from scattering 155

    4.7.3. Mechanism 14: Color from interference 158

    4.7.4. Mechanism 15: Color from diffraction 161

4.8. Color occurrences in many fields 164

    4.8.1. Color in pigments and dyes 164

    4.8.2. Color in biology 164

    4.8.3. Color in minerals and gems 165

    4.8.4. Color in glass, glazes, and enamels 166

    4.8.5. Color in the atmosphere 166

    4.8.6. Sources of white and colored light 166

4.9. Further reading 167

References 167

4.1. Introduction

In Chapter 1 were covered the basics of color, including Newton's discovery of the spectrum. Based on this, and on the color perception concepts covered in Chapter 3, we can deduce that we perceive color when some of the wavelengths emitted by a source of white light are selectively absorbed, reflected, refracted, scattered, or diffracted by matter on their way to our eyes; alternatively, a non-white distribution of light may have been emitted in some way.

Fifteen specific color-producing mechanisms can then be distinguished, originating in a variety of physical and chemical mechanisms, collected into five groups with the color originating in: simple excitations and vibrations; ligand (crystal) fields; molecular orbitals; energy bands; and geometrical and physical optics. The five groups of this admittedly somewhat arbitrary classification break down in this way: three of them, simple excitations and vibrations, energy bands, and geometrical and physical optics are normally part of the physics curriculum; molecular orbitals are normally part of chemistry; and ligand (crystal) fields may be covered in either discipline. It is unusual, however, for even a book-length treatment of any of these topics to concern itself with the colors produced as such.

An outline of these fifteen causes of color is given in table 4.1, based on the book "The Physics and Chemistry of Color" by K. Nassau (1983), from which all illustrations are taken. More details can be found there, together with many references for further reading; only some selected references are given at the end of this chapter. In addition to being organized by mechanisms, the book includes further discussions of the color causes involved in a variety of fields including pigments and dyes – here covered in Chapters 10 and 11; the preservation of color in our artistic heritage – here covered in Chapter 12; as well as details of the occurrences of color in many different fields, here outlined in Section 4.8.

There are many widely repeated erroneous color cause attributions. As one example, the blue color of the ocean is frequently attributed to reflection from the sky and that of ice to a structural change involving cold. Again, simplistic color attributions, such as that blue and green are caused by copper, dark blue by cobalt, etc., are only occasionally correct. Color plate 10 shows some of the many colors given by chromium. Even in a highly restricted field such as the dark blue gemstones of color plate 12, cobalt can indeed be one of the causes of a deep blue, yet six other mechanisms not involving cobalt can also produce a similar color; I regret that I could not borrow the dark blue Hope diamond (colored by Mechanism 10 of table 4.1, involving a boron impurity acting as an electron acceptor and also giving electrical conductivity) for inclusion in this figure. Detailed and lengthy studies may be needed to establish the cause of an unknown color.

TABLE 4.1
Examples of the fifteen causes of color.

Simple excitations and vibrations
1. Incandescence:
    Hot objects, the sun, flames, filament lamps, carbon arcs, limelight, lightning*, pyrotechnics*.
2. Gas excitations:
    Vapor lamps, neon signs, corona discharges, auroras, lightning*, pyrotechnics*, lasers*.
3. Vibrations and rotations:
    Water, ice, iodine, bromine, chlorine, blue gas flame.

Transitions involving ligand field effects
4. Transition metal compounds:
    Turquoise, malachite, chrome green, rhodochrosite, smalt, copper patina, fluorescence*, phosphorescence*, lasers*, phosphors*.
5. Transition metal impurities:
    Ruby, emerald, alexandrite, aquamarine, citrine, red iron ore, jade*, glasses*, dyes*, fluorescence*, phosphorescence*, lasers*.

Transitions between molecular orbitals
6. Organic compounds:
    Dyes*, biological colorations*, fluorescence*, phosphorescence*, lasers*.
7. Charge transfer:
    Blue sapphire, magnetite, lapis lazuli, ultramarine, chromates, Painted Desert, Prussian blue.

Transitions involving energy bands
8. Metals and alloys:
    Copper, silver, gold, iron, brass, "ruby" glass.
9. Pure semiconductors:
    Silicon, galena, cinnabar, vermillion, cadmium yellow and orange, colorless diamond.
10. Doped semiconductors:
    Blue and yellow diamonds, light-emitting diodes, lasers*, phosphors*.
11. Color centers:
    Amethyst, smoky quartz, desert "amethyst" glass, fluorescence*, phosphorescence*, lasers*.

Geometrical and physical optics
12. Dispersion, polarization, etc.:
    Rainbows, halos, sun dogs, photoelastic stress analysis, "fire" in gemstones, prism spectrum.
13. Scattering:
    Blue sky, red sunset, blue moon, moonstone, blue eyes, blue skin, blue butterflies*, blue bird feathers*, other blue biological colors*, Raman scattering.
14. Interference:
    Oil slick on water, soap bubbles, coatings on camera lenses, biological colors*.
15. Diffraction:
    Aureole, glory, diffraction gratings, opal, liquid crystals, biological colors*, diffraction spectrum.

---

* Only in part.

It may seem remarkable that so many distinct causes of color should apply to the small band of electromagnetic radiation to which the eye is sensitive: less than one "octave" in an electromagnetic spectrum of more than 80 "octaves" (see Section 1.5 and fig. 1.4). Yet much happens in this narrow band because this is the energy range where radiation begins to interact with the outer electrons on atoms. At energies lower than those in the visible range, radiation only induces small motions of atoms and molecules which we sense as heat; at energies higher than those of the visible range, radiation ionizes atoms by removing one or more electrons and can therefore destroy molecules. This is further discussed in the next section. Only in the narrow optical region to which the human eye is sensitive is the energy of light well attuned to non-destructive interactions with the electrons in matter with the diversity of colorful results summarized in table 4.1. Electrons are involved in essentially all of these color causes and we "see" electrons in action whenever we perceive color.

4.2. Some fundamentals

A concept that is basic to much of what follows is the energy diagram. Figure 4.1 shows the relationship between energy in electron volts eV, wavelength in nanometers nm, frequency in hertz Hz, and the perceived color for that part of the electromagnetic spectrum relevant to this chapter. (For these units, their interconversion, and related matters, see Section 1.5 and tables 1.1 and 1.2.)

Quantum theory applies when the energy of a photon (or quantum), which is the smallest quantity of light that can exist, is absorbed or emitted by an electron, atom, or group of atoms. Such a system usually can only exist at certain energy levels. It can therefore absorb or emit only certain quantities of energy and we can represent such

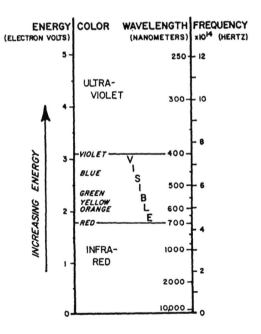

Fig. 4.1. The spectrum with several descriptions.

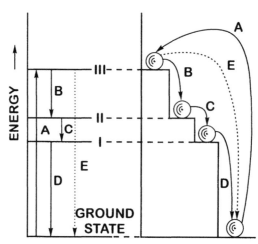

Fig. 4.2. Energy level scheme (left) of a quantum system that can only exist at discreet energy levels; some transitions are allowed (solid arrows), others are forbidden (dotted arrows); schematic equivalent (right) with a ball on a flight of steps.

a system by an energy diagram as at the left in fig. 4.2. It takes energy to raise the system up to a higher energy level; energy is again released when the system returns downwards. An easily visualized equivalent is the energy diagram of a ball being carried up a flight of irregular height steps and rolling back downwards as at the right in fig. 4.2. In the gravitational field of the earth, the ball gains energy when it is raised and loses this energy when it falls downward.

There may be "selection rules" which "allow" or "forbid" certain of these steps in a quantum system, so that absorption A and emission C in this diagram may be allowed and will occur strongly, while emission B may be forbidden and will not occur at all or only weakly or slowly.

It should be noted that when a band of wavelengths is removed from white light, e.g., by absorption, then the color perceived will be complementary to the color(s) removed, as readily deduced from the color triangle as described in Section 1.6.1 and fig. 1.7. Thus if the red end of the spectrum is removed, then a blue color is seen; if the blue end of the spectrum is removed, then a yellow is seen; removing green produces a reddish purple; and so on.

A consideration of the dispersion curve may also be helpful in what follows. Such a curve shows how the refractive index of a transparent material varies with the wavelength or energy as in fig. 4.3. The increase of the refractive index with increasing energy (decreasing wavelength) which leads to Newton's spectrum in a prism is only the small central region of the full dispersion curve, shown for a colorless crown glass in this figure. At low energies in the infrared (left) there are features derived from the absorption of infrared energy; this produces excited lattice vibrations, originating in the molecular framework derived from the bonding between atoms. At high energies (right) there are features originating from the unpairing and excitation of previously paired electrons on

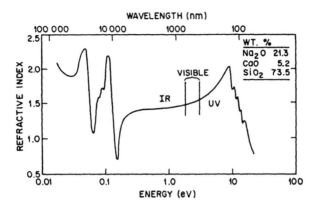

Fig. 4.3. The extended refractive index curve of a colorless crown glass.

individual atoms which leads to absorptions in the ultraviolet. The infrared absorptions of fig. 4.1 can be shifted into the visible region with light-weight, strongly-bonded atoms, as in the vibrations and rotations of Mechanism 3 in table 4.1. Similarly, the ultraviolet absorptions of fig. 4.1 can be shifted into the visible region in a wide variety of situations leading to color from Mechanisms 1, 2, and 4 through 11 of table 4.1.

4.3. Color from simple excitations and vibrations

4.3.1. Mechanism 1: Color from incandescence

We are used to speaking of "red hot", "white hot", and so on; these are parts of the color sequence black, red, orange, yellow, white, and bluish white seen as an object is heated to successively higher temperatures. The light produced consists of photons given off by electrons, atoms, and molecules when various parts of their heat vibration energy are released as photons. Max Planck found in 1900 that the quantization of energy was necessary to explain the idealized "black body" radiation, thus initiating the quantum theory. Planck's equation for the energy E (in $W\,cm^{-2}$) radiated into a hemisphere at wavelength λ (in μm) and in interval $d\lambda$ at temperature T (in K) is given by:

$$E\,d\lambda = \frac{37\,415\,d\lambda}{\lambda^5}\left[\exp\left(\frac{14\,338}{\lambda T}\right) - 1\right]^{-1}. \tag{4.1}$$

The total radiated energy E_T (in W) is given by:

$$E_T = 5.670 \times 10^{-12}\,T^4. \tag{4.2}$$

At any given temperature there is a peak in the intensity of the emitted radiation and the peak shifts toward shorter wavelengths (higher energies) with increasing temperature as can be seen in the curves of fig. 4.4. Wien's law gives the peak wavelength λ_m as:

$$\lambda_m = 2\,897/T. \tag{4.3}$$

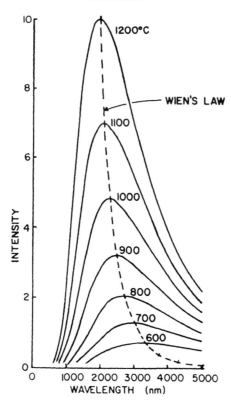

Fig. 4.4. Planck's blackbody equation curves for different temperatures, also showing Wien's law.

Our definition of "white" is derived from our evolution on earth, exposed to light from the 5700° C temperature of the surface of the sun: its radiation peak near 550 nm (2.25 eV) is paralleled in the maximum sensitivity of our eyes in the same region. The sequence of incandescence colors is shown on a color triangle in fig. 4.5. No matter how high the temperature, a blue-white is the "hottest" color.

The color of incandescence radiation can be used to measure temperature in radiation pyrometers of several types. Lighting sources from the primitive candle through limelight, arc lamps, incandescent filament lamps, and flash bulbs are based on incandescence. The usual aim here is to avoid color. Part of the light from pyrotechnic devices is derived from incandescence, originating from high temperatures produced by chemical reactions, such as the burning of magnesium metal.

4.3.2. Mechanism 2: Color from gas excitation

Every substance when heated gives the incandescence of Mechanism 1. Specific chemical elements, present as a vapor or a gas, can have their electrons excited into higher energy levels by the addition of energy in one of several forms. Light is then produced when part or all of the excess energy is emitted as a photon in the process of gas excitation.

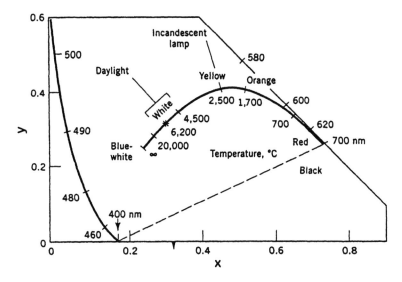

Fig. 4.5. Black body colors shown on the CIE chromaticity diagram.

Examples of various forms of energies used for this include: electrical excitations as in arcs, sparks, lightning, neon tubes, and sodium and mercury vapor lamps; chemical reactions in a flame to produce high temperature as in the chemist's flame test for sodium, potassium, copper, and a few other elements; and high energy particles as in the northern aurora borealis and southern aurora australis displays.

Unusual is triboluminescence, as when we crunch a wintergreen "lifesaver" candy in the dark in front of a mirror (after letting our eyes accommodate to the dark for a few minutes). The high voltage field produced during the mechanical creation of electrically-charged sugar crystal surfaces accelerates electrons to high energy. These electrons then excite nitrogen gas molecules in the air by collision to produce the ion N_2^+ which has a weak blue luminescence, also seen in auroras. Some ultraviolet is produced and causes the oil of wintergreen (methyl salicilate) vapor to fluoresce with a particularly intense blue color.

A sodium atom in the gaseous state has energy levels as shown in fig. 4.6. In a sodium vapor lamp a high voltage produces ionization from the initial "ground state" up to the sodium ion Na^+ plus one electron at or above the top line of this figure. As the electron recombines with the ion in a series of steps, the system passes along allowed transitions, the arrows in this figure, only some of which are shown. These downward steps produce heat and/or photons. The position of the various levels shown, as well as the specific transitions allowed are given by quantum theory. All paths terminate in the lowest two arrows, corresponding to the emission of the well-known yellow "sodium doublet" at 589.0 nm and 589.7 nm.

Sodium and mercury vapor lamps produce yellow and blue illumination, respectively, with high efficiency per unit of electricity consumed. They are often used in parking lots, where their unaccustomed colors may disguise the color of our autos. Mercury

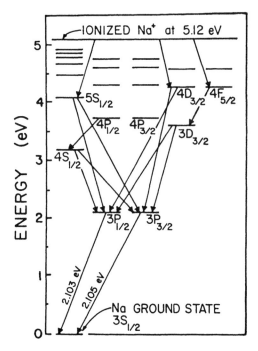

Fig. 4.6. Energy level scheme of an isolated atom of sodium, showing some of the allowed transitions.

vapor lamps also produce much ultraviolet. This can be converted by a phosphor coating in fluorescent tube lamps into some heat plus lower energy yellow, orange, and red light to yield a better color balance, closer to white, for indoor lighting.

Gas lasers also employ light emitted by gas excitations. In gas lasers, such as the helium–neon laser, the electrical excitation produces coherent light with optical feedback from mirrors at each end of the tube; all the photons have the same frequency and are in step both in space and in time.

4.3.3. Mechanism 3: Color from vibrations and rotations

As discussed in Section 4.2 and shown in fig. 4.3, most vibrations between the atoms in molecules absorb only at very low energies in the infrared. In a stringed musical instrument the pitch (sound frequency) is raised if the mass of the string is reduced or if the tension applied to the string is increased. Analogously, the highest vibrational frequencies between atoms occur with the lightest atoms, as with hydrogen, when most strongly bonded, as among water molecules strengthened by an additional interaction called hydrogen bonding.

The isolated water molecule is bent and has three fundamental vibrations as in fig. 4.7. When individual water molecules are trapped in channels in the gemstone emerald (beryl containing both water and chromium $Be_3Al_2Si_6O_{18} \cdot xCr_2O_3 \cdot yH_2O$), the absorption spectrum of fig. 4.8 shows at the left in the 800 to 7000 nm range a series of sharp

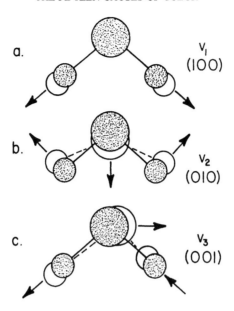

Fig. 4.7. The three fundamental vibrations of the water molecule H_2O: the symmetrical stretch (a), the symmetrical bend (b), and the antisymmetrical bend (c).

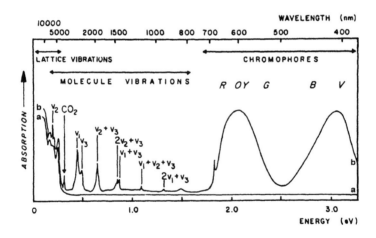

Fig. 4.8. The absorption spectra of colorless beryl (a) and emerald (b), the latter showing vibrational absorptions derived from water molecules and carbon dioxide CO_2. After D.L. Wood and K. Nassau (1968).

absorptions. These can be identified with the three vibration frequencies ν_1, ν_2, and ν_3 of fig. 4.7, as well as with combinations of two or more vibrations at somewhat higher frequencies (higher energies, shorter wavelengths) as labeled in fig. 4.8. Note that the spectrum of pure beryl also given in this figure does not show any of these water absorptions.

The additional hydrogen bonding between adjacent H_2O molecules in liquid water or solid ice raises the energies of these vibrations and leads to very weak combination absorptions at the red, low-energy end of the visible spectrum. As a result, pure water and ice have a complementary very pale blue color. This is best seen at tropical white-sand beaches and in ice caves in glaciers; a green, color in water or ice is usually derived from the presence of algae. That reflection from a blue sky is not an important factor in the blue of the ocean can be seen when the blue remains even when the sky is covered by white clouds. Also, the blue is not polarized, which it would be if it were derived from reflection. With dark clouds and wind, it is the wind-produced surface roughness and air bubbles in the foam that scatter light and prevent it from penetrating deeper into the water and producing the blue color.

Excited vibrational states as well as the related excited rotational states of a molecule can be superimposed on the electronic excitation states of the previous section. Such combinations are involved in the violet color of iodine vapor, the reddish-brown of bromine liquid and vapor, and the green of chlorine gas. The blue-green color emitted from the lower parts of an oxygen-rich gas flame, as seen on a kitchen range or a Bunsen burner, also involves such combined vibrational, rotational, and electronic excitations in unstable molecules such as CH and C_2.

4.4. Color from ligand field effects

A large energy is required to excite one of a paired set of electrons in most inorganic substances. It follows that electronic absorptions generally occur in the ultraviolet as in fig. 4.3. Substances containing only paired electrons are therefore usually colorless (exceptions occur in semiconductors and color centers, see Sections 4.6.2 and 4.6.4). Unpaired electrons occur in transition metal compounds, usually in the so-called d or f atomic orbitals. These electrons become excited at smaller energies and cause light absorptions in the visible region of the spectrum. Such unpaired electrons are present in salts of the so-called d transition elements such as chromium (see color plate 10), iron, cobalt, nickel, and copper; of the 4f lanthanides such as cerium and neodymium; and of the 5f actinides such as uranium.

Consider a crystal of the aluminum oxide corundum Al_2O_3, also known as colorless sapphire when in gem quality form. In sapphire, each aluminum ion Al^{3+} is surrounded by six oxygen ions O^{2-} in the form of a slightly distorted octahedron, as shown in fig. 4.9. All electrons are paired and there is no absorption in the visible region. Consider what happens if one out of every hundred aluminums is replaced by a chromium. Trivalent chromium Cr^{3+} has 18 paired and three unpaired electrons; the latter are situated in the 3d orbitals which have the capacity to hold ten electrons, two in each of the five orbitals customarily designated d_{xy}, d_{yz}, d_{xz}, $d_{x^2-y^2}$, and d_{z^2}. In the free trivalent chromium ion all the 3d electrons occupy energy levels having the same energy as in the central part of fig. 4.10, so that again there can be no light-absorbing transitions.

Such energy levels are, however, perturbed by the existence of the six neighboring oxygens, the "ligands", in fig. 4.9. The spacing of the levels as at the left and right in fig. 4.10 is affected by two factors: the geometrical distribution (here distorted octahedral) and the strength of the bonding, i.e., the ligand field (or the equivalent size of the

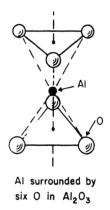

Al surrounded by
six O in Al₂O₃

Fig. 4.9. The distorted octahedral oxygen ligand environment around an Al ion in sapphire Al_2O_3.

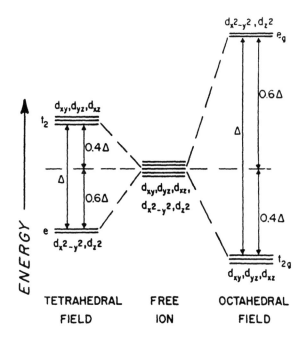

Fig. 4.10. The splitting of the five 3d orbitals in tetrahedral and octahedral ligand fields.

electronic field produced by the oxygen ions – the crystal field), here a very large 2.23 eV. This subject is covered by ligand field theory. An earlier, somewhat less sophisticated version was crystal field theory. Ligand field theory can also be viewed as a special case of the molecular orbital theory discussed in Section 4.5.

Consider now a crystal of ruby, which is Al_2O_3 containing 1% or so chromium in the form of trivalent chromium Cr^{3+} replacing Al^{3+}. The symmetry of the ligand field and

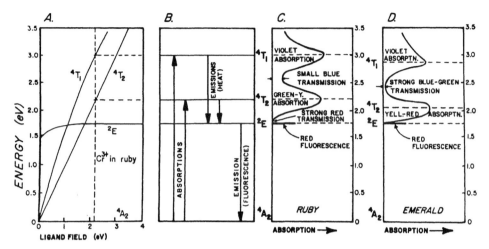

Fig. 4.11. The term diagram of Cr^{3+} in a distorted octahedral field (A), the energy levels and transitions in ruby (B), and the resulting absorption spectra and fluorescence of ruby (C) and emerald (D).

its strength give the energy level scheme at A, called the term diagram, and the energy level transition scheme at B in fig. 4.11 for the 2.23 eV ligand field in ruby. Here the violet and yellow-green regions of white light passing through ruby are absorbed as in the two upward arrows in B, broadened into bands by thermal vibrations and other effects. The result is the absorption spectrum shown at C, giving ruby its deep red color. The Cr^{3+} in ruby must pass through the energy level labeled ^2E in losing its energy again, with the emission of some heat. When it returns from ^2E back down to the ground state, a photon in the red region of the spectrum is emitted, giving ruby a red fluorescence, also seen under ultraviolet illumination. The red fluorescence in ruby is harnessed in the ruby laser.

Now consider a Cr^{3+} present in a crystal of beryllium aluminum silicate, the emerald discussed in connection with fig. 4.8 and shown in color plate 10. Here the symmetry is also distorted octahedral, but the ligand field is a little weaker, being 2.05 eV instead of the 2.23 eV of ruby. Although this is a relatively small change, it produces a significant shift in the absorption bands as shown at D in fig. 4.11 (and as also seen in the upper curve of fig. 4.8). The result is a change from the red color of ruby to the deep green color of emerald! Interestingly enough, the position of the ^2E level does not change significantly with the ligand field as can be deduced from A in fig. 4.11, so that the same red fluorescence is present in both red ruby and green emerald.

Let us now ask: what might be the color produced by chromium in a ligand field intermediate between that of green emerald and that of red ruby? Nature provides an answer in the form of the extremely rare and valued gemstone alexandrite, demonstrating how she can confound our expectations and yet turn out to be perfectly reasonable. The absorption bands are indeed intermediate between those of ruby and those of emerald. However, they are so positioned in alexandrite that in daylight or the similar quality light from a fluorescent tube lamp (both rich in blue) our eyes perceive a blue-green color,

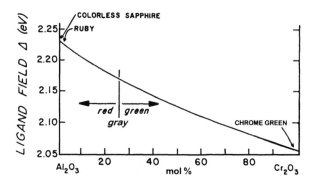

Fig. 4.12. The variation of the ligand field and the color in the solid solution system between colorless sapphire Al_2O_3 and chrome green Cr_2O_3. After D. Reinen (1969).

somewhat resembling emerald, while in the light from a candle or from an incandescent lamp (both rich in red) we see a reddish color, somewhat resembling a ruby, as show in color plate 11.

As the chromium concentration of pure, colorless sapphire is increased to beyond that found in ruby, there is a change from colorless to red, next to gray, and finally to the dark green of pure chromium oxide Cr_2O_3, the pigment chrome green. This sequence is a result of the weakening of the ligand field as shown in fig. 4.12. A specimen on the red side of gray (including ruby itself) can be turned green by heat, which causes a reduction of the ligand field as the atoms move apart from thermal expansion in "thermochromism". In the reverse process, a specimen a little on the green side of gray can be turned red by the application of pressure in "piezochromism" by moving the atoms together and strengthening the ligand field.

4.4.1. Mechanism 4: Color in transition metal compounds

Transition metal colors in pure transition metal compounds is Mechanism 4. This mechanism provides the colors of many minerals and paint pigments. Such colors are sometimes called "idiochromatic", i.e., self-colored. They include many pigments such as the chrome green mentioned above, green viridian $Cr_2O(OH)_4$; blue smalt glass $K_2CoSi_3O_8$, and Thenard's blue Al_2CoO_4; gemstones such as pink rhodochrosite $MnCO_3$ and green malachite $Cu_2(CO_3)(OH)_2$; and minerals and ores such as brown manganite $MnO(OH)$, red iron ore Fe_2O_3, yellow goethite $FeO(OH)$, and green bunsenite NiO. The idiochromatic mineral shattuckite is colored by cobalt and is shown in color plate 12.

4.4.2. Mechanism 5: Color from transition metal impurities

Mechanism 5 covers occurrences of color where the transition metal is only an impurity, typically present in an otherwise colorless substance at a level of a few percent or less. Such colors are sometimes called "allochromatic", i.e., other-colored. Examples among minerals and gemstones include the chromium-based gemstones ruby, emerald, and alexandrite discussed in Section 4.4; the manganese-colored pink morganite form of beryl; the iron-colored blue to green aquamarines and tourmalines, yellow citrine, green jade (in part); and so on. The allochromatic mineral spinel can be colored by cobalt as shown in color plate 12. Allochromatic colors in ceramics, glass, glazes, and enamels include purple from manganese (but also see Section 10.4), green from iron, and an intense violet-blue from cobalt.

4.5. Color from molecular orbitals

4.5.1. Mechanism 6: Color in organic compounds

In this mechanism color derives from molecular orbital effects in organic compounds. Here the electrons involved in the color-causing transitions belong to several atoms within the molecule; the closely related charge transfer colors are discussed separately in Section 4.5.2.

A "conjugated" organic compounds is one that contains alternating single and double bonds in chains and/or rings of carbon atoms; occasionally a different atom, such as a nitrogen, may also be involved. In such an arrangement there are "pi-bonded" electrons located in molecular orbitals which belong to the whole chain and/or ring system. The excited states of such electrons occur at energies similar to those of the unpaired electrons in transition metal compounds and can therefore absorb and emit photons in the visible region.

The longer the conjugated chain, the more do the transition energies move from the ultraviolet into the visible region. The absorptions of the conjugated cyclic benzene C_6H_6 or the linear 2,4-hexadiene C_6H_{10}, CH_3–$CH = CH$–$CH = CH$–CH_3 are still in the ultra-violet, but with the conjugated linear ten carbon chain 2,4,6,8-decatetraene $C_{10}H_{14}$, the absorption has moved into the blue end of the spectrum and produces a complementary pale yellow color. These desired "bathochromic" shifts of the absorptions to lower energies can be obtained in several ways in addition to extending the length of the conjugated chain.

Such shifts are produced by the presence of electron donor groups which push electrons into the conjugated system, such as the $-NH_2$ group in the dye crystal violet $C(C_6H_4NH_2)_3$ shown in fig. 4.13, or by electron acceptor groups which pull electrons out of the conjugated system, such as the $-NO_2$ group. Large bathochromic shifts tend to

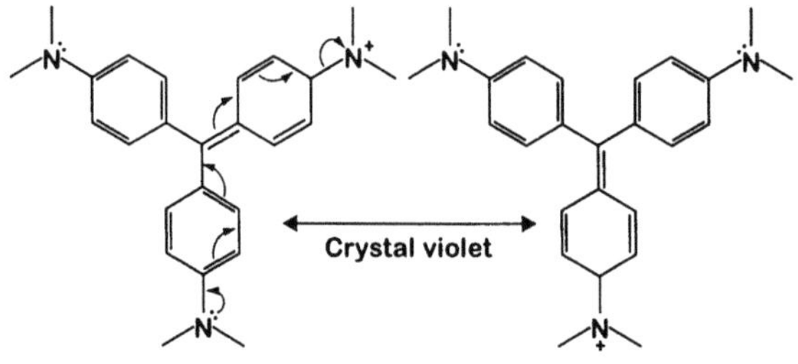

Fig. 4.13. Two resonance structures of the dye crystal violet.

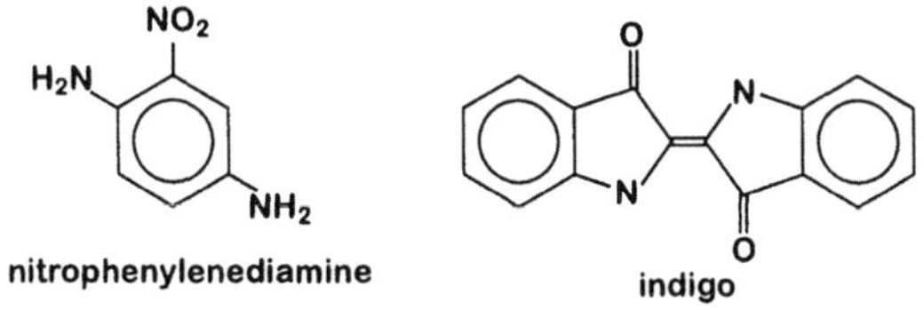

Fig. 4.14. The structures of a nitrophenylenediamine hair dye (left) and indigo (right).

occur in molecules which have many resonance structures; two of the resonance forms of crystal violet, produced by the shifting of electrons, are shown in fig. 4.13.

An example of a molecule containing both donor and acceptor groups is the nitro-phenylenediamine dye shown at the left in fig. 4.14. This dye absorbs in the blue part of the spectrum and gives a complementary yellow to brown color; it is used in hair dyes because its small size permits it to penetrate into the hair. An imitation of blue sapphire, consisting of a colorless spinel "doublet" held together with a layer of cement containing an organic blue dye, is shown in color plate 5.

A useful conjugated "chromophore" (color bearing) group is present in the blue dye indigo, shown at the right in fig. 4.14. This dye has been used since antiquity, from the woad of the "Picts" (painted people) whom Julius Caesar fought in Britain in 58 B.C. up to today's all-pervading blue jeans. With two bromine atoms added, the result is Tyrian purple, a dye laboriously extracted from certain sea shells and used exclusively by Roman emperors as a status symbol.

Molecular orbital dye colors occur widely in plants (see color plate 19) and animals (see color plate 18) as well as in the products of the modern synthetic dye and pigment industry. Exactly as with ligand field energy levels, some of the absorbed energy may be re-emitted in the form of fluorescence. This is used in dye lasers, where absorption of radiation energy leads to the fluorescence; tuning of the wavelength of the laser light is possible because of the broad nature of the fluorescence peaks.

Fluorescence is also used in the "fluorescent brighteners" added to laundry detergents; ultraviolet present in daylight is converted to blue fluorescence, making the laundered fabrics look whiter. Chemical energy can also excite fluorescence in organic substances and leads to fluorescence (or the much slower phosphorescence) as in the bioluminescence of fireflies and angler fishes and in the chemoluminescent "lightsticks" of color plate 13; here a slow chemical reaction emits light over a period of one half to several hours.

If the conjugated framework of an organic colorant molecule is destroyed, then the color will be lost. This can happen in bleaching by the chlorine or peroxide in chemical bleaches or by fading from oxidizing polutants in the atmosphere. These substances add to both ends of a double bond, converting it to a single bond. Light fading usually derives from a direct break-up of the conjugated system from the energy in ultraviolet or even blue light present in sunlight and other illuminations. In photochemical bleaching or fading, ultraviolet provides the energy to permit even the oxygen in the air to add across a double bond. These are the reasons why museums take care to avoid ultraviolet in their illumination. More details are given in Chapter 12.

4.5.2. Mechanism 7: Color from charge transfer

A crystal of sapphire Al_2O_3 containing a few hundredths of one percent of titanium is colorless. The presence of a similar amount of iron by itself shows at most a very pale yellow color. When both impurities are present together, the result is a magnificent deep blue color, that of blue sapphire, seen in color plate 12. The mechanism at work is intervalence charge transfer, the motion of an electron from one transition metal ion to another, induced by the absorption of the energy of a photon. This results in a temporary change in the valence state of both ions, with reversion to the original state soon thereafter, usually with the emission of heat. This mechanism also causes the black or dark colors of many variable valence transition metal oxides such as the iron oxide magnetite. It is sometimes also called electron hopping or cooperative charge transfer.

Consider two adjacent aluminum sites in corundum occupied by divalent iron Fe^{2+} and by titanium Ti^{4+} as illustrated in fig. 4.15. The motion of an electron from the Fe to the Ti changes the valence state of both ions:

$$Fe^{2+} + Ti^{4+} \longrightarrow Fe^{3+} + Ti^{3+}. \tag{4.4}$$

This process requires energy as shown in fig. 4.16; since the energy of 2.1 eV corresponds to the absorption of a photon of yellow light, the result is the complementary color blue.

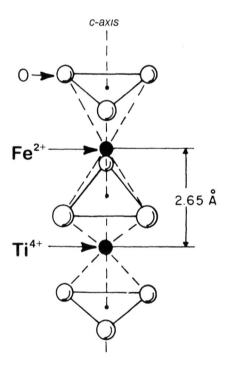

Fig. 4.15. Two adjacent distorted octahedral sites containing Fe^{2+} and Ti^{4+} in blue sapphire; compare with fig. 4.9.

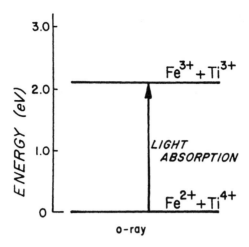

Fig. 4.16. Transition from the ground state to the excited state in the blue sapphire of fig. 4.15.

Blue sapphire is an example of "heteronuclear" intervalence charge transfer with two different transition metal ions involved. "Homonuclear" intervalence charge transfer involves two valence states of the same metal in two different sites A and B, as in the ferrous-ferric iron oxide magnetite Fe_3O_4 or $Fe^{2+}O \cdot (Fe^{3+})_2O_3$:

$$(Fe^{2+})_A + (Fe^{3+})_B \longrightarrow (Fe^{3+})_A + (Fe^{2+})_B. \tag{4.5}$$

The right-hand side of this equation has a higher energy than the left-hand side; the result is a broad-band light absorption and a black color. In blue sapphire this mechanism is also present, but here it absorbs only in the infrared and does not contribute to the color.

This same mechanism gives the "carbon-amber" brown beer-bottle color in glass made with iron sulfide and charcoal, and the brilliant blue color of the pigment Prussian blue $(Fe^{3+})_4[Fe^{2+}(CN)_6]_3$. The brown to red colors of many rocks, e.g., in the USA Painted Desert, derive from homonuclear charge transfer involving iron impurities.

Charge transfer can also involve electrons on a transition metal and on a ligand atom. One example is the charge transfer from oxygen to chromium in the yellow chromate $KCrO_4$ and the orange dichromate $(NH_4)_2(Cr^{6+})_2O_7$ of color plate 10. The formal valence of 6+ on the Cr does not permit unpaired electrons and therefore prevents ligand field colors as occur in the trivalent chromium colors of the upper two rows of color plate 10. Ligand-metal charge transfer is also present in blue sapphire, but its absorption occurs only in the ultraviolet.

A final example of charge transfer is the deep blue gemstone lapis lazuli of color plate 12, which is the massive, crystalline form of the pigment ultramarine, approximately $CaNa_7Al_6Si_6O_{24}S_3SO_4$. This color derives from charge transfer within groups of three sulfurs (S_3^-).

Charge transfer transitions are "allowed" by the selection rules and therefore give intense colors, produced by as little as $1/100$ percent each of Fe and Ti in blue sapphire;

by contrast the "forbidden" transitions in the ligand-field colors are so weak that one percent or more of Cr is required to produce intense colors in ruby and emerald.

4.6. Color from band theory

4.6.1. Mechanism 8: Color in metals and alloys

Consider two hydrogen atoms coming together and bonding to form the molecule H_2 as shown in fig. 4.17. Two equal energy atomic orbitals, one from each hydrogen atom, interact to form two molecular orbitals. If the electron spins are opposed, there is formed a bonding molecular orbital which has a lowered energy. With like spins, an antibonding orbital results with a raised energy. These molecular orbitals can accommodate exactly the same number of electrons as the sum of the atomic orbitals from which they were formed, namely two per orbital. For the normal bonding distance d between the two atoms in fig. 4.17, the two equal energy atomic orbitals at A in fig. 4.18 convert to the two molecular orbitals at B in this figure; the two electrons, one each from the two hydrogens, now both occupy the lower level.

If we consider four atoms interacting to form molecular orbitals, the result will be four separate energy levels as at C in fig. 4.18. Extrapolate this approach to a piece of metal containing some 10^{23} strongly interacting atoms per cubic centimeter. We can now visualize the bonding in a cubic centimeter of metal to involve 10^{23} molecular orbitals in an electron energy band (or energy band, or just band) as at D in fig. 4.18. If this band has an energy range of 1 eV, then the spacing between adjacent levels would be 10^{-23} eV, a negligibly small quantity for all practical purposes, so that the band can be viewed as being quasi-continuous in energy values.

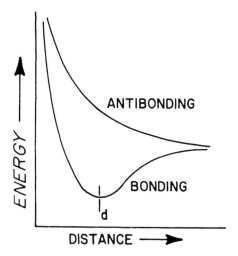

Fig. 4.17. The energy of bonding and antibonding molecular orbitals of two hydrogen atoms interacting to form the hydrogen molecule H_2.

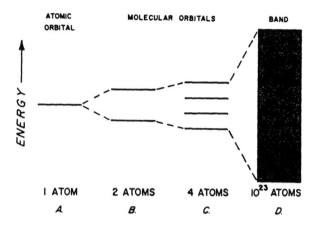

Fig. 4.18. The conversion of atomic orbitals (A) into molecular orbitals (B and C) and into a band (D).

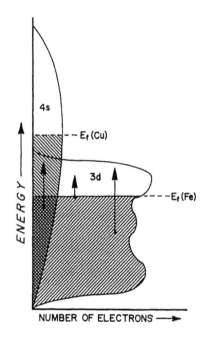

Fig. 4.19. Schematic density of states diagram for the metals iron Fe and copper Cu. After J. Slater (1951).

Band theory gives a full explanation of the thermal, electrical, optical, and many other properties of the metallic elements and of the alloys formed by mixtures of metals. The exact shape of the energy band depends on the element, on the atomic orbitals involved (3d and 4s for the first row transition elements as in fig. 4.19), on the geometrical packing arrangement of and spacing between the atoms, and on the relative numbers of atoms present in alloys.

The total number of electrons involved in a band is the sum of the valence electrons for all the atoms present, that is the sum of those electrons outside the full shells and therefore available for bonding. These electrons now occupy the band from the bottom upward, just as if one were pouring water into an oddly-shaped container, as in fig. 4.19. This density of states diagram shows that the capacity to hold electrons varies at different energies within the band. The highest energy filled level is called the Fermi surface, usually designated E_f, and is illustrated for iron Fe and copper Cu in the metallic state in this figure. The outer bonding electrons involved in the band no longer belong to individual atoms but they belong to the piece of metal as a whole; they are delocalized. In copper, which has three more conduction electrons than iron, the band is filled to a higher level.

The good electrical and thermal properties of metals immediately result from this arrangement. An electric field raises the energy of an electron from below the Fermi surface to a high energy level in the band, as indicated for iron by three vertical arrows in fig. 4.19; this creates a negatively-charged electron above E_f and a positively-charged hole, which is a missing electron, in the otherwise continuous ocean of electrons below E_f. In the applied electric field these two species move in opposite directions, the result being an electric current. A similar excitation by heat also produces electrons and holes, both of which diffuse away from the hot region thus producing a flow of thermal energy, without any net movement of electrical charge; this flow of energy is the thermal conductivity.

When light falls onto a metal, an electron at or below the Fermi surface can also become excited into a higher energy level in the band by absorbing the energy from a photon, producing an electron–hole pairs, again as indicated by any of the arrows in fig. 4.19. The light is so intensely absorbed that it can penetrate to a depth of only a few hundred atoms, typically less than a single wavelength. Since the metal is a conductor of electricity, this absorbed light which is, after all, an electromagnetic wave, will induce alternating electrical currents in the metal surface. The theory of electromagnetism shows that these currents immediately re-emit the photon back out of the metal, thus providing the strong reflection of polished metal surfaces.

The efficiency of this reflection process depends on the selection rules which apply to the atomic orbitals which had formed the energy band. If the efficiency of the absorption and re-emission process is approximately equal at all optical energies, then the different colors in white light will be reflected equally well. This condition leads to the silvery colors of metals such as iron, chromium, and silver. If the efficiency decreases with increasing energy, however, as is the case for gold, brass, and copper, the result is a slightly reduced reflectivity at the higher energy blue end of the spectrum, leading to some absorption. This gives the observed complementary yellow colors of gold and brass and the reddish color of copper.

The direct light absorption of a metal in the absence of reflection can be observed under special circumstances. Gold is extremely malleable and can be beaten into gold leaf less than 100 nm thick. This is less than the thickness necessary to support fully the electric currents which produce metallic reflection. Under these circumstances the weakly reflected color is still that of gold but a bluish-green color is seen when light is transmitted through the gold leaf. When gold is in a colloidal form, however, as in the 10 nm diameter particles which give the color to ruby glass, a complex scattering theory originated by Mie is required to explain the unexpected red color. A much poorer red color in glass derives from Mie scattering from copper oxide particles and a yellow color from metallic colloidal silver particles.

The colors of alloys, combinations of metals, sometimes with small quantities of non-metallic elements present, follow the same general pattern but are difficult to predict *a priori*. For example, the addition of 25 percent copper to pure gold produces an alloy with a not unexpected reddish color, while a similar amount of silver added to copper produces a greenish color.

4.6.2. *Mechanism 9: Color in semiconductors*

In some materials to which band theory applies, it is possible for a gap, the band gap, to occur within the energy band, with important consequences for color. This happens when there are exactly 4.0 outer valence electrons per atom (per atom on average in compounds or alloys) available for entry into the band. The result is that there are now formed two bands separated by the energy band gap (or energy gap or band gap or just gap). The lower energy band, called the valence band, is exactly filled to capacity, and the upper band, called the conduction band, is exactly empty, as shown at the left in fig. 4.20. The size of the energy spacing between the two bands, the band gap energy, is usually designated E_g.

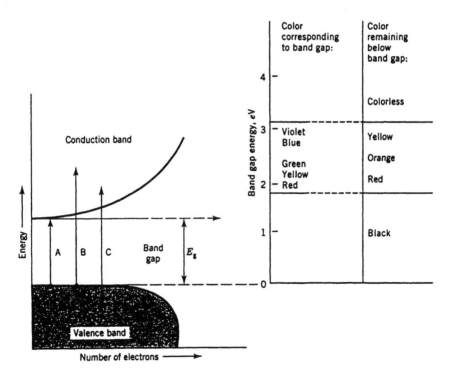

Fig. 4.20. The absorption of light in a band-gap material (left), the variation of the size of the band gap (center), and the resultant color sequence (right); also see color plate 15.

TABLE 4.2
Colors of some band-gap semiconductors.

Substance	Mineral name	Pigment name	Band gap (eV)	Color
C	Diamond	–	5.4	Colorless
ZnO	Zincite	Zinc white	3.0	Colorless
CdS	Greenockite	Cadmium yellow	2.6	Yellow
$CdS_{1-x}Se_x$	–	Cadmium orange	2.3	Orange
HgS	Cinnabar	Vermillion	2.0	Red
HgS	Metacinnabar	–	1.6	Black
CdSe	–	–	1.6	Black
Si	–	–	1.1	Black
PbS	Galena	–	0.4	Black

Consider the absorption of light as represented by the vertical arrows A, B, and C in fig. 4.20. There are no electron energy levels within the band gap, so the lowest energy light that can be absorbed corresponds to arrow A. This involves the excitation of an electron from the top of the valence band to the bottom of the conduction band; the energy of this light of course corresponds to the band gap energy E_g. Light of any higher energy can also be absorbed as, for example, by the arrows B and C.

If the substance represented by this figure has a large band gap, that is large on the scale of optical energies, such as the 5.4 eV of pure diamond or the similar value of pure sapphire, then none of the energies of the visible spectrum can be absorbed; these substances are indeed colorless. Such large band-gap semiconductors are excellent electrical insulators. While physicists view such materials as large band-gap semiconductors, they can also be viewed as covalently to ionically bonded insulators, as is usually done by chemists.

Consider now a medium band-gap semiconductor, a material with a somewhat smaller band gap, where the band gap energy lies within the visible spectrum photon energy range. An example is the compound cadmium sulfide CdS, which is also known as the pigment cadmium yellow and the mineral greenockite. The 2.6 eV band-gap energy listed in table 4.2 results in absorption of violet and some blue, but of none of the other colors; this leads to the complementary yellow color as can be deduced from the color scales at the right of fig. 4.20.

At a somewhat smaller band gap of 2.3 eV, absorption of violet, blue, and green results. This produces the complementary orange color as deduced from fig. 4.20. At a yet smaller band gap as in the pigment vermillion, also known as the mineral cinnabar HgS, a band-gap of 2.0 eV results in the absorption of all visible energies except those corresponding to red and thus leads to a red color. All visible energies are absorbed if a material has a band-gap energy less than the 1.77 eV (700 nm) limit of the visible spectrum. Such narrow band-gap semiconductors therefore can absorb all colors to give black, as in the last four materials of table 4.2.

An illustration of this change in the band-gap energy is shown by mixed crystals of yellow cadmium sulfide CdS, $E_g = 2.6$ eV, and black cadmium selenide CdSe,

$E_g = 1.6$ eV. These compounds have identical structures and form continuous solid solutions. Color plate 15 illustrates the color sequence yellow, orange, red, black of these mixed crystals as the band-gap energy decreases from 2.6 eV to 1.6 eV, following the sequence as shown at the right in fig. 4.20. Mixed crystals of this series, such as Cd_4SSe_3, the painter's pigment cadmium orange, are also used to provide a range of yellow to orange to red colors in glass and plastic.

Mercuric sulfide HgS exists in two different crystalline forms as shown in table 4.2: cinnabar (the pigment vermillion) with $E_g = 2.0$ eV has a deep red color but can transform on exposure to light in an improperly formulated paint to black metacinnabar with $E_g = 1.6$ eV. This has happened in a number of old paintings and can occur in as little as five years.

4.6.3. Mechanism 10: Color in doped semiconductors

If a small amount of an added impurity, here called the dopant, forms an extra energy level within the band gap, then the color can change. Photons can now be absorbed by or emitted from a wide band-gap semiconductor at energies less than the band gap.

A pure diamond crystal is composed of a single molecule of interconnected carbon atoms, each of which contributes four valence electrons to exactly fill the valence band. Now consider a diamond in which just a few carbon atoms out of a million have been replaced by nitrogen atoms, each containing five valence electrons. The structure of the diamond is not significantly disturbed and there is no room for the extra electrons in the valence band. These electrons enter a new energy level inside the band gap. This is called a donor level because electrons from it can be donated into the empty conduction band during the absorption of energy, as shown by the upward arrow at the left in fig. 4.21. This donor level is broadened by thermal vibrations and other factors, as at the right in this figure. The resulting absorption at the blue end of the spectrum produces the complementary yellow color seen in both natural and man-made nitrogen-containing diamonds.

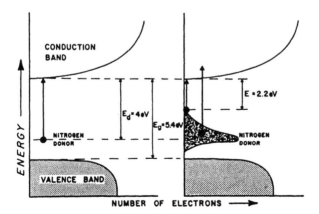

Fig. 4.21. The nitrogen donor energy level in the band gap of diamond (left) forms a broadened band and is involved in the absorption of light (right) to give a yellow color.

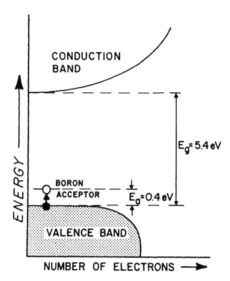

Fig. 4.22. The boron acceptor energy level in the band gap of diamond is involved in the absorption of light, resulting in a blue color as in the Hope diamond.

Boron has one less electron than carbon, and the presence of just a few boron atoms per million carbon atoms in diamond produce an analogous energy level inside the band gap, but now containing holes, as shown in fig. 4.22. The resulting energy level is called an acceptor level since it can accept electrons from the valence band, which remains exactly full. The energy required for this change in boron-doped diamond is very small and, because of broadening, absorbs only at the red end of the spectrum, leading to the complementary blue color. Since the acceptor level energy is so small, even the thermal energy at room temperature can produce this excitation. As a result, the holes produced in the valence band can now move in the presence of an electric field and blue boron-doped diamonds, including the famous Hope diamond, conduct electricity at room temperature.

Some semiconductor materials can be doped to contain both donor and acceptor levels, as in fig. 4.23. They can absorb ultraviolet or electrical energy to produce the transition 'a' from the valence band up to the conduction band. If the return path proceeds via steps 'f', 'g', and 'd', then light may be emitted corresponding to the energy release from one of these steps, such as 'g'; this is then fluorescence or electroluminescence, respectively. Such a fluorescence occurs in phosphors, for example in zinc sulfide ZnS containing Cu and other additives, used as an internal coating in fluorescent tube lamps. This phosphor converts the ultraviolet produced by the mercury arc (see Section 4.3.2) into visible light, particularly into red light, so as to produce a warmer color approximating daylight. Phosphors are also used inside a television screen, activated by a stream of electrons (cathode rays) in cathodoluminescence. Electroluminescence can use a similar powder deposited onto a thin sheet of metal and covered with a transparent conducting electrode to produce lighting panels, often used for nightlights.

Some phosphors contain specific impurities which form trapping levels as at the right in fig. 4.23. An excited electron may land in a trap, as by steps 'a' and 'j' in this figure,

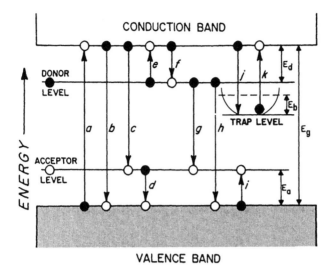

Fig. 4.23. Possible absorption and emission transitions in the band gap of a semiconductor phosphor containing an acceptor, a donor, and a trap level.

and can then only be released when additional energy is added to permit step 'k' and subsequent light emission via 'f', 'g', and 'd' as above. If the trap level is close to the conduction band, then even the thermal energy of room temperature may be able to supply the energy required for a slow release from the trap by step 'k', thus resulting in phosphorescence, lasting seconds to many minutes. If the energy required for release is a little larger, then infrared light may permit the escape, so that higher energy visible light is then produced as in an activated infrared detecting screen.

Finally, there is injection luminescence which can occur in a crystal containing a contact region between differently doped semiconductor regions. An electric current can then produce recombination between electrons and holes in the junction region, giving light from just a few volts in tiny light-emitting diodes (LEDs); these are widely used in alphanumeric display devices (usually red) in electronic equipment, in fiber-optic communication systems, in compact disc players, and so on. With a suitable geometry, the emitted light can be controlled to function as a semiconductor laser.

4.6.4. Mechanism 11: Color from color centers

Expose a century old glass bottle in the desert to the ultraviolet radiation present in the strong (because of unpolluted desert air) sunlight for ten years or so; the glass will acquire an attractive purple color. Or expose such a bottle to an intense source of energetic radiation, such as in a cobalt-60 gamma-ray cell in the laboratory; an even deeper purple color appears within a few minutes, as shown at the upper left in color plate 17. The color disappears on heating the bottle in a medium-hot baking oven.

The color in this desert "amethyst" glass derives from a color center associated with a manganese impurity, used at one time to decolorize greenish iron-containing glass. Similar color centers explain the colors of the gemstones amethyst, smoky quartz, and blue topaz. Many other materials, both natural and man-made, can be irradiated to produce color centers, including irradiated diamonds of various colors. All the specific color centers mentioned so far have colors that are stable on exposure to light; they lose color only when heated. Other color centers exist that are unstable and fade when exposed to light, while yet others fade even during storage in the dark.

The term color center is sometimes used so loosely that even transition metal and band gap colorations are so designated. This rare usage ignores the unique characteristics of color centers; the conventional restricted meaning is used here.

Consider an alkali halide crystal, such as ordinary table salt, sodium chloride NaCl; this consists of a three-dimensional array of Na^+ and Cl^- ions. A single Cl^- can be missing in two different ways: if a compensating Na^+ is also missing, then the crystal remains electrically neutral and color is not produced. If, however, a Na^+ is not missing, then electrical neutrality must still be maintained; one way is for a free electron, designated e^-, to occupy the site vacated by the Cl^-. This entity is a color center, in this instance also called an F-center, after the German word Farbe, color. One can view the lone electron as if it were part of an imaginary negatively-charged transition metal ion located in the ligand field of the surrounding K^+ ions, or one can view this electron as forming an energy level trap within the band gap of this transparent wide band-gap semiconductor material, exactly as at the right in fig. 4.23.

Some form of energy, such as irradiation by ultraviolet or by high energy electrons, X-rays, or gamma rays, can now excite an electron from the valence band into the conduction band and from there into the trap by steps 'a' and 'j' in fig. 4.23. Usually there exist excited energy levels such as E_b within the trap itself, a level at 2.7 eV above the bottom of the trap for NaCl, which can absorb a photon, leading to a yellow/brown color in irradiated NaCl. Note that the electron is still within the trap even when in this excited energy level. The excess energy can be lost as fluorescence or as heat, with the electron still remaining in the trap: another photon can then again be absorbed.

Only by supplying energy corresponding to step 'k' in fig. 4.23, more than 3 eV for NaCl, can the electron leave the trap and return via the conduction band to the valence band. This can happen if the crystal is heated and results in destruction of the color center and the bleaching of the color. If the energies for absorption and for destruction are about the same size, then bleaching can occur merely while the material is being illuminated, leading to optical bleaching or fading.

If the energy of step 'k' is sufficiently small, the color may even fade at room temperature. This occurs in self-darkening sun glasses, where the ultraviolet present in sunlight produces darkening and room temperature leads to fading as soon as there is no ultraviolet to maintain the darkened color. Other color centers are possible in the alkali halides, designated F' (involving two electrons trapped at a Cl^- vacancy), M (two adjacent F vacancies), V_K, etc.; these may absorb in the visible, ultraviolet, or infrared regions. Some of these color centers show fluorescence and some are used as laser materials. As an alternative to irradiation, growth in the presence of excess metal or solid state electrolysis have also been used to generate color centers.

The general description of a material capable of supporting a color center is given in fig. 4.24, where the colorless state is shown above and the colored state below. Two

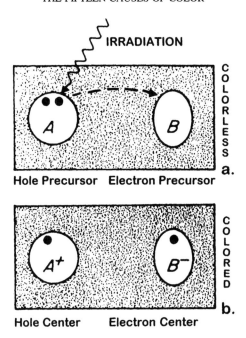

Fig. 4.24. The irradiation of hole and electron precursors (a) forms hole and electron centers (b), either or both of which could be color centers.

kinds of precursors are necessary in the colorless state. The hole precursor A can lose an electron, e.g., when absorbing energetic radiation as shown, and form a hole center A^+. The electron precursor B can accommodate the electron ejected from A and then forms the electron center B^-. Either the A^+ or the B^- in the lower part of fig. 4.24 can be the color center that absorbs light to give color; even both do so in some materials. On heating, the electron is released from B^- and returns to A^+, with the material returning to the colorless state of A plus B.

Several gemstone materials derive their colors from color centers. Colorless rock-crystal quartz, shown center above in color plate 17, is composed of silicon oxide SiO_2, shown schematically at (A) in fig. 4.25. All natural and synthetic quartz contains some aluminum as an impurity; an Al^{3+} typically replaces one in $10\,000$ Si^{4+}. To maintain electrical neutrality either a hydrogen ion H^+ or a sodium ion Na^+ is also present for each Al. Such quartz is colorless. When irradiated, either naturally in the ground over many thousands of years or in a few minutes in the laboratory, e.g., with X-rays or in a cobalt-60 gamma-ray cell, the result is smoky quartz, shown at upper right in color plate 17. As illustrated at (B) in fig. 4.25, irradiation ejects an electron from an oxygen adjacent to the Al^{3+}; the whole $[AlO_4]^{5-}$ entity acts as the hole precursor and converts to the hole center $[AlO_4]^{4-}$. The electron is trapped by the H^+ or Na^+ electron precursor, producing the neutral H or Na atom electron center. Here it is the hole center that is the light-absorbing color center and provides the gray to brown to black color of smoky quartz seen in the color plate.

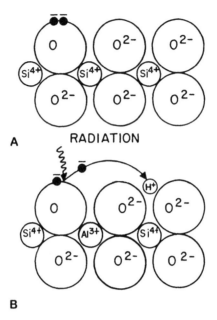

Fig. 4.25. Schematic representation of the structure of quartz (A) and the formation by irradiation of the smoky quartz color center (B).

Also shown in color plate 17 at lower left is the yellow citrine form of quartz (often erroneously called smoky topaz) which contains Fe^{3+} instead of Al^{3+} as an impurity. On being irradiated, some citrine changes to the purple amethyst form of quartz, shown at lower right in the plate; the color derives from light absorption in the hole color center $[FeO_4]^{4-}$, analogous to that in smoky quartz.

The color centers in both amethyst and smoky quartz are stable to light, these colors being lost only on heating from 300 to 500° C; if not overheated, the color centers and the colors can then be restored by another irradiation.

Natural yellow to orange to brown so-called "precious" topaz contains a color center, but the color may be stable to light or it may be unstable and will then fade in a few days in light, depending on the nature of the material. Colorless topaz can be irradiated to produce these same colors, but they will usually fade on exposure to light. Blue topaz contains a stable color center; much colorless topaz can be turned into this same stable blue color by an appropriate irradiation treatment. The color center in deep blue Maxixe beryl, shown in color plate 12, is unstable and fades very slowly when exposed to light at room temperature.

The precise chemical nature of most color centers is unknown. Interestingly, the irradiation of colorless diamonds can produce color centers giving yellow, blue, orange, brown, green, or rarely red colors. Although the first two of these are similar in appearance respectively to the nitrogen-caused yellow and the boron-caused blue band-gap-impurity colors discussed above, they represent much less valued materials which can be distinguished by spectroscopic and other examinations.

4.7. Color from geometrical and physical optics

4.7.1. Mechanism 12: Color from dispersion

The discovery by Newton of dispersion, also known as dispersive refraction, and its dependence on the uv electronic and ir vibrational features was covered in Section 4.2 and fig. 4.3.

In a colorless transparent substance, the refractive index n at wavelength λ in the visible region is given by the Sellmeier dispersion formula

$$n^2 - 1 = a\lambda^2(\lambda^2 - A^2)^{-1} + b\lambda^2(\lambda^2 - B^2)^{-1} \ldots, \tag{4.6}$$

where $A, B, \ldots$ are the wavelength of the individual infrared and ultraviolet absorptions seen in fig. 4.1 and $a, b, \ldots$ are constants representing the strengths of these absorptions. Only two or three terms, corresponding to the absorptions closest to the visible region, are required for an excellent fit in this region.

Anomalous dispersion results when there is a light absorption in the visible region of an otherwise transparent medium. It is now necessary to use the complex refractive index $N = n + ik$ instead of n in the Sellmeier dispersion formula, where i is the imaginary $(-1)^{1/2}$ and k is the absorption coefficient. The variation of n and k in a glass having a violet color derived from an absorption in the green part of the spectrum is shown in fig. 4.26. In the absorption region, the natural resonating frequency of the absorbing entities interacts with the vibration of the light in a complicated manner, involving the phase velocity and the phase angle. This produces a speeding up of the light, thus giving a lower n, on the short wavelength side of the absorption and a slowing down, with a higher n, on the other side. In the central region of the absorption, the refractive index increases with the wavelength, instead of decreasing as usual; this region may be difficult to observe since it occurs exactly where the light is most strongly absorbed.

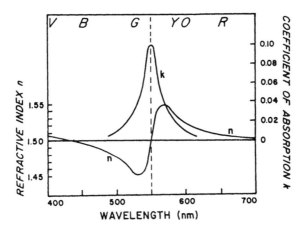

Fig. 4.26. Refractive index and absorption coefficient variation in a violet crown glass having an absorption band centered in the green at 550 nm.

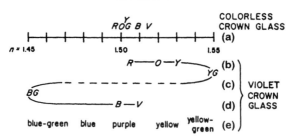

Fig. 4.27. Color sequences produced by dispersion in the colorless crown glass of fig. 4.1 (a) and in the violet crown glass of fig. 4.26 (b to e).

If a beam of light is passed through a prism cut out of the colorless glass of fig. 4.3, then the sequence of colors seen is the normal spectral sequence shown at 'a' in fig. 4.27. For the green-absorbing glass of fig. 4.26, which therefore has the complementary color violet, the sequence in the lower half of fig. 4.27 applies. The red to yellow-green sequence at 'b' follows normal behavior as at 'a' in this figure, as does the blue-green to violet sequence at 'd'. The yellow-green to blue-green sequence at 'c' is reversed; note that green itself does not appear here since it is absorbed. The overall color sequence that is observed from this prism is shown at 'e' in fig. 4.27; compared to the normal spectrum at 'a' this is truly anomalous, both in the sequence of colors as well as in the width of the spectrum.

If either the refractive index variation or the coefficient of absorption variation is known for all wavelengths for a material, then the other one can be calculated by use of the Kramers–Kronig dispersion relationships. Usually we think of absorption as the cause of dispersion, but the two are inextricably interdependent: one cannot exist without the other.

In addition to the spectrum produced by a prism, there are a variety of dispersion-produced color phenomena. A ray of light passing into the top of a faceted gemstone follows a path controlled by total internal reflections and is seen returning from the top of the stone as the brilliance. Since the geometry of the path corresponds to that in a prism, these transmitted rays are also dispersed into a spectrum, leading to flashes of color, the fire in a stone. The amount of brilliance depends on the refractive index and the amount of fire depends on the dispersion; diamond is paramount in both refractive index and dispersion among naturally-occurring gemstones; therefore it has the strongest brilliance and fire.

The refracted paths through a raindrop produce the primary and secondary rainbows as in fig. 4.28; higher orders can be demonstrated in the laboratory but cannot be seen in nature. The refracted paths through hexagonal ice crystals produce the 22° and the 45° halos around the sun and moon, the parhelia such as sundogs or moondogs, as well as a variety of other white and colored arcs.

Finally, there is the often discussed but rarely seen green flash, occurring at the last moment of the setting of the sun. Here the density gradient of the atmosphere acts as a prism, separating the colors as shown in fig. 4.29. Since the violet and blue rays are

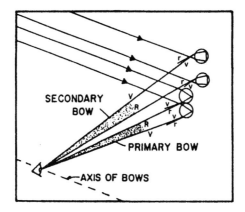

Fig. 4.28. Formation of the primary and secondary rainbows inside raindrops by one and two reflections, respectively.

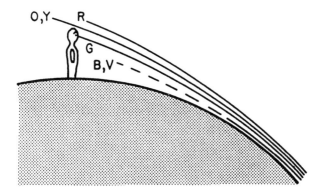

Fig. 4.29. The formation of the green flash at sunset from a combination of dispersion and scattering.

scattered from the beam as described in Section 4.7.2, a green image is seen just at the final setting of the sun under favorable circumstances.

A more complicated case than dispersive refraction is double refraction, which provides color when certain anisotropic crystals are viewed in polarized light between crossed polarizers. Here the polarized light on entering the crystal is resolved into the ordinary (or o or ω) ray and the extraordinary (or e or ε) ray as in a calcite crystal of fig. 4.30. These rays move at different velocities through the crystal and are recombined in the second polarizer to produce color by interference; interference is discussed in Section 4.7.3.

4.7.2. Mechanism 13: Color from scattering

Perfectly clean air does not seem to scatter light. Sunbeams and laser beams reveal themselves in the presence of dust, mist, or smoke. Yet even the purest substances, including gases, liquids, crystals, and glasses, do scatter light when carefully examined in the laboratory.

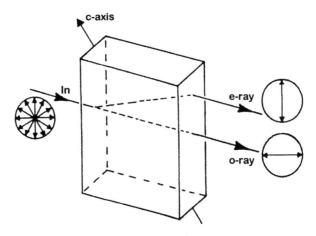

Fig. 4.30. The separation of a beam of light in a calcite crystal into polarized ordinary and extraordinary rays.

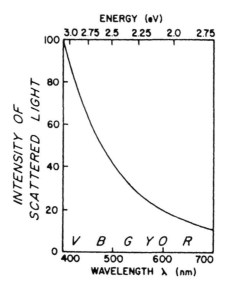

Fig. 4.31. The intensity of Rayleigh scattered light varies with λ^{-4}.

Leonardo da Vinci had observed that a fine water mist produced light scattering. For many centuries misleading ideas produced only confusion. The English experimental scientist John Tyndall (1820–1893) first showed that the amount of scattering from particles small compared to the wavelength of light depends on the wavelength, with blue being much more strongly scattered than red.

The English scientist Lord Rayleigh was the first to realize that scattering particles were not necessary; even the purest substance contains variations in its refractive index which can scatter light. He showed that the intensity of the scattered light I_s is related to that of the incident light I_0 by the inverse fourth power of the wavelength λ

$$I_s/I_0 = \text{const} \cdot \lambda^{-4}. \tag{4.7}$$

If we take the intensity of scattered violet light at the 400 nm limit of visibility to be 100, then red light at 700 nm is scattered only at an intensity of $100(700/400)^{-4} = 10.7$, as shown in fig. 4.31. The terms Rayleigh scattering and Tyndall blues are often applied to the scattered blues.

The atoms and molecules in a gas, liquid, crystal, or glass, are evenly distributed on a macroscopic scale; however, there is considerably non-random distribution at a scale just above the atomic level. As one example, thermal effects cause individual molecules, as well as small clusters of a few molecules, to come together in collisions for a brief instant before dispersing again in gases and liquids; this produces fluctuations in density which scatter light just as do particles of dust or impurities. In a glass there will be similar density and refractive index variations from imperfect mixing of the various ingredients of the glass as well as from density fluctuations frozen-in from the liquid when the glass was formed. Even in an apparently perfectly ordered single crystal, there usually will be a variety of point defects (impurity atoms, vacancies, and clusters of these) and line and plane defects (dislocations, low angle grain boundaries, stacking faults, and the like), all of which produce scattering. There are local density fluctuations from the thermal vibrations of the atoms or molecules even in a perfect crystal, were it to exist.

The blue of our sky derives from sunlight being scattered by dust and other refractive index variations in the atmosphere. A dark background permits an intense blue scattered color to be perceived, as in the dark of outer space backing our blue sky. The removal of the longer wavelengths by scattering also gives the orange to red color of the rising and setting sun, the latter being more intense because surface soil dries out during the day and is then stirred up as dust by wind and by the activities of humans and animals. Sunrise and sunset colors are particularly spectacular when a volcanic eruption has added large amounts of dust to the upper atmosphere.

The Rayleigh scattering process itself involves units that are very small compared to the wavelength of light; photons are absorbed by them and then re-emitted, some as scattered light. Since light is a transverse oscillation, the scattered light is polarized as indicated in fig. 4.32. In a direction perpendicular to the beam the scattered light is completely polarized in an orientation perpendicular to the incident beam, while in other directions, such as at the $\Theta°$ angle shown in this figure, there is an additional component of the polarization parallel to the incident beam direction of $\cos^2 \Theta$. The combination $(1 + \cos^2 \Theta)$ gives the total light-scattering intensity. The blue of the overhead sky at sunset, corresponding to $\Theta = 90°$, is however not completely polarized as might be expected because some of the initially scattered light is scattered again with a resulting partial randomizing of the polarization. As much as one-fifth of the light from a clear sky has undergone such multiple scattering.

If the size of scattering particles approaches 400 to 700 nm, the visible wavelength range, or exceeds it, then the very complicated Mie scattering theory applies and colors other than blue can occur. Mie theory also applies to scattering particles which are

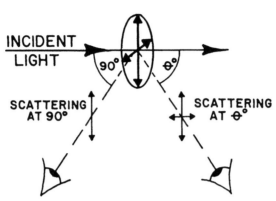

Fig. 4.32. Rayleigh scattered light at 90° is fully polarized; at other angles it is only partly polarized.

electrically conducting as previously mentioned for colloidal metals in Section 4.6.1. All colors are scattered equally by large particles, giving white as in fogs and clouds.

Several atmospheric phenomena derive from scattering. In addition to the blue of the sky, the red of the sunset, and the white of clouds, there is also that epitome of rare occurrences, the blue moon, involving oil droplet clouds from forest fires. Most blue and green bird feather colors and many butterfly wing blues (see color plate 18) derive from scattering, as do many other animal and some vegetable blues. The same scattering phenomenon is seen in the blue color of the iris of our eyes. This is particularly intense in most infants, where dark pigments such as melanin have not yet been formed and where the blue scattering in the iris is seen against the dark interior of the eye. Combined with a little yellow-to-brown melanin pigment, a green iris color results; if melanin dominates, the iris is brown to black. Melanin is absent in albinos, in whose eyes the weak blue scattering adds to the weak red of the underlying blood vessels to produce a pink iris. The pink skin color of light-skinned people similarly derives from weak scattering in the skin combined with the weak red from subsurface blood vessels.

Rayleigh and Mie scattering are called elastic scattering, because there is no change of the wavelength. Inelastic scattering mechanisms, involving a shift in the wavelength of the scattered light, include Raman scattering and Brillouin scattering, both of which can be used for laser operation, as well as several other forms of scattering.

4.7.3. Mechanism 14: Color from interference

Two light waves (or photons) of the same wavelength, i.e., having the same energy, can interact under appropriate circumstances in two extreme ways: they can add so as to reinforce if the waves are in phase as shown at A in fig. 4.33; or they can subtract and cancel if the waves are out of phase, as at B in this figure. Intermediate conditions give intermediate intensities. In this section are covered only those causes of color which involve interference without diffraction; the combination of interference with diffraction is covered in the following Section 4.7.4.

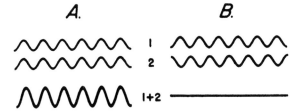

Fig. 4.33. Equal intensity, equal wavelength light waves 1 and 2 produce constructive reinforcement if they are in phase (A) or destructive cancellation if they are out of phase (B).

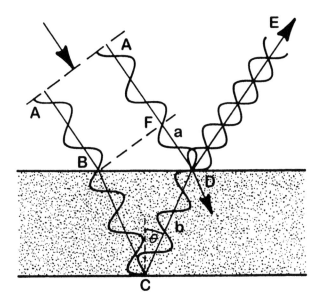

Fig. 4.34. Interference of light reflected from the front and back surfaces of a thin parallel film.

Interference without the occurrence of diffraction was first observed by the French scientist A. Fresnel about 1815. A monochromatic beam of light was directed onto two mirrors made of black glass so that light was reflected only at the front surfaces. The mirrors were inclined at a small angle to each other to produce two overlapping beams of light on a screen. Fresnel observed a series of alternating bands of light and dark, the interference fringes. If either mirror was covered or removed, the fringes disappeared and he saw only the uniform illumination derived from the remaining mirror.

Interference also occurs within thin films. When a plane, coherent, monochromatic beam of light A–A in fig. 4.34 falls at an angle onto a thin sheet of glass or plastic or onto the thin film of a soap bubble, then part of the wave will enter as shown at B in this figure. Part of this beam will be reflected at the back surface at C and a part of this reflected beam will leave at D in direction E. Part of a second wave F in beam A–A will be reflected at the upper surface at D so that it also leaves in direction E. As drawn in fig. 4.34, there is an extra path length of exactly five wavelengths while beam B traverses the distance $2b$ inside the film to reach D; beam F travels only one wavelength a in the air to reach the same point. The net path difference between the two beams is thus four wavelengths, so that they might be expected to be in phase with each other. However, a reflection at a medium of higher refractive index as that of beam F at the top surface at D produces a phase change equivalent to one-half wavelength, whereas this does not happen at the lower surface at C, this being reflection at a medium of lower refractive index. The two beams appearing in direction E are therefore out of phase as shown. There will therefore be destructive cancellation as at B in fig. 4.33 with no light if the intensities are equal.

As either the angle, the thickness, or the wavelength changes, alternating bands from cancellations and from reinforcements will be seen, e.g., as a series of dark and light bands in a tapered film with monochromatic light. With white light, however, the sequence of overlapping light and dark bands from all the spectral colors leads to Newton's color sequence. Starting with the thinnest film there appear black, gray, white, yellow, orange, red (end of the first order); violet, blue, green, yellow, orange-red, violet (end of the second order); and so on. These colors are seen in the tapered air gap between touching non-flat sheets of glass, in cracks in glass or in transparent crystals, in a soap bubble, in an oil slick on a water surface, and in the petrological microscope.

Direct sunshine or a bright light results in the most spectacular colors. Iridescence is a term applied to some interference colorations, implying that multiple colors are seen as in the rainbow (Latin iris), that the colors are intense, and that there is a change of color as the viewing angle is changed. Since we usually see such intense reflection only from mirrors or polished metal, iridescence seems metalliclike to us.

Antireflection coatings on camera lenses employ interference in a reverse manner. A coating is applied that has a refractive index which is the geometrical intermediate between the refractive indexes of glass and air and that has a thickness of one quarter the wavelength of light. Such a coating reduces the overall reflected light to less than one half, while multiple layers can reduce this to less than one tenth. Such layers improve the amount of light transmitted and reduce extraneous reflected light; they usually appear purple to the eye in reflected light.

Many structural colorations in biological systems owe their color to thin film interference, usually involving multiple layered structures. The layers may be composed of keratin, chitin, calcium carbonate, mucus, and so on. There is frequently a backing layer of a dark pigment such as melanin which enhances the color by absorbing nonreflected light. Examples of interference colors in nature include pearl and mother of pearl, the transparent wings of house and dragon flies, iridescent scales on beetles and butterflies (see color plate 18), and iridescent feathers on hummingbirds and peacocks. The eyes of many nocturnal animals contain multilayer structures that improve night vision and give a green iridescent (metalliclike) reflections (see color plate 16); these are often seen at night as the roadside reflections of automobile headlights from animal eyes.

The observation and study of interference has been greatly facilitated by the availability of monochromatic light from lasers. This has led to the widespread use of a variety of interference-based devices, including Twyman–Green and multiple-reflection interferometers, such as Fabry–Perot etalons used for precision measurements, as well as interference filters.

Interference of polarized white light in an optically anisotropic substance, such as crumpled cellophane viewed between crossed polarizers, derives from the double refraction discussed in Section 4.7.1 and also leads to color. Photoelastic stress analysis uses this effect to check glass for strains and to study the stresses in machinery and in deformed plastic models of medieval cathedrals.

4.7.4. Mechanism 15: Color from diffraction

Diffraction in the strict sense refers to the spreading of light at the edges of an obstacle; it ultimately supplied the incontrovertible proof for the wave nature of light. Diffraction was first described in detail in 1665 in a posthumous book by the Italian mathematician F. Grimaldi. He used a small opening in a window shutter, just as did Newton to obtain the spectrum, but studied the shadows of small objects. Grimaldi observed that the shadows were larger than could be explained by the geometry of the light beam. He also saw colored fringes as in fig. 4.35, not only outside, but even inside the shadow under certain conditions. He used the word diffraction for this effect. It remained for the three contemporaries Thomas Young (1773–1829), Joseph von Fraunhofer (1787–1826), and Augustin Fresnel (1788–1827) to provide adequate descriptions and explanations.

Diffraction involves the spreading of a wave into the geometrical shadow behind an obstacle as seen in an ocean wave passing a rock as in fig. 4.36. Consider sunlight passing through a large opening in a shutter on its way to a screen. As the opening is made smaller, so the patch of light on the screen becomes smaller and its edges become sharper. Beyond a certain point, however, the edges become indistinct and begin to show colored fringes. Parts of the light wave spread into the geometrical shadow region, just as in fig. 4.36, but on a much smaller scale. The spreading part of the wave can now interfere with the undisturbed part of the wave. An edge in monochromatic light produces

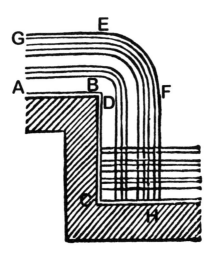

Fig. 4.35. Drawing of diffraction given in Grimaldi's 1665 book; H is the geometrical shadow and the thin lines represent colored fringes.

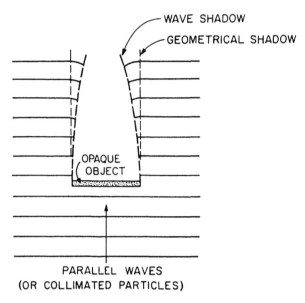

Fig. 4.36. The geometrical and wave shadows produced by a rock in the ocean.

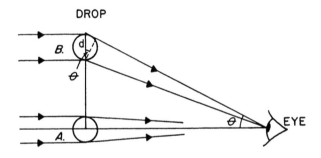

Fig. 4.37. Diffraction from the edges of a drop of water leading to the corona.

a sequence of light and dark bands, while white light gives a sequence of colors similar to the Newton color sequence discussed in Section 4.7.3.

The result from the interference of light diffracted from opposite sides of small particles is a sequence of colored rings called the corona (not to be confused with the solar corona seen only during an eclipse). The corona can frequently be seen surrounding the sun (using very dark sunglasses), derived from diffraction at small particles in a cloud or fog. In fig. 4.37 the path difference between the two rays diffracted from opposite edges of the particle at A is zero, thus producing reinforcement. At B the path difference is $d \sin \Theta$; when this is a whole wavelength, that is when $n\lambda = d \sin \Theta$, where n is an integer, reinforcement again occurs. Half-way between these reinforcements there will be cancellation. The size of the cloud particles can be deduced from measuring the diameter of the corona. For uniformly-sized particles there is the usual interference color

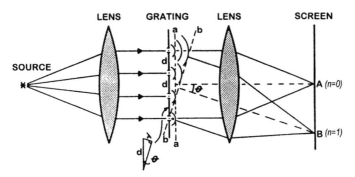

Fig. 4.38. Schematic diagram for a diffraction grating spectroscope showing constructive reinforcement for the undeflected beam with $n = 0$, and for the first order spectrum with $n = 1$.

sequence. With varied sizes of particles present, merely a bluish disk, the corona aureole, is seen surrounding the sun. This same mechanism occurs in Bishop's ring, also known as the glory and the specter of the Brocken, surrounding the shadow of ones head on a bank of fog or the shadow of one's airplane on a cloud, as seen in color plate 14. Transluscent clouds having uniform particle sizes when viewed near the sun (with very dark sunglasses or two sets of sunglasses) may show a range of pastel colors, then called iridescent or luminescent clouds.

A regular array of scattering objects or openings forms a diffraction grating; this can be two- or three-dimensional. This effect is used in the grating spectroscope as in fig. 4.38, where the path difference is again $n\lambda = d \sin \Theta$. Some colors in the animal kingdom also derive from diffraction gratings, as in the beetle Serica sericae and the indigo or gopher snake Drymarchon corais couperii, both of which show spectral colors in direct sunlight. This effect can also be seen by viewing at a glancing angle across a phonograph record or compact disc, or by looking at a distant streetlamp or flashlight through a cloth umbrella.

The gemstone opal (see color plate 20) is the epitome of diffraction gratings, showing on a white or a black background flashes of varied colors called the play of color (opalescence is merely a milky-white translucent appearance). At one time the colors in opal were believed to involve thin film interference, but electron microscope photographs revealed the secret of opal, demonstrating a regular three-dimensional array of equal-size spheres as shown in fig. 4.39. The actual composition of the spheres is amorphous silica, SiO_2 containing a small amount of water, cemented together with more amorphous silica containing a different amount of water so that a small refractive index difference exists between the spheres and the spaces between them.

Finally, there are liquid crystals, which are organic compounds with a structure intermediate between that of a crystal and that of a liquid. Some of these have twisted structures which can interact with light in a manner similar to diffraction to produce color, as in contact fever thermometers. Some beetle colors are derived from liquid crystal structures on the outer layers of their cuticles (also see Chapter 15, Section 15.2.3).

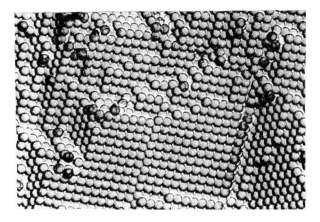

Fig. 4.39. Electron microscope view of synthetic opal; individual spheres are about 250 nm (0.00001 in.) across. Photograph courtesy Ets. Ceramiques Pierre Gilson.

4.8. Color occurrences in many fields

4.8.1. Color in pigments and dyes

Inorganic pigments are described in detail in Chapter 10, organic dyes in Chapter 11, and the preservation of color in Chapter 12. The use of color in printing and in photography, using many of these same pigments and dyes, is covered in Chapter 13.

The majority of inorganic pigments (see color plates 15, 29 and 30) derive from Mechanism 4, transition metal compounds; examples are cobalt blue $CoAl_2O_4$, and the chrome green Cr_2O_3 of color plates 10 and 29. Some also derive from Mechanism 7, charge transfer: e.g., Prussian blue and the iron oxide browns of color plate 30 (both involving Fe^{2+} and Fe^{3+}); the ultramarine (lapis lazuli) of color plate 11 (S_3^-); and chrome yellow (Cr^{6+} and O). Mechanism 8, metals and alloys, is involved in metallizing pigments, usually flakes of aluminum, brass, or copper.

Some pigments involve Mechanism 9, pure semiconductors: zinc white ZnO; the cadmium yellow to oranges Cd(S, Se) of color plate 15; and vermillion HgS all have colors derived from band gap absorptions. Fluorescent and phosphorescent pigments usually involve Mechanism 10, doped semiconductors, frequently (Zn, Cd)S activated with Cu, Ag, or Mn. Pearlescent pigments use Mechanism 11, interference in thin films, involving ground-up fish scales; layers of TiO_2 on mica flakes; and the like.

Almost all dyes are based on Mechanism 6, organic compounds (see color plates 28 and 32). There are, however, the so-called pigment or padding dyes, where a dyeing-type process is used to precipitate inorganic pigments onto a fabric. A typical example is military khaki, using iron and chromium salts.

4.8.2. Color in biology

A general discussion is given in Chapter 8. While most of biological colorations derive from the organic compounds of Mechanism 6, three other mechanisms provide significant additional colors, often called structural colors.

Mechanism 13, scattering, provides many animal blues, as well as the blue component in many greens, pinks, and purples: the iris in blue eyes and the blue component of pink skin in humans; blue skin colors in monkeys, turkeys, etc.; blue as well as the blue component in the green colors of many butterflies (see color plate 18), birds, fishes, snakes, lizards, etc.

Sometimes difficult to distinguish are biological structural colors derived from Mechanisms 14, interference, and 15, diffraction. When particularly intense, the term iridescent is appropriate. Interference colors are the most common, occurring on the scales of butterfly wings, the transparent wings of flies, in the eyes of nocturnal animals (see color plate 16), and on hummingbird feathers and butterfly wings (see color plate 18); they also add to the luster of our hair and nails. Quite rare are diffraction grating colors, e.g., on some beetles, such as Serica sericea and snakes, such as Drymarchon corais couperii.

4.8.3. Color in minerals and gems

All but the first two of the 15 causes of color are active in minerals and gemstones; a summary with examples is given in table 4.3.

TABLE 4.3

Causes of color in minerals and gemstones.

Color cause No.	Examples	Color plates
3: Vibrations	Blue water and ice	–
4: Transition metal compounds	Many minerals and ores, e.g., Cu, in green malachite, turquoise, and blue azurite	10, 12, 30
5: Transition metal impurities	Many gemstones, e.g., Cr in ruby, emerald, and alexandrite	10, 11, 12
6: Organic compounds	Amber; coral	–
7: Charge transfer	Blue sapphire (Fe–Ti); black magnetite (Fe–Fe); blue lapis lazuri (S_3^-)	10, 12, 30
8: Metals and alloys	Gold; copper; silver; electrum	–
9: Pure semiconductors	Colorless diamond; yellow sulfur and greenockite; red cinnabar; black galena	15
10: Doped semiconductors	Yellow and blue diamonds	–
11: Color centers	Amethyst; smoky quartz; yellow and blue topaz	12, 17
12: Dispersion	"Fire" in a faceted diamond	–
13: Scattering	Moonstone; star ruby; body color in white opal	–
14: Interference	Labradorite; bornite (peacock ore); iris agate	–
15: Diffraction	"Play of color" in opal	20

4.8.4. Color in glass, glazes, and enamels

Many glass-related colors derive from Mechanism 5, transition metal impurities, particularly the iron-produced green and cobalt-produced intense violet blue. An important glass color, the carbon-amber beer-bottle brown, is based on Mechanism 7, charge transfer, from Fe^{2+} to Fe^{3+} in the presence of sulfur. Mechanism 8, charge transfer, gives yellow to orange to red to black glass, based on particles of $Cd(S,Se)$; see color plate 15. Mechanism 11, color centers, gives the purple color of desert amethyst glass (see color plate 17), based on ultraviolet-exposed or irradiated glass containing Mn.

Two different types of colorations are provided by Mechanism 13, scattering. Translucency to opacity, as in opal glasses, are based on light scattering from relatively large particles of substances not soluble in glass, such as fluorides. Very small colloidal particles of gold give a brilliant red to purple color in gold ruby glass; a less attractive red is similarly produced by copper oxide and a yellow by silver.

4.8.5. Color in the atmosphere

Mechanism 1, incandescence, gives the color to stars, including our sun, ranging from red through yellow and white to a bluish white. It also contributes to the light emitted by lightning strokes. Mechanism 2, gas excitations, gives the color to the auroras and to St. Elmo's Fire, an electrical discharge from high voltage power lines or from pointed objects, such as the tips of ships' masts, during electrical storms. Mechanism 12, dispersion, gives many color effects, including rainbows, halos, arcs, and sundogs. This mechanism is involved in the green flash of the setting sun, and also gives mirages.

Mechanism 13, scattering, gives the blue of the sky, the red of the sunset, the white of clouds, mists, and fogs, and the rare blue moon. Mechanism 14, interference, gives the corona aureole, iridescent clouds, and the glory (also known as Bishop's ring and the specter of the Brocken) as in color plate 14.

4.8.6. Sources of white and colored light

The incandescence of Mechanism 1 provides us with yellow to white light sources ranging from the candle to incandescent light bulbs; there are also photo flash bulbs and pyrotechnic fireworks, usually using the burning of zirconium and magnesium, respectively. Mechanism 2, gas excitation, is used in fluorescent tube lamps (combined with the phosphors of Mechanism 10; also see below), in the neon tubes used in advertising signs, in blue alphanumeric gas discharge displays, and in some lasers such as the red light produced by the helium–neon laser.

Mechanisms 4 and 5, transition metal compounds and impurities, are used in lasers such as in some neodymium salts and in the ruby laser, respectively. Mechanism 6, organic compounds, is used in organic dye lasers and in lightsticks, where an organic reaction gives off a cold light, as in color plate 13. Mechanism 10, doped semiconductors, is widely used in LEDs, light emitting diodes, and in semiconductor lasers, usually red but also available in other colors.

The term *luminescence* covers many types of light emissions, including: fluorescence, frequently stimulated by ultraviolet (black light) illumination; cathodoluminescence in television tubes and in vacuum fluorescence alphanumeric displays; thermoluminescence, on raising the temperature; chemiluminescence, as in lightstick; and bioluminescence, as in the firefly *Photinus*, in some fish, jellyfish and Jack-o-lantern fungi.

Phosphors are widely used in fluorescent tube lamps to convert the otherwise wasted ultraviolet into visible light by fluorescence. On the inside of television tubes they provide cathodoluminescence in three colors for color TV (see color plate 3); these phosphors are typically based on Mechanism 5, transition metal impurities (e.g., yttrium vanadate containing Eu for red) and Mechanism 10, doped semiconductors (e.g., Ag-activated zinc sulfide for blue and Mn-activated zinc silicate for green). Also see Chapter 15, Sections 15.2.2 and 15.3.3.

4.9. Further reading

Apart from my own extended text on the subject (Nassau 1983), there appear to be no comprehensive modern discussions of the causes of color; well over 100 recommendations for further reading are given in Appendix G of this book (Nassau 1983). The following selected items are accordingly representative rather than comprehensive.

A general source for many color-related topics is the multivolume *Kirk-Othmer Encyclopedia of Chemical Technology* (Howe-Grant, 1991). Light and optics, usually including topics such as incandescence, dispersion, scattering, interference, and diffraction are covered in optics textbooks (Wood 1934; Ditchburn 1976; Jenkins 1976; Born and Wolf 1980). Ligand-field (and crystal-field) theory is covered in advanced inorganic chemistry texts (Cotton and Wilkinson 1988) as well as in specialized treatments (Figgis 1966; Reinen 1969; Burns 1970). Color in organic molecules is covered to some extent in advanced organic chemistry texts (Streitweiser and Heathcock 1981) and also in specialized treatments (Fabian and Hartmann 1980; Gordon and Gregory 1983). There are few detailed discussions of color derived from charge transfer (Burns 1970; Smith and Strens 1976; Cotton and Wilkinson 1988).

Band theory is covered in many solid state physics texts (Bube 1974; Kittel 1986), although color produced by this mechanism is not usually included. There are several treatments of color centers (e.g., Farge and Fontana 1979; Marfunin 1979); most contain some discussion of color caused by this mechanism. There are specific discussions of color in minerals and gemstones (Burns 1970; Marfunin 1979; Nassau 1978, 1980; Fritsch and Rossman 1987), in glass (Bamford 1977; Volf 1984; Nassau 1985), in plastics (Nassau 1986), in the atmosphere (Minnaert 1954; Greenler 1980) in particular the recent *Color and Light in Nature* (Lynch and Livingstone 1995), and six early but still definitive studies of animal coloration (Mason 1923, 1924, 1926, 1927).

References

Bamford, C.R., 1977, Color Generation and Control in Glass (Elsevier, New York).

Born, M., and E. Wolf, 1980, Principles of Optics, 6th edn (Pergamon Press, New York).

Bube, R.H., 1974, Electronic Properties of Crys-

talline Solids (Academic Press, New York).

Burns, G.R., 1970, Mineralogical Applications of Crystal Field Theory (Cambridge Univ. Press, Cambridge, UK).

Cotton, F.A., and G. Wilkinson, 1988, Advanced Inorganic Chemistry, 5th edn (Wiley, New York).

Ditchburn, R.W., 1976, Light, 3rd edn (Academic Press, New York; reprinted by Dover, New York, 1991).

Fabian, J., and H. Hartmann, 1980, Light Absorption in Organic Colorants (Springer-Verlag, New York).

Farge, Y., and M.P. Fontana, 1979, Electronic and Vibrational Properties of Point Defects in Ionic Crystals (North-Holland, New York).

Figgis, B.N., 1966, Introduction to Ligand Fields (Wiley, New York; reprinted by Krieger, Melbourne, FL, 1986).

Fritsch, E., and G.R. Rossman, 1987, An update on color in gems. Gems and Gemology **23**, 126–139; **24** (1988), 3–15 and 81–102.

Gordon, P.F., and P. Gregory, 1983, Organic Chemistry in Color (Springer-Verlag, New York).

Greenler, R., 1980, Rainbows, Halos, and Glories (Cambridge Univ. Press, New York).

Jenkins, F.A., 1976, Fundamentals of Optics, 4th edn (McGraw-Hill, New York).

Howe-Grant, ed., 1991, Kirk-Othmer Encyclopedia of Chemical Technology (Wiley, New York; 4th edn).

Kittel, C., 1986, Introduction to Solid State Physics, 6th edn (Wiley, New York).

Lynch, D.K., and W. Livingstone, 1995, Color and Light in Nature (Cambridge Univ. Press, New York).

Marfunin, A.S., 1979, Spectroscopy, Luminescence and Radiation Centers in Minerals (Springer-Verlag, New York).

Mason, C.W., 1923, Structural colors in feathers. J. Phys. Chem. **27**, 201–251 and 410–447.

Mason, C.W., 1924, Blue eyes. J. Phys. Chem. **28**, 498–501.

Mason, C.W., 1926, Structural colors in insects. J. Phys. Chem. **30**, 383–395.

Mason, C.W., 1927, Structural colors in insects. J. Phys. Chem. **31**, 321–354 and 1856–1872.

Minnaert, M., 1954, The Nature of Light and Color in the Open Air (reprint) (Dover Publications, New York).

Nassau, K., 1978, The origin of color in minerals. Am. Mineral. **63**, 219–223.

Nassau, K., 1980, Gems Made by Man (Chilton, Radnor, PA; reprinted by Gemmological Institute of America, Santa Monica, CA, 1987).

Nassau, K., 1983, The Physics and Chemistry of Color: The Fifteen Causes of Color (Wiley, New York).

Nassau, K., 1985, The varied causes of color in glass, in: Proceedings of the 1985 Defects in Glass Symposium, Vol. 61 (Materials Research Society, Pittsburgh, PA).

Nassau, K., 1986, Color in plastics: The varied causes, in: Proceedings 1986 ANTEC (Society of Plastics Engineers, Stamford, CT).

Reinen, D., 1969, Ligand field spectroscopy and chemical bonding in Cr^{3+} containing oxidic solids. Struct. Bonding **6**, 30–51.

Slater, J., 1951, Quantum Theory of Matter (McGraw-Hill, New York).

Smith, G., and R.G.J. Strens, 1976, Intervalence transfer absorption in some silicate, oxide and phosphate minerals, in: R.G.J. Strens (ed.), Physics and Chemistry of Minerals and Rocks (Wiley, New York) p. 583.

Streitweiser, A., and C.H. Heathcock, 1981, Organic Chemistry, 2nd edn (MacMillan, New York).

Volf, M.B., 1984, Chemical Approach to Glass (Elsevier, New York).

Wood, R.W., 1934, Physical Optics, 3rd edn (MacMillan, New York; reprinted by Optical Society of America, Washington, DC, 1988).

Wood, D.L., and K. Nassau, 1968, The characterization of beryl and emerald by visible infrared absorption spectroscopy. Am. Min. **53**, 777–800.

chapter 5

COLOR IN ABSTRACT PAINTING

SANFORD WURMFELD

Hunter College of the City University of New York
New York, NY, USA

Color for Science, Art and Technology
K. Nassau (Editor)
1998 Elsevier Science B.V.

CONTENTS

5.1. Introduction 171

5.2. Color models and art 171

    5.2.1. Color mixture and assimilation 172

    5.2.2. Color contrast 174

5.3. Emergent properties in paintings 175

5.4. Figure–field organization 177

5.5. Spatial organization 181

5.6. Temporal organization 183

5.7. Film color 185

    5.7.1. Luminous film color 190

5.8. The aesthetics of color painting 192

References 193

5.1. Introduction

The subject of this chapter is the development of the use of color as a means of expression in abstract painting[1]. The expressive use of color in painting can be divided into two broad categories – the aesthetic and the semantic[2]. Aesthetic color expression refers to the meaning derived from the use of color intrinsic to the structuring of the image and to its visual experience. Semantic color expression refers to the meaning derived from all the possible relationships colors or groups of colors might have to ideas external to the painting either through association, or metaphor, or convention. This chapter attempts to analyze only the aesthetic category of color expression in abstract painting. (Chapter 6 deals with some semantic aspects of color.)

Such an investigation requires a review of a variety of subjects. A selective history of color theory, prior to the beginnings of abstraction but relevant to its development, is necessary to understand the state of color ideas available to painters in the early twentieth century. This short history is expanded to include contemporary color ideas important to artists working today, followed by an analysis of how color is used to stimulate and control various sensations and perceptual experiences for the viewer of abstraction. The analysis offered includes many examples of paintings over the course of the history of abstraction. Finally, the importance of these ideas about color expression relative to the aesthetics of painting today is discussed. The goal is to offer not only the specifics of the complexity of color ideas developed in abstract painting, but also some perspective on the importance of these developments in art.

5.2. Color models and art

Abstract color painters select color from a defined set of relationships within an overall set of limits. In this way individual colors are understood to be elements in a system which allows these colors to be used in ensembles. Such use of color creates an experience of more than a simple aggregate of separate pieces and becomes the structure of a pictorial experience. Painters may come to this use of color through investigations of artistic problems, or from the analysis of visual experiences, or by the study of color theory.

Most current artists derive their notions about color order systems from the theoretical work of Albert Munsell (1916) and Wilhelm Ostwald (1931). These color scientists

---

[1] The reader is referred to the following recent volumes which contain discussions of color in representational paintings: Gage (1993); Kemp (1990); Birren (1965). More specific periods of color in painting are covered in: Hills (1987); Hall (1992); Homer (1964); Ratliff (1992).
[2] These categories were developed in Moles (1966).

started developing their two different color systems over one hundred years ago, just at the time that artists were beginning to experiment in the development of abstraction in painting. Ostwald, already a Nobel Prize winner in chemistry by the time he turned his attention to color, devised a three-dimensional color system in the shape of a double cone (much like Runge's sphere and influenced greatly by Ewald Hering's idea of four psychological primary hues). Ostwald constructed his symmetrical model based on color relationships that one would discover through color mixing particularly by using a rotating disc containing variable area sectors of hue, black and white. Munsell, an artist and art educator in Boston, devised an irregularly shaped color model based on the visual parameters of how the human eye/brain system sees equal increments in discriminating the differences between colors (also see Section 2.3.3 of Chapter 2).

What is true for both the Ostwald and Munsell color systems, and for many others that have been developed, is that they explain color relationships in terms of visual categories of like and unlike elements. There are no one-to-one physical equivalents of color measurement that create these visual color orders. Rather they assume that all human observers with so-called normal vision will be able to group colors in scales of order – the most fundamental being the achromatic or grey scale of brightness (value) from white to black. Another order common to all normal observers is the hue scale which is circular in arrangement so that alike colors are close together and colors least alike are opposite each other on the circle (also see Section 1.7.5 of Chapter 1). The third dimension of color, which is commonly called saturation, or what Munsell refers to as chroma, defines the relative purity of a hue measured by its amount of difference and therefore distance from the achromatic scale of grays on the Munsell model. Ostwald's system differs by organizing color in this dimension according to the relative degree of whiteness (tints), greyness (tones), or blackness (shades) contained in colors of specific hue.

5.2.1. Color mixture and assimilation

The development of three dimensional models of color was partly in response to one of the most basic of artistic problems: the need to predict and control color mixture, whether in pigments or optically. Artists needed to understand how to predict the possible variety of visual colors they might make from the limited palette they had available. By learning from experience, they were able to formulate simple pragmatic rules of color mixture. The ancients, for example, must have found such rules in order to construct mosaics effectively and to paint the murals we can still see in the Naples Museo Nazionale. What follows is not intended as a complete history of color theory[3], but rather a selective one touching on only those developments most relevant to the specific understanding of mixture and assimilation.

Our present understanding of color order, when considered in the larger context of the history of art, is fairly recent. Though a color sphere from 1611 by Aron Sigfrid Forsius, a Finnish mathematician, was recently rediscovered in a Swedish manuscript and one Robert Fludd apparently printed the first color circle diagram in an English

---

[3] In addition to the works on color in painting cited in footnote 1, the reader is referred to Halbertsma (1949) and Boring (1942).

medical manuscript in 1629–1631 (see the article by Parkhurst and Feller 1982), one of the earliest documentations of a color ordering to predict mixtures for artists was by a contemporary of Peter Paul Rubens, Franciscus Aquilonius, in 1613. His diagram showed in a linear order three primary hues blue, red and yellow in between black and white. Through connecting arcs he shows the predicted possible mixtures to make secondary hues as well as tints and shades.

But our modern conception of a color order system starts with Isaac Newton's color circle published in his *Opticks*. This of course stemmed from Newton's famous experiments refracting light through a prism to create a spectrum (see Section 1.4 of Chapter 1). His recreating white light by recombining only selected pairs of wavelengths allowed Newton to understand that white light, like other hue experiences, was a subjective experience stimulated by an infinite number of different wavelength combinations of light. He further understood that not all hues were the result of a specific stimulus found in the spectrum. This led him to compose his hue circle (see fig. 1.3) and thus to explain the experience of purple (or those extra-spectral hues experienced by combinations of the shortest visible wavelength, blue-violet, and the longest visible wavelength, red). Newton's color circle had the advantage of ordering hues by similarity, those close to each other on the circle, and by complementary, those opposite each other on the circle. Newton used his circular diagram to predict that a hue located on the circle could be mixed from the two adjoining colors, and that two hues opposite each other on the circle when mixed in light could produce white. This established the basic laws of the additive mixture of color (see Section 1.6.1 of Chapter 1).

In the eighteenth century J.C. Le Blon published a treatise explaining his method of three color printing. This seems to be the first such example of a reproductive process that makes use of both optical mixing and pigment mixing principles (Le Blon 1980). Most impressively, in order to reproduce a full color print, he was able to visualize individually the varying separate quantities of red, yellow, and blue required to make a full color picture – not having available to him the photographic color separation process that we now have. Later in the century Moses Harris in England devised a color circle overprinted by progressive densities of black lines, thus making use of the principle of optical mixture in order to show gradients of each hue toward dark (Harris 1963).

In the early nineteenth century in England, Thomas Young postulated his trichromatic theory of vision (see Section 1.7.2 of Chapter 1), which used the basic ideas of predictive color mixture from three so-called primary colors as a partial explanation of color vision (Young 1802). Two contemporaries of Young in England, Mary Gartside and James Sowerby, published treatises which added to the artists' understanding of color mixture. Gartside also based her ideas on three primary colors and showed possible mixtures through a blotting technique. Sowerby proposed three primary colors and showed through painted diagrams the possible hues resulting from the optical mixture of these, as well as the physical mixtures resulting from a transparent overlay of these primaries.

Thirty years later Michel Eugene Chevreul, as head of the Gobelin tapestry works in Paris, formulated and published a set of rules to elucidate the effects from the optical mixture of adjacent colors on one another in various mediums. Before the end of the century, Charles Lacouture in France had published a complete color atlas based on the principles of optical mixture from three primary hues and black on a white ground; and

Wilhelm von Bezold, in Germany, had further developed the principles of the "spreading effect" that Chevreul had earlier documented.

The process of chromo-lithography pioneered in the 19th century, made much of this color theory available to painters in a visual format[4]. The effect of this progress on painters of the nineteenth century is already well documented (Gage, Kemp, Ratliff). All this led in this century to the color solids of Munsell and Ostwald which have become well known tools for the artist to understand the possible mixing of color both optically and in pigments, since mixed colors could now be envisioned as occupying positions in the color solids between the parent colors.

5.2.2. Color contrast

The visual and artistic problems which arise from seeing two color forms side by side with a common edge, led observers and artists to investigate the issue of simultaneous color contrast. Keen observers must have noticed long ago how color changes both in varying light conditions and as a result of the proximity of one color to another. Again what follows is only a selective history of color investigations which developed the awareness of color contrast.

According to Edwin Boring (1942), Aristotle noted that the edge of a black next to a white appears blacker than the rest of the color, and much later the great Islamic scholar Alhazen postulated that one's ability to see the stars at night and not during the daytime is the result of the increased contrast offered by the dark sky. During the Renaissance in Italy, Leonardo had written much in his notebooks about the apparent colors one sees in varying light and shadow and a variety of other visual conditions (Richter 1970). Certainly the most direct history of these investigations is seen in the development of his paintings and those of other artists of the period as they describe such color phenomena within the paradigm of Renaissance perspective.

The Comte de Buffon, the French naturalist, recorded before the mid-eighteenth century the phenomena of what were called "accidental colors" or what we now know as afterimages. Many others did comparable research later in the century including the already mentioned author of the color circle, Harris, and Robert Waring Darwin, the father of Charles Darwin. In the nineteenth century Johann Wolfgang von Goethe and Chevreul popularized the knowledge of simultaneous contrast and the negative afterimage. Chevreul codified the occurrence of an enhanced contrast along the edge of any color toward its complement – now defined visually according to the rules of color contrast as that color seen in response to the original color stimulus. Harris had already noted that the "most contrary" colors are placed opposite each other on the color circle. Late in the nineteenth century Ernst Mach explained in more detail the spatial contrast effects at an edge that psychophysicists now refer to as Mach bands.

Later, Ewald Hering explained hue organization according to four principle hues and the achromatic pair of black and white. He believed the experience of color was the result of an excitation of these three opponent process pairs: red and green, blue and yellow, black and white (also see Section 1.7.1 of Chapter 1). This was in opposition to Young's trichromatic vision theory, adopted and expanded by Hermann von Helmholtz,

---

[4] See my catalog for the exhibition (Wurmfeld 1985).

which had seemed ideal to explain the phenomenon of color mixing. Hering's model, however, appeared to offer a better explanation of color contrast phenomena. Giving form to Hering's color theory, Hermann Ebbinghaus in 1902 created an octahedron or double pyramid color model with Hering's opponent process colors occupying the six vertices of the solid.

Simultaneous contrast is the reverse color effect to mixture. Colors seen next to each other tend to separate visually by increasing their color difference over viewing duration in a predictable manner. This phenomenon is also diagrammed on the color circle and the more complete color solids. Colors are visually altered over viewing time to appear like the complementary color found at the opposite side of the color solid. Just as the color models of Munsell and Ostwald function to diagram the properties of color mixture, they also predict by their arrangement of colors the properties of color contrast.

5.3. Emergent properties in paintings

Thus the color solids of Ostwald and Munsell provide systems to understand and predict color phenomena. Color painters can make use of these systematic relations to order paintings. Colors are understood either to group together because of their degree of similarity or to appear separate because of their degree of difference. The tendency to group together is called assimilation and the tendency to separate is called contrast (see Jameson and Hurvich 1975). Color elements in a painting may be made to assimilate or to contrast with each other depending on a number of form factors: size, shape, relative location, and number of elements. And so color painters select each element in a painting by its relative size and shape in relation to its hue, saturation and brightness, and then place it in a location on the canvas in relation to other elements in order to stimulate the desired experiences for the viewer of assimilation or contrast in various parts of a picture.

These are the fundamental building blocks for the color painter in the making of a picture; but the magic of color painting entails much more: the emergent visual properties that occur through the artist's manipulation of a palette in the above-described manner. Without these emergent visual properties, representation in painting would not have been possible and color abstraction would have been a dead end. What I am referring to is the endless variety of organizational possibilities that can be made to appear in order to create an abstract color picture[5]. Paintings are physically just colors placed on a flat opaque surface and yet they are perceived as having a dual visual quality; the same opaque two dimensional surface of visual elements can also be organized by the viewer into a variety of surface pattern, spatial, and temporal experiences.

---

[5] The methodology to describe these ideas was developed together with Prof. Stanley Novak, a psychophysicist in the Department of Psychology, Hunter College. With N.E.H. funding we jointly developed a team-taught course first offered in 1976 bridging the relationship between the parallel developments over the last 150 years of experimental visual psychology and art – especially abstract painting of this century.

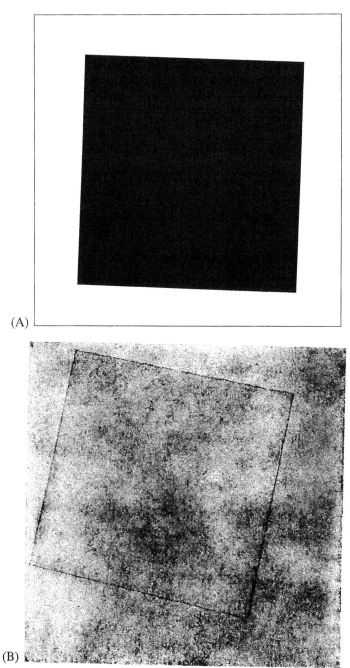

Fig. 5.1. Kasimir Malevich: (A) "Black Square", 1929, oil on canvas, $80 \times 80\,\mathrm{cm}$, State Tret'iakov Gallery, Moscow, redrawn by Elizabeth Pugh; (B) "White Square on White", 1918, oil on canvas, $31\frac{1}{4} \times 31\frac{1}{4}$, reproduced with permission from the Museum of Modern Art, New York, NY.

Fig. 5.2. Edgar Rubin "Synsopleuere Figurer", 1915, redrawn by Elizabeth Pugh.

5.4. Figure–field organization

As one color or one assimilated group is seen in contrast to another color or assimilated group, the viewer tends visually to divide the picture into a two dimensional organization or figure and field. Crucial to this experience is the acceptance of the physical extent of the picture as the artificially defined limit of the field. The seminal Russian artist Kasimir Malevich's early "Suprematist" compositions in black and white (see fig. 5.1(A))[6] are clear examples of paintings displaying this type of figure–field organization in a pictorial format. The Dutch painters Piet Mondrian and Theo Van Doesberg also painted works by

---

[6] In much of the abstract art described in this chapter, surface texture and other subtle effects may not survive in a reproduction and may well prevent the various intended emergent properties from being seen. This includes size, which also may be critical.

Fig. 5.3. Wladyslaw Strzeminski, "Architectural Composition 2a", 1927, 73×78 cm, reproduced with permission from the Museum Sztuki, Lodz, Poland.

1918 that could be understood as abstract figural elements on a field. In 1915 Edgar Rubin, the Danish gestalt psychologist, published examples of reversible figure and field pictures (see fig. 5.2) which explicate this surface organizational visual phenomenon. Following these same principles using simple black and white elements, the Polish Constructivist, Wladyslaw Strzeminski, made paintings during the 1920s which stimulated reversible figure–field relationships (see fig. 5.3).

Painters can control even more complex multiple organizations of figure–field relationships. Just as the viewer reverses one figure–field organization, the viewer can create visually a number of such alterations, each different from the others. A simple example is the juxtaposition of two different patterns of black and white so that the viewer sees, on the one hand, a figure–field reversal between the two patterns and, on the other hand, a figure–field reversal between the black and the white. The painting "Mechano-Faktur" (1923) by Henryk Berlewi, the Polish Constructivist artist, is a good early example of this effect.

Jackson Pollock did another variation in such paintings as "Autumn Rhythm" (Metropolitan Museum of Art, New York) which can be understood to explore the tension between a multiplicity of elements seen as figures on a field or as a cohesive visual group experienced as an all-over field. By involving more colors to control specifically figure–field relations, ever more complex organizations can be effected. During the 1950s

Fig. 5.4. Frank Stella, "Jill", 1959, enamel on canvas, $90\frac{3}{4}''\times 78\frac{3}{4}''$, reproduced with permission from the Albright–Knox Art Gallery, Buffalo, NY.

Mark Rothko selected colors to create paintings with such multiple visual organizations; in one way the color forms are seen grouping according to value, in another by hue, and in a third way by saturation. A number of other American Abstract Expressionists – Barnett Newman, Franz Kline, and Robert Motherwell, for example – investigated the issue of scale in relation to figure and field. They made black and white paintings which stimulate figure–field reversals on a large enough scale to be experienced as a wall. This affects more directly the actual space of the viewer.

The two-color works by Frank Stella done in the early 1960s create even more of these same effects by using simple means. His use of $2''$ wide concentric bands of a

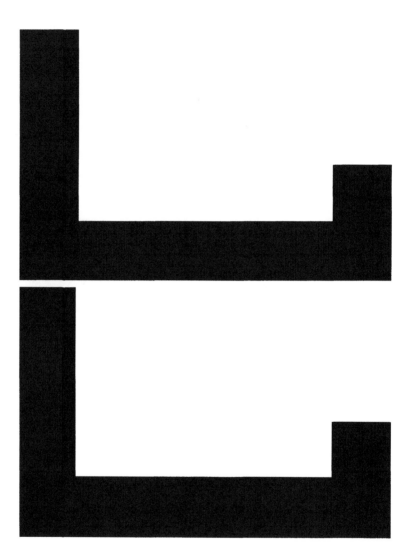

Fig. 5.5. David Novros, "2 : 16", 1965, Vinyl lacquer on canvas, $121\frac{1}{4}'' \times 78\frac{1}{2}''$, Park Place Gallery, New York, NY, reproduced with permission from the artist.

color leaves a 1/4″ band of raw canvas in between to form the figure–field relationship (see fig. 5.4). We can visually organize these works as a figure–field reversal based either on the color contrast or on the different patterns caused by lines of different orientation. These images can also each be seen as an all-over pattern or what is referred to as all-field. Further to the problem of color, we notice that the level of contrast between the colors in these paintings by Stella alters, depending on which pattern the viewer attends to as figure or as field. That which is seen as figure exhibits more color contrast within the pattern; that which is seen as field shows greater assimilation of color.

Further exploring the limits of figure–field organization, a number of artists in the 1960s, including Stella and David Novros, broke with the convention of accepting a rectangular-shaped canvas as the accepted limit of the field. By using varying shaped canvases, these artists made more apparent the inherent figure–field relationship between the painting and the wall (see fig. 5.5). These new shaped canvases made the wall or environment color and pattern an integral part of the visual experience offered by the artist, a relationship previously separated by the convention of the frame. Failing the opportunity to work on large cooperative architectural-scale projects in which they could control entire spaces, most artists retreated to the convention of the rectangle, but with a renewed understanding of the role of the wall as an environmental color.

5.5. Spatial organization

Simultaneous to the viewer's experience of the organization of color elements into figure and field is the emergence of the perception of three-dimensional or spatial relationships. Very simply, assimilated colors are seen as occupying the same spatial plane, while contrasted colors are seen as located in a different spatial plane. In the extreme condition of total assimilation or no contrast, such as we experience with the lights out or in a thick fog, there is no spatial perception; rather there is spatial disorientation. Thus in a painting total assimilation would produce an all-field organization. Color painters manipulate spatial organization through the control of the degree of contrast between colors or assimilated groups in relation to size differences, or shape changes, or location changes of the elements on the canvas. Often, as with figure–field organizations, there occurs visually a spatial reversal from the same stimuli. Since colors are always affected by the context in which they are seen, the spatial organization is dependent on the color context presented; there is no absolute rule; as is often thought, that certain colors are always seen as advancing and others receding.

Parallel to the investigations of the gestalt psychologists, abstract color painters developed a number of approaches to the expression of space in paintings. In its simplest manifestation, space is sensed in a painting from the degree of color contrast between the color of the figure and the color of the surround or field. Spatial control by color contrast can be enhanced by the perception of apparent overlap of forms. This in turn can be further complicated by controlling color to make an apparent transparency of elements.

In the development of abstraction early in this century, one sees that the ideas to create space by overlap, size and shape difference, as well as relative contrast of color to an established field color, were all investigated by Malevich. In his simplest compositions a single element of black is placed on a white field to create visual space from color contrast (see fig. 5.1(A)). Malevich varied the sense of space by varying the amount of color contrast to the field – making paintings with red squares and near white squares on a white field (see fig. 5.1(B)). He also varied the shape of the square to effect a warped space. Another composition with two elements presents a vertical rectangle of red apparently overlapped by a horizontal rectangle of black to create the impression of space. These compositions were followed by slightly more complex ones using three or more colors in varying shapes and sizes which create a visual space in the experience of the painting. In the work "Yellow Quadrilateral", Malevich experimented further and presented a trapezoidal form with a color progression from yellow to almost white on a white field to create a visual recessional spatial experience.

Other Russian artists of this period, seemingly following Malevich's example, developed different approaches to creating the appearance of space in an abstract picture. Liubov Popova ("Architectonic Composition", 1917) used an altered edge condition by painting progressional colors juxtaposed with possible overlap of forms in order to make ambiguous the relative spatial location of the depicted planes. Ivan Kliun introduced images of apparent transparency, a technique also extensively used later by Laszlo Moholy-Nagy and Wassily Kandinsky in their years at the Bauhaus.

The illusion of transparency in an opaque medium on a two dimensional surface is described as controlled by the gestalt idea of good continuity of color and form (Metelli 1974; Metelli has fully described the parameters controlling the effect of transparency). Josef Albers defined transparency as a spatial effect in practice for the artist in his 1963 book *Interaction of Color* (Albers 1963). Derived from his teaching practices at Black Mountain College in the 1940s and at Yale University School of Art in the 1950s, Albers demonstrated the control of transparency through numerous silk screen prints. His basic approach was the use of a mid-color between two parent colors to create the illusion. Albers also showed how alternative color selections could effect spatial perception by weighting the color toward one side or the other. If a color appearing as transparent was in less contrast to one of the parent colors, then it would be seen as closer to that parent color spatially.

Albers created a number of series of paintings in which he controlled color to create apparent transparencies stimulating ambiguous figure–field and spatial visual readings. Many other artists used the effects of apparent transparency including Max Bill, Richard Lohse and Kenneth Noland. Morris Louis and Piero Dorazio made paintings that employed the transparent quality of the particular paint medium they were using in order to effect transparency in their pictures.

By the simplicity of his format, Rothko also created a sense of space in his paintings by controlling the amount of color contrast between the rectilinear elements and the perceived field. In her early work in the 1960s, the British artist Bridget Riley filled rectilinear canvases with an overall pattern on a single color field that progressively changed in size, shape and especially color contrast to create strong evocations of a warped spatial plane.

The result of all these artists' pictures was to show that abstraction could effectively create a spatial experience by using a wide variety of all the above techniques of color and form control.

5.6. Temporal organization

Beyond the surface and spatial experiences of pictures, abstract color painters learned to stimulate a temporal organization in painting as well. This can occur even though a painting is physically a single fixed image, since the visual organization of that image might change over the duration that a viewer looks at the painting. Figure–field reversal, for example, by necessity requires duration since it is impossible for the viewer to see both figure–field organizations simultaneously – rather they can only be seen sequentially.

It is also known that color is especially sensitive to the effects of viewing duration and so, as color changes visually over time, this will likely alter the viewers' figure–field and spatial perceptions. A prime example of this is seen in Albers "Homage to the Square" series. As the viewer fixates on the center of the picture, the colors change through simultaneous contrast. What at one moment would appear to be flat colors making a multiple transparency which has ambiguous figure–field and spatial readings, can seemingly become, with greater fixation time, a series of apparently progressional colors continually changing from edge to edge. The so-called "fluting effect", discussed thoroughly in Albers' book, is a name for the viewer's experience of those afterimages seen along the boundaries of colors when viewed simultaneously – mimicking the spatial effect of a "fluted" classical column.

If it can be readily accepted that an artist organizes a picture in such a way that a viewer can elicit figural and spatial experiences, then it is possible to understand that the artist can also structure the viewer's sense of time for expressive purposes. Ad Reinhardt's so-called black paintings of the 1950s and 1960s are particularly obvious examples. Looking at the painting in the collection of the Museum of Modern Art, New York, the initial experience is of a five foot square of a single black which reads as all field and no space. As the viewer fixates for a time on the painting, some contrast can be discerned and thus a number of colors are seen which yield a figure–field and spatial organization that, at times, is based on overlap and then, through simultaneous contrast, can appear as a nine square grid. Thus the abstract color painter can select color so that through the viewer's duration of fixation on the image, changing color and form experiences are created over time.

In the early 1960s, Larry Poons created a series of large paintings using small colored dots or oval shapes spaced so as to be seen both as multiple figures on a field and as an all-over pattern on a single color field. With time the viewer elicits afterimages of these dots or ovals and so the information of the painting is increased by the simultaneous viewing of the directly seen dots along with the afterimages. Moreover, the afterimages appear to float and wander around the canvas from the natural eye movements of the viewer, further complicating the surface and spatial readings of the painting and creating a virtual sense of movement. Ellsworth Kelly in the mid 1960s produced his "Spectrum" series which presented vertical rectilinear panels each of one color and combined in a horizontal spectral hue order. Through the interaction of the colors along the meeting edges, the viewer experiences the "fluting effect" increasing dramatically with viewing duration. Curiously, both Poons and Kelly professed dismay at seeing these so-called "optical effects" in their work and quickly moved to rid their paintings of such experiences. Poons reacted by making a series of similarly formed paintings which, because of his new color selections, inhibited the appearance of afterimages. Kelly opted to separate the panels leaving wall space between the colors which prevented any color interaction between them. This newer work of Poons and Kelly thus fails to stimulate the viewers' durational experience of the painting.

Starting in the late 1960s and early 1970s, Robert Swain painted large canvases covered by a one foot square grid of colors in a progressional sequence (see color plate 21), much like a section from a Munsell or Ostwald color chart. The effects of these paintings were much more than ordinary. Because the colors share edges, a prolonged fixation time produces for the patient viewer undulating experiences of fluting between the rows of colors, alternating between horizontal and vertical organizations. The colors presented in visual sequences also make for changing figure–field and spatial readings over the entire surface of these paintings caused by the shifting apparent transparencies.

But the temporal organization of the picture can also be effected by a different viewing technique. Rather than fixating on the entire image, viewers can scan surfaces part by part for information. Such scanning creates sequences of viewing and so establishes differing contexts for the color and form in any part of the picture, depending on the scanning sequence. In this way a painting can offer different sensory experiences of what is in fact a single fixed image. Though the perceptual psychologist Rudolph Arnheim (1954) in the 1950s proposed a popular theory that surfaces are usually scanned in a particular sequence, more recent psychophysical evidence and anecdotal experience seem to show that scanning is not so specifically ordered. Rather the scanning order is affected by the attitude of the observer and the attraction of points of increased visual information[7].

Pollock and Willem de Kooning, as well as their contemporaries Kline and Motherwell, created images which seem to demand scanning as well as full-field fixation viewing. By scanning the Pollock "drips", the viewer effects a constantly changing spatial and figure–field experience of the painting quite different from the earlier described all-field view. Looking at deKooning's "abstract landscapes" of the 1950s, while one spatial color experience is possible with fixation on the whole work, another is possible by the scanning of the parts of the painting. DeKooning seemed to perfect this idea through what John Canaday, the New York Times critic, mockingly called the "New York stroke", in which all brush strokes which appear to be first on top eventually are in turn overlapped by a different color stroke – a fact only verified through scanning.

Kelly offers another approach to making scanning a part of the experience of the painting as demonstrated in his small work "Seine" from the early 1950s. This is a two color black and white painting divided into a grid of small rectangles with the amount of each color changing arithmetically in the vertical columns from left to right. The painting can be viewed as a whole from a distance and through assimilation appears as a progression from left to right of light to dark to light; but when seen close-up, so as to focus on the contrast of black and white elements, one can scan the picture to experience a progressive change from black figure on white ground, to figure–field reversal, to white figure on black ground, back to figure–field reversal, and finally back to black figure on white ground. Whether the viewer fixates on it as a whole or scans the surface from part to part, either or both experiences of the painting tend to effect a sense of time.

---

[7] Goldstein, E.B., Sensation and Perception (Wadsworth Publishing Company, Belmont, California, 1980) pp. 180–183.

In the 1960s Guido Molinari started a series of paintings composed of sequences of vertical stripes of color. The experience of these paintings is only completed as the viewer scans the color bands, discovering repetitions and alterations of the color patterns.

More recently Doug Ohlson has created large paintings which use a series of horizontal or vertical rectilinear shapes in relation to a field color (see color plate 24). Because the colors selected are frequently of close value and hue to each other or to the field, by fixating on the whole painting the viewer stimulates the appearance of bars of afterimages which considerably complicate the experience. We tend to see extended edges and whole bars of color in afterimages simultaneously with the surface color. The result is a shifting composition which makes the figure–field and spatial readings ambiguous. In the process of reworking the painting, Ohlson paints the edge of the bars or the field color at some places with a specific contour and at other places a less specific one. The result is that the scanning experience of the color relationships challenges the viewers' ability to organize the image spatially much in the way that deKooning pioneered. These multiple experiences of fixation and scanning place the viewer in a reverie where time is no longer physically measurable, but part of a continuum of subjective experience.

Just as it is known that any single physical form or color has distinctly different sensory experiences which are dependent on the visual context, it is also known that the same physically measured duration can be affected by the visual context to change the subjective experience of time. In color painting this consideration can be a major expressive element in the making of a picture. Indeed, it has become a major new creative problem in the most advanced abstract painting since the 1950s. As other hot media like film and video are developed which control the duration of each image, painting continues to function as a passive image which allows active viewers to effect their own subjective sense of time through their color experience.

5.7. Film color

As has been outlined, the contrast and assimilation of colors create the surface, spatial, and durational experiences of a painting which bring about the emergence of a picture. But how is it that some paintings achieve a greater sense of picture than others – an image floating as if detached from the limit of a physical surface? This quality requires further investigation of color experience.

David Katz, the gestalt psychologist working early in this century, seems to have been the first to describe color experience as divided into modes of appearance (Katz 1935). He started by defining the mode of "surface color", which he characterized as color seen attached to the perception of an object. Surface color associated with viewing an object is, therefore, seen specifically located in space. Seeing a painting in the surface color mode infers perceiving the painting as a discrete object with the material quality of paint colors on a flat canvas. Katz's analysis seems particularly relevant since the emergence of a picture quality from a painted surface might then be understood to be simultaneous with an alternative to experiencing the perception of surface color.

The fostering of an alternative mode of color perception – "film color" – when looking at a painting, is a primary method for seeing a picture and not just a surface. Katz defined film color as a color sensation detached from the experience of any object. When color is seen in the film mode, Katz stated that it appears as frontal in a bi-dimensional plane of indefinite spatial localization and with an apparent spongy texture. Katz's prime example of the experience of film color is what he called the subjective grey: the color impression one gets with eyes closed. This is a very specific example of Katz's idea of pure film color but, interestingly, an illusion of the film mode can be effected in painting. This can be achieved by a number of techniques.

One such example is related to the idea that the viewer can perceptually organize a painting into figure and field. Rubin, responding to Katz's work, noted that when figure–field relationships are seen, the figure tends to display the quality of surface color and the field takes on the film mode. Katz had documented that film color is experienced more in peripheral vision and surface color more centrally, in the fovea. This tends to reinforce the notion that what is perceived as field has a more film-like quality, while what is visually organized as figure has a more surface quality. While attending to the visual field, the viewer is aware of the specificity of things seen in the center and the progressively blurred quality of color and contours as the eye attends to the periphery. A simple approach which attempts to achieve some film color in painting is seen in pictures clearly organized to elicit figure–field relationships. Moreover, when the painting effects a figure–field reversal, it is understood that at some moment in a durational sequence of viewing, each part of the painting is seen as surface and each part as film color. In addition, non-referential color, that is color which appears to lack all information about the material and specific spatial location of an object, helps the appearance of the film mode. As a result, completely abstract paintings which have little or no apparent paint texture can appear more film-like in color. The work of the aforementioned Strzeminski done in the 1920s displays non-referential color organized in a figure–field reversal.

Even more specific to the idea of effecting film color through figure and field control is the approach to create an all-over or all-field painting pioneered by Pollock. When viewing a Pollock work such as "Autumn Rhythm" (previously described), one can induce a film quality by fixating on the whole and not focusing specifically on any of the drip elements. This is of course only one of many perceptual constructs for the active viewer while experiencing this painting. All-over or all-field painting has been a particular compositional approach investigated by many artists since Pollock, both singularly and in combination with other ideas (see cover illustration and color plate 22, both painted by Wurmfeld).

In the late 1960s Brice Marden combined the approach of non-referential color with an all-field composition. Marden made paintings which present an all-over single color on a surface that is completely matte and without texture, achieving some quality of film color for the attentive viewer. Because the viewer can also perceptually construct such a painting as a figure on the entire wall as the field, this returns the color in such a painting to a surface quality. And thus it must be noted that such simple examples are more dependent on the attitude of the viewer to achieve any quality of film color than some other, more complex, approaches which will be discussed further.

The illusion of transparency, earlier described as a technique to control visual space, also tends to effect a partial film color illusion in painting. This is because the viewer senses "seeing through" the color which appears transparent, evoking a film-like appearance. As has been pointed out previously, many artists made use of transparency as an

organizing principle for their paintings. Albers used the idea to great effect in his series of paintings, "Homage to the Square" begun in 1949. The constructivists and early
abstractionists had created transparency effects, but limited to one part of the painting.
Albers painted a concentric square format which created a possible impression of film
over the entire surface, as any of his colors could appear transparent. Noland, who had
been Albers' student at Black Mountain, made a number of paintings in the 1960s using rectilinear planes in a color sequence which also presented all-over transparency. In
his "Veil" series, Louis poured physically transparent paint in layers on large canvases
which, because of the actual transparency of the paint, created some film-mode color.
Louis, however, reestablished the physical quality of the paint as surface color by leaving
the edges of each applied color unmixed and visible at the periphery of the painting.

Compounding the use of transparency in Albers' work in order to create film color, are
the afterimages stimulated by the specific contours of each color. The so-called "fluting
effect" addressed earlier allows the viewer to experience, with fixation time, afterimages
which alter the seen color and the mode of appearance of the underlying surface color.
Colors seen as afterimages appear in the film mode because the viewer cannot localize
them to any surface or spatial plane and they appear to have Katz's described "spongy"
texture. In those "Homage to the Square" paintings where Albers used a progression of
colors of the same hue range, the fluting effect is extremely apparent when the paintings
are fixated upon over a period of time. Each apparent concentric square then appears to
be a progressional color area rather than a single flat color. With such viewing time, the
identity of the three or four physical pigments painted is displaced by a curious confusion
of many more or even fewer colors visually floating in the film mode.

Afterimages stimulated by seeing the work of Albers, Poons, Kelly, and Swain (see
color plate 21) referred to earlier, are particularly effective in creating the impression
of film color. That the sharp contours in the work of all these artists should assist the
appearance of film color is counter to the normal rule for general experiences described
by psychophysicists which states that fuzzy contours of form increase the appearance of
film, while sharp contours bring on surface color. Ralph M. Evans, the color scientist,
showed how a circle with a specific edge stimulates much more simultaneous contrast
and surface mode color than the same circle with a fuzzy or out of focus edge which tends
to assimilate more with the surround color and thus appear film-like[8]. A painting by Bill
from early in his career can be viewed as a study of the relationship between the edge
condition of a dot and the amount of color contrast with its surround color. The result
is not only a painting with a strong spatial organization, but also one in which the color
progresses from being seen as surface to being experienced as film through a number
of stages in between. Rothko's paintings also make use of this effect. The so-called
atmospheric quality of his paintings, which is now understood to be color in a film-like
mode, is due in some measure to the lack of clarity in the contours of the color shapes.

The issue of focus seems fundamental to the possible experience of film color. In his
original investigations, Katz postulated the impossibility of matching colors that were
in different modes of appearance and so he devised the "reduction screen". A small

---

[8] Evans (1974). A similar illustration appears in Ratliff (1992) p. 63. Ratliff uses the experiment to compare
relative simultaneous contrast. Also see Beck (1972).

aperture in a larger surface which is held at a distance to mask the test color was so-called because it could "reduce" any color seen through the aperture to the film mode. This is accomplished because the viewer tends to focus on the frame or aperture, and the test color seen at some undefined distance is out of focus and so always perceived as film color. Though Katz's analysis was highly original, the idea of viewing through an aperture had precedents. In the mid-nineteenth century, the English painter and critic John Ruskin, had described the practice of isolating a scene by looking at it through a frame-like aperture as a drawing aid. Hering and a number of other nineteenth century psychophysicists had used such screens as a way of isolating a particular color experience in a variety of experiments. Much more recently James Turrell, an artist from California, has created room sized experiences of film color by using the reduction screen on an environmental scale and carefully controlling the ambient light.

Related to the question of focus, but an even more effective way to create the experience of film color throughout a painting, is the use of color and form assimilation. If the elements of color are small repeated units placed throughout the surface of the canvas so as to be seen to assimilate but not to blend in total optical mixture, there emerges an all-over sense of film color from this vaguely out of focus image (see cover illustration and color plate 22, both painted by Wurmfeld). The level of assimilation is a function of the inverse relationship between the size of the color elements and the degree of contrast between the colors used (see color plate 23, painted by Wurmfeld). This is now understood as one of the major aesthetic results of Seurat's pointillist technique. In a painting from 1931 ("Unistic Composition 9"), Strzeminski presented narrow repeated horizontal bands of a color juxtaposed with two other colors and thus brought this technique to abstraction.

Varying the degree of visual texture seems natural to the linear foundation of the intaglio process (see Section 13.3.5 of Chapter 13). Starting in the 1950s, Vincent Longo, a painter and printmaker, presented the idea of film-mode color in the graphic medium. Using the line inherent to the various intaglio processes as a basic small repeated unit, he created fields of grey film color varying in value by the density of the marks on the surface (see fig. 5.6). Though it may be argued that earlier representational etchings display some of these qualities, Longo makes the film color integral to the subject of his abstract work.

In her 1964 painting "Current", Riley painted black and white patterns in small units which tend to assimilate in the manner described and to produce, coincidentally, some induced hue experience where physically only black and white existed. When there is an appearance of hue from the presentation of certain frequencies of a black and white pattern, referred by color scientists as the Benham effect (or Fechner's colors), this hue exhibits a film-like quality.

Richard Anuszkiewicz has created several series of paintings using near assimilation. Even more effective in producing film quality color has been his simultaneous use of partial assimilation with the apparent transparency of one assimilated group with another, exemplified in his well known painting "All Things Do Live in the Three" (1963). Anuszkiewicz has also done a number of works that juxtapose a perception of film mode color in one part of the painting with direct surface color in another. By leaving one part of the painting in a single flat color, untextured by the near assimilation of two or more

Fig. 5.6. Vincent Longo, "Regular", 1969, etching, $23\frac{1}{2}'' \times 17\frac{1}{2}''$, reproduced with permission from the artist.

colors, the viewer sees the flat color in the surface mode while the visually textured area is seen in the film mode. But in fact, Katz had associated the perception of surface color with seeing texture, although he was referring to the material quality of the surface of specific objects in ordinary vision. It can be inferred from the example of the paintings cited, however, that visual texture in painting caused by the near assimilation of details can indeed serve to aid the apparent film color experience.

Julian Stanczak, a fellow student of Albers together with Anuszkiewicz at Yale in the early 1950s, has used a number of grids of small repeatable units to effect film color in his painting through incomplete assimilation. A favored technique of Stanczak is to use the assimilation of color to make a progression from a sharp contour at one edge of an assimilated group to an out of focus or fuzzy edge at its opposite side, which increases the viewer's sense of film color all the more.

5.7.1. Luminous film color

Furthering the attempt to counterpoint the reality of the physically opaque surface of the canvas, color painters have tried to make pictures emit a sense of light by creating a luminous film color experience. Some of this effect may be the result of the phenomenon of color constancy. This is the tendency for the viewer to see changes in the lightness or hue of an object as apparent changes in illumination of a single object color, rather than as different or separate surface colors. Attempts by representational painters[9] to instill in the viewer a sense of light or luminance is a successful way to destroy surface color perception in a painting. Because Katz described color constancy in ordinary visual experience as an attribute of surface color, it seems paradoxical that the ability of the representational painter to foster an acceptance of color constancy is what tends to shift the viewer's attention from seeing merely a flat object to seeing a picture.

In abstract color painting, artists have used parallel techniques. In the late 1960s, Riley painted canvases which established a sense of an all-over grid through repetition of black and grey forms on a constant white ground (for examples see reproductions in de Sausmarez 1970). Because she then sequentially altered the elements in shape and color so that they assimilate more with the ground, there emerges for the viewer an effect not only of warped space, but also of luminous film-mode color.

Luminous film color appears in painting when the viewer dissociates a change in color of the paint on the canvas from the surface paint color and perceives this variable of light and dark as a change in brightness or luminance. This is not to say that abstract color paintings appear as a light source per se, but rather that such paintings can at times manifest a sense of their own mysterious luminosity. Such luminosity is seemingly accomplished by establishing an average level of lightness in the painting and then shifting part of the picture toward a lighter level of brightness beyond the established adaptation level. Katz pointed out that peripheral vision contributes to a large extent to the viewers' impression of the quality and intensity of prevailing illumination. He went on to say that "the color of a small area signifies nothing ... only when it is grouped as part of the whole ... is a sense of illumination imparted" (see Katz 1935, p. 227).

---

[9] One might think immediately of the work of the English artist Turner, but even more specific to what follows is the experience the viewer has from the "American Luminist" painters, especially Sanford Gifford.

Referring more specifically to the experience of painting, Katz states that when we stand "close to the canvas the effect we get is that of the reduction of the colors of the picture to the illumination of the room" (see Katz 1935, p. 227). It would seem, therefore, that some viewing distance is helpful to the experience of luminosity that can be elicited from a painting. Perhaps that is why certain color painters have found more success with especially large color field painting; the overall perceived field still takes on this function of the periphery when the painting is viewed from some distance.

A painting by Kliun – "Red Light, Spherical Composition", 1923, presenting a color progression from a dark surround to a light and more saturated circular red center with an out-of-focus contour, is an early example of luminous film color in abstraction. Kliun, in another work, uses the technique of a darker surround with lighter transparency toward the center of the picture to create some sense of luminosity. Moholy-Nagy also used a similar method in a painting from 1922 (Museum of Modern Art, New York). The apparent luminosity one sees in some paintings by Rothko is the result of his having selected colors for the floating rectangles that are lighter and more saturated than the darker and less saturated surround. In some of Albers' "Homage to the Square" paintings the viewer may perceive a sense of luminous color as a result of his arranging the color in a progression from a darker surround to a lighter center. Poons also created some luminous effects by his selection of lighter colors for his dots or ovals in relation to the single field color. Anuszkiewicz and Stanczak have used this technique in some of their work, painting peripheral field areas with darker assimilated groups and progressively lightening the colors toward the center so as to emit a sense of brilliance from the opaque canvas surface (see color plate 25, painted by Wurmfeld).

Another technique to effect a sense of illumination in an opaque painting is also experienced in nature. When we are in a dense fog, there exists low contrast throughout the visual field, creating for the spectator a sense of overall light and an ensuing definite reduction of all objects in the field of vision from surface color toward film color[10]. This luminous film experience caused by low contrast and a dominant color is true as well for very light experiences as one might have when being "whited out" in a snow storm, or for very dark experiences as one might have in an extremely low ambient light, or even for hue-controlled situations such as seeing under the red safe-light in a dark-room environment. The tendency to see luminous film color in such circumstances is related to apparent transparency as well as focus and assimilation effects. More importantly, the possibility exists for artists to use this phenomenon in color painting to create a more luminous film color experience (see the yellow area of color plate 25, painted by Wurmfeld).

The origin of this phenomenon in abstract painting is evident from a comparison of Malevich's famous "white on white" paintings of 1918 with his earlier black and white work (see figs 5.1(A) and (B)). Though the sense of luminous film color is hardly strong in Malevich's low contrast pictures, the difference evoked in these works is striking in comparison to his high contrast paintings. Much later in the 1950s Reinhardt worked on three series of paintings which were all in extremely low contrast: a red group, a blue group, and a black group. Perhaps the most curious results of this effect are seen

---

[10] Actually, this example is what Katz uses to specify "volume color", but the visual quality of volume color and its illusion in painting is like film color.

in Reinhardt's work because of the sensation of a dark light that the viewer draws out from the so-called black paintings (referred to earlier). This is due to the combination of extremely low contrast, the matte and entirely textureless surface, and the subsequent durational experience of hue changes in the paintings caused by the simultaneous contrast of these closely-valued hues.

Albers "Homage to the Square" series includes many examples which use closely related colors to add to the experience of luminous film color. Some of his paintings in this format are constructed from three or four reds or yellows and, because of the dominance of just one hue range, tend to present the overall impression of a luminous hue. More recently other artists, such as Molinari in Canada and Johannes Geccelli in Germany, have painted extremely low contrast work to create a sense of luminous film color.

Abstract artists create an experience of a picture with film color and ultimately luminous film color by controlling contrast and assimilation of color elements. By establishing figure–field organization, visual spatial experience, and subjective time experience, color painters have moved their work to a new level of complexity with the creation of these picture qualities.

5.8. The aesthetics of color painting

The structuring of a picture by the artist through the related selection of color and form in order to offer for an active viewer varying pictorial experiences from a single physical image is a paradigm for abstract color painting. How can we understand the significance of this development? A comparison to an earlier era will be instructive.

Erwin Panofsky used the term "symbolic form" to describe the perspective practice of the Renaissance (Panofsky 1991). He had borrowed the term from Ernst Cassirer in order to establish the idea that perspective itself was not only a construct for the making and ordering of a painting, but also expressive of the ideas of that era. Samuel Edgerton has elaborated on this more recently in his book on *The Renaissance Rediscovery of Linear Perspective* (Edgerton 1975), arguing that perspective was the paradigm of the Renaissance. Paradigm was a term Edgerton had taken from Thomas Kuhn's influential book *The Structure of Scientific Revolutions* (Kuhn 1962). Though Kuhn certainly used the word paradigm in a number of different ways, one important meaning for the word is schema or world-view. For the artist the paradigm is a schema or a set of assumptions from which one draws inferences and within which one functions.

Though perspective has been taught over the centuries since the Renaissance and has been understood as a technique for visually ordering images, both Panofsky and Edgerton took a different look at the issue and saw that perspective, the structuring principle in the making and the viewing of painting, is more importantly the major expressive element of the work as well. It is curious that in a similar way, color has often been relegated to mere technical status in the analysis of abstract painting.

An emphasis on color was crucial to that revolutionary impulse of the late 19th and early 20th century to reemphasize the flatness of the canvas while still constructing representational images. Color again became the focal issue with the realization that a flattening of the picture is no longer expressive of the same meaning with the advent of non-representational painting. Rather the problem is reversed, because the physical limitations of the painting become all the more apparent with constructed abstraction. By investigating color, the abstract artists discovered new possibilities for stimulating more complex visual experiences in order to further contest the physical limits of the canvas and thus to alter the nature of their expression.

The definition and evolution of this paradigm for color painting is evident from the examples referred to in the history of twentieth century painting. As in other eras, more than one paradigm for painting has existed and has been developed, and so many artists have followed competing paradigms. Moreover, many practitioners followed hybrids of different paradigms in their work, so that one cannot categorize all the artists referred to here as adhering totally to this new paradigm of color painting.

But there is no doubt that this paradigm functions as the symbolic form for constructed color abstraction. The color organization of contrast and assimilation, leading to various pictorial structures of pattern, space, duration, film, and luminosity, are not just a set of technical rules convenient to the artist, but as the structuring principle for the making and viewing of abstract painting, they are the major expressive elements as well. This approach to color is a paradigm for painting in our time.

References

Albers, J., 1963, Interaction of Color (Yale Univ. Press, New Haven).

Arnheim, R., 1954, Art and Visual Perception (University of California Press).

Beck, J., 1972, Surface Color Perception (Cornell Univ. Press).

Birren, F., 1965, History of Color in Painting (Van Nostrand Reinhold, New York).

Boring, E.G., 1942, Sensation and Perception in the History of Experimental Psychology (D. Appleton-Century Company).

de Sausmarez, M., 1970, Bridget Riley (New York Graphic Society).

Edgerton, S.Y., 1975, The Renaissance Rediscovery of Linear Perspective (New York).

Evans, R.M., 1974, The Perception of Color (Wiley, New York).

Gage, J., 1993, Colour and Culture (Thames and Hudson).

Goldstein, E.B., Sensation and Perception (Wadsworth Publishing Company, Belmont, California) pp. 180–183.

Halbertsma, K., 1949, A History of the Theory of Colour (Amsterdam).

Hall, M., 1992, Color and Meaning: Practice and Theory in Renaissance Painting (Cambridge Univ. Press).

Harris, M., 1963, The Natural System of Colours (London, 1766; reprint by the Whitney Library of Design, New York).

Hills, P., 1987, The Light of Early Italian Painting (Yale Univ. Press).

Homer, W.I., 1964, Seurat and the Science of Painting (MIT Press).

Jameson, D., and L. Hurvich, 1975, From contrast to assimilation: In art and in the eye. Leonardo 8, Spring 1975, pp. 125–131. Reprinted in: M. Henle (ed.), Vision and Artifact (Springer, New York, 1976).

Katz, D., 1935, The World of Colour (Kegan Paul, Trench, Trubner & Co., London). This is the English translation of the second German edition "Der Aufbau der Farbwelt", published in 1930 and reprinted by Johnson Reprint Corp., London and New York, 1970. The first German edition published in 1911 was titled "Die Erscheinungsweisen der Farben und ihre Beeinflussung".

Kemp, M., 1990, The Science of Art (Yale Univ. Press).

Kuhn, Th., 1962, The Structure of Scientific Revolutions (University of Chicago Press).

Le Blon, J.C., 1980, Coloritto (London, 1723–26; reprint ed. F. Birren, Van Nostrand Reinhold, New York).

Metelli, F., 1974, Scientific American, April.

Moles, A., 1966, Information Theory and Esthetic Perception, transl. J.E. Cohen (University of Illinois Press).

Munsell, A., 1916, A Color Notation and Color Atlas (Geo. H. Ellis, Boston).

Ostwald, W., 1931, Color Science (2 vol.), transl. J. Scott Taylor (Winsor and Newton, London; orig. German edition 1919).

Panofsky, E., 1991, Perspective as Symbolic Form (Zone Books, New York; original German edition published in 1927).

Parkhurst, Ch., and R. Feller, 1982, Who invented the color wheel. Color Research and Application 7(3).

Ratliff, F., 1992, Paul Signac and Color in Neo-Impressionism (Rockefeller Univ. Press).

Richter, J.P. (ed.), 1970, The Notebooks of Leonardo Da Vinci, Vol. 2 (Dover, New York).

Wurmfeld, S., 1985, Color Documents: A Presentational Theory (Hunter College Art Gallery).

Young, Th., 1802, On the theory of light and colours, Royal Philosophical Society Transactions, Vol. 92.

chapter 6

COLOR IN ANTHROPOLOGY AND FOLKLORE

JOHN B. HUTCHINGS

Bedford, England
(Unilever PLC, retired)

Color for Science, Art and Technology
K. Nassau (Editor)

CONTENTS

6.1. Introduction 197

6.2. Earliest use of color 197

6.3. Color in anthropology and folklore 199

    6.3.1. Traditions, beliefs, legends, and customs 201

6.4. Total appearance 204

6.5. Discussion 205

    6.5.1. Usage 205

    6.5.2. Colors and their meanings 206

6.6. Bibliography 207

References 208

6.1. Introduction

For 300 000 years *Homo sapiens* and their predecessors have used color in ceremony and art. In particular, they used the colors black, white and red, a triad familiar to 20th century anthropologists and folklorists. In archaeology, color is a fossil indicator of human behaviour, but for more recent times we can attempt to achieve a greater understanding.

Color and appearance are essential to the well-being of many living organisms, essential to the lowliest of insects, to plants, to animals and to mankind. The ability of the human species to see color has probably been inherited from ancestors who needed to see and select food that was safe and wholesome to eat. In common with that of other animal species, our total appearance, that is our color, color patterning, design and behavioural display, have adapted to physical, geographical, climatological and sexual environments (Hutchings 1994a). However, we act not only as animal organisms, we are creative. Our capacity for belief and power to imagine, coupled with our understanding of materials, have enhanced our ability to create profound effects on our fellows and on the appearance of our environment. The inheritance of color vision has been exploited and color now forms a highly significant part of modern life. Cosmetics, clothing, household pets, food, decoration inside and outside the house, landscape architecture are all deliberately colored and patterned in attempts to satisfy personal tastes and apparent needs for colorful surroundings. Examples of this essential nature of color and appearance are abundant in anthropology and folklore.

The major preoccupations of the living can be classed as *pressures*, *concerns*, and *enjoyment*. Our *pressures* include problems of staying alive, of satisfying needs for food, shelter and clothing, and desires to appear acceptable as a member of our group. Our *concerns*, on the other hand, are longer term and include the desire for good health, a long life, our hope for a continuation of food source, our desire that the sun will return and bring the season of spring with it, of increasing our power. Our *enjoyment* includes celebration which may be concerned with marking stages in life, such as birth or marriage, or celebrating the actual return of spring. In this category also fall artistic and aesthetic aspects of color and design. Color and decoration touch every aspect of our drive to continue life, before and after death.

This chapter contains an outline of the archaeological evidence concerning the earliest uses of color, then considers color within the context of anthropology and folklore.

6.2. Earliest use of color

Archaeological literature contains many references to the finding of ochre at ancient sites. The terms *ochre* or *red ochre* cover a wide range of oranges, browns and reds. They may also include a number of other red-colored materials of mineral origin. What is certain is that in Palaeolithic sites such materials were deliberately collected because of their color. Pieces of red, brown and yellow ochre have been found in shelters used by *Homo erectus* in France perhaps 300 000 years ago. In Czechoslovakia, 50 000 years later, tools used to abrade red ochre were discovered in a shelter in which a wide area of floor was colored with ochre powder. That is, there was early cultural tradition of color use in an adult nonsubsistence context (Marshack 1981). Such collections are fossil indicators of human behaviour, as opposed to tools, which are fossil indicators of human skill and development (Wreschner 1980).

After this early Acheulian period, there is little evidence of the use or collection of ochre until the Middle Mousterian, approximately 70 000 years ago. Neanderthal man created a new dimension in behaviour that became an integral part of culture. The long gap of hundreds of thousands of years might have been caused because early hominids appeared not to have the power of abstraction, of not being able to produce a recognizable picture. Such a conclusion is at best based on problematic and severely limited evidence. What is significant is not the cranial capacity or intellectual level but the evidence of a lack of interest or ability to engage in symbolic or representative activities with color as a criterion in decoration. In Mousterian sites, *Homo sapiens neanderthalensis* used ochre in burials. As in all periods, practice differed among groups. Some used ochre, others deposited funerary gifts of products of the hunt and tools, and others deposited nothing with the body.

Later, within the Upper Palaeolithic period (10–30 000 BP), members of *Homo sapiens sapiens*, whose intelligence was comparable to ours, continued these practices. Marine shells, animal bones and teeth were included in burials and so became part of the symbolism. Ochred human figurines carved from stone have been found in campsites. In Russia, carvings in bone were filled with black paste, while further west highlighting was achieved using ochre. Cave paintings (see color plate 26) using yellows, browns and reds from ochres, black from charcoal and iron and manganese earths, and white marl may be seen, for example, in the French Dordogne and Altamira in Spain. During this period, use of ochre was widespread. Ochre burials have been found in Spain, France, across Europe south of the ice sheet, into the Russian steppes, and further south in the Mediterranean islands, across North Africa to the countries of the Near East, in Southern Africa and North America. In later Mesolithic burials, inclusion of ochre continued, but the use of shells and animal bones declined (Wreschner 1980). The subjects of cave painting changed from the Upper Palaeolithic polychromic bison, horses, deer and bovids to the Mesolithic (10 000 BP) vivacious, action-packed scenes of dancers and the hunt.

In pre-history, sources of these colors are in the main mineral and durable. Discovery of colors derived from biological sources, such as plants, is rare. However, evidence for vegetable reds has been found in Poland. Gromwell (Lithospermum officinale L.) was found near the head of a corpse in an Upper Palaeolithic period tomb, and hawthorn fruits (Crataegus sp.) have been found in several Early Iron Age cremation burials (Malinowski 1980).

Unfortunately, we know nothing of the ideas behind the paintings and their colors. However, color was used in different ways. In certain contexts, such as burials, red was used because it was red, other durable colors presumably seemed unsuitable. Again, practice was not uniform, the whole or part of the body may have been smeared with the pigment, or the ochre may have been scattered over the body and grave goods, or it may have been placed in a bowl beside the body. Ochre may have been considered to be symbolically related to blood and therefore needed to sustain life beyond the grave. Wrenscher quotes the Maori legend of the woman who went to the nether world, found a bowl of ochre, ate the ochre, became strong and was restored to life. However, there are other reasons why a supply of ochre might have been given to the dead (Wreschner 1980). It would be surprising if the use of color had not extended to decoration of the body.

Later discovery of exotic materials, such as lapis lazuli and amber, and a number of colored minerals and pigments, as well as the developing technology of metal extraction provided the raw materials which humans have used to decorate themselves, their homes and their buildings ever since.

6.3. Color in anthropology and folklore

There can be more certainty about the motives for the use of color and decoration in historical and more recent times. Colors, designs and pictures may provide ways of giving pleasure, or be means of communication. In communication color identifies or symbolises. Colors in and on buildings can be used in a practical sense, that is as markers of domestic or religious space. They can also be used in a symbolic sense, for example, red to denote life or danger, and black, death. Body decoration (see color plate 27) may be used, for example, to denote status, age or sex.

There are three types of body decoration. Scarifying and tattooing are permanent and more typical of rigid societies and social structures (Brain 1979; Ebin 1979). Scarifying or cicatrizing is a process by which the flesh is cut and encouraged to scar. A hooked thorn is used to lift up the skin, which is cut with a small blade to produce a protruding scar. The application of substances such as ashes, charcoal or indigo can be used to make the scars more permanent and stand out in color as well as in physical form. Tattooing, in which a pigment is pricked into the dermis with a sharp point, is also permanent. The simplest tattoos involve the use of carbon, but a more colorful display can be achieved using metallic pigments. It is possible to make delicate markings. This has enabled tattooing to be developed into a fine art form in some societies, notably Japan and New Zealand. Painting is impermanent, so offers a great range of design and usage possibilities. Different colors and designs can be used for different occasions.

There are seven motives for decorating the body. First, protection can be gained from high winds, sunburn, glare and flies. Second, some people enjoy decorating themselves and do it as a hobby. Third, a major motive for decoration is celebration of the milestones of life. Fourth, people decorate themselves, or are decorated, because of many beliefs. These include placation of the gods, indications of status, or belief that decoration will help cure an ailment. Fifth, colors are used to create a frightening or awe inspiring presence, say in battle where power is contested by display. Sixth, decoration is used to disguise or impersonate. The seventh class involves decoration aimed at increasing acceptability. This has led to the foundation of the world's huge cosmetic industry (Ebling 1981). The main functions here are to denote fertility and conceal the ravages of time.

TABLE 6.1
The basic anthropological color triad (constructed from data in Brain 1979).

		Black	Red	White
Availability		charcoal, soot	iron oxide, blood, henna, red camwood	white clay
Examples of belief	Thompson Indians	death, bad luck, opposite to red	good life, heat, earth	ghosts, dead people
	Tchikrin S. America	transition, danger, death	health, energy, sensitivity	transcendence over normal world, purity
Psycho-biological Significance		excreta, decay, death or transitional state	blood, mother/child tie, war, hunting	semen, mother's milk, reproduction

Use of color and design varies significantly with culture. The traditional color worn by brides in western Europe and Japan is white, in eastern Europe red and black, in China red, and in African villages any colors providing they are bright.

Anthropologists have identified a basic *color triad* consisting of black, white and red (Turner 1967). Other colors may be used when they are available, but the triad colors tend to be found in all body decoration. It has been suggested that the triad is among the earliest symbols used by man and that the three colors represent the products of the body. These products are, in turn, symbols of important social relationships as indicated in table 6.1.

Basic senses are given to triad colors among the Ndembu of northwestern Zambia. Blackness symbolises badness or bad things, to have misfortune or disease, witchcraft (a person having a black liver can kill someone), death, darkness, or sexual desire. Only one of these linkings, the last, will be out of place to individuals of the first world. However, among the Ndembu there is a direct link. Women with very black skins are most desirable as mistresses. Sooty black bark of certain trees is rubbed on the vulva to increase sexual attractiveness.

Whiteness is life or goodness, making healthy, purity, to be without bad luck or tears, chieftainship or huntsmanship, when people meet together with or bring gifts to ancestral spirits, to bring into the world infants, animals or crops, to wash oneself or to sweep clean, to do no wrong or not to be foolish, and to become mature or an elder, that is when white hair is becoming visible.

Red things are of blood or of red clay. Categories of blood are animal blood (huntsman-ship or meat), mother's blood of parturition, menstrual, the blood of murder, wounding, or circumcision, the blood of witchcraft. Red things have power to act for both good and ill. For example, living creatures have blood or they will die. Wooden figurines have no blood so cannot take part in life's activities, but if they are given blood by sorcerers they can move and kill people. Red (or black) semen is ineffective, good semen is white good blood.

The Ndembu also use red, white and black in a number of binary systems. Examples of the senses of white and black are goodness/badness, life/death, and to make visible/to create darkness, the good blood shed by the hunter/the bad blood of menstruation or murder. Such arrangements tend to be direct opposites. White and red in ritual, however, tend to be complementaries. They may denote life activities such as the continuance and taking of life, male and female, milk and flesh, peace and war. In these cases black remains hidden, perhaps the unknown or potential as opposed to the actual (Turner 1967).

Approaches to body decoration can be very different. Thompson Indians, who inhabited a plateau between the Rocky Mountains and the Thompson River in North America, had formal patterns and colors that were intimately linked with cultural institutions. There were guardian spirits, each of whom had designs, and ritual specialists, who could cure, or perhaps cause, illness. The specialists prescribed the designs to be used for each particular person and/or occasion. Formal decoration was rigidly specified. In wartime the warrior had an individual pattern but was not allowed to wear black unless he had killed an enemy. Red was used to bring good luck to the warrior, black (the opposite to red) to bring bad luck to the enemy. Pattern variations included half the face red, half black, divided along the line of the nose. In another, stripes of black and red covered the chin and jaw bone, the number possibly indicating the number of the enemy he had killed. The feet and lower legs were camouflaged with black. A painted right arm denoted blood while an amulet was painted on the left. A typical warrior pattern was to have alternate ribs painted red and black.

The Thompson warrior's hair was red on black and his necklace and head band were made from the skin of the grizzly bear. Eagle tail feathers were daubed with red and attached to the head band. On his face a raven may have been painted; other symbols used were two black snakes drawn on the left cheek and the sun, also in black, on the right. Other formal designs were related to other aspects of Thompson Indian life. Shamans were ritual experts who protected the village during the winter season, and were themselves painted and prescribed painting for others. Also, public ceremony and dance required the use of specific colors and patterns, each seemingly having a particular symbolism.

Among the Sudanese Nuba, however, body decoration is less formal. Each individual uses design to complement particular features of the body, the aim being to achieve pure beauty of form. Both approaches can apply. For example, the Bangwa of Cameroon sometimes decorate purely to enhance the body, while at other times the form of decoration is dictated solely by the type of occasion being celebrated. Status within the Bangwa is marked by more permanent forms of decoration, such as scarification or cicatrization (Brain 1979).

6.3.1. Traditions, beliefs, legends, and customs

Different cultures have different meanings for colors. This section contains the results of a study carried out in Britain. A search of folklore archives and collections housed in

TABLE 6.2
Frequencies of use of specific colors. A summary of data collected from British people between 1960 and 1990 (Hutchings 1991).

white – 219	black – 176	red – 161
green – 83	blue – 48	silver – 24
yellow – 17	brown – 15	grey – 12
dark – 11	pink – 11	gold – 8
purple – 5	fair – 4	
cream – 1	khaki – 1	orange – 1

TABLE 6.3
Contribution frequency classifications. A summary of data collected from British people between 1960 and 1990 (Hutchings
1991).

Omens and indications	312	Cures and amulets	59
Clothes and dressing up	192	Food	29
Flowers	188	Weather lore	26
Rites of passage	161	Fishing	26
Calendar customs	91	Animals	18

the British Isles, and contributions from members of the British public have produced a total of 547 items concerning Custom and Belief, 110 Folk Narrative, and 51 Language. The archive searches include items noted since 1960, so this search into oral tradition may include memories from the last years of the nineteenth century.

An analysis of the contributions by color name is shown in table 6.2. All general color names have been included as well as the more specific descriptors dark, fair, silver and gold. With the exception of cream, khaki and orange all colors listed play significant roles in folklore. The predominance of the triad colors mirrors its importance in anthropology. Contributions in the Custom and Belief, and Folk Narrative categories can be classified as shown in table 6.3.

Hence, color plays an important part in many areas of folklore. The following principles can be noted.

A widespread use of white metal (silver), black and red for cures was found. It is reported that epilepsy, appendicitis and warts could be alleviated, respectively, by the burial of a black cock under the bed, a soot and water mixture, and a black snail. Red is used for protection, a red coral necklace is worn to protect children from disease, red wool or cloth keeps rheumatism, smallpox or a sore throat at bay. Silver water prepared using a silver ring or coin has healing properties for humans and animals.

Faces are blackened for calendar customs such as Plough Monday, May Day, Halloween and New Year. The reasons given for use in the first three of these ceremonies are to frighten or to be a disguise – the customs may have, in the past, involved damage to property if money was refused, a type of trick or treat. At New Year its use is probably a reinforcement of the general black and white nature of the event. This is because good (or bad) luck is attributed to dark (or fair) haired men in New Year and wedding ceremonies. Before the wedding, particularly in Scotland, black is used to identify and disfigure the bridegroom-to-be. Where the color of human hair is concerned, the trio dark, fair and red substitute for black, white and red. Dark and fair men may be lucky, but a fair-haired man is unlucky when fishing. It may be unlucky to meet a dark-haired person on the way to a wedding, but lucky to have one present at the ceremony. Today there is money in this custom for those chimney sweeps who hire themselves out for attendance at weddings. Possibly this custom arises from the old belief that perfection brings bad luck. The sweep's job is to dirty the bride's beautiful dress. Red headed people are unlucky on land and on water, they are bad-tempered, mad or are bad for milk and butter, and cause weakness in cattle.

A widespread belief still is that white flowers are funeral flowers, and unlucky if brought into the house. In the hospital and at home the presence of red and white flowers in the same vase is to many families an ill omen. This belief is said to arise from the resemblance of the arrangement to blood and bandages. Another reason given is that red and white are Crucifixion colors.

Many beliefs concern black, white and even red animals. Presence of a black cat can bring good or bad luck. On the other hand, sight of a white horse will bring good luck if some form of action is taken, such as saying "White horse, white horse bring me good luck!", or by spitting on your shoe and marking it with a cross. The first of the month should be greeted by saying "White rabbits!". Milk from a red cow has healing properties.

In this section it is tempting to note the many references to salt, soot or coal, and rowan, which has red berries. Perhaps these act as tangible symbols of their respective colors. For example, black food, such as the traditional Scottish black bun and black pudding, appear to reinforce the black or dark-haired-men tradition of Christmas and New Year. Then it is good luck for the coming year that the first person entering the house on New Year's Eve should be a dark-haired man bearing something black such as a lump of coal. A rowan tree in your garden is said to keep witches away.

The principle of curing like with like holds for color. Cures for yellow jaundice may incorporate an infusion of the yellow sap of the barberry, or gin or beer containing saffron. Red amulets help cure ailments such as fevers and rheumatism. Brown paper also features in many cures. Used with goose grease it can be put on the chest for coughs as well as featuring in the relief of lumbago, stomach pains and toothache. However, the physical properties of brown, as opposed to any other color, paper provide the good practical reasons for its suitability for such applications.

Certain activities make their own particular demands of color. Fishermen do not like brown and green. Orange-brown jerseys made from wool dyed with crotal, a rock lichen colorant, are not appreciated. This is because brown and green are earth colors and therefore tend to "attract" the boats.

Green cuts across all folklore category boundaries and is widely regarded with apprehension. Sayings include "Wear green, wear black" and "Marry in green, you'll be ashamed to be seen". Many contributors to the survey never wear anything green, and some will have nothing green in the house. The reasons given include: "Green gowns are worn by witches", and "It is the fairies' (or piskies') color". Fishermen regard green as unlucky. ". . . the Almighty had clothed His world in green, it is presumptuous of us to wear it – a kind of irreverence". In some families in the Aberdeen area of Scotland, an older unmarried daughter wears a green garter at her sister's wedding "out of envy". However, green can have its uses. One correspondent, when a baby, had a green ribbon tied around her wrist by her east European grandmother, so that she should have "second sight". Green also brings good luck to those who wear it on Saint Patrick's Day (Hutchings 1997).

Color symbolisms occurring in narrative appear to be of two types; those giving comfort and those that are more frightening. Again there is an overwhelming predominance of the triad colors. The comforting group consists of associations with bible stories, such as the robin who received his red breast as he tried to comfort Jesus on the Cross, the white blotches of the lily are Mary's milk, and the red spots on the purple orchid are Christ's blood. There are more of these "bible of the folk" color beliefs (Bowman 1991). Various mechanisms, all associated with whiteness or brightness, are used to guide the soul to the light of heaven. These include seagulls, a golden butterfly, candle flames, meteors and white birds.

Frightening stories more frequently contain black as a color than white or red; stories of black dogs are widespread. In Northern Ireland the dog tends to appear after a death, in Wales it normally either guards a traveller from danger or is an omen of death, while in Norfolk Black Shuck tends to be friendly.

6.4. Total appearance

In all this emphasis on color we must not forget that color forms only part of a series of images that can be termed *total appearance* (Hutchings 1994b). There are two classes of appearance image, the impact or gestalt image, and the sensory, emotional and intellectual images. The impact image is the initial recognition and hedonic judgement. The descriptors sensory, emotional and intellectual can be used to prompt questions of the image. Attempting to put ourselves in the shoes of ancient man, when we see in the cave a painting of a bison, do we prostrate ourselves (because it is a holy image), or laugh (because we admire it as a cartoon sketch), or wonder at the dangers that face us outside the cave (because it is used for training young people to survive in the real world), or die of fright (because this is not the sign of our tribe). Such is the background to consider when thinking of how people respond to images and behave.

These images rely on a number of factors. The first involves factors present in our immediate environment, such as geographical, including climate and landscape, social factors, such as whether we are in a crowd or alone, and medical factors, that is how we are feeling at the time of viewing, whether we need to seek for shelter or food.

The second series of factors concerns those inherited and learned responses to specific colors, scenes or events. Factors contributing include our culture, tradition, memory, preference and fashion. The third series concerns our visual capabilities, whether the viewer's color vision is "normal", whether light intensity is sufficient for the object or scene to be interpretable. All these factors are personal and govern the ways in which we as individuals respond to a well defined scene. All these factors integrate to form images in our mind where *all* colors and images exist.

The fourth series concerns the definition of the object or scene. Included are physics and chemistry properties as well as form and temporal or movement properties. With whatever lighting is available, these building blocks are put together to form the object or scene.

In archaeological times only some of this information is available for us to examine; in present day anthropology and folklore times we can attempt to examine them all.

6.5. Discussion

6.5.1. Usage

Certain events are vital to our continued existence, and there are general principles of human behaviour by which we can ensure that these events do or do not occur. These principles held in the historical past, they may be current, they may even have held in the time of the earliest use of color, but we cannot be certain. For example, there are two ways in which we might secure a continuation of food supply. This can be done by using food preservation techniques known to our group, and by performing such rites that will ensure the return of migrating herds or the growth of the year's crops. Similarly we not only seek protection from everyday elements such as the cold and wet (by finding a dry south-facing cave), but also from those longer term elements such as, perhaps, disease or the return of spring (by conducting the appropriate rites). We may also secure continuation of human life by conducting burial rites. During these activities we must look out for signs, which may be good or bad. For example, we may look for the reversal of the movement on the horizon of the rising sun that will herald the fact that spring will once again return. If, on the other hand, we see a bad sign, such as to see a red-haired woman on our way to the fishing boat, then we must do something positive to counteract its effect, such as turn round and go home. Color and decoration play major parts in many of these practices and beliefs.

None of these traditions, whether it was the Neanderthal use of red or the more recent red hair belief, is universal within a large population. They may apply to small regions, to some families, or just individuals. For example, one of the widest spread traditions of the late 20th century world, is that of a bride marrying in white. However, it is doubtful whether within any one small area, *all* brides will be so married. Oral tradition belongs to the family or group.

The importance of red, white and black is widespread. This applies throughout archaeology, anthropology and folklore. Studies of less advanced languages have indicated that color words are added to languages in a specific sequence. The first two words are black and white, the third, red (Berlin and Kay 1979). This sequence is discussed further in Chapter 1. These three colors happen to be those most frequently used worldwide in folktales (Bolton and Crisp 1979), and in "within living memory" folklore in the British Isles. The use of the triad colors in body decoration is not exclusive; other colors are added to patterns of black, white and red when they become available. Wrenschner has observed that biology lies at the root of human color behaviour (Wreschner 1980). This is demonstrated through the root of ritual, which tends to be the most highly symbolic or cultural activity in which humans engage (Bolton 1980). Much of the evidence is provided by color.

6.5.2. Colors and their meanings

Color is used for different purposes. It is used for aesthetic purposes; for communication using color coding; for identification, such as the delineation of ritual areas; and for symbolism. Colors used for making pictures and for decorating stone figures, bone carvings or live bodies may have to "mean" something, although this is not essential. However, colors used for symbolic purposes, such as for amulets and cures, do have to "mean" something. It seems reasonable that "meanings" must be important for us as human beings. It might hardly be worth the trouble otherwise.

A color applied as part of a painting to a wall, an artifact or to a human body is a signal, a communicator of information. To be effective it must be visible, and to be visible it must have a contrast color against which it is easily observed. The larger the color difference the more visible the signal. Such are the principles behind early heraldic and flag designs. The largest contrast available to us is that existing between black and white, between dark and light. A black surface normally reflects very little of the light radiation incident upon it; a white surface reflects a high proportion of the incident radiation. So, the anthropological "meanings" of black and white seem sensible. That is, the wide availability of blackish colorants from fire, and whitish colorants from clays, and their equation with two of the three body fluids may be a happy coincidence. In this case the choice for a third color to be "red for blood" is straightforward.

The choice of the third color for purely aesthetic purposes rests on its suitability as an effective color-contrasting component of the picture, one that will contrast with other colors used and with the background. The latter is not essential as other colors can be used alongside to make the necessary contrast. Color choice therefore lies not only in its availability, but also in the artist's preference and its necessity for the desired picture.

Red is a nominal description of the third triad color. The actual color range includes paler and darker reds, browns, oranges and some yellows. This range occupies a large part of color space, only greenish, bluish and purplish hues are excluded. Indeed, dark tones of these colors can appear black under certain circumstances. It might be argued that it is the very wide availability of "red" from minerals, blood, burned earth, as well as from vegetable sources, that renders its choice automatic. That is, if a third color is required then it has to be red because of its availability.

There is evidence, however, that the choice of red is more than coincidental. Australian aborigines use red for body decoration. They travel hundreds of miles to fetch the red ochre used from the "blood-stained battle-grounds of their ancestors". Similarly, in the Upper Palaeolithic period in Poland, haematite has been found in hunting camps 400 km from the mining site (Malinowski 1980). Hence great effort is involved in the supply of red pigment.

This importance of color to humans occurs throughout the world and throughout historical times. From ancient Babylon to modern Europe, there are very many examples in which color forms an *essential* part of a protection or cure. For example, red forms an *essential* element of protection against the evil eye in Italy, while in Christian Greece the protection *must* be the Islamic blue. Similarly, there are many examples of monuments and buildings of different historical periods and civilizations in which the deliberate use of traditional colors and color patterns are indicated.

Colors can have many "meanings" even within one culture, hence each color must be studied within its context. For example, in the matrilineal cultures of Central Africa the color red may symbolise many things. They include woman, man, in-laws, rainbow, morning, desire, birth, etc., according to the situation and context in which they are used (Jacobson-Widding 1980). This is a learned process. It is the same learned process of which advertisers make good use when they train potential purchasers to associate a particular color or emblem with their product.

Color in amulets and cures act as visible and tangible declarations of belief and psychological reinforcements of purpose. Color and appearance in custom and belief makes distinct and sets apart an individual on an occasion (as a bride at a wedding), or makes distinct the occasion (as Christmas decorations), or site (as a laying-out area). A change in the look of a person or space immediately identifies the leading players and areas to participants in the ceremony or occasion. In some rites color is seen to be vital and every effort must be made to comply with custom. This extends even to the bride from a very poor family who was dressed in old lace curtains. The black faces of the Plough Monday revellers identifies those seeking to frighten a victim into giving money. Soot is an effective way to identify and disfigure the Scottish bridegroom before his wedding, black lead and graphite grease are more difficult to remove. However, in many customs color is irrelevant, it is appearance that is important. For example, providing her workmates identify the bride-to-be by decorating her with ribbons and pieces of colored paper, the colors themselves are not important. The identification of the chief figure becomes the excuse for and is the focus of the party (Hutchings 1991).

Color is a powerful stimulus and motivator that can be used in different ways to control our actions, direct our lives and to make life a joy, or a misery. This appears to have been the case ever since early man learned to hold a piece of ochre.

6.6. Bibliography

1. Berlin, B., and P. Kay, 1979, Basic Color Terms (University of California Press, Berkeley and Los Angeles).

2. Brain, R., 1979, The Decorated Body (Harper and Row, New York).

3. Ebin, V., 1979, The Body Decorated (Thames and Hudson, London).

4. Ebling, F.J.G., 1981, The role of colour in cosmetics, in: Natural Colours for Foods and Other Uses, ed. J.N. Counsell (Applied Science, London).

5. Hutchings, J.B., 1997, Folklore and symbolism of green. Folklore **108**, forthcoming.

6. Hutchings, J.B., 1997/8, The colours in folklore survey – a progress report. Color Research and Application, forthcoming.

7. Hutchings, J.B., and J. Wood, eds, 1991, Colour and Appearance in Folklore (The Folklore Society, London).

8. Hutchings, J.B., M. Akita, N. Yoshida, and G. Twilley, 1996, Colour in Folklore, with Particular Reference to Japan, Britain and Rice (The Folklore Society, London).

9. Turner, V., 1967, The Forest of Symbols (Cornell Univ. Press, Ithica).

10. Wreschner, E.E., 1980, Red ochre and human evolution, a case for discussion. Current Anthropology **21**, 631–644.

References

Berlin, B., and P. Kay, 1979, Basic Color Terms (University of California Press, Berkeley and Los Angeles).

Bolton, R., 1980, A note published with Wrenschner's (1980) paper, pp. 633–635.

Bolton, R., and D. Crisp, 1979, Color terms in folktales – a cross-cultural study. Behavior Science and Research **14**, 231–253.

Bowman, M., 1991, The colour red in "the Bible of the Folk", in: Colour and Appearance in Folklore, eds J.B. Hutchings and J. Wood (The Folklore Society, London) pp. 22–25. (Available from The Folklore Society, University College, London.)

Brain, R., 1979, The Decorated Body (Harper and Row, New York).

Ebin, V., 1979, The Body Decorated (Thames and Hudson, London).

Ebling, F.J.G., 1981, The role of colour in cosmetics, in: Natural Colours for Foods and Other Uses, ed. J.N. Counsell (Applied Science, London).

Hutchings, J.B., 1991, A survey of the use of colour in folklore – a status report, in: Colour and Appearance in Folklore, eds J.B. Hutchings and J. Wood (The Folklore Society, London) pp. 56–60.

Hutchings, J., 1994a, Living Colour – Appearance in Nature, in: Color, ed. Nassau K. (Elsevier, Amsterdam).

Hutchings, J.B., 1994b, Food Colour and Appearance (Blackie Academic and Professional, Glasgow).

Hutchings, J.B., 1997, Folklore and symbolism of green. Folklore **108**, forthcoming.

Hutchings, J.B., and J. Wood, eds, 1991, Colour and Appearance in Folklore (The Folklore Society, London).

Hutchings, J.B., M. Akita, N. Yoshida, and G. Twilley, 1996, Colour in Folklore, with Particular Reference to Japan, Britain and Rice (The Folklore Society, London).

Jacobson-Widding, A., 1980, A note published with Wrenschner's (1980) paper. Current Anthropology **21**, 637.

Malinowski, T., 1980, A note published with Wrenscher's (1980) paper. Current Anthropology **21**, 637–638.

Marshack, A., 1981, On Palaeolithic ochre and the earliest uses of colour and symbol. Current Anthropology **21**, 188–191.

Turner, V., 1967, The Forest of Symbols (Cornell Univ. Press, Ithica).

Wreschner, E.E., 1980, Red ochre and human evolution, a case for discussion. Current Anthropology **21**, 631–644.

chapter 7

THE PHILOSOPHY OF COLOR

CLYDE L. HARDIN

Syracuse, NY, USA
(Syracuse University, retired)

Color for Science, Art and Technology
K. Nassau (Editor)

CONTENTS

7.1. Introduction 211

7.2. The philosophy of color 211

7.3. Bibliography 218

References 219

210

7.1. Introduction

Are colors part of the physical world, or do they depend upon perceiving minds? How can we reconcile the picture of the world presented to us by science with the view of the world that naturally suggests itself to common sense? Are color sensations identical with brain states, or different from them? These questions are at the intersection of science and philosophy. Several possible answers are proposed; all of them face serious difficulties.

7.2. The philosophy of color

A problem that has long preoccupied philosophers is whether color qualities are to be located in the physical world, independent of the consciousness of perceivers, or whether they are mind-dependent phenomena. This is not merely a matter of the terminology that we choose to employ, as in the moldy riddle, "If a tree falls in the forest and nobody is around, does it make a sound?" That question is readily answered once the questioner is obliged to specify whether 'sound' is taken to mean "vibration of the air" or "auditory sensation."

What intrigues philosophers is quite different. It may be expressed by a pair of related questions: First, how can we reconcile the picture of the world presented to us by science with the view of the world that naturally suggests itself to common sense? Second, are color sensations (and other sensations and feelings) identical with brain states, or different from them?. The first of these issues was addressed by Galileo in his essay of 1623, *The Assayer*:

> "Whenever I conceive any material or corporeal substance, I immediately feel the need to think of it as bounded, and as having this or that shape; as being large or small in relation to other things, and in some specific place at any given time; as being in motion or at rest; as touching or not touching some other body; and as being one in number, or few, or many. From these conditions I cannot separate such a substance by any stretch of my imagination. But that it must be white or red, bitter or sweet, noisy or silent, and of sweet or foul odor, my mind does not feel compelled to bring in as necessary accompaniments. Without the senses as our guides, reason or imagination unaided would probably never arrive at qualities like these. Hence I think that tastes, odors, colors, and so on are no more than mere names so far as the object in which we place them is concerned, and that they reside only in the consciousness. Hence if the living creature were removed, all these qualities would be wiped away and annihilated."

Galileo's sentiments had been voiced some 2000 years earlier by the atomist Democritus, and repudiated by Aristotle, who preferred to locate colors outside of living creatures, in the interaction of light and matter. Medieval as well as ancient thinkers had followed Aristotle in this. For them, the qualities that objects are seen to have are by and large the qualities that they do have; they held that the business of our senses is to reveal Nature's finery rather than to clothe things with raiment of our own devising. But Galileo's words heralded the birth of a new view of the Book of Nature: it is, he was fond of saying, written in the language of mathematics. The objects of mathematics are *quantitative*: number, structure, and motion. The powerful new mathematical physics seemed not to allow for the *qualitative* features of experience. Colors, sounds, and odors were thus swept from the physical world of matter in motion into the dustbin of the mind, where they joined thinking, purposiveness, and consciousness, which the new physics had likewise dispossessed from the world of matter. This "bifurcation of nature", as philosopher-mathematician Alfred North Whitehead (1925) later dubbed it, served physics well, but left obscure the nature of mind and its relationship to the physical world.

Science and philosophy have both become far more sophisticated than they were in the 17th century, yet the basic picture of a bifurcated nature remains with us. As some of its early critics remarked, this picture offends common sense, for nothing seems so evident as that color is an inherent feature of the world outside our bodies: grass is green, the sky is blue, coal is black. How could it be that shape is an inherent feature of the physical world, whereas color is a construction of the mind? Can we even imagine a shape with no color whatever? Furthermore, as we shall shortly see, there are grave difficulties in understanding in what sense colors could be "in the head". For these reasons, several philosophers in recent years have tried to show how colors could in fact be identified with items in the physical world. If they were successful, color would be detected by the senses rather than being created by them. So let us briefly consider some of the proposed physical candidates for identification with the colors (Hardin 1988).

The first of these is that colors are to be identified with certain wavelengths of light. Thus, an object is said to be yellow if it predominantly reflects light of about 580 nm. Many physicists use the terms 'red light' or 'green light' in such a way as to suggest their acceptance of this point of view. This is commonly traced back to Isaac Newton, who is said to have shown that white light is really made up of "light of all colors". Newton himself was far more cautious, remarking in the *Opticks* that "the rays, to speak properly, are not colored. In them there is nothing else than a certain power and disposition to stir up a sensation of this or that color" (Newton 1704). The problem with the proposed identification of light with wavelength is, of course, that it flies in the face of metamerism (Chapter 1, Sections 1.7.6 and 1.8). Metamerism occurs whenever two stimuli with different spectra look the same because they stimulate the same responses in the cones of the retina. A monochromatic light of 580 nm, for example, can be precisely matched by a mixture of monochromatic lights of 540 nm and 670 nm, neither of which is seen as yellow. Similarly, indefinitely many pairs of monochromatic lights can match a reference white. If a perfectly good white can be "made up of two colors" and in indefinitely many ways, it cannot be true that white light is identical with "light of all colors".

A second proposal acknowledges that colors are not natural physical kinds, but are, instead, heterogeneous classes of spectral reflectances or emissions grouped together by how they affect the human visual system. By this account, an object is red just in case it reflects light that produces a certain ratio of cone responses. This obviously meets the challenge of metamerism, but one might seriously question whether it captures the central features of color as we experience it. We have learned from the physiologist and psychologist Ewald Hering (the opponent theory, Chapter 1, Section 1.7.1) that there is an elementary red that contains no perceptual trace of any other color, whereas a color such as orange, for example, is never elementary, but is always seen as a perceptual mixture of red and yellow. There are four elementary, or *unitary* hues – red, yellow, green, and blue – and innumerable perceptually mixed, or *binary* hues, such as orange, purple, or chartreuse (Hering 1920). However, it makes no sense to say that a spectral reflectance or emittance is elementary rather than mixed. The unitary–binary hue structure, so central to the colors as we know them, has no counterpart in the domain of spectral reflectances or emittances, therefore spectral reflectances and emittances cannot be identical with colors, since they lack an essential property that the colors have. Spectral reflectances and emittances are of course causally essential to our perception of color, but that does not entitle us to say that they *are* colors.

A third proposal makes no attempt to identify colors with any particular physical property or process. Rather, it says that colors are identical with whatever causes color sensations in normal observers under standard conditions. Thus, a bird's feathers would be blue just in case they would look blue to a normal observer under standard conditions. Although this works well enough as a rough-and-ready account of when we are entitled to attach color words to objects, it fails when made to perform more exacting tasks. Normal conditions will vary considerably with the type of object that is observed (compare the conditions for seeing the blue of a rainbow, a star, and a Munsell sample, respectively), and normal observers do not in fact have identical visual responses. As professional color-matchers know only too well, there can never be such a thing as a metameric match that will satisfy all normal observers, and it is an established fact in visual science that the spectral locus of a unique hue is a statistical construct taken over populations of normal observers, populations which show wide variance in their locations of unitary blue and green (also see Section 1.8 of Chapter 1).

A final line of defense by some who wish to locate colors outside the head is to urge that colors are indeed properties of the physical world even though they correspond to nothing in the world that is described by contemporary physics. This simply shows, say they, that contemporary physics is incomplete. This is likely to seem to most readers to be a council of desperation, for although no competent person would claim that today's physics is able to capture all of the phenomena of nature, it seems highly unlikely that the phenomena now left out include the colors. Unlike the situation in the 17th century, the physics and chemistry of color stimuli are well understood. The physical mysteries of color that are left to explore all seem to turn around what goes on once light is absorbed by the photoreceptors of eyes.

Faced with these serious objections to any attempt to locate colors in the physical world outside of nervous systems, our search to find the locus of color must next turn to the processes that transpire within living creatures, just as Galileo had supposed. Unfortunately, our task has now become even more difficult. It is obvious that colors are not, literally speaking, *in* brains as raisins are *in* cookies. The question is what relationship color perceivings have to brain processes. This is in fact the second of the two questions that were raised at the beginning of this essay. Those who hold that subjective sensations of color, or pain, or feelings of envy are reducible without remainder to neural processes and biochemical events are mind–body *materialists*, and those who deny this are mind–body *dualists*. Although psychophysicists and physiologists commonly call themselves materialists and indeed look like staunch materialists when they putter about their laboratories, catch them in the off-hours, get them to use such words as 'subjectivity' and 'consciousness', and many of them will sound like closet dualists. True, few if any of them suppose, as did some dualists of old, that the bits of mind-stuff that compose sensory qualities are fragments of an autonomous domain, with causal powers independent of brain function, but many of them are persuaded that the event of sensing a patch of red in their visual field, although it may be *caused* by brain processes, could not be *identical with* a brain process, or at least that we could never have satisfactory grounds for deciding that it is identical with a brain process.

From the 17th century to the present day, arguments on behalf of dualism have ultimately involved appeals to a small underlying set of intuitions which commonly take the form of little pictures of how the world must be. All of them purport to show that materialist accounts of subjective experience must inevitably leave something out. Let us undertake our examination of the issue between dualism and materialism by looking at three such pictures, along with their associated arguments, and see how well they survive analysis.

We shall call the first picture *Blind Mary*, the second *Leibniz's Mill*, and the third *Chromatic Inversion*. In each case, we are to suppose, for the sake of argument, that neuroscience has attained a utopian state of perfect knowledge, so that there are no scientific issues left to be resolved and only logical and philosophical questions remain. According to the supposition, we not only know how the whole physics and chemistry of the nervous system works, but just how its functional architecture enables the organism to extract and utilize information in the environment. We can predict just how someone will respond to whatever stimulus is thrown at her, no matter how complex that stimulus might be.

In the first instance we are to imagine a physiologist – call her Mary – who is in possession of this prodigious utopian understanding, but unfortunately suffers from having been blind from birth. She knows all about the neural states that sighted people are in when they see red, she knows in the fullest detail how people react to seeing red, she knows all of the things that people associate with seeing red, but does she know what it is to see red? Or, as it is sometimes put, does she know what it would be *like* to see red? It seems that we know something that she does not, and so her knowledge of seeing, however complete it may be from a physiological standpoint, necessarily leaves out something that is central to seeing. Expressed another way, if Mary were to gain her sight, would she come to know something that she did not and could not have previously known? (Jackson 1982).

This argument trades heavily on the idea that having visual experience is being in possession of a certain item of knowledge. But we need not grant this. Knowledge, we might insist, is a conceptual matter, not a sensory one, even though our knowledge is gained through sensory stimulation and must constantly be subjected to sensory test. By hypothesis, Blind Mary does not have the experiences that we have, because she cannot be in the neural states that we can be in. And as a consequence, she cannot imagine, as we can, the look of a patch of red. But conceiving, thinking, and knowing are one sort of thing, imagining quite another. As the famous philosopher-mathematician Reneé Descartes, the founder of mind–body dualism, long ago pointed out (Descartes 1641), we can readily conceive of a polygon of a thousand sides, easily distinguishing it from a polygon conceived as having a thousand and one sides, or from one of 999 sides. But can we *imagine* a thousand-sided polygon, distinguishing it in imagination from one of 999 sides? Furthermore, we can see and remember and imagine things that seem to us to be singular and indescribable. Notice that this doesn't mean that these things are not in principle categorizable or describable, just that we at the particular moment lack the capacity to do so. So we may readily agree that Mary can't be in a sensory state that we can be in, but we need not grant that we know something about the state that she doesn't. Indeed, unless we possess her utopian science, we will know much less about our states than she does. Furthermore, she will likely understand more than we do about what, in the literal sense of the word, seeing red is *like*, that is, what other sensory states the state of red-seeing most nearly resembles.

If you were attracted by the Blind Mary argument in the first place, you will doubtless not be persuaded by this response to it. The response turned upon an unwillingness to use the word 'knowledge' in a broader rather than a narrower sense. You may feel that there is a deep fact of some sort to which we have so far simply refused to give expression.

Perhaps this feeling can be captured more successfully by our second image, taken from the 17th century philosopher–scientist G.W. Leibniz (1973):

> "It must be confessed, moreover, that perception and that which depends on it are inexplicable by mechanical causes, that is, by figures and motions. And, supposing that there were a machine so constructed as to think, feel, and have perception, we could conceive of it as enlarged and yet preserving the same proportions, so that we might enter into it as into a mill, and this granted, we should only find on visiting it, pieces which push one against another, but never anything by which to explain a perception."

It was Leibniz who proposed a universal computing language, and devised an early calculating machine, so it is not difficult for us to transpose his image into suitably modern electronic or electrochemical terms. Some of us may remember reacting to the overzealous claims of the Artificial Intelligence hucksters, especially in the earlier days of computers, by imagining the machines for which they had written their programs allegedly for "thinking" and "seeing", and responding, "That hunk of junk couldn't really think or see, no matter how complicated we made it and its program. All it could ever do is imitate a thinking or seeing organism". Since then, it has become rather less clear that thinking need be a consciously accessible process, and somewhat clearer that whether a program should be regarded as a program for thinking has more to do with the detailed character of the program, its scope and capabilities and the manner of its realization, than with any questions of basic principle. Seeing has seemed a different matter though, because in at least some of its aspects, particularly the perception of color, qualitative character plays a central role; wavelength discrimination is simply not enough. If we were to walk through an enlarged digital computer, electrons whizzing past our ears, we might still ask, "How could this thing experience red?"

But let us be careful here. We might equally well suppose ourselves walking within an enlarged brain, dodging depolarizing neurons and squirts of neurotransmitters, and be equally tempted to ask, "How could *this* thing experience red?" Cells do not *look* to be any more capable of supporting sensory episodes than silicon chips do, but the fact is that networks of cells do just that. So the question is not *whether* they support sensory episodes, but *how*. And there are only three answers that we need consider. The first two are dualist answers. They agree that the processes occurring in neural networks cause sensory events, and that those events are not identical with the neural processes.

According to the first answer, which philosophers have called *interactionism*, mental events can affect other mental events as well as neural events. From the point of view of a visual scientist, this is an odd suggestion indeed, for it requires not only that our utopian neurophysiological pictures be incomplete insofar as it leaves out the domain of subjective phenomena, but that it be *causally* incomplete, so that some neural events downstream from certain perceptual episodes could not be explained by means of *any* previous physical and chemical happenings, but would require that some subjective episodes be invoked. But how would a neuroscientist react if in the course of tracing out a complex biochemical sequence someone told her that a particular missing step would be forever missing because its causal role is filled by a nonphysical subjective event?

This sort of consideration has made many scientists of a dualist persuasion look with favor upon our second answer, which philosophers have called *epiphenomenalism*. According to this view, brain events cause subjective sensory events, but these sensory events are causally idle. The subjective experiences occupy the stage of the mind, but it is the nervous system backstage that pulls the strings. To change metaphors, sensory events are the shadows on the wall, or the foam on the beer. But this position won't do either, for we are quite persuaded that those sensory events play a real role in our mental lives. Colors, for example, delight, excite, annoy, and depress us. They affect our beliefs and incite our behavior.

So we want to have our subjective sensory experiences make a difference, but we do not want them to usurp the hegemony of physical and chemical events in the brain. How can we have it both ways? By abandoning both of our first two dualist alternatives and adopting our third, materialist alternative, that the subjective sensory events are nothing but the neural happenings themselves. But here Leibniz's Mill returns to haunt us. How can a bunch of depolarizing neurons be *identical with* the event of my experiencing red rather than merely occasioning it? Who, on seeing a neural network in action could ever guess that it is generating an experience of red, rather than an experience of green, or a toothache, rather than being nothing more than a bunch of neurons acting up? We shall elaborate this objection more fully in a moment, but for now let us content ourselves with observing that if it has force against materialism, it has force against dualism as well, since nobody who simply saw the neural network could have guessed that it was causing nonphysical subjective experiencings either. So the most that it could show is that the mind–body relationship is ultimately and intractably mysterious. But does it actually show this much?

Let us look at the matter more closely by considering Chromatic Inversion, our third picture of the relationship between sensory states and neural states. Although the possibility of chromatic inversion was raised in the 17th century by John Locke (1689), the version that we shall consider here is by a contemporary philosopher J. Levine (1983), who invites us to imagine that we have before us the utopian accounts of red-seeing and green-seeing:

> "Let's call the physical story for seeing red 'R' and the physical story for seeing green 'G'....
> When we consider the qualitative character of our visual experiences when looking at ripe McIntosh
> apples, as opposed to looking at ripe cucumbers, the difference is not explained by appeal to G and R.
> For R doesn't really explain why I have the one kind of qualitative experience – the kind I have when
> looking at McIntosh apples – and not the other. As evidence for this, note that it seems just as easy to
> imagine G as to imagine R underlying the qualitative experience that is in fact associated with R. The
> reverse, of course, also seems quite imaginable."

Such an appeal to imagination is characteristic of many philosophical arguments. The reason for the appeal is that philosophers use imaginability as a tool for separating out the necessary features of a situation from the the merely contingent ones. For instance, even though it may be false that there is life on Mars, it is nevertheless possible that there should have been life on Mars, whereas it is not only in fact false but necessarily false that on Mars, $2 + 2 = 5$. One way of seeing the difference in status between the propositions that life exists on Mars and that $2 + 2 = 5$ on Mars, is to notice that one can readily imagine the state of affairs in which the first would be true but that a state of affairs in which the second would be true is quite literally unimaginable. In the case at hand, what is being argued is that since one can imagine that someone sees red when G obtains rather than R, seeing red is not a necessary feature of R's occurring. But if it were true that seeing red is nothing but R's occurring, that seeing red is identical with being in brain state R, then to see red is necessarily to be in state R, and to conceive the one is to conceive the other. The contrary would not be imaginable since it would be impossible, and impossible states of affairs are not imaginable.

If the argument that appeals to Chromatic Inversion is correct, there can never be empirical grounds that can justify a claim that subjective sensory processes are identical with neural processes. All that an empirical investigation could establish is that the one is correlated with the other.

But if this argument proves anything it proves too much. Could one not say that heat of a monatomic gas, thermodynamically conceived, is not identical with, but only correlated with the kinetic energy of its constituent molecules? After all, we can imagine that the course of the history of science was different, and that heat turned out to be a very subtle mechanical fluid, the caloric fluid, just as the inventor of thermodynamics, Sadi Carnot, had initially supposed.

Many of us would be strongly inclined to reject such an argument, and urge that heat really is nothing but random kinetic energy. For once we come to understand the properties of matter, and the laws of thermodynamic phenomena, it becomes clear to us that there are many reasons why the two preclude the possibility of caloric mechanisms of heat. Such mechanisms never really were possible, any more than it was ever really possible that trees should speak or witches fly. The absence of adequate concepts of these matters gave rise to an appearance, or, if you will, an *illusion* of their possibility. Once the details are spelled out, accounts of caloric fluid, speaking trees and flying witches collapse from their own incoherence. These things can be imagined, but only as long as they are imagined *schematically*.

In the case at hand, what we have been given is just a schematic account, and it is far from evident that, were the **R** and **G** stories of neural functioning to be spelled out, that we could just as easily imagine red-seeing associated with **G** as with **R**. Needless to say, we don't have in our possession anything that approaches proper utopian **R** and **G** stories of chromatic processing. Visual scientists are in no position now to make assertions about identities in this domain that are comparable to what physicists can say about thermodynamics and statistical mechanics. However, visual science is far enough along now for us to say useful things about the neural basis of color qualities.

For instance, suppose that instead of talking about red and green and **R** stories and **G** stories, we had talked about red and orange on the one hand, and **R** stories and **O** stories on the other. Could a proper neural **O** story – one given in terms of central rather than peripheral processes – have been just as fit to be an account of red seeing as of orange seeing? Most of us would be inclined to deny it. The reason is of course that we demand of a proper **O** account that it have a structure appropriate to a binary process, whereas a proper **R** account must have a structure appropriate to a unitary process. Of course unitary and binary processes do not come abstractly labeled as such, but one must give an account of the hues all at once, and within such an account the relative simplicity and complexity of the processes ought to be salient. There is of course no guarantee that our best neural explanations of orange seeing and red seeing will exhibit this sort of structural differentiation, but should they not, we would be disinclined to think that we had accounted for the phenomenally unique character of the one and the binary character of the other. If the future efforts of visual scientists should have one result, we would have evidence for a claim that the phenomenal event is nothing but the neural process, whereas if they should have another, we would not be able to support such a claim.

The challenge of Chromatic Inversion has not been fully met. Nevertheless, using the example of how the unitary and binary colors might be modeled by appropriately structured neural processes, we can at least gain a glimpse of how future science might meet such a challenge, and in meeting it, might provide solid reasons for accepting a materialist view of how color qualities are realized by brain states. On the other hand, it might not; nature might indeed be bifurcated, or else we might be constitutionally incapable of understanding why it is not. In any event, these are now open questions, and are apt to remain so for quite a long time.

7.3. Bibliography

1. Dennett, D.C., 1991, Consciousness Explained (Little Brown, Boston).
2. Hardin, C.L., 1988, Color for Philosophers: Unweaving the Rainbow (expanded edition) (Hackett Publishing Company, Indianopolis, IN, and Cambridge, MA).
3. Hilbert, D., 1987, Color and Color Perception: A Study in Anthropocentric Realism (Center for the Study of Language and Information, Stanford, CA).
4. Hurvich, L.M., 1981, Color Vision (Sinauer Associates, Sunderland, MA).
5. Thompson, E., 1995, Colour Vision: A Study in Cognitive Science and the Philosophy of Perception (Routledge, London and New York).
6. Westphal, J., 1991, Colour: A Philosophical Introduction (Basil Blackwell, Oxford).

References

Dennett, D.C., 1991, Consciousness Explained (Little Brown, Boston).

Descartes, R., 1641, Meditations on First Philosophy. In: Haldane and Ross (transl. and eds), The Philosophical Works of Descartes, Vol. 1 (Cambridge Univ. Press, Cambridge, MA, 1931).

Galileo, G., 1623, The Assayer. In: S. Drake (transl. and ed.), Discoveries and Opinions of Galileo (Doubleday, New York, 1957) pp. 231–280.

Hardin, C.L., 1988, Color for Philosophers: Unweaving the Rainbow (expanded edition) (Hackett, Indianapolis, IN, and Cambridge, MA) Ch. 2.

Hering, E., 1920, Outlines of a Theory of the Light Sense, transl. L.M. Hurvich and D. Jameson (Harvard Univ. Press, Cambridge, MA, 1964).

Hilbert, D., 1987, Color and Color Perception: A Study in Anthropocentric Realism (Center for the Study of Language and Information, Stanford, CA).

Hurvich, L.M., 1981, Color Vision (Sinauer Associates, Sunderland, MA).

Jackson, F., 1982, Epiphenomenal qualia. Philosophical Quarterly 32, 127–136.

Leibniz, G.W., 1973, Monadology. In: G.H. Parkinson (ed.), Leibniz: Philosophical Writings (Dent and Sons, London) Sec. 17.

Levine, J., 1983, Materialism and qualia: The explanatory gap. Pacific Philosophical Quarterly 64, 354–361.

Locke, J., 1689, Essay Concerning Human Understanding, ed. P.H. Nidditch (Oxford Univ. Press, Oxford, 1975) Bk II, Ch. 32, Sec. XV.

Newton, I., 1704, Opticks (Dover Publications, New York, 1952) prop. 2, th. 2, def.

Thompson, E., 1995, Colour Vision: A Study in Cognitive Science and the Philosophy of Perception (Routledge, London and New York).

Westphal, J., 1991, Colour: A Philosophical Introduction (Basil Blackwell, Oxford).

Whitehead, A.N., 1925, Science and the Modern World (Macmillan, London).

chapter 8

COLOR IN PLANTS, ANIMALS AND MAN

JOHN B. HUTCHINGS

Bedford, England
(Unilever PLC, retired)

Color for Science, Art and Technology
K. Nassau (Editor)

CONTENTS

8.1. Introduction 223

8.2. The greens 223

    8.2.1. Adaptation of organisms to the earth's environment 223

    8.2.2. Why leaves are green 225

    8.2.3. Greenery under stress 229

8.3. Color and appearance of animals and flowering plants 231

    8.3.1. The three driving forces 231

    8.3.2. Coloring mechanisms 232

8.4. Color and appearance in plants 234

8.5. Color and appearance in animals – the principles 235

    8.5.1. Animals that do not want to be seen 235

    8.5.2. Animals in the open 236

    8.5.3. In the sea 237

    8.5.4. Color and appearance in animals – applications of the principles 238

8.6. Color of humans 239

    8.6.1. Melanin skin pigmentation 241

8.7. Bibliography 243

References 243

8.1. Introduction

Biological organisms must live in harmony with their environment. If they do not, they perish. Three driving forces lead to the continuing optimization of color and appearance in biological organisms. An organism may be colored by default, that is, through the reflection of unwanted radiant energy. This is the driving force behind greenness in energy-absorbing systems such as leaves. Color has evolved in many organisms through a combination of the light reflected or transmitted from one organism and the color vision of that or another organism. Examples include animals and flowering plants. Lastly, color can arise from a biochemical substance that has been optimized for a purpose other than coloring; blood is probably an example (Hutchings 1986a, 1986b, 1986c).

These driving forces are discussed in terms of three convenient areas of biological color and appearance. First, the greens of leaves; second, the color and appearance of animals and flowering plants; third, the color of mankind.

8.2. The greens

8.2.1. Adaptation of organisms to the earth's environment

Energy absorption is the basis of greenery color, and pigments are the basis of energy absorption. The color we perceive arises from unwanted energy that is reflected or transmitted. Life on earth, it is said, began in a warm organic soup which had been chemically formed through the action of high-energy ultraviolet radiation passing through a reducing atmosphere. Such an atmosphere might have existed over much of the planet or locally near regions of high volcanic activity. Before life could emerge from the seas, ultraviolet radiation present in sunlight had to be prevented from reaching the earth's surface. It is argued that this condition was achieved when, through the photosynthetic action of primitive bacteria, the atmosphere became oxidizing and ozone formed in its higher layers (Scientific American 1978).

The soup contained all the organic matter necessary to provide energy needed for single-celled organisms to flourish. This is how the present day fermenting bacteria *Clostridia* survive. Later, using single stage photosynthesis, bacteria developed that could manufacture food within their own cell walls by using energy absorbed directly from the sun. Such a biological system, relying on absorption of electromagnetic energy for its survival, must have a means of absorbing energy. That is, it must have an absorption spectrum.

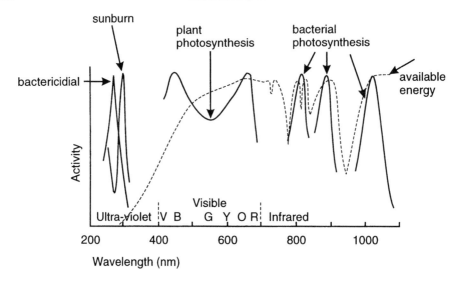

Fig. 8.1. Action spectra of various biological functions compared with the intensity available in terms of photons at each wavelength at the earth's surface.

From the organism's viewpoint it is the action spectrum, the wavelengths that supply energy to drive a particular biological mechanism, that is important. Figure 8.1 gives examples of action spectra (solid lines) of energy absorbing systems against a background of intensity available at the earth's surface (dotted line). The action spectra for photosynthesis to occur in bacteria living in an aqueous environment depend upon optimizing absorption of the infra-red radiation occurring between the water absorption bands. Only this radiation can penetrate through the water to the bacteria (Norbert 1978). Hence, specific organic molecular structures have evolved to respond to specific wavelength bands. Note that short wavelength energy is largely absorbed by the atmosphere, so no survival mechanism has evolved to absorb in this ultraviolet region. The bactericidal action spectrum, fig. 8.1, indicates that bacteria could not survive such conditions.

Photosynthesis results in the production of carbohydrates. Single stage photosynthesis involves an electron pump which increases the energy of electrons obtained from an electron donor, commonly hydrogen sulphide, to such a level as to initiate the reaction. Energy is obtained directly from bacteriochlorophyll pigments and indirectly from antenna pigments, which can be a mixture of bacteriochlorophyll and carotenoids. Both groups of pigments are optimized to suit the specific environment of the particular bacteria (Bjorn 1976). Examples of this photosynthesis can be seen in the volcanic pools of Yellowstone Park. These contain many brilliantly colored species of bacteria.

Later, single cell organisms such as cyanobacteria (blue-greens), extracted from water the hydrogen needed for carbohydrate production. A double-stage photosynthesis based on green chlorophyll, the green-blue biliverdin and antenna pigment systems evolved to provide the higher energy necessary for this reaction to occur. The resulting action spectrum is more complex. Double-stage photosynthesis, necessary for plant life to begin on land, involves a double electron pump consisting of the old pump (photosystem 1, or PS1), responsible for delivery of electrons leading to the fixation of carbon dioxide for conversion to carbohydrate, and PS2, which removes electrons from water for delivery to PS1. Plants redistribute energy between photosystems according to irradiation wavelength and time. This is done by switching the chlorophyll molecules between states PS1 and PS2 (McTavish 1988). These processes enable carbon dioxide to be fixed, and carbohydrate to be produced without help from a soup environment.

Cyanobacteria, such as the blue-green algae species *Spirulina*, provide the red astaxanthin carotenoids that contribute to the red, black and white colors of the lesser flamingo, the firebird, family *Phoenicopteridae*. *Spirulina* die-off turns the African alkali-rich volcanic lakes, such as Lake Natron where these flamingo feed, into a thick green sludge. When this occurs every few years the birds, being filter-feeders, must travel elsewhere to survive. Green pigments in the sea interact with the environment and affect climate. Rapid growth of cyanobacteria in calm water produces greenish toxic scums (Codd et al. 1992). These release dimethyl sulphide, the "bracing" smell of the seaside. This compound provides condensation nuclei and leads to an increase in cloud cover, and thus may contribute to a cooling of the earth (Fell and Liss 1993). Chlorophyll pigments in plankton also help to control water surface temperatures. Monsoon rains flowing into the Arabian Sea result in an upswelling of nutrients to the surface, where there is a marked increase in the growth of phytoplankton. This results in more solar radiation being absorbed and an increase in surface temperature. Thus, biological mechanisms can contribute to seasonal and regional modification of heat exchange interactions between ocean and atmosphere (Sathyendranath et al. 1991).

8.2.2. Why leaves are green

Light is a source of energy (see Section 1.5 of Chapter 1) and a source of information. Leaves are green, like surface seaweed, because absorption (and reflectance, fig. 8.2(a)) is based on the chlorophyll molecule. However, efficient energy absorbers ought to be black. It has been suggested that leaves are not black because independent individual controls (operated by the two peaks of the chlorophyll absorption spectrum, fig. 8.2(b)) might be needed to control the diverse needs of leaf and plant, fig. 8.2(c) (Hutchings 1986a). This usefulness of the independence of different parts of the absorption spectrum has been demonstrated. Switching of chlorophyll molecules between states PS1 and PS2 enables plants to photosynthesize with maximum efficiency under the different lighting conditions prevailing at different times of day (McTavish 1988).

Goldsworthy, however, has proposed that the red and blue peaks of the chlorophyll absorption spectrum are directly linked to the environment of the first chlorophyll-based organisms. The prokaryotic organism, *Halobacterium halobium*, adapted to the high salt content of the primeval soup and was able to survive through primitive photosynthesis by the formation of the purple pigment bacteriorhodopsin. This has a broad absorption peak and utilizes energy from much of the daylight spectrum not absorbed by water, fig. 8.3. Chlorophyll-based photosynthesis probably evolved well after the bacteriorhodopsin-based photosynthesis systems had become established. The more efficient chlorophyll-driven system would, it is argued, stand a greater chance of surviving and flourishing if

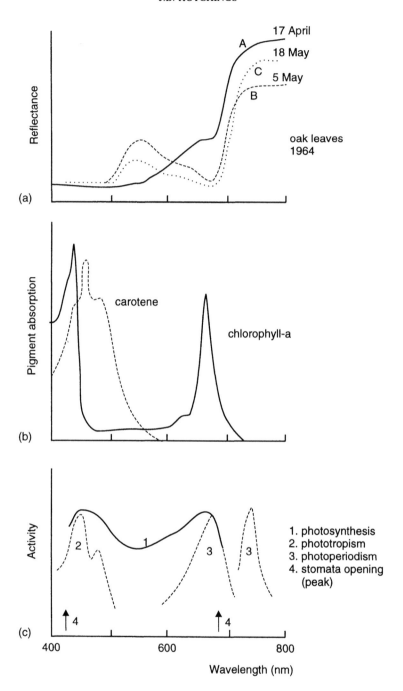

Fig. 8.2. Factors contributing to leaf color: (a) Variations of oakleaf reflectance over one spring month. (b) Absorption spectra of major pigments in leaves. (c) Leaf-function action spectra.

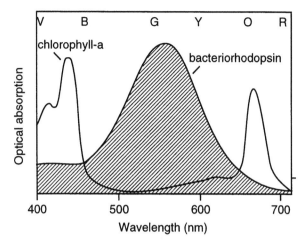

Fig. 8.3. The absorption spectra of chlorophyll-a and membranes containing bacteriorhodopsin.

it absorbed energy not utilized by the *halobacterium*. Hence the green chlorophyll-a pigment evolved preferentially to absorb either side of the bacteriorhodopsin peak, fig. 8.3. The eventual reduction of suitable environments for the halobacterium meant that more light became available in the middle of the spectrum. This, in turn, allowed the development of other chlorophylls and accessory pigments such as carotenoids (Goldsworthy 1987). Absorption at the 670 nm peak raises the chlorophyll molecule from the ground to the singlet state, while absorption at the 440 nm peak raises it to the doublet state. The dangerous triplet state may expose the molecule to oxidation, and carotenoids may be present in leaves to quench it.

These arguments have not remained unchallenged. When a system has opted for a metalloporphyrin structure to perform a function, the system is limited by the chemical properties arising from the high symmetry of its conjugated carbon-skeleton molecular structure (see Section 4.5.1 of Chapter 4 for this and other aspects of colored organic molecules). Quantum mechanics arguments suggest that the design of the chlorophyll molecule is accidental, a compromise between its ability to absorb energy maximally and its ability to act as a lipid-soluble electron carrier (Symmons 1987).

Adaptation of photosynthetic organisms to the environment is also important in the sea. Seawater containing organic materials preferentially absorbs incident daylight at wavelengths at both the blue and red ends of the visible spectrum, fig. 8.4(a) (Bjorn 1976). Surface living seaweed, like other chlorophyll-based systems, is green, and the absorption spectrum has the characteristics typical of such systems, fig. 8.4(b) (Fenchel and Staarup 1971). Deeper living seaweeds tend to be brown, an increase in absorption compensating for loss of incident energy, fig. 8.4(c) (Van Norman et al. 1948). Seaweeds living still deeper, such as the alga *Porphyridium cruentum*, are red, fig. 8.4(d) (Bjorn 1976). They have developed the pigment phycoerythrin, which specifically absorbs radiation at about 550 nm, the wavelength of peak transmission through seawater. Figure 8.4(e) (Bjorn 1976) shows how this has arisen. The action spectrum of one electron pump has retained

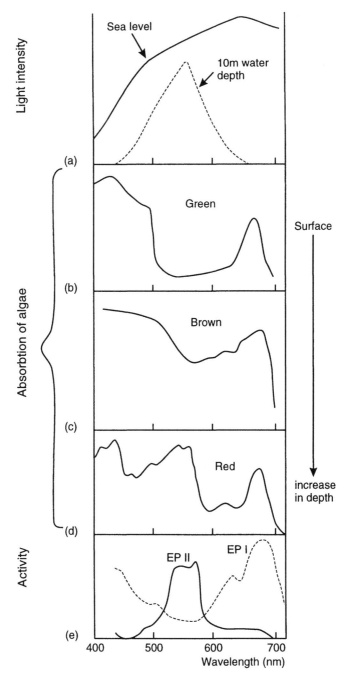

Fig. 8.4. Adaptation of the absorption spectrum of algae to energy availability in the sea. (a) Light intensity available at sea level and at 10 m depth; (b)–(d) Absorption spectra of seaweeds living at increasing depths: (b) green, (c) brown, (d) red; (e) Action spectra of electron pumps (EP) I and II of the red algae.

the green chlorophyll absorption characteristic, while the other pigment produces an action spectrum peak around 550 nm.

Photosynthesis is not the only function vital to the success of a plant. Light, or lack of light, controls the life of a plant from seed to death. Other plant functions controlled by light are phototropism, photoperiodism and stomata opening. The action spectra of these functions are compared in fig. 8.2(c).

Phototropism has its action spectrum maximum at about 430 nm and hence is affected by blue light. It is the mechanism that detects changes in light quantity and encourages plants to grow tall or to manoeuvre their leaves into an appropriate mosaic for light gathering (heliotropism). There are two types of phototropism. Some plants have leaves whose laminae face east-west. This decreases light interception at midday and reduces leaf temperature by 5°C thus limiting water loss. With this orientation, photosynthetic rates are high in the morning and evening when water-transfer stresses are lower. The mechanism which fixed this type of leaf orientation is light induced, as plants grown in the shade have randomly orientated leaves. Another strategy of maximum light interception all day is exhibited by leaves whose laminae are maintained perpendicular to the direction of solar radiation. This orientation appears to be of particular advantage in seedling establishment of annual vegetation constrained to complete its life cycle quickly. These plants capture approximately 38% more energy than plants having fixed horizontal leaves. Solar tracking elevates the temperature, and thus significantly increases the weight of seeds produced (Smith 1984). Specific leaf cells enabling plants to do the tracking have been isolated (Anon 1986).

Photoperiodism has two overlapping action spectra which absorb maximally at 665 nm (red) and 730 nm (far red), fig. 8.2(c). Plant phytochrome pigments sense shade as an increase in the ratio of far red to red light. Light reflected from or transmitted through leaves is rich in far red light. When this is detected by the stem of a neighbouring plant, more sugars are directed into new shoots so that it can compete and outgrow its neighbour. This response occurs well before shading by the neighbouring plant becomes important (Szabo 1993; Ballare et al. 1990). The far red to red ratio also determines the habitat of the plant and controls germination and flowering, which often depend on the relative lengths of day and night. Plants can be induced to flower at the wrong time of year by suitable manipulation of the incident illumination. Stomata regulate the flow of gases and water vapour into and out of the leaf and their action spectrum peaks at approximately 432 nm and 675 nm. The stomata control is also sensitive to levels of carbon dioxide and drought.

In early spring, there is greater absorption at the phototropism maximum, photosynthetic activity is vigorous and leaves are light yellow-green (fig. 8.2(a), curve A). At this time the mosaic-directing function may be more important than the photoperiodism function. Increase in chlorophyll content causes the green color to deepen through later spring (curves B and C) through an increased absorption in the 600–680 nm region. Leaf chloroplasts, which contain the pigment, are more densely located on the upper surface and are darker there (Gates et al. 1965).

8.2.3. Greenery under stress

Over-absorption of infrared wavelengths can lead to overheating, and particular problems arise in desert conditions. The thick fleshy leaves of water-storing desert plants do not transmit, and only have reflectance to control the degree of coupling to incident radiation. Hence, these plants tend to reflect substantially more energy at all wavelengths. A covering of fine thorns offers no advantage over thornless desert plants from the heat transfer standpoint. However, long thorns cast shadows on the cactus surface, and reduce heat absorption without obstructing the flow of air (as well as protecting the plant against grazing). Some desert plants exude a white secretion, develop a crust of wax, or grow white hairs. These mechanisms seem to be designed to maximize light reflection and minimize water loss (Simons 1982). At the other end of the temperature scale are the algae/fungal symbiotic endoliths that live in the Antarctic through winter temperatures of $-60°C$. These organisms live inside rocks, are red to absorb very low light levels, and are active for only a few hours each year.

Leaves may contain eight chlorophylls, all of which have similar absorption spectra, and many more carotenoids. The absorption spectra of two of these pigments, chlorophyll-a and carotene, are shown in fig. 8.2(b). Together these pigments normally yield a green material. In the autumn, chlorophyll destabilizes, the protein to which it is attached breaks down, and sugars and nitrogen-rich compounds are transported to seeds and underground tubers for storage. When greens fade, carotenoids are revealed and the leaf becomes yellow-brown. Sometimes in dying leaves, carotenoids combine with anthocyanins to yield red and purple colors (particularly in North American trees – see color plate 19), but it is not known why these are formed in autumn (Thimann 1950). The inherent instability of the 1.2 billion tons of chlorophylls destroyed annually is important ecologically. If a significant quantity were to find its way into the oceans before breakdown, the seas would become green, energy absorbed would increase, and the seas would heat up (Hendry 1988).

There is probably no single general reason why some plants are evergreen. In the Galapogos evergreens are found more frequently in areas of prolonged drought. These are capable of taking advantage of short periods of rain or warmth during otherwise hostile climatic conditions. In Mediterranean and oceanic areas, plants that retain their leaves can continue photosynthesis through the winter. Further north such plants are able to commence photosynthesis much earlier in the spring. Other plants use their leaves as overwinter reservoirs for minerals and nutrients (Moore 1984). Decline, in the form of needle bleaching and loss of photosynthetic capacity, of trees in European evergreen forests has been blamed directly on air pollution. However, it seems that this decolorization may result from nutrient deficiencies through changes in soil environment and/or root damage (Lange et al. 1987).

Leaves at the top of a forest canopy tend to be a lighter color than those lower down. This helps to control leaf temperature. Approximately two percent of light falling on the canopy reaches the plants that inhabit the forest floor. Leaves in an herbaceous layer of a structured community tend to be larger, and when the canopy layer is denser there is an increase of leaf area to weight (Twidell 1988). Under these conditions rainfall and leaf size tend to be positively related (Moore 1984). Some plants living in low light conditions possess absorption reinforcement mechanisms and are an iridescent blue or blue-green color. The iridescence is caused by interference structures that decrease reflection at longer wavelengths in the visible spectrum. Hence these leaves more efficiently utilize available energy (Lee and Lowry 1975).

Much of this low energy reaching the forest floor is in the form of flecks of unfiltered sunlight. These may illuminate the plant for a period of seconds or perhaps up to 30 minutes. Rainforest plants have adapted to this by shortening the photosynthesis initiation response time of their leaves to illumination. Also, when the light intensity falls, the photosynthesis rate declines more slowly than in plants that demand higher light intensities. When such rainforest understorey species are exposed to greater light intensities, their leaves are bleached and the plant dies (Brown and Press 1992).

Plants that live close to the forest floor sometimes possess variegated or patterned leaves. Total chlorophyll content of the light areas of such leaves may only be 2% that of the green areas. Hence photosynthesis is less efficient than leaves that are wholly green (Araus et al. 1986). However, variegation destroys the plant outline and makes leaves more difficult to be seen by vertebrate herbivores (Timson 1990).

On land, growth areas of seedlings are situated below ground in the dark, whereas light needed to fuel growth is available only to the tips of seedlings. Light transfer to the underground stem can be achieved by fibre-optic light-pipes (Mandoli and Briggs 1984). In the sea, the generally accepted level of light necessary to cause growth is one percent of the maximum intensity at the surface. However, *Johnson-sea-linkia profunda*, an aquatic photosynthetic fleshy alga, has been found growing at a depth of 268 metres where the light level is 0.0005% of the value at the surface (Littler et al. 1985).

The need to absorb light energy in order to survive is the basis of a great deal of the color in nature. Adaptation to different environments leads to much subtlety. The total interaction of organisms with their environment involves a complex consideration of energy flow, organism temperature, diffusion theory, chemical rate processes and molecular biology (Gates 1963).

8.3. Color and appearance of animals and flowering plants

8.3.1. The three driving forces

Examples of all three coloration driving forces occur in plants and animals. Sunlight is essential to many cold-blooded animals. To survive they must develop a balance between energies received, absorbed, and reflected. Absorption-basking butterflies often have melanin-pigmented darker markings on wing areas near to the body. From here, heat can be transferred with minimum loss to where it is needed. Reflection-basking butterflies, which bask with their wings set at a "V" shape, are of lighter colors. These reflect solar radiation from wing to wing until it reaches the insect's body (Anon 1985). Warm blooded animals can use heat absorption to restore internal temperatures to thermoneutrality, but sunbathing at higher temperatures leads to heat stress (Horsfall 1984). Desert animals counteract the extreme heat of day in various ways. They shelter where possible, and undergo daily color or form change. The latter includes exposing skin or subsurface insulation layers, which have different reflection characteristics, at different times of the day (Hamilton III 1973).

A major explosion of color on earth came with the coevolution of flowering plants and insect flight, the second driving force. Insect, animal and fish coloration cannot be discussed without consideration of vision, either of their own species or that of a predator.

Human trichromatic color vision probably coevolved with the color of fruit (Mollon 1989). On the other hand, if it is not required for survival, the mechanism for color vision can diminish. For example, there is a relatively high incidence of color defective vision among industrialized Caucasians, and greater incidence of normal, trichromatic, color vision among peoples whose survival depends on direct working with the land (Cruz-Coke 1970). In a shorter timeframe, photopigment characteristics of fish (Muntz 1975) and man (Hill and Stevenson 1976) can change in response to changes in spectral absorption of the surroundings.

Light detection mechanisms are useful for their ability to form images, and as a means of initiating a meaningful response (such as a color change) to incident radiant energy. Animal colors change according to background, to season, to clearly identify male and female, or, like the cichlid fish, to frighten other males in battle. Both males darken during fighting, the loser signalling his defeat by becoming lighter again. The color of butterfly pupa can be changed by altering the wave-length of incident light. Yellow light, associated with a leafy environment, results in green pupa while blue wavelengths, associated with a non-leafy environment, result in brown pupa (Smith 1980). Detection of the light which initiates these changes might be made by the eye, or by dermal light sensors (Nelson 1981).

Porphyrins form an example of the third driving force, that of coloration by coincidence. Nature has adapted one structural form to be the basis of respiration of vegetation, animals and humans. Included are the green chlorophyll in leaves, brown pinnaglobin in bivalves, blue haemocyanin in some molluscs and red haemoglobin in humans. These pigments occur within the organism and are normally not a major factor in outward appearance. Indeed, from a survival viewpoint, they would be unsuitable. Carrots produce red carotenoids for apparently neither driving forces one or two. Corals and sea anemones tend to be highly colored and some fluoresce, but the biological function remains obscure.

8.3.2. Coloring mechanisms

Colors in nature are caused by organic pigments and/or physical structures (see color plates 16, 18 and 19, and Sections 4.5.1 and 4.7.2 to 4.7.4 of Chapter 4). Table 8.1 contains a list of the seven major pigment groups, their function or suspected function in plants, and an indication of their value to herbivores and carnivores further along the food chain.

Animals cannot synthesize vitamins A, B_2, K_1, and K_2, so must obtain them from plants, hence the value of the pigments in biological terms is indisputable. Animals can synthesize other pigments; an example is sclerotin which is used for photoprotection.

Melanins in animals are used for concealment, communication or protection from radiation (Kettlewell 1973; Prota and Thomson 1976). Mammalian coloring is determined mainly by two related melanins. Eumelanins are black or brown, phaeomelanins range from yellow to reddish brown. Some animals are able to change color according to their background by expanding or contracting melanophores in the skin. Melanophores are connective-tissue cells containing melanin that allow the underlying skin color to come through to different extents. Animals such as the chameleon can change color quickly. This animal is a variegated green when the melanophores are contracted, but when caught out in the open, these expand, and the animal turns brown.

TABLE 8.1

Major organic plant pigment groups, and their value to herbivores and carnivores (adapted from Hutchings (1986b)).

Pigment group	Function in plant	Value to herbivores and carnivores
Betalains	Pollination? Virus protection?	No physiological function known
Carotenoids	Photosynthesis Pollination	Vital, precursors of vitamin A which cannot be synthesized
Chlorophylls/ Porphyrins	Photosynthesis	Not essential? Haemoglobin can be synthesized
Flavins	Quench molecular oxygen	Vital, vitamin B_2 cannot be synthesized
Flavonoids	Growth control pollination	No physiological function known
Indoles	Unknown	Melanins can be synthesized
Quinones	Respiratory enzymes	Vital, includes vitamins K_1, K_2 which cannot be synthesized

Many animal species exhibit seasonal changes in color. Some types of deer go through light-controlled seasonal moulting and color-change cycles. An extreme example is the Northern Hemisphere weasel, *Mustela ermina*. In winter it is called an ermine and is mainly white, but in summer it has a brown coat and is called a stoat. Others, such as cattle and mink, have seasonal moulting cycles without accompanying dramatic color changes. The earless lizard exhibits long-term adaptation. In the southern USA it is grey after adapting to the light-colored gypsum dunes, elsewhere it is variegated.

The octopus achieves a lightness match to its environment by regulating a series of broad-band reflectors. These reflect incident light whatever its wavelength (Messenger 1979). Cephalopods such as cuttlefish can produce controlled regular patterns, waves and moving clouds of colors. They may do this to communicate with other individuals, perhaps to mesmerise prey or to confuse predators (Wells 1983).

Albinism is caused by genetic disorders of the melanin pigmentary system. Animals, such as the white mouse, have pink eyes and extremities. This color is derived from the red of blood vessels and blue from light scattering. Albinism should not be confused with leukemism or whiteness; for example, the white weasel has dark eyes and extremities.

The plants' use of chlorophyll has forced some insects and other animals to become green. Methods include modifying ingested pigment to color the blood, allowing the green color of ingested food to show through a transparent body wall, and synthesizing green pigments that are unrelated to chlorophyll (Bjorn 1976).

Colors can also be formed physically by a number of mechanisms (Hendry 1988). Iridescent interference colors (see Section 4.7.3 of Chapter 4) are caused by regular structural arrays, such as occur in the Morpho butterfly (Wright 1969) (see color plate 18) and hummingbirds (Greenewalt et al. 1960). The Tyndall effect is caused by light-scattering fine particles situated over a light-absorbing melanin layer. If the particles are smaller than the wavelength of light Rayleigh scattering occurs and the light scattered looks blue. This scattered light can combine with pigment colors. For example, the color of some green birds and green irises in humans arise from Tyndall scattering combined with a yellow pigment. Light scattered from larger particles is white. This scattering, caused by the separation of finely divided materials by air spaces, leads to white in feathers and some blossoms (Simon 1971). The various types of scattering are described in Section 4.7.2 of Chapter 4. Diffraction describes the spreading of light at the edge of an obstacle (see Section 4.7.4 of Chapter 4). A diffraction grating consists of a regular two or three dimensional array of scattering obstacles or openings. These arrays occur in the beetle *Serica sericae* and the indigo or gopher snake *Drymarchon corais couperii*. Both exhibit spectral colors in direct sunlight (Nassau 1983).

8.4. Color and appearance in plants

Plants such as hazel, the pollen of which is dispersed by the wind, are usually dull in color. Insects and birds act as very efficient transport systems for other flowers. Flower petals, which are modified leaves, surround egg, pollen, and sometimes, as an extra bribe for the insect, nectar. Obviously the visual characteristics of the pollinator and the reflectance properties of the flowers must be compatible. For example, many flowers efficiently reflect ultraviolet wavelengths. Some of these are pollinated by honey bee workers which have four sets of visual cells. One of these has a peak sensitivity in the ultraviolet at about 340 nm (Bjorn 1976). On the other hand, birds tend to have good red but poor blue vision, hence they are attracted to red flowers. However, not all red flowers are pollinated by birds; for example the European poppy is pollinated by insects. This flower also reflects ultraviolet light and it is this that attracts the bees, not the redness.

Color pattern and flower shape also have vital roles in pointing the way to the pollen. The strategy of using "nectar guides" is fulfilled in various ways through the use of different colors (including ultraviolet), color intensities, or different shapes of petal as in the magnolia.

Flowers possessing reduced color contrasts tend to be rarer. For example, the small perennial larkspur, *Delphinium nelsonii*, which is native to the mountains of western USA, exists in blue and in the much rarer white forms. The latter produce fewer seeds because they are partially discriminated against by the bumblebee and hummingbird pollinators. This is because the white form has inferior ultraviolet-reflecting nectar guides, which slows down nectar collection. Hence the natural selection of albinos results neither from their rarity nor from their overall color but from the reduced color contrast within the flower (Waser and Price 1983).

Color and appearance mimicry are used to attract pollinators. The spider orchid has the colors, form and scent of a female bee; the male copulates with the flower and pollen is carried away. Several species of orchid use neither floral mimicry nor olfactory deception to attract pollinating insects. These also have no nectar and are not self-pollinating. Their strategy appears simply to be among other species that when in flower attract bees (Anon 1983).

Flower petals of the scarlet gilia become lighter in color during their season. Darker colors attract hummingbirds, and, when these emigrate from the system, pollination is continued by hawkmoth (Paige and Whitham 1985). Many flowers change color after pollination. An example is the petunia which changes from pink to blue-purple. The evening primrose flower signals that its nectar has been withdrawn by changing color from yellow to orange-red (Anon 1987b).

The food color/vision argument can probably be extended to humans. Yellow and orange tropical fruits may have co-evolved with the trichromatic color vision of Old World monkeys. Fruit of yellow to orange colors is consistently taken, and forms a substantial part of the diet. The monkey needs trichromatic color vision to find the fruit amongst the foliage (Mollon 1989). In contrast, fruits predominantly taken by birds are red and purple, while fruits taken by ruminants, squirrels and rodents are dull colored, green or brown. Each animal species has evolved its own retinal cone pigments, that enable it to search for food efficiently. As well as providing food for the animals the plant is also helped. Fruit is usually carried away to be eaten, and the seeds are dispersed where they are spat out or excreted.

A similar argument can be used to reason why blue fruit is so rare. Just as the Old World monkeys have mechanisms to enable them to isolate yellow and orange fruits against a contrasting background of green, they also must isolate them against a background of blue sky. Blue fruit would be difficult to pick out against such a background, hence possible driving forces for a mature blue or green fruit are reduced (Hutchings 1994).

Cases in which pollination helps both plant and insect are forms of cooperative coevolution. However, there is also antagonistic coevolution, such as the interaction between the *Passiflora* vine and the *Heliconius* butterfly. The vines have evolved effective chemical defences against insects, but the *Heliconius* has evolved the ability to circumvent them. The *Passiflora* has in turn evolved leaf patterns which resemble the butterfly eggs. *Heliconius* larvae are voracious predators of the passion flower and only one set of eggs is usually found on any one plant. Hence, the butterflies do not lay on plants that display dummy eggs (Gilbert 1982).

8.5. Color and appearance in animals – the principles

Humans respond to their environment according to images of total appearance. Appearance images are derived from a combination of the color, pattern and behaviour of whatever is out in front (Hutchings 1994). Similar considerations seem relevant to animals. That is, appearance as a whole must be considered. The principle governing the appearance of animals, and the visual characteristics of the possible predator, is simple. The animal does not want to be seen, or it does want to be seen, or it doesn't mind being seen (Cott 1957).

8.5.1. Animals that do not want to be seen

Whether it is for defence or aggression, animals that do not want to be seen have three alternatives: to be camouflaged, in which case the animal blends with its environment like a baby ostrich or a leopard; to be disguised, in which case it looks like an inanimate object, like a mantis which can look like leaves; or to remain hidden (Ward 1979).

The four methods used for camouflage tend to hold for all types of creature. Obliterative or counter shading is that in which the underside of an animal is a lighter color than the upper side. Hence when the belly is in shadow it does not look darker than the back. Disruptive coloration allows the eye to focus on a large patch of color (as on the giraffe) not allowing the eye to concentrate on the shape of the object bearing the patch. An example of the third method is the disruptive marginal pattern of the tiger. (The zebra's stripes are now believed to be a form of display, not of camouflage (Anon 1988).) The fourth method involves the boundary concealment of shadow. Some twig-like creatures have fleshy tubercules or hairs that interrupt the boundary between the insect and the twig, thus eliminating shadow (Cott 1957).

There are two types of mimicry. Batesian mimicry is that in which a harmless and edible species sufficiently resembles a harmful non-edible species to escape the attentions of the predator. Müllerian mimicry occurs when two distasteful or harmful species resemble each other. There can be switches between the two types of mimicry with season; perhaps food becomes more or less abundant. Some mimicry is widespread. *Danaus chrysippus* is a widely distributed butterfly which is large conspicuous and poisonous, and thirty eight other species are known to mimic it (Cregory and Gombrich 1973). Only a few generations are needed for some butterfly species to substantially change their wing pattern to match that of a different species (Anon 1979a).

When concealment is required, patterns tend to be detailed and delicate. Very often there is a link between pattern and the anatomy in adaptive coloration, and this results in the disruption of the surface or obliteration of contour (Cott 1957).

8.5.2. Animals in the open

Animals that stay out in the open have no enemies, or rely on large numbers, or appear unpalatable. Members of the last group may taste nasty or look as if they taste nasty. Other members are Batesian mimics; harmless creatures, like the European hoverfly, which copies the warning colors of the well-defended wasp. Some creatures use combinations of defence strategy. There can be conflict in intent, such as between remaining undetected by predators but being conspicuous enough to find a mate. An example is the female crab spider, which can change body color to either yellow or white depending on the flowers beneath which she lurks waiting for prey. Her body carries two bright red eye spots that act as warning colors to frighten bird predators. Hence, this creature can remain both concealed (its prey cannot see red) and conspicuous (bird predators can see red) at the same time (Lythgoe 1979).

When visibility from a distance is required, either to act, say, as a warning threat in poisonous frogs or recognition in butterflies, then coarse, bold, and simple patterns have been developed. Avoidance of highly-colored unpalatable food can be a learned response. Free-ranging crows, which ate chicken eggs that were painted green and contained a nonlethal toxin, subsequently avoided green eggs at various locations whether or not they contained toxin (Anon 1980). However, avoidance can also be innate. Chicks ate green painted worms but avoided prey painted other colors (Roper 1987).

Brightly colored (aposematic) prey are discriminated against when present at low frequencies, but, when they are relatively common, aposematism is selectively favoured. This tends to be confirmed by the observation that, among British butterflies, all species which live in families of 100 or more are aposematically colored, compared with only a small proportion of the solitary species (Harvey 1983).

Pagel discusses the design of animal signals in terms of *honest* and *conventional* signals. In the former, only the best quality males can afford to handicap themselves by energy-wasteful displays. An example is the peacock, who perhaps sacrifices everything it needs for defence merely for the sake of appearances. *Conventional* signals are those in which the costs of the signal is not part of the message. These include signals in the form of, say, a colored mark on the animal's coat (Pagel 1993).

It is very likely that there are a number of causes, acting singly or in combination, for such showy coloration. Kirkpatrick and Ryan (1991) discuss proposed mechanisms for the evolution of mating preferences under three headings: direct selection of preferences, arising when the preferences affect survival or fecundity of the female; the runaway process of indirect selection of preferences, arising when a character genetically correlated with mating preference becomes exaggerated; and the parasite hypothesis of indirect selection of preferences. The parasite hypothesis links the male's ability to display with his health and vigour. From the male's viewpoint the display may be used: to attract attention and advertise the fact that he can escape attack easily (Anon 1979b); and to use features such as an exaggerated eye stimulus and the concentric circle pattern, say of the golden pheasant, to act on the female through a form of hypnosis (Anon 1981a).

Incidental to the efficient light utilization of nocturnal animals is the bright green reflection from interference seen from their eyes at night (see color plate 16 and Section 4.7.3 of Chapter 4).

8.5.3. In the sea

The above principles hold for aquatic as well as land creatures, but there are others that concern the distribution and availability of photosynthetic pigments in the sea. Chlorophyll-rich phytoplankton live near the surface of the sea. At night, transparent marine zooplankton become green after grazing on them. After feeding, the zooplankton must swim down a considerable depths to prevent the chlorophyll becoming energized. This would release lethal oxygen radicals (Hendry 1988).

There are high concentrations of carotenoids in small oceanic animals living near the surface, where they are produced, and at depths of 500 to 700 metres (Herring 1972). Whereas the carotenoids at the surface tend to be blue and inconspicuous, those at greater depth tend to be red, like those coloring the beautiful bright red shrimp. No daylight can penetrate to this depth. However, some fish, like the pilot light fish which live here, are bioluminescent. The majority of these fish emit their peak radiation at about 450 nm to 510 nm, that is, around the wavelength of minimum reflectance of the shrimp. Brittlestars which live at great depths (400–5000 metres) vary in color according to the season. These creatures feed on detritus (rotting algae) that falls to the sea bed in the summer months. Hence these animals are colored incidentally (Anon 1982b).

8.5.4. Color and appearance in animals – applications of the principles

Species identification is important, especially within crowded populations. Near warm reefs, food is rich, life is abundant and many of the brilliantly colored fish that live there have both rods and cones (Bruce 1975). Such fish are brightly colored and patterned to aid identification (Lythgoe and Northmore 1973). Sexual dimorphism contributes to the variety of color. For example, the female blue-head wrasse, *Thalassoma bifasciatum*, is yellow, while the male is royal blue with a dark zebra-striped saddle. The female can change sex and, with it, her color (Warner 1982). Mates of the jewel fish, *Hemichromis bimaculatus*, recognize each other solely on the basis of visual cues, and parents of crested tern chicks (*Sterna bergii*) similarly recognize their eggs and their young (Rowland 1979).

Patterns signal status. The great tit is a common Eurasian bird with yellow and black underparts and a black and white head. Dominant great tits are those with widest black breast stripes. Tits possessing narrower stripes are frightened away (Roper 1986). Brighter colored species of warblers, genus *Phylloscopus*, live in darker habitats. They can also make themselves temporarily more conspicuous by flashing bright color patterns. Experimentally increasing or decreasing this conspicuousness of males within a given habitat leads to an increase or decrease in territory size (Marchetti 1993). The female of the blue-head wrasse selects a male on the basis of the area of his white band and his body length (Warner and Schultz 1992).

When the females of some species of baboon and monkey are receptive they display a patch of bright red skin on their hindquarters. However, color also acts as a reproductive isolating mechanism. For example, eye/head color contrast is a major factor in species recognition of gulls. A male gull shows little interest in females if this contrast is not characteristic of their species (Rowland 1979). Display contributes to the formation of the bond occurring between male and female. The brightly-colored male wire-tailed manakin, a tropical forest bird, repeatedly brushes the female's chin with a twisting movement of his tail filaments prior to copulation (Anon 1981b).

Patterns induce parents to feed nestlings. The red throat of the baby cuckoo excites the step-parent to feed the intruder instead of its own chicks. The red gape of the kittiwake and the contrasting spot on the beak of the gull are exaggerated feeding stimuli directing the chick's attention to where food may be obtained (Tinbergen 1957).

Specific patterns are used in distractive display. For example, eyespots, such as those on butterfly wings, frighten birds (Anon 1981b). The thecla, a tropical American butterfly, has bold lines that converge onto false eyes. When these "eyes" are attacked, the butterfly takes off in the opposite direction (Anon 1982a). Parent jewel fish possess brightly colored medial fins which can be alternately raised and lowered during an alarm. This signal acts as a distractive display and also directly on the young, who approach the parent or sink to the bottom where they can be less easily seen by predators (Rowland 1979).

Some animals attempting to attract a mate use color and pattern in the construction of elaborate environments. The male satin bowerbird, *Ptilonorhynchus violaceus*, decorates his bower with brightly colored objects, giving a preference to dark blue and yellow-green. Some species of stickleback, family *Gasterosteidae*, build nests initially using green and yellow, but change to red during the final stages. Whether this is a specific entrance-marking mechanism, or because red body coloration plays a significant part in courtship, or because red happens to be available is not known (Rowland 1979).

Kortmulder (1972) has attempted to unravel the close links between color, pattern and behaviour in a study of seven species of Asiatic *Barbus* fish. Conclusions involving just two species are used as illustration. The *B. cochonius* is of subdued color, its black and white markings are blurred and its fins transparent, there is sexual dichromatism, and its swimming movements are smooth. The species has group social structure that contains both males and females. The males are not vicious, the behavioural accent being a threat rather than a bite. Kortmulder concludes that, as their reproduction activities occur daily throughout the season, the male does not want to waste time courting so it is a positive advantage for males and females to look different. On the other hand, the *B. tetrazona* is vividly colored, has contrasting black and white markings and its fins are like bright red flags. The males have the same appearance as the females, and their movements are jerky. They are a strongly territorial species and any fish, male or female, that invades the territory is attacked. Because territorial species are loners, their reproductive activities are irregular. Hence the female does not have to look different because the male has more time to distinguish the female, who must make herself known by her behaviour.

Kortmulder has also observed that logical adaptive behaviour would make it desirable that territorial males should advertize. Hence the largest fins of *B. tetrazona* are highly contrasted bright red and black, and jerky movements increase the advertizement. Mobbing behaviour in males and females which gang together to challenge invaders would again make it an advantage if the females have a red flag too. On the other hand, group behaviour species have no need to advertize so positively, and colors shade into each other (Kortmulder 1972). Schooling seems to reduce the risk of being eaten; most fish that school are prey rather than predators (Partridge 1982). For prey, the adaptive value is in reducing the probability of being eaten once the school has been detected. When attacked, the school closes up and the actions of individuals are synchronized to cause greater confusion to the predator. The more black, contrasty, and jerky of movement, the fiercer the species – the more gentle fish cannot afford to be conspicuous. Model experiments have shown that black patterned models have a lower probability of being attacked by males (Kortmulder 1972).

Color, pattern, shape and behaviour generally represent nature's best attempt to adapt a creature to and help it survive in its environment.

8.6. Color of humans

The appearance of human skin is governed by the nature of the surface, by the epidermis, dermis, subcutaneous tissues and pigmentation in the hair shaft and follicle. Keratin lamellae are produced on the skin surface, melanin is produced beneath this horny layer, and yellow carotenoids are deposited deeper in the dermis. Also influencing the outward color of the skin are the number of blood vessels, the extent to which they are filled, the degree of blood oxygenation, and any bile pigments that happen to be in the blood. A bluish color component of the skin is produced by scattering (see Section 4.7.2 of Chapter 4).

Subjective color assessment of the skin is a critical weapon in the medical armoury as skin color is an indicator of our well-being (Roberts 1975). Pallor is caused by a diminished blood flow, by blood which has a low oxygen-carrying capacity, and by fluid beneath the skin layers. Hypochromia, or reduced skin color, is a condition exhibited, for example, by African patients suffering from tuberculosis (Pearson 1982). Local depigmentation is a symptom of conditions such as vitiligo, piebaldism and onchocerciasis, or river blindness. Clinicians can judge the prevalence of the tropical disease onchocerciasis by getting villagers to sit so that the lower part of their legs are lined up. The proportion of those affected with "leopard skin" can then rapidly be estimated (Murdoch 1983).

Redness is caused by too much oxyhaemoglobin, by increased blood flow, or by high temperature as heat is dissipated from the surface capillaries. Jaundice or icterus results from plasma which has been discolored yellow, or sometimes green, by bile pigments. This condition also causes yellowness of the eyes and mucous membranes. Yellowness of the skin is also caused by high serum carotene levels resulting from over-ingestion of, for example, tomatoes, carrots and foods cooked in red palm oil. Yellow-brown or black patches on the skin are caused by exposure to the sun, dyspepsia, or certain specific diseases. Chlorosis, or greensickness, an iron-deficiency anemia most commonly found in pubescent females, results in a green tinge of the skin. This can also be caused by some toxic chemicals.

Cyanosis, a blue appearance, is caused by cold as capillary vessels on the skin surface close to conserve heat, or by the presence in the blood of reduced haemoglobin, which is blue, rather than the red of oxygenated haemoglobin. There is one other cause of blueness. The "blue people" of Troublesome Creek, Virginia, USA, owe their color to lack of the enzyme diaphorase. In normal people, red haemoglobin is converted very slowly to blue methaemoglobin, and the enzyme is needed to reverse the reaction. This condition, inherited as a simple recessive trait, is alleviated by a daily methylene-blue pill, which has the same effect as the enzyme (Trost 1982).

Humans have adapted over a long period of time to local conditions, and they can now be divided into races on the basis of geographical distribution. Although structural features of the body differ, only color differences are considered here. People who live in the tropics are generally very dark, but some equatorial forest dwellers can be paler or yellow. Perhaps these peoples optimized carotene and keratin, as have the yellow Mongols, rather than melanin production as skin coloring. Alternatively they may have adapted to life in the dark forest rather than the brightly-lit plains inhabited by black equatorials. The major pigment is, however, melanin, and the skin color of all other races, American Indian, Australoid, Caucasian and Pacific, can be related to the size, aggregation properties, and speed of response of the melanocytes, the cells which produce the pigment melanin.

Members of paler races attempt to adapt to stronger than normal sunlight conditions by tanning, which occurs in two parts. First, the existing melanocytes manufacture melanin, yielding a temporary tan; second, more melanocytes are produced allowing the manufacture of more melanin to produce a more permanent tan. The fairest skin occurs in Northern Europe, but there is a distinction between Scandinavian fair skin, which is capable of tanning, and Celtic fair skin, which has lost the ability to produce melanin. The latter people burn when exposed to sunlight.

Generally speaking, darkness of hair and eyes goes with darkness of skin. Melanin concentration accounts for the color of hair from blonde to black, although its total appearance is affected by variations in surface structure, reflecting light in different ways, and to its coarseness or fineness. The highest incidence of blonde hair can be found among Scandinavian females aged 15–21 years. The pigment causing red hair, of which the Welsh have the highest incidence, is also thought to be melanin-derived. Greying of hair is a progressive condition accompanying aging and involving the replacement of pigmented by non-pigmented hair. Sudden whitening of the hair through emotional shock is rare. Middle-aged people tend to possess a mixture of pigmented and non-pigmented hair, and sudden whitening appears to involve the selective loss of the former.

Melanin concentration and structural light scattering account for the wide range of eye colors. The iris is colored because it is necessary to limit the amount of light entering the eye, albinos having an intense visual sensitivity to sunlight. Scattering (see Section 4.7.2 of Chapter 4) backed by very dark melanin gives blue irises. Combined with yellow melanin, scattering gives green. Melanin dominates in brown to black irises. In albinos, blue scattering and red from blood vessels gives pink eyes.

In the animal kingdom there is a relationship between eye darkness and self-paced and reactive skills. Creatures such as cats, that have the self-paced hunting techniques of stalking and surprise, are light-eyed. Those that prey in a reactive manner, such as swifts that feed on insects while in flight are dark-eyed. Dark-eyed human subjects have significantly faster reaction times (to visual *and* auditory stimuli) than those who are light-eyed. As yet there is no satisfactory explanation for these findings (Robins 1991).

8.6.1. *Melanin skin pigmentation*

A number of theories have been advanced to account for man's color via melanin pigmentation. These include its usefulness as camouflage, as an aggressive social signal, as resistance to sunburn and cancer, and as optimization of vitamin D biosynthesis.

The development of a natural camouflage would be a useful adaptation and a significant survival factor for an early man engaged in tribal wars and hunting. It would also protect women and children from large predators. However, man could be more effectively camouflaged if he matched his background more precisely, as do many forest animals. Modern man has recognized this by designing an optimum jungle-fighting uniform that is patterned green and brown. Also, as a relatively large animal, man should not be homogeneously colored. The camouflage hypothesis for skin coloration has been rejected (Hamilton III 1973; Robins 1991).

Skin coloration might act as an aggressive social signal. In many populations of man and animals, males tend to be darker and more aggressive. Hence, the hypothesis that the degree of pigmentation indicates the level of threat, and that depigmentation, like hairlessness and smooth, soft skin, is an adaptation to minimize threat. Hamilton has rejected this because selection pressures resulting in differentiation between populations are usually based on environmental rather than social characteristics (Hamilton III 1973).

Naturally dark skin gives protection against ultraviolet radiation absorption which causes skin cancer. It is interesting to contrast Northern America and Northern Europe in this context. In the former, the incidence of cancer increases towards the south. This is an expected response of a mixed immigrant population to the effect of the earth's curvature. In Europe however there is a higher incidence of skin cancer in the north. This is caused by two factors. First, the Mediterranean peoples are protected by having a moderately dark skin, which can tolerate increased exposure. Second, more of the

Northern fair peoples are holidaying south for the sun, and it is among these that the disease is increasing. It does not help if the fair-skinned tan, because they do not have the extra protective mechanisms possessed by naturally dark-skinned peoples (Durie 1980). This protection arises from the many and large melanosomes present in the keratinocyte cells beneath the stratum corneum. The melanosomes appear to be more effective in protecting the keratinocyte nucleus (Robins 1991). The evidence that dark human skin is an adaptation to prevent solar-radiation damage is unconvincing, Hamilton argues. In the tropics the occurrence of sunburn and the development of cancer in unpigmented man cannot be used logically as evidence to support the role of black pigment as radiation protection in indigenous tropical races (Hamilton III 1973).

Some UV absorption within UV-B wavelengths is essential for vitamin D production and the consequent avoidance of rickets. Overuse of barrier creams for absorbing ultra-violet radiation can result in low levels of vitamin D in the bloodstream, particularly of older people (Anon 1987a). Cases of reduced vitamin D production can also be found in dark-skinned Asians who have emigrated to Northern Europe. The theory was that dark skin protected those living in the tropics from over-exposure to UV radiation and vitamin D poisoning. Also, that white skin was an adaptation that could take advantage of lower levels of UV present in higher latitudes. Eskimo are dark-skinned but live further north than most Europeans. They eat large quantities of vitamin D-containing food from fish, seal and whale liver, and therefore no selection for lighter skin pigmentation was necessary (Wagner 1974). More positively, Eskimo adaptation to high UV reflection from ice and snow may have led to their dark skin. However, Negroids can apparently synthesize sufficient vitamin D while living in the Arctic. Also, Caucasoids are incapable of such synthesis during winter months in northern regions, such as Boston and Edmonton, North America. At such times they use up vitamin D stored in the body from the previous summer. There is little archaeological evidence of rickets, which is a relatively modern disease of urbanization and caused by under-exposure to sunlight. Hence, optimization of vitamin D synthesis is not a major driving force for skin color (Robins 1991).

Other driving forces for skin color have been suggested. For example, that depigmentation is a protection against cold injury; that black skin is a positive disadvantage in hot climates because it absorbs heat better than white skin; that Negroid skin was the successful adaptation to tropical conditions and resistance to infection and parasites. However, all these postulates have been shown to be ill-founded. Skin color and latitude (hence sunlight and UV tolerance) are linked. The specific reasons for this are unknown (Robins 1991).

We have adapted not only to a physical, climatological environment but also, as with other living species, to a reproductive sexual environment. The instinct and drive to reproduce is powerful and we have evolved accordingly in our color and design. For example, areas of sexual interest are accentuated by higher concentrations of melanin. When looking at an object, our eyes do not remain on one fixed area but tend to scan over the surface. When we look at a face we seek to communicate and gain information; that is, we look more at the eyes and mouth, those parts which contain higher melanin concentrations. We assist nature in emphasizing particular regions of sexual interest. For example, the larger fleshier lips of the female are further accentuated by the use of lipstick. This routine has been practiced for at least 4000 years. From time to time in different cultures, emphasis has been placed on different features as releasers of sexual interest. An example is the evolution of protruding buttocks among the Hottentot and Bushmen females (Morris 1980).

It is sometimes said that it is possible that an individual's family background can be deduced from their appearance. If this is true, then a person's appearance can be affected by a natural selection mechanism based on some aspect of appearance. Japan has an isolated population which has a strong social class distinction. Skin color has long been regarded there as a criterion for evaluating physical attractiveness, especially in the female. The hypothesis has been tested using reflectance measurements made on the inside of the upper arm, and these have shown that although males are rather darker than females in all areas, reflectances from men are higher in the higher social class than in the middle and lower classes. It is concluded that social selection for light skin color has had some genetic effect and, even in a nominally stable population, there are forces that drive apart groups on the basis of appearance (Hulse 1967).

8.7. Bibliography

1. Attenborough, D., 1980, Life on Earth – A Natural History (Reader's Digest Association, British Broadcasting Corporation, London).

2. Bjorn, L.O., 1976, Light and Life (Hodder and Stoughton, London).

3. Burt Jr., E.J., 1979, The Behavioural Significance of Color (Garland STPM Press, New York).

4. Gates, D.M., 1963, The energy environment in which we live. American Scientist **51**, 327–348.

5. Evolution and Design of Animal Signalling Systems, 1992, Proceedings of a conference, 28–29 October 1992, published in Philosophical Transactions of the Royal Society of London.

6. Nassau, K., 1983, The Physics and Chemistry of Color (Wiley, New York).

7. Roberts, Sh.L., 1975, Skin assessment for color and temperature. American Journal of Nursing **75**, 610–613.

8. Robins, A.H., 1991, Perspectives on Human Pigmentation (Cambridge Univ. Press, Cambridge).

9. Scientific American **239**(3) (1978). The whole issue is devoted to evolution.

References

Anon, 1979a, Two way evolution in butterfly genes. New Scientist **83**(1164), 191.

Anon, 1979b, Birds assume colours to avoid being eaten. New Scientist **84**(1199), 110.

Anon, 1980, Predators learn to avoid bright food. New Scientist **86**(1211), 285.

Anon, 1981a, How the Peacock got his tail. New Scientist **91**(1266), 398–401.

Anon, 1981b, South American birds like to twist. New Scientist **92**(1276), 236.

Anon, 1982a, Falseheads fool butterfly biters. New Scientist **93**(1288), 81.

Anon, 1982b, Bathybius – so Huxley wasn't seeing things. New Scientist **95**(1322), 690.

Anon, 1983, Orchids fool bees. New Scientist **99**(1369), 345.

Anon, 1985, Black ladybirds warm up fast. New Scientist 106(1418), quoting from P. Brakefield and P. Willmer, Heredity 54(1418), 9.

Anon, 1986, Leaves follow the sun. New Scientist 112(1535), 21.

Anon, 1987a, New Scientist 115(1574), p. 24, quoting M. Hollick et al., J. Clinical and Endocrinological Metabolism 64, 1165.

Anon, 1987b, Why flowers blush. New Scientist 115(1579), p. 31, quoting from D. Eisikowitch and Z. Lazar, Botanical J. of the Linnean Soc. 95, 101.

Anon, 1988, Stripes before your eyes. New Scientist 117(1598), p. 40, quoting from Biological J. Linean Soc. 32, 427.

Araus, J. L., J. Sabido and F.J. Aquila, 1986, Structural differences between green and white sectors of variegated Scindapsus aureus leaves. J. Amer. Soc. Hort. Scientist 111, 98–102.

Attenborough, D., 1980, Life on Earth – A Natural History (Reader's Digest Association, British Broadcasting Corporation, London).

Ballare, C., A.L. Scopel and R.A. Sanchez, 1990, Far-red radiation reflected from adjacent leaves: An early signal of competition in plant canopies. Science 247, 329–332.

Bjorn, L.O., 1976, Light and Life (Hodder and Stoughton, London).

Brown, N., and M. Press, 1992, Logging forests the natural way? Scientist 133(1812), 25–29.

Bruce, A.J., 1975, Coral reef shrimps and their colour patterns. Endeavour 34, 23–27.

Burt Jr., E.J., 1979, The Behavioural Significance of Color (Garland STPM Press, New York).

Codd, G.A., C. Edwards, K.A. Beattle, W.M. Barr and G.J. Gunn, 1992, Fatal attraction to cyanobacteria. Nature 359, 110–111.

Cott, H.B., 1957, Adaptive Colouration in Animals (Methuen, London).

Cregory, R.L., and E.H. Gombrich (eds), 1973, Illusion in Nature and Art (Gerald Duckworth, London).

Cruz-Coke, R., 1970, Colour Blindness – an Evolutionary Approach (Charles C. Thomas, Springfield, IL).

Durie, B., 1980, Why bother with a suntan? New Scientist 87, 516–517.

Fell, N., and P. Liss, 1993, Can algae cool the planet? New Scientist 139(1887), 34–38.

Fenchel, T., and B.J. Staarup, 1971, Vertical distribution of photosynthetic pigments and the penetration of light in marine sediments. Oikos 22, 172–182.

Gates, D.M., 1963, The energy environment in which we live. American Scientist 51, 327–348.

Gates, D.M., H.J. Keegan, J.C. Schleter and V.R. Weidner, 1965, Spectral properties of plants. Appl. Optics 4, 11–20.

Gilbert, L.E., 1982, The coevolution of a butterfly and a vine. Scientific American 247(2), 102–107.

Goldsworthy, A., 1987, Why did Nature select green plants? Nature 328, 207–208.

Greenewalt, C.H., W. Brandt and D.D. Friel, 1960, Iridescent colors of humming bird feathers. J. Opt. Soc. Am. 50, 1005–1013.

Hamilton III, W.J., 1973, Life's Color Code (McGraw-Hill, New York).

Harvey, P., 1983, Why some insects look pretty nasty. New Scientist 97(1339), 26–27.

Hendry, G., 1988, Where does all the green go? New Scientist 120(1637), 38–42.

Herring, P.J., 1972, Depth distribution of the carotenoid pigments and lipids of some oceanic animals. Parts 1 and 2. J. Mar. Biol. Ass. U.K. 52, 179–189, and 53 (1973) 539–562.

Hill, A.R., and R.W.W. Stevenson, 1976, Long-term adaptation to ophthalmic tinted lenses, in: Modern Problems in Ophthalmology, Vol. 17 (Karger, Basel and London) pp. 264–272.

Horsfall, J., 1984, Sunbathing – is it for the birds? New Scientist 103(1420), 28–31.

Hulse, F.S., 1967, Selection for skin colour among the Japanese. Am. J. Phys. Anthrop. 27, 143–155.

Hutchings, J.B., 1986a, Colour and Appearance in Nature, Part 1. Color Research and Application 11, 107–111.

Hutchings, J.B., 1986b, Colour and Appearance in Nature, Part 2. Color Research and Application 11, 112–118

Hutchings, J.B., 1986c, Colour and Appearance in Nature, Part 3. Color Research and Application 11, 119–124

Hutchings, J.B., 1994, Food Colour and Appearance (Blackie, London).

Kettlewell, B., 1973, The Evolution of Melanism (Clarendon Press, Oxford).

Kirkpatrick, M., and M.J. Ryan, 1991, The evolution of mating preferences and the paradox of the lek – review article. Nature 350, 33–38.

Kortmulder, K., 1972, A Comparative Study in Colour Patterns and Behaviour in Seven Asiatic Barbus Species (A.J. Brill, Leiden).

Lange, O.L., H. Zellner, J. Gebel, P. Schrammel, B. Köstner and F.-C. Czygan, 1987, Photosynthetic capacity, chloroplast pigments, and mineral content of the previous year's spruce needles with and without the new flush: Analysis of the forest-decline phenomenon of needle bleaching. Oecologia (Berlin) 73, 351–357.

Lee, D.W., and J.B. Lowry, 1975, Physical basis and ecological significance of iridescence in blue plants. Nature **254**, 50–51.

Littler, M.M., D.S. Littler, S.M. Blair, and J.N. Norris, 1985, Deepest known plant life discovered on an uncharted seamount. Science **227**, 58–59.

Lythgoe, J.N., 1979, Ecology of Vision (Clarendon Press, Oxford).

Lythgoe, J.N., and D.P.N. Northmore, 1973, Colours under water, in: Colour 73 (Adam Hilger, London) pp. 77–98.

Mandoli, D.F. and W.R. Briggs, 1984, Fiber optics in plants. Scientific American **251**(2), 80–92.

Marchetti, K., 1993, Dark habitats and bright birds illustrate the role of environment in species divergence. Nature **362**, 149–152.

McTavish, H., 1988, A demonstration of photosynthetic state transitions in nature. Photosynthesis Research **17**, 247.

Messenger, J.B., 1979, The eyes and skin of octopus: Compensating for sensory deficiencies. Endeavour (N.S.) **3**, 92.

Mollon, J.D., 1989, "Tho' she kneel'd in that place where they grew...", the uses and origins of primate colour vision. J. Exp. Biol. **146**, 21–38.

Moore, P.D., 1984, Why be evergreen? Nature **312**, 703.

Morris, D., 1980, Manwatching, a field guide to human behaviour (Triad Panther, St. Albans).

Muntz, W.R.A., 1975, Visual pigments and the environment, in: Vision in Fishes, ed. M.M. Ali (Plenum Press, New York) pp. 565–578.

Murdoch, M., 1983, lecture (Wellcome Research Institute for parasitic Infections).

Nassau, K., 1983, The Physics and Chemistry of Color (Wiley, New York).

Nelson, W.H., 1981, Resonance Raman spectroscopy in the study of carotene-containing biomolecules and microorganisms. International Laboratory **11**(2), 12–20.

Norbert, Ph., 1978, in: The Photosynthetic Bacteria, eds R.K. Clayton and W.R. Sistrom (Plenum Press, New York).

Pagel, M., 1993, The design of animal signals. Nature **361**, 18–20.

Paige, K.N., and T.G. Whitham, 1985, Individual and population shifts in flower color by scarlet gilia, a mechanism for pollinator tracking. Science **227**, 315–317.

Partridge, B.L., 1982, The structure and function of fish schools. Scientific American **246**, 90–99.

Pearson, C.A., 1982, Face colour as a sign of tuberculosis. Color Research and Application **7**, 31–33.

Prota, G., and P.H. Thomson, 1976, Melanin pigmentation in mammals. Endeavour **35**, 32–38.

Roberts, Sh.L., 1975, Skin assessment for color and temperature. Am. J. Nursing **75**, 610–613.

Robins, A.H., 1991, Perspectives on Human Pigmentation (Cambridge Univ. Press, Cambridge).

Roper, T., 1986, Badges of status in avian societies. New Scientist **109**(1494), 38–40.

Roper, T., 1987, All things bright and poisonous. New Scientist **115**(1568), 50–52.

Rowland, W.J., 1979, The use of colour in intraspecific communication, in: The Behavioural Significance of Color, ed. E.J. Burt Jr. (Garland STPM Press, New York) pp. 381–418.

Sathyendranath, Sh., A.D. Gouveia, S.R. Shetye, P. Ravindran and T. Platt, 1991, Biological control of surface temperature in the Arabian Sea. Nature **349**, 54–56.

Scientific American, 1978, **239**(3). The whole issue is devoted to evolution.

Simon, H., 1971, The Splendor of Iridescence (Dodd, Mead, New York).

Simons, P., 1982, The touchy life of nervous plants. New Scientist **93**(1296), 650–652.

Smith, A.G., 1980, Environmental factors influencing pupal colour determination in lepidoptera. Proc. Roy. Soc. London B **207**, 163–186.

Smith, H., 1984, Plants that track the sun. Nature **308**, 774.

Symmons, M., 1987, Why plants are green. Nature **329**, 769–770.

Szabo, M., 1993, Plant sugars move in mysterious ways. New Scientist **140**(1902), 18.

Thimann, K.V., 1950, Autumn colors. Scientific American **83**(4) 40–43.

Timson, J., 1990, How plants with patterned leaves compete. New Scientist **127**(1732), 25.

Tinbergen, N., 1957, Defence by color. Scientific American **97**(4), 48–54.

Trost, C., 1982, The blue people of Troublesome Creek. Science **3**(9), 35–39.

Twidell, J., 1988, Green plants. New Scientist **117**(1594), 76 (letter).

Van Norman, R.W., C.S. French and F.D. Macdowall, 1948, The absorption and fluorescence spectra of two red marine algae. Plant Physiol. **23**, 455–466.

Wagner, R.H., 1974, Environment and Man (Norton, New York).

Ward, P., 1979, Colour for Survival (Orbis Books, London).

Warner, R., 1982, Metamorphosis. Science **3**(12), 43–46.

Warner, R.R., and E.T. Schultz, 1992, Sexual selection and male characteristics in the bluehead wrasse, *Thalassoma bifasciatum.* Evolution **46**, 1421–1442.

Waser, N.M., and M.V. Price, 1983, Pollinator behaviour and natural selection for flower colour in *Delphinium nelsonii.* Nature **302**, 424–433.

Wells, M., 1983, Cephalopods do it differently. New Scientist **100**(1382), 332–338.

Wright, W.D., 1969, The Measurement of Colour (Adam Hilger, London).

chapter 9

THE BIOLOGICAL AND THERAPEUTIC EFFECTS OF LIGHT

GEORGE C. BRAINARD

Department of Neurology
Jefferson Medical College
Philadelphia, PA, USA

Color for Science, Art and Technology
K. Nassau (Editor)
1998 Elsevier Science B.V.

CONTENTS

9.1. Introduction 249

9.2. The science of photobiology 249

9.3. Human responses to light 251

9.4. Regulation of biological rhythms by light 252

9.5. Response of the pineal gland to different wavelengths of light 255

9.6. The effects of color on blood pressure and general arousal 260

9.7. Light treatment of winter depression and other disorders 262

9.8. The effects of different wavelengths in light therapy. 265

9.9. Color therapy (chromotherapy) 267

9.10. Placebo response concerns 268

9.11. Conclusion 269

Acknowledgements 270

9.12. Addendum to Chapter 9 by K. Nassau

   Double blind testing for biological and therapeutic effects of color 270

   9.12.1. Test and control light sources 271

   9.12.2. Survey light sources 274

   9.12.3. In conclusion 276

References. 276

9.1. Introduction

Light can be a potent biological and therapeutic force. Light is fundamental to life on the planet earth. Without radiant energy from the sun, the earth would be a frozen rock in space unable to support sophisticated life processes. It stands to logical reason that specific physical qualities of light, wavelengths or their color appearance, for example, would be intricately involved in life processes. While this seems to be a very basic idea, unanswered questions abound. What physiological responses are most responsive to the spectral quality of light? Which specific wavelengths or colors have the greatest biological impact? How can the colors of light be harnessed to treat human disease and illness? These and other questions have been pondered since antiquity and many theories have been suggested.

Color thoroughly permeates our lives – it is embedded in our culture, our religion, our very language. Virtually everyone has feelings and opinions about color – what it means, what is does for us and what it does to us. Given that mankind has used color in ceremony and art for some 300 000 years as discussed in Chapter 6 by Hutchings (1997), it is obvious that color would be employed for protecting health and curing ailments since the earliest times. It is appropriate to ask: What forms the basis for using color to influence human health and to treat disease: superstition? faith? opinion? logic? clinical observation? empirically proven fact? A comprehensive review of the world's literature on color philosophy, color psychology or color therapy is beyond the scope of this chapter. The principal focus of this chapter will be to examine what has been discovered empirically about the capacity of light and color to regulate human biology and treat disease. In addition, some of the avenues about color therapy which await empirical confirmation will be considered.

9.2. The science of photobiology

On the every-day human level, light and color enrich our surroundings, provide stimulation and information, create atmosphere, and add to the quality of our lives in myriad ways. There is little doubt that we humans respond strongly to light and color in our surroundings. How are such effects achieved? Quite simply, all human responses to light and color can be understood through the basic principles of photobiology (Smith 1989).

The science of photobiology is based on the interaction of light energy with living organisms. More specifically, photobiology involves the study of how the ultraviolet, visible and infrared portions of the electromagnetic spectrum influence biological processes (Smith 1989). As table 9.1 illustrates, light and color can be considered from different

TABLE 9.1

The photobiologically active portion of the electromagnetic spectrum: spectral bandwidths, color appearances, wavelengths and photon energies*.

Spectral band width color appearance	Wavelength range	Photon energy (eV)
Far infrared (IR-C)	1000–3 μm	0.001–0.41
Middle infrared (IR-B)	3–1.4 μm	0.41–0.89
Near infrared (IR-A)	1.4–0.76 μm	0.89–1.63
Visible range	760–380 nm	1.63–3.26
Red	760–610 nm	1.63–2.03
Orange	610–585 nm	2.03–2.12
Yellow	585–575 nm	2.12–2.16
Green	575–490 nm	2.16–2.53
Blue	490–440 nm	2.53–2.82
Violet	440–380 nm	2.82–3.26
Near ultraviolet (UV-A)	400–315 nm	3.10–3.94
Middle ultraviolet (UV-B)	315–280 nm	3.94–4.43
Far ultraviolet (UV-C)	280–100 nm	4.43–12.4

* The infrared, visible range and ultraviolet limits are defined by CIE (1987). Specific color limits are approximate.

viewpoints. Light can be understood on an experiential level of color-appearance, or in more physical terms of photon energies and wavelengths. Photobiological responses are induced by the absorption of light by molecules in living cells or tissues. As photons are absorbed by specific molecules in living organisms, these molecules undergo physical-chemical changes which ultimately lead to larger scale reactions in cells, tissues or the whole organism.

In general, a relatively restricted portion of the electromagnetic spectrum mediates photobiological responses. This does not mean that other portions of the electromagnetic spectrum cannot induce changes in living organisms. For example, wavelengths below 200 nm can destroy living cells by ionizing cellular molecules, while wavelengths above 1.4 μm can influence living tissues by the simple process of heating or burning. Almost all of the species which inhabit the earth have evolved specific photobiological mechanisms which allow them to use ultraviolet, visible or infrared energies from the sun to survive. For example, single-celled organisms and higher order plants capture light energy to produce food. In contrast, fish, birds and mammals do not use sunlight to generate nutrients directly, but instead use light to mediate visual sensation and to regulate daily and seasonal rhythms.

There is a wide diversity of mechanisms by which living systems use visible and near-visible electromagnetic energy for survival. Despite this diversity, it is important to note that almost all living species share a common feature in their ability to respond to light stimuli. All photobiological responses are driven by the singular process of specific molecules within the cell or tissue absorbing photons and then undergoing physical–chemical changes which lead to an overall physiological change in the organism (Smith 1989; Coohill 1991). Such light absorbing molecules are called chromophores or photopigments. This chapter will primarily be concerned with human photobiology and empirical studies that demonstrate human responses to selected wavelengths or "colors" of light. Where appropriate, reference will be made to animal photobiological studies which lay the foundation for human photobiology.

In considering human responses to visible and near-visible wavelengths from the photobiological viewpoint, it is important to understand the definition of an action spectrum. An action spectrum is the relative response of an organism to different wavelengths (Smith 1989; Coohill 1991). The concept of action spectra dates back to the 19th century when plant biologists observed that plant growth depends on the spectrum of light to which the plant is exposed. Over the past four decades, photobiologists have evolved a refined set of acceptable practices for determining action spectra which are applicable in all organisms from simple plants to complex animals such as humans.

In general, an action spectrum is determined by first measuring the light dose required to evoke the same biological response at different wavelengths. The action spectrum is then formed by plotting the reciprocal of the quantity of incident photons required to produce the given effect versus wavelength. The action spectrum for a given biological response usually corresponds to the wavelength absorption spectrum for the chromophore or photopigment that absorbs the radiation responsible for the effect. Generally, it is not an easy or short task to develop high quality action spectra. It is not advisable to attempt formulating action spectra without referring to the specific conditions and guidelines that were developed by the photobiological community (Smith 1989; Coohill 1991). Despite the potential difficulty of developing action spectra, it is a fundamental photobiological technique with high utility. Knowledge of the action spectrum for a given biological response helps identify the physiological mechanism that mediates the effects of light and can guide the development of optimum artificial light sources for that biological effect.

9.3. Human responses to light

As is true for almost all species which inhabit the earth, humans evolved under environmental light provided principally by the sun, moon and stars. In addition, mankind has probably used various forms of firelight since prehistoric times and lamplight for more than 5000 years (Schivelbusch 1988). Artificial illumination powered by electricity which is used pervasively in homes, workplaces and other institutions is relatively recent – slightly more than 100 years. What are the various mechanisms by which light can influence humans? Photobiological responses in humans can be divided into two broad categories: surface effects and internal effects (see table 9.2). Photobiological effects which occur on the body surface are mediated by light being absorbed by the external skin and ocular tissues. In contrast, internal responses to light, for the most part, are mediated by photoreceptors in the retina which are connected to nerves that project to specific regions of the brain. The internal effects of light can be further divided into visual and nonvisual responses.

The breakdown of human photobiology into surface and systemic physiology is useful for organizational purposes but somewhat arbitrary. The human body is a thoroughly integrated system. Events occurring on the body surface can influence internal processes, while changes in internal physiology most certainly affect the status of surface tissues. For example, the photobiological regulation of vitamin D synthesis takes place in the dermal layer of the human skin when it is exposed to ultraviolet radiation and therefore is a "surface" photobiological effect. Specifically, far- and middle-ultraviolet radiation between 250 and 315 nm from sunlight is absorbed by molecules in the epidermis which initiates the formation of vitamin D from 7-dehydrotachysterol (MacLaughlin et al. 1982; Holick 1989). Once active vitamin D is synthesized, however, it acts more like a hormone than a vitamin. It leaves the skin, enters the blood stream, and is responsible for regulating the intestinal absorption of calcium and the normal mineralization of the skeleton. For most children and adults, up to 90% of their daily vitamin D requirements are met by casual exposure of skin to sunlight. The human skeleton can become seriously weakened from vitamin D deficiency (rickets in children and osteomalacia in adults) due to lack of sunlight exposure or insufficient dietary vitamin D. Thus, vitamin D regulation which is driven by a "surface" photobiological event influences widespread internal physiology and overall human health (MacLaughlin et al. 1982; Holick 1989).

It is useful to point out that photobiological effects can include both normal regulatory processes as well as destructive processes. For example, the same wavelengths of ultraviolet radiation which are beneficial in vitamin D regulation or the natural tanning response of the skin can seriously harm the body, including sunburn (erythema), immune suppression, induction of skin cancer and premature aging of the skin when one is over-exposed to sunlight (Jagger 1985; Urbach 1992). Thus, wavelengths or colors of light may be beneficial or essential to human health at one dosage, toxic or destructive at a higher dose. The remainder of this chapter will be concerned with the "internal" photobiological responses. Since there is a thorough discussion of vision in Chapter 3 by Krauskopf (Krauskopf 1997), the focus here will be further narrowed to nonvisual responses pertaining to circadian and neuroendocrine regulation and the related therapeutic use of light. Numerous resources are available to the reader interested in exploring other areas of photobiology and photobiology listed in table 9.2 (Jagger 1985; Wurtman et al. 1985; Häder and Tevini 1987; Smith 1989; Rey 1993).

9.4. Regulation of biological rhythms by light

Biological rhythms have been observed in all species from simple single-celled organisms to highly complex human beings (Aschoff 1981; Binkley 1990). Some biological rhythms have short cycles lasting only seconds or minutes in length, while other rhythms consist of cycles lasting up to a year or longer. In animals and humans, the best studied rhythms are those which are approximately 24 hours in length. These daily rhythms are called circadian rhythms and are controlled by neural mechanisms in the brain. All mammals studied, including humans, exhibit numerous circadian rhythms such as cycles of hormone secretion, body temperature, muscle strength, and blood pressure. The timing of these rhythms involves coordinating or entraining external time cues with an internal pacemaker in the brain.

TABLE 9.2

Major photobiological effects and therapeutic applications of radiant energy in humans.

Effects or applications	Ultraviolet (200–400 nm)	Visible light (380–760 nm)	Infrared (0.76–3 μm)
Surface effects			
Skin	Vitamin D synthesis	Drug photosensitivity	Burns
	Protective tanning	Chemical photosensitivity	
	Immune stimulation		
	Immune suppression		
	Erythema		
	Carcinogenesis		
	Accelerated aging		
	Drug photosensitivity		
	Chemical photosensitivity		
Cornea	Photokeratitis		Burns
	Pterygium		
	Droplet keratophy		
Lens	Cataracts		
	Coloration		
	Sclerosis		Infrared cataracts
Internal effects			
Retina-thalamus	Retinal lesions	Vision	
		Retinal lesions	
		Solar retinitis	
		Macular degeneration (?)	
Retina-brainstem		Ocular movement	
		Ocular reflexes	
Retina-hypothalamus	Circadian regulation	Circadian regulation	
	Neuroendocrine control	Neuroendocrine control	
Phototherapy			
	Psoriasis	Bilirubinemia	Radiant heating
	Herpes simplex	Photodynamic therapy	Low level laser (?):
			Wound healing
			Pain alleviation
	Vitiligo	Low level laser (?):	
		Wound healing	
		Pain alleviation	
	Dentistry	Seasonal depression	
		Non-seasonal depression	
		Sleep disorders	
		Menstrual cycle disturbances	
		Circadian disruptions:	
		Shiftwork	
		Jet lag	
		Syntonic optometry (?)	
		Chromotherapy (?)	
*Photosurgery			
		*Laser surgery:	*Laser surgery:
		Retinal detachment	Retinal detachment
		Glaucoma	Glaucoma
		Tattoo removal	Tattoo removal

* These therapeutic applications of radiant energy are not based on classical photobiological mechanisms since they occur at higher than normal physiological temperatures. Items followed by a question mark indicate uses or effects of light that are controversial or uncertain.

Over the past thirty to forty years, animal studies have confirmed that environmental light is the primary stimulus for regulating circadian rhythms and some seasonal cycles (Aschoff 1981; Binkley 1990). Initially, the effects of light on rhythms and hormones were observed only in nonhuman species. Research during the last ten to fifteen years, however, has confirmed that light entering the eyes is a potent stimulus for controlling the human circadian system (Lewy et al. 1980; Czeisler et al. 1986; Wetterberg 1993). Across the span of evolution, daily cycles of light and darkness produced predominantly by the rising and setting of the sun, have provided a stable environmental time cue. Generally, circadian rhythms in humans and animals are strongly synchronized with the 24 hour cycle of light and darkness in the surrounding environment. When humans are not exposed to daily cycles of light and darkness (as has been done in long term cave experiments and other isolation studies) they still express physiological rhythms, but the rhythms "free-run", that is they follow a cycle period different from 24 hours. Free-running rhythms are not entrained or synchronized to the outside environment (Aschoff 1981; Binkley 1990). To understand how light regulates daily physiological rhythms in humans, it is useful to consider the neural structures that process photic stimuli.

Humans are highly dependent on visual perception of the environment and, consequently, the scientific study of vision and visual mechanisms is a centuries old endeavor (Krauskopf 1997). In humans, as in most vertebrate species, light enters the eyes and stimulates the retina. Nerve signals are sent from the retina to the visual centers of the brain permitting the sensory capacity of vision. In addition, signals are sent from the retina by a separate nerve pathway into the hypothalamus, a non-visual part of the brain (see fig. 9.1). The hypothalamus is an important part of the brain that influences or controls many basic functions of the body including hormonal secretion, body temperature, metabolism and reproduction, as well as higher cognitive functions such as memory and emotions (Morgane and Panksepp 1979). Information about environmental light is sent from the retina to a specific area of the hypothalamus, the suprachiasmatic nucleus (SCN) (Pickard and Silverman 1981; Moore 1983; Klein et al. 1991). The SCN is considered to be a fundamental part of the "biological clock", or circadian system, which regulates the body's 24 hour rhythms. This small part of the brain is thought to be a key pacemaker for synchronizing circadian rhythms such as sleep and wakefulness, body temperature, hormonal secretion and other physiological parameters including cognitive function. Although it is clear that light is the primary stimulus for regulating the circadian system, other external stimuli such as sound, temperature and social cues may also influence the body's timing functions (Aschoff 1981; Binkley 1990; Klein et al. 1991).

As indicated in fig. 9.1, the neural pathway responsible for vision is anatomically separate from the pathway responsible for circadian regulation. There is a functional connection, however, between the primary visual pathway and circadian neuroanatomy by way of the lateral geniculate nuclei (Aschoff 1981; Binkley 1990; Klein et al. 1991). Although the retinohypothalamic tract projects most densely to or around the suprachiasmatic nuclei, this pathway also projects in to other hypothalamic nuclei (preoptic nuclei, anterior and lateral hypothalamic areas, retrochiasmatic area and dorsal nuclei) as well as nuclei outside the hypothalamus (thalamic intergeniculate leaflets, midbrain periaqueductal grey) (Aschoff 1981; Binkley 1990; Klein et al. 1991). Whereas the retinal projection to the suprachiasmatic nuclei has been studied extensively for its involvement in regulating circadian physiology, the functional roles of the other nonvisual projections from the retina are unknown.

It is important to understand that the biological clock extends a broad influence to nearly every tissue and cell in the human body. Hence, light and the physical qualities of light such as wavelength or color may impact many biological processes in humans by way of circadian physiology. For simplicity, the next section will focus on a single aspect of how the biological clock responds to specific wavelengths of light.

9.5. Response of the pineal gland to different wavelengths of light

The SCN and the circadian system in general relay retinal information to many of the major non-visual control centers in the nervous system including other areas in the

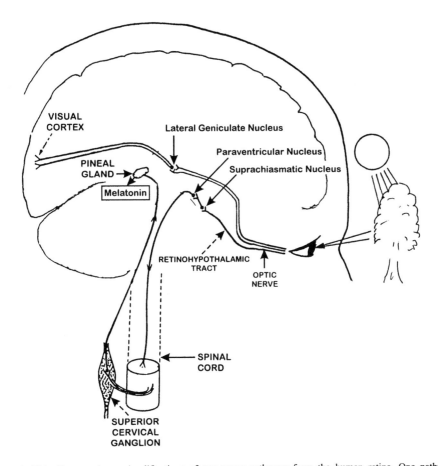

Fig. 9.1. This diagram shows simplifications of two nerve pathways from the human retina. One pathway supports the sensory function of vision (also see Chapter 3 and fig. 3.1) and the second pathway mediates the non-visual control of the pineal gland.

hypothalamus, the midbrain, thalamus and limbic system (Klein et al. 1991). As shown in fig. 9.1, one nerve pathway that carries nonvisual information about light extends from the SCN to the pineal gland via a multi-neuron pathway (Pickard and Silverman 1981; Moore 1983; Klein et al. 1991). Cycles of light and darkness relayed by the retina entrain SCN neural activity which, in turn, synchronizes physiological rhythms in the pineal gland. Within the pineal, a rhythm of production and secretion of the hormone melatonin is synchronized to cycles of light and darkness in the environment. In humans and all other vertebrate species studied to date, high levels of melatonin are produced and secreted during the night while low levels are released during the day (Reiter 1981, 1991). In addition to entraining the daily rhythm of melatonin secretion from the pineal gland, light can have an acute suppressive effect on this hormone. Specifically, exposure of the eyes to light during the night can cause a rapid decrease in the high nocturnal synthesis and secretion of melatonin (Klein and Weller 1972; Brainard et al. 1982; 1988). As is true for all photobiological phenomena, the intensity (Brainard et al. 1982, 1983, 1988; Lynch et al. 1984; Nelson and Takahaskhi 1991), wavelength (Brainard et al. 1984a, 1986b, 1994a; Podolin et al. 1987; Benshoff et al. 1987), timing and duration (Brainard et al. 1984b) of the light stimulus are important in determining the regulatory influence on the pineal gland.

Studies done with hamsters, rats and mice suggest that wavelengths in the blue and green portion of the spectrum have the strongest impact on melatonin secreted from the pineal gland (Brainard et al. 1984b, 1994a; Benshoff et al. 1987; Podolin et al. 1987) as well as the circadian system in general (Takahashi et al. 1984; Milette et al. 1987; Holtz et al. 1990). Although the highest sensitivity may be in the blue-green range, this does not preclude other wavelengths from regulating rhythms and hormones. For example, if the intensity is sufficiently high, short wavelengths in the "non-visible" ultraviolet region of the spectrum (Brainard et al. 1988, 1994a; Benshoff et al. 1987; Podolin et al. 1987) and longer wavelengths in the red portion of the spectrum (McCormack and Sontag 1980; Vanecek and Illnerova 1982; Nguyen et al. 1990) are quite capable of modulating melatonin or other biological rhythms in rodents. Some of the published data can be used to support the hypothesis that the retinal photopigment rhodopsin (commonly found in rod cells) is the primary receptor for circadian and hormone regulation (Cardinali et al. 1972a, 1972b; Brainard et al. 1984a, 1988, 1994a; Takahashi et al. 1984; Benshoff et al. 1987; Bronstein et al. 1987; Podolin et al. 1987). In contrast, it has also been suggested that one or more cone photopigments may be involved in these regulatory effects (Brainard et al. 1984a, 1988, 1994a; Benshoff et al. 1987; Milette et al. 1987; Podolin et al. 1987; Thiele and Meissl 1987; Holtz et al. 1990; Provencio et al. 1994). In mice, for example, there is now strong evidence for two cone type photoreceptors (middle wavelength sensitive and UV sensitive photoreceptors) mediating circadian regulation (Provencio et al. 1994). Further studies are required to conclusively identify which specific photoreceptors and photopigments are involved in regulating hormones and rhythms in animals.

Unlike the studies with animals, initial human experiments did not demonstrate that light stimuli could entrain daily rhythms or suppress melatonin (Vaughan et al. 1976; Jimerson et al. 1977; Lynch et al. 1978; Wetterberg 1978). As a result of these early studies, many investigators suggested that the human species did not retain the sensitivity to light in terms of nonvisual biological regulation as had been demonstrated in animals.

A pivotal study published in 1980, however, reversed this preliminary thinking. This study, done at the National Institutes of Health, demonstrated that exposing the eyes of normal volunteers to 2500 lux of white light during the night induced an 80% decrease in circulating melatonin within one hour. In contrast, exposure of volunteers to 500 lux of white light, an illuminance typical of indoor offices or residential lighting, did not suppress melatonin (Lewy et al. 1980). Normally, typical room lighting levels would be sufficient for suppressing melatonin in many animal species and would be adequate for human vision (Klein and Weller 1972; Brainard et al. 1982, 1983, 1984a, 1986a, 1994a; Lynch et al. 1984; Benshoff et al. 1987; Podolin et al. 1987; Nelson and Takahashi 1991; Rey 1993; Krauskopf 1997). The discovery that much higher levels of light are needed to suppress melatonin in humans provided the groundwork for numerous studies on the circadian and hormonal responses of humans to bright artificial light.

Shortly after it was discovered that light at 2500 lux can suppress melatonin in humans, a study was done to determine more precisely the dosages of light needed to suppress melatonin in normal volunteers (Brainard et al. 1988). In that study, healthy male subjects were exposed to carefully controlled intensities of light at 509 nm for one hour during the night. As indicated in table 9.1, this monochromatic light normally appears green to the human eye. The volunteers were continuously exposed to this green light between 2:00 AM and 3:00 AM when their blood levels of melatonin would be typically high. During the light exposure the volunteers had their pupils dilated, their heads held steady relative to the light source by an ophthalmologic head holder, and their eyes covered by translucent white integrating spheres as shown in fig. 9.2. This procedure produced a constant and uniform illumination of the whole retina during the entire light exposure. The monochromatic green light was produced by an optical apparatus shown in fig. 9.2. The results of this experiment (table 9.3 and fig. 9.3) demonstrated that light affects a human hormone in a dose-response fashion: i.e., the brighter the photic stimulus the greater the suppression of melatonin (Brainard et al. 1988). Thus, light irradiance is an important factor in determining the biological potency of a photic stimulus.

In addition to light irradiance being critical in determining if a light stimulus will suppress melatonin, the spectral quality of light is important in determining its relative biological impact. After completing the dose-response work illustrated in fig. 9.3, a preliminary test was done to compare the effectiveness of six different wavelengths or colors of light for suppressing nocturnal plasma melatonin (Brainard et al. 1988). These monochromatic wavelengths were 448, 474, 509, 542, 576 and 604 nm and had the color appearances of blue, blue, green, green, yellow and orange, respectively (see table 9.1). The results of this small study suggested that the peak sensitivity for melatonin suppression is in the blue-green range as seems to be the case in some lower mammals. It is premature, however, to put much emphasis on such a conclusion since relatively few subjects were tested in this pilot study.

Interestingly, *all* of the stimuli used in the complete dose-response study and the pilot work with different wavelengths activated the visual system: both the volunteers and the experimenters saw each of the different light intensities and accurately described them according to their color appearance. In the dose-response study, the lower intensities of green light, however, did not change hormone levels whereas the higher intensities induced a 60–80% decrease in this hormone. Thus, light that activates vision does not

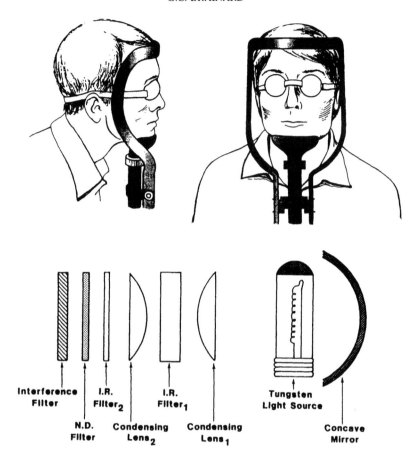

Interference | I.R. | I.R. | Tungsten
Filter | Filter₂ | Filter₁ | Light Source

N.D. | Condensing | Condensing | Concave
Filter | Lens₂ | Lens₁ | Mirror

Fig. 9.2. The top illustration shows how a volunteer's head was held motionless and the eyes were covered with translucent white integrating spheres during the experimental light exposures. The monochromatic light stimulus was produced by the optical arrangement shown at the bottom (Brainard et al. 1988). The data shown in fig. 9.3 were derived from the experimental technique shown in this drawing.

necessarily cause hormonal change. It appears to be generally true in both animals and humans that much more light is needed for biological effects than for vision.

It is misleading to state that only bright light can induce nonvisual biological changes. The term "bright" refers to a subjective visual sensation and is thus a relative descriptor (Thorington 1985; Rey 1993). A white 2500 lux light indoors indeed appears bright relative to typical indoor levels. In contrast, 2500 lux of light outdoors is relatively dim compared to daylight at high noon which reaches 100 000 lux (Thorington 1985). Furthermore, high illuminances are *not* always necessary for changing rhythms and hormones. For example, in the dose-response study with monochromatic green light, only 5 to 17 lux was needed to significantly suppress melatonin (Brainard et al. 1988). Similarly, only 100 to 300 lux of white light is needed to significantly suppress melatonin (Bojkowski et al. 1987; McIntyre et al. 1989; Brainard et al. 1993, 1994b; Gaddy et al. 1993, 1994).

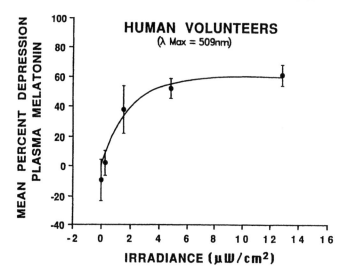

Fig. 9.3. This graph shows the dose-response relationship between green light exposure of normal volunteers eyes and suppression of the hormone melatonin. The light stimulus in this study was monochromatic light at 509 nm, with a 10 nm half-peak bandwidth. The radiometric and photometric values for the test stimuli in this study are shown in Table 9.3 (Brainard et al. 1988).

TABLE 9.3

Scotopic lux refers to the measurement of light that is stimulatory in night vision or vision mediated by the rod system of the eye. Photopic lux refers to the measurement of light that is stimulatory to day vision or vision mediated by the cone system of the eye.

Irradiance $\mu W/cm^2$	Quanta photons/cm^2	Illuminance	
		photopic lux	scotopic lux
0.01	9.2×10^{13}	0.03	0.17
0.30	2.8×10^{15}	1.03	5.25
1.60	1.5×10^{16}	5.50	27.98
5.00	4.6×10^{16}	17.18	85.90
13.00	1.2×10^{17}	44.66	227.37

Several studies have made it clear that the specific characteristics of the light stimulus and the exposure technique are critical in determining how much illuminance is necessary for inducing a biological change in healthy humans (Brainard et al. 1988, 1993, 1994b; Gaddy et al. 1993, 1994).

Interestingly, it has been shown that in some blind subjects with loss of pupillary reflex, no conscious perception of the light stimuli, and no outer retinal functioning as determined by electroretinographic testing, melatonin suppression can be induced by exposing the eyes to 10 000 lux of white light (Czeisler et al. 1995). Such data suggest that light detection for the circadian and neuroendocrine systems may not rely on the rods or cones used for vision. Other recent experiments have begun to examine whether a retinal deficiency in the color perceptual system could affect light-mediated melatonin suppression. In one study, volunteers were screened for color vision defects using the Ishihara, Farnsworth–Munsell, and Nagel anomaloscope tests (Ruberg et al. 1996). Fourteen volunteers with color vision deficiencies were compared to seven volunteers with normal color vision for light-induced melatonin suppression. Subjects with color vision deficiencies were diagnosed as follows: five subjects had protanopia (functionally lacking the "red" cone photoreceptor system), six subjects had deuteranopia (functionally lacking the "green" cone photoreceptor system), three subjects were anomalous trichromatic observers (having milder color confusion in the red–green axis) and one subject was an unspecified dichromat (with an unknown cone deficiency). Each subject was studied on two nights separated by at least one week: 1) a control night when they were exposed only to darkness, and 2) a night when they were exposed to 200 lux of white light in a full visual field with their pupils dilated from 2:00 AM to 3:30 AM. This study showed that melatonin was suppressed after light exposure both in subjects with normal color vision and subjects with color vision deficiencies ($p < 0.001$), with no significant difference in melatonin suppression between the two groups. These results suggest that a normal trichromatic photopic system is not necessary for light regulation of the human pineal gland. A separate study done on melatonin regulation in volunteers with color vision deficiencies further reinforces the hypothesis that a normal trichromatic visual system is not necessary for light-induced suppression of melatonin or regulating short term circadian changes in the melatonin rhythm (Rubery et al. 1996).

Whereas centuries of experimental work has helped determined the nature of photoreceptors that deliver specific color information to the visual system of animals and humans, there has been only three or four decades of work on the nonvisual control of biological rhythms and hormones by light. Given the current paucity of data available on how wavelengths of light regulate melatonin or the circadian system in general, it is premature to draw any conclusions as to what photoreceptors mediate these nonvisual effects of light in humans. Thus, there are many open questions about how light and color stimuli influence the brain and body. Are biological rhythms controlled by one or more of the same photoreceptors that support vision? Does the circadian system have its own unique photoreceptors? How do different combinations of wavelengths or colors influence nonvisual rhythms and hormones? Are all nonvisual biological effects of color mediated through the circadian or neuroendocrine systems?

9.6. The effects of color on blood pressure and general arousal

The effects of light and wavelength on the human circadian system with specific emphasis on the pineal gland have been reviewed above. For the most part, those discussions are based on extensive controlled, empirical studies that have been published in the peer-reviewed biomedical literature. Does that represent all there is to know about the influence of color on human biology? There is a broad popular literature (books, magazines, newspaper articles) which suggests that color has other specific physiological effects on the human body and influences emotions, mood and feelings. Occasionally, popular writers will refer to one or more scientific references to support their contentions.

More frequently, however, this literature is anecdotal or without a controlled scientific basis. Is there sufficient evidence to support this popular wisdom?

An important article by Kaiser reviews the diverse literature on the nonvisual physiological effects of color in humans with emphasis on how different studies have shown that color can influence electrical patterns of the brain, galvanic skin response, blood pressure, heart rate, respiration rate, eyeblink frequency and blood oxygenation in humans (Kaiser 1984a). Kaiser's article concludes that "there are reliably recorded physiological responses to color in addition to those generally associated with vision. However, it may be that some are indirect effects mediated by cognitive responses to color" (Kaiser 1984a). This raises a fundamentally important point. Are the physiological effects of color a *direct* biological response to the color itself or are they *indirect* – are they mediated by mental associations connected to the color. For example, as discussed above, red wavelengths can influence biological rhythms but they are weaker than blue or green wavelengths. Such physiology is due to a direct response to color; there is no cognitive intermediary – an animal or human subject does not have to experience or learn anything about red for the biological response to occur. In contrast, an indirect physiological response to color would be based on mental associations with the color. For example, an individual may have strong cognitive associations such as sex, blood, fire or danger for the color red. Associations like these may be intellectually and emotionally stimulating which, in turn, may lead to a physiological arousal such as raised blood pressure or accelerated heart rate. In general, when considering the effects of color on humans, it is important to discern if the color has a direct physiological impact on a particular biological function or if the color effect depends on cognitive learning or emotional associations which lead to a more generalized biological response. The Addendum to this chapter outlines one strategy of controlled testing to determine if a given biological or psychological response is regulated specifically by light wavelength versus color appearance.

One of the ideas more frequently expressed is that color can influence blood pressure in humans – more specifically, red raises blood pressure while blue lowers blood pressure. A careful study published in 1958 by Gerrard, tested a variety of objective physiological and subjective mental responses of 24 healthy male volunteers to different colored light stimuli. In that study, systolic blood pressure, skin conductance, respiration, heart rate, eyeblink frequency and electrical activity of the brain were measured relative to red and blue light exposure. The results showed that with the exception of heart rate, red light produced a significantly greater arousal in all physiological responses compared to blue light. In that study, the subjective preferences, attitudes and associations to the different color stimuli were assessed in each of the volunteers. The subjective responses generally matched the physiological responses. In short, red tended to be more emotionally arousing or intellectually stimulating than blue and, similarly, red was associated with increased physiological arousal. Gerrard reached two major conclusions about this study: 1) "The response to color is differential, i.e. different colors arouse different feelings and emotions and activate the organism to a different degree" and 2) "The differential response to color is a response of the whole organism, involving correlated changes in autonomic functions, muscular tension, brain activity and affective-ideational responses" (Gerrard 1958).

A more recent study attempted to replicate the findings of Gerrard's study with a specific focus on the effects of color on cardiovascular function (Yglesias et al. 1993). In that study, 16 healthy, male volunteers had their heart rate, mean arterial blood pressure, systolic blood pressure and diastolic blood pressure measured automatically while they were exposed to equal photon densities of patternless monochromatic red (620 nm) or blue (460 nm) light. Prior to the experiment, each volunteer completed a test for their expectations of how the color stimuli might affect their blood pressure and heart rate. Relative to pre-test expectations, the subjects "liked" the blue color significantly better than the red, predicted that blue would be significantly more relaxing than red, and predicted that red would raise blood pressure significantly more than blue. In sharp contrast to these mental expectations, none of the cardiovascular responses varied significantly relative to the color stimuli (Yglesias et al. 1993). These results demonstrate that the subjects had very significant psychological expectations that the color would influence their blood pressure. Despite this positive mental attitude, however, the red and blue stimuli did not induce any significant changes in blood pressure or heart rate. Thus, Gerrard's original observation (Gerrard 1958) that red and blue light differentially effects blood pressure was not repeated in this study.

Do color stimuli influence blood pressure in normal humans? If color does influence blood pressure, why were the effects of red and blue light not similar in the two studies discussed above? Was some unidentified test parameter different between the two studies causing different outcomes? Are the blood pressure effects of color simply not repeatable phenomena? Unfortunately, there are no definitive answers to these questions. There is, however, one striking consistency between the two studies. The subjective mental responses of the subjects in both experiments had very similar attitudes about the color stimuli (Gerrard 1958; Yglesias et al. 1993). Specifically, they found the red to be more emotionally arousing or stimulating than the blue. Such a consistent cognitive response to the colors coincides with the broad popular belief that color influences blood pressure and heart rate. Scientifically this conclusion is premature since there is not a strong, consistent empirical base of studies that supports the idea. It would be a mistake to say that the hypothesis that color influences blood pressure has been disproven. At this time, it simply remains an open question that needs further study.

9.7. Light treatment of winter depression and other disorders

While research proceeded on the effects of light in controlling hormones and rhythms in humans, concurrent studies tested the use of light as a therapeutic tool for improving mood and psychological status of patients diagnosed with winter depression. It has been known since antiquity that some individuals are adversely affected by the changing seasons. More recently, the specific condition of fall and winter depression or Seasonal Affective Disorder (SAD), has been formally described in the scientific literature (Lewy et al. 1982, 1987; Rosenthal et al. 1984, 1988; Terman et al. 1989) and been included in the American Psychiatric Association's diagnostic manual, DSM-IV-R (American Psychiatric Association 1994). Individuals affected with this malady often experience a dramatic decrease in their physical energy and stamina during the fall and winter. As daylengths become shorter, individuals with SAD often are not able to meet their job responsibilities and/or are not able to cope with the needs of routine family life. In addition to a general decrease in physical energy or stamina, they experience emotional depression along with feelings of hopelessness and despair. Other symptoms of winter depression or SAD may include increased sleepiness and need for sleep, increased appetite (particularly for sweets and other carbohydrates), and a general desire to withdraw from others. People afflicted with this malady often feel compromised in meeting the ordinary demands and responsibilities of everyday life. Fortunately, for many of the patients who are accurately diagnosed with SAD, daily light therapy has been found to effectively reduce symptoms (Lewy et al. 1982, 1987; Rosenthal et al. 1984, 1988; Terman et al. 1989).

There are now numerous clinics which specialize in diagnosing winter depression and treating people with light therapy (Society for Light Treatment and Biological Rhythms 1993). The majority of light therapy for SAD is done on an outpatient basis with individuals receiving their therapy at home or in their workplace. Although specific treatment recommendations vary somewhat between clinics, a procedure that has been used frequently over the past ten years involves a patient sitting at a specific distance from a fluorescent light panel which provides a 2500 lux exposure when looking directly at the lamp. The patient is told not to gaze steadily at this bright light, but rather to glance directly at the unit for a few seconds each minute over a two hour period. During the therapy period, patients can chose to read, work at a computer, watch television, or do other hand work. Patients often respond to light therapy after two to seven days of treatment and continue to benefit as long as the treatment is repeated daily throughout the months that the individuals experience winter depression (Lewy et al. 1982, 1987; Rosenthal et al. 1984, 1988; Terman et al. 1989).

There are now a variety of light devices available for treating SAD. One of the newer devices is configured as a work-station with a fluorescent light panel angled over and in front of the head. The device delivers an illuminance of approximately 10 000 lux at eye level. Since the illuminance provided by this unit is four times brighter than the 2500 lux panels, the therapy session is reduced to thirty minutes (Terman et al. 1990). A different approach is to make the light therapy unit wearable. In this case, a compact fluorescent or incandescent light source is mounted on the head and powered by a small battery pack. These easily portable devices permit patient mobility around the household or workplace during the light therapy session and appear to effectively reduce symptoms with daily treatments as short as thirty minutes (Stewart et al. 1990; Rosenthal et al. 1993). Finally, one of the most recent innovations in light therapy is the automatic dawn simulator. This treatment unit is placed in the bedroom near the sleeping patient and consists of a light source connected to a programmable timing device. The dawn simulator gradually raises the illumination in the room during the patient's wakeup period and has been indicated as an effective substitute or supplement to the other forms of light therapy (Avery et al. 1992, 1993).

The newer light therapy devices are configured to shorten therapeutic time, increase patient mobility or permit therapy during the wakeup period. Although such devices are readily available to clinicians and their patients, testing is continuing to determine if these innovative designs have the same therapeutic potency as the light panels used in the earlier studies on treating winter depression. Doubtless there will be continued development, diversification and improvement of light therapy devices and strategies in the coming years.

Over the past decade, the majority of studies on light therapy have been concerned with treating winter depression. Other research, however, has begun to extend the applications of light therapy. Investigators have had some success in treating certain sleep disorders with phototherapy (Rosenthal et al. 1990; Terman et al. 1995). In addition, studies have indicated that individuals with either non-seasonal depression (Yerevanian et al. 1986; Kripke et al. 1989), menstrual disorders (Parry et al. 1987, 1989; Lin et al. 1990; Kripke 1993) or the eating disorder, bulimia (Berman et al. 1993; Lam et al. 1994) may benefit from light therapy. Much more work needs to be done in determining the utility of light for treating these disorders. It is clear, however, that we are entering a frontier of medicine in which man's biological response to light is being studied for its potential to alleviate specific illnesses. Such medical developments have encouraged investigators to explore the possibilities of using light for various domestic or non-medical applications.

One area of study involves the function and disfunction of the human biological clock under more challenging situations. Jet lag is a condition that results from rapid transport over several time zones. For example, a plane flight from Philadelphia to Paris involves crossing six time zones which means that the new environment has a cycle of sunlight and darkness that is shifted six hours compared to the home environment. While readapting the circadian system to a new social and geophysical time zone, many people experience uncomfortable symptoms such as insomnia at night, daytime sleepiness, gastric distress, irritability, depression and confusion (Winget et al. 1984). Such symptoms can pose serious problems for the business traveler and can be an unwelcome adversity to the leisure traveler. Preliminary studies have tested the use of bright light exposure to prevent or ameliorate jet lag (Daan and Lewy 1984; Wever 1985; Boulos et al. 1995). The initial findings from these studies are generally positive and investigators are optimistic that light will be a useful tool for quickly resetting the traveler's biological clock and overcoming some of the problems associated with jet travel (Society for Light Treatment and Biological Rhythms 1991).

Like intercontinental travelers, shiftworkers may suddenly alter their sleep and waking times as a consequence of their work schedules. Currently, there are an estimated 20 million individuals in the United States who are employed in shiftwork circumstances (US Congress 1991). In fact, most modern industrialized nations have a significant segment of their workforce doing some form of night work. There are often strong economic incentives encouraging businesses to use shiftwork. Some employees prefer night work over day work and adapt well to shift schedules. Generally, however, there are drawbacks of decreased production, increased accidents and increased health problems among those who are working when the body has a natural tendency to be at rest or asleep (Moore-Ede et al. 1982, 1983; Czeisler et al. 1990; US Congress 1991; Moore-Ede 1993). Researchers believe that poor adaptation of the circadian system to night work and shift schedules causes some of these ailments (Moore-Ede et al. 1982, 1983; Czeisler et al. 1990; US Congress 1991; Moore-Ede 1993). Some recent studies have shown that bright light exposure of shiftworkers at the proper time can speed up the adaptation of their biological clock to a new schedule (Czeisler et al. 1990; Eastman 1990b, 1991; Dawson and Campbell 1991). Other studies have demonstrated that bright light exposure during night work can enhance performance and alertness while producing profound biological adjustments (French et al. 1990; Campbell and Dawson 1990; Badia et al. 1991; Brainard et al. 1991; Hannon et al. 1991; Campbell et al. 1995). Clearly, the new data on light treatment for night workers and intercontinental travelers holds great promise. Most scientists, however, agree that there is currently insufficient data for a set prescription on how best to use light for shiftwork and jet lag applications (Society for Light Treatment and Biological Rhythms 1991).

9.8. The effects of different wavelengths in light therapy

Considerable research has been directed at determining the optimum light intensity, exposure duration and time of day for light treatment of winter depression (Rosenthal et al. 1988; Terman et al. 1989; Terman and Terman 1992; Rosenthal 1993). In addition to those physical parameters of light therapy, the spectral characteristics of light are important in determining its antidepressant effect. Most of the clinical trials on treating winter depression have employed white light emitted by commercially available lamps. The white light used for treating SAD has been effectively provided by a range of lamp types including various incandescent lamps (such as halogen, nitrogen and krypton) as well as different fluorescent sources (such as cool white, tri-phosphor, and "sunlight simulating") (Lewy et al. 1982, 1987; Rosenthal et al. 1984, 1988; Yerevanian et al. 1986; Terman et al. 1989, 1990; Stewart et al. 1990; Terman and Terman 1992; Avery et al. 1992, 1993; Rosenthal et al. 1993; Rosenthal 1993).

As illustrated more thoroughly in Chapter 1 and in the Addendum to this chapter, it is important to note that there is an infinity of wavelength combinations that appear "white" to human observers. For example, two complimentary narrow band wavelengths at 440 nm (violet-appearing monochromatic light) and 560 nm (yellow-appearing monochromatic light) can be mixed to produce a light that appears white to an observer. There are many such complimentary pairs of colors. Alternatively, many sets of three or more visible wavelengths can be mixed together to produce white-appearing light. Among all the possible sources currently available, relatively few "white" light types have been tested for SAD therapy. The effective color temperature, expressed in absolute Kelvin (K) units, varies from approximately 2900 K to 6300 K for the different white light sources that have been tested in SAD therapy – a range which indicates how very different these white light sources are from one another in spectral balance. Studies which employed these different white light sources all showed that, *at a sufficiently high illuminance*, they could effectively reduce SAD symptoms (Lewy et al. 1982, 1987; Rosenthal et al. 1984, 1988; Yerevanian et al. 1986; Terman et al. 1989, 1990; Stewart et al. 1990; Terman and Terman 1992; Avery et al. 1992, 1993; Rosenthal et al. 1993; Rosenthal 1993) It is critical to note that these findings do not support the notion that these different white light sources are equivalent therapeutically, or that "white-appearing" light represents the optimum therapy for winter depression. By raising the dose (illuminance or intensity) of white light, the differences in spectral balance are obscured.

Current evidence supports the idea that light therapy for SAD works by way of light shining into the eyes as opposed to light on the skin (Wehr et al. 1987). As with the regulation of biological rhythms and hormones, it is not known which ocular photoreceptors mediate the therapeutic benefits of light in winter depression. To date, three studies have specifically compared different portions of the spectrum for clinical efficacy in treating SAD (Brainard et al. 1990; Oren et al. 1991; Stewart et al. 1991). In one of these studies, 14 patients were treated with an equal photon dose of green or red fluorescent light

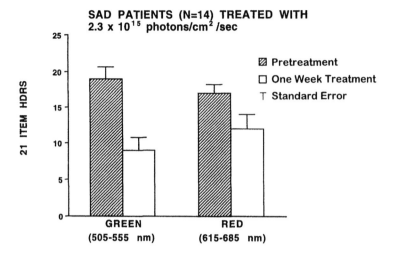

Fig. 9.4. The data illustrated here demonstrate that green light was significantly more effective than red light for treating winter depression (Oren et al. 1991). The bars in this graph give mean (and standard error of the mean) depression rating values for patients before treatment (hatched bars) and after one week of treatment with equal photon densities of green or red light. Numbers in parentheses at the base of the top graph indicate the half-peak bandwidth of the light source. The chart below provides the radiometric and photometric characterization of the light sources used in this study (Brainard 1998).

TABLE 9.4

Color appearance	Quanta photons/sec/cm²	Irradiance µW/cm²			Illuminance lux	
	315–800 nm	315–400 nm	400–700 nm	700–800 nm	photopic	scotopic
Green	2.3×10^{15}	0	802	6	4426	8869
Red	2.3×10^{15}	0	600	94	686	43

for a period of one week. The photon dose of 2.3×10^{15} photons/cm²/sec was selected because many previous studies had shown this particular photon density of white light to be clinically effective in one week of therapy (Rosenthal et al. 1984, 1988; Terman et al. 1989). The green and red light sources used in that study were not monochromatic but had half-peak bandwidths of approximately 505–555 nm and 615–685 nm, respectively. Patients' clinical status before and after light therapy was followed by means of the 21-item Hamilton Depression Rating Scale (HDRS), a standard scale for measuring symptoms associated with depression (Hamilton 1967). As illustrated in table 9.4 and fig. 9.4, one week of light therapy with both green and red light sources produced an improvement in depression symptoms in the groups of patients tested. Depression symptoms were reduced 51% by the green light source and 30% by the red source. Hence, green light was significantly more potent than the red light for treating winter depression at equal photon densities (Oren et al. 1991).

Together, the three studies which compared different portions of the spectrum for treating SAD form the ground work for defining an action spectrum of light therapy for winter depression (Brainard et al. 1990; Oren et al. 1991; Stewart et al. 1991). As discussed earlier, the traditional approach to defining a complete action spectrum, however, requires substantially more testing (Smith 1989; Coohill 1991). A thoroughly defined action spectrum can guide the development of light treatment devices that emit the optimum balance of wavelengths for treating SAD and will yield important information about the photosensory mechanism responsible for the beneficial effects of light therapy. Currently, it is premature to predict what photopigment(s) or photoreceptor(s) mediate the antidepressant effects of light.

An important issue debated among SAD researchers concerns the role of ultraviolet radiation (UV) in light therapy. Many of the early studies on SAD therapy successfully utilized white fluorescent lamps that emitted a portion of UV wavelengths (Rosenthal et al. 1984, 1988; Lewy et al. 1987; Terman et al. 1989). The early studies led to the suggestion that UV wavelengths are necessary for successful therapy. The literature, however, now shows that SAD symptoms can be reduced by lamps which emit little or no UV (Lewy et al. 1982; Yerevanian et al. 1986; Terman et al. 1990; Stewart et al. 1990, 1991; Rosenthal et al. 1990, 1993; Oren et al. 1991; Avery et al. 1992, 1993; Terman and Terman 1992; Rosenthal 1993). Thus, UV wavelengths do not appear to be *necessary* for eliciting beneficial effects of light in treating depression. Does this rule out UV having any role in relieving winter depression? Among all the living creatures, it has been shown that many species of insects, fish, birds have specific UV photoreceptors in their eyes (reviewed in Vision Research (1994)). Even some mammalian species appear to have UV photoreceptors (Jacobs 1992; Jacobs and Deegan 1994) and in some animals UV wavelengths can regulate seasonal reproduction, melatonin production, and circadian rhythms (Brainard et al. 1986b, 1992, 1994a; Benshoff et al. 1987; Podolin et al. 1987; Provencio et al. 1994). Furthermore, in normal, healthy humans up to the age of at least 25 years, near ultraviolet radiation (UV-A) can be visually detected (Brainard et al. 1992; Sanford et al. 1996). Thus, though the latest studies show positive therapeutic responses when UV is excluded in SAD treatment, they do not demonstrate that UV is necessarily noncontributory. Whether or not UV wavelengths contribute to the optimum balance of wavelengths for SAD therapy remains an open question.

9.9. Color therapy (chromotherapy)

The preceding section on light therapy for winter depression and other disorders is based predominantly on controlled empirical studies done by investigators in medical clinics, hospitals or universities. Beyond the biomedical literature, there is a rich and long history of light and color being used for treating human disorders (Birren 1961, 1978; Ott 1973; Gimbal 1988; Lieberman 1991). Kaiser has developed an extensive academic overview of these color therapies for the interested reader (Kaiser 1984b). For the most part, such color therapy or chromotherapy has not been systematically investigated by conventional scientific methods and has not been included in mainstream modern medicine. That is not to say that there is no potential value in color therapy or no basis for what has been claimed about the healing power of color – it simply indicates that the level of scientific evidence is lacking. Hence, the views of scientists about such color therapies range from skeptical to scathing. In contrast, individuals with less stringent need of empirical proofs often embrace color therapy and are excited about its potential.

There is a form of color therapy which is currently used by some professional optometrists which has a modicum of scientific support and documentation. During the 1920s and 1930s, an optometric method was developed to treat patients with a variety of visual problems by exposing their eyes to selected wavelengths or colors of light (Henning 1936, 1939, 1940; Spitler 1941). The current therapeutic practice is termed syntonic optometry and is based on the concept that correcting visual problems with colored light stimuli could lead to a broader improvement in the patient's behavior, performance and academic achievement (Spitler 1941). For this therapy, the practitioner derives diagnostic information from the patient's case history, symptoms, pupillary response, ocular-motor skills, ocular examination, and visual field plotting. If the patient is determined to have blurred vision, a crossed or lazy eye (strabismus or amblyopia), double vision, reduced peripheral vision, or eye strain, then treatment of these primary conditions is thought to improve secondary problems such as headaches, poor attention and memory, reading problems, loss of coordination or academic underachievement. Once the syntonic practitioner determines that syntonic therapy is appropriate, then the patient is treated by exposing the eyes to selected visible wavelengths which are experienced as colors. Often a series up to twenty colored light treatment sessions is given over a four or five week period. Each treatment session is about 20 minutes in length and the color stimuli are varied according to the diagnosis of the patient. The progress of the patient is monitored during the series of syntonic treatments.

Unlike other forms of color therapies, the practice of syntonic optometry is discussed here for two reasons. First, this practice was developed by professional optometrists and thus there has been a more formal documentation of case studies compared to other forms of chromotherapy (Spitler 1941; Otto and Bly 1984). Secondly, two experimental studies have been done to document the efficacy of syntonic therapy (Kaplan 1983; Lieberman 1986). Despite these efforts, most optometrists are skeptical and do not advocate this therapy. Considerable research is still required to verify the clinical efficacy of the syntonic method and determine its related physiological mechanisms. Given the potential therapeutic value that is claimed for syntonic optometry, a series of strictly controlled experiments which evaluate the clinical efficacy of this therapy would appear to be warranted and timely.

9.10. Placebo response concerns

In general, the various forms of colored light therapy or chromotherapy still need to be studied more thoroughly in controlled settings before the scientific community will take such treatments seriously. Not having the support of the conventional medical community, however, has not prevented some therapists from continuing to treat patients since they see positive responses. The critical question is: do patients get better because of a specific therapeutic value of this treatment or are they responding to the general circumstance of being clinically treated? More to the point, does color therapy work by way of a specific biological mechanism or are the patient improvements due to placebo responses?

Discerning generalized placebo responses from specific clinical responses is pertinent not only to color therapy but to all forms of medical treatment. As an example, examine the treatment response data shown in fig. 9.4. In this study, each light treatment was associated with some therapeutic improvement. Does this indicate that each light was at least partially effective in treating depression symptoms, or are *some* of the therapeutic benefits of this therapy due to a non-specific or placebo response? Since patient expectations of treatment outcome are thought to contribute significantly to the placebo effect, evaluation of expectations before treatment is one strategy for approaching this question. Prior to any light treatment, subject expectations were systematically probed in the study shown in fig. 9.4. In general, all subjects had positive expectations about the success of light therapy and there were no significant differences between the expectations for the different light spectra in that study (Oren et al. 1991). This evidence supports the idea that *some* of the therapeutic benefits of the different light spectra *may* have been due to placebo responses. Patients generally expected light therapy to help and, indeed, depression symptoms were reduced by both red and green treatments. Reduction in depression symptoms, however, were significantly greater after treatment with green light source compared to treatment with the red source. The patients, however, did not indicate an expectation that green light would be more potent in reducing their symptoms compared to red light. This supports the idea the differential therapeutic response to the different light spectra was not merely an extension of the patients' preconceived beliefs and are thus less likely to be attributed to general placebo responses.

It has been well documented that patients with a wide range of disorders – depression, schizophrenia and anxiety as well as cancer, diabetes and ulcers – can successfully respond to inactive or placebo treatments (Ross and Olson 1981; Eastman 1990a; Turner et al. 1994). Hence, it would be remarkable if patients receiving treatment for winter depression did not show some level of placebo response to light therapy. In fact, some therapeutic improvements are almost always observed with light treatments regardless of light intensity, wavelength and duration (Lewy et al. 1982, 1987; Rosenthal et al. 1984, 1988; Terman et al. 1989; Terman and Terman 1992; Rosenthal 1993). Although it is obvious that light therapy will reduce patients' depression symptoms, the critical question is how much of the patients' response to light therapy is due to a non-specific placebo response versus a genuine clinical response? This remains an open question in the SAD field and has been discussed most insightfully by Eastman (Eastman 1990a; Eastman et al. 1993). Similar questions can be generalized to all forms of clinical practice including drug therapy and surgery. Indeed, unless the issue of placebo responses are addresses in controlled studies, any new therapy will be treated skeptically by the medical community.

9.11. Conclusion

Light can be a potent regulator of human biology. It is well established that light can regulate physiology both on the body surface (skin or dermis) and internally (the circadian and neuroendocrine systems). As is expected from the science of photobiology, the intensity and wavelength are important in determining the capacity of a photic stimulus to regulate human physiology. Throughout history, there have been many claims about the use of light and color as therapeutic agents. In many cases the claims for color therapy, or chromotherapy, are not supported by controlled scientific studies and thus await empirical confirmation. Over the past 15 years, however, light has been used successfully for treating winter depression (SAD) and numerous controlled studies on this application of light therapy have appeared in the biomedical literature. The success of light therapy in benefiting patients with winter depression has led to explorations of using light to treat other medical conditions. The modern study of light as a therapeutic tool has opened an exciting new chapter in medicine and in the understanding of the nature of man.

Acknowledgements

The author appreciates the dedicated support of John P. Hanifin for the overall editing and preparation of this chapter. Sincere thanks to John Georgiou, Rick Ruberg, Betsy Hancock, Ray Gottlieb and Hal Kelly for their contributions to the manuscript. The wavelength research on light therapy for winter depression illustrated in fig. 9.4 was supported, in part, by NIMH Grant #MH-44890.

9.12. Addendum to Chapter 9
Double blind testing for biological and therapeutic effects of color

KURT NASSAU

As briefly discussed in Sections 9.6 to 9.10 above, there have been surprisingly few satisfactorily controlled experimental trials of color therapy. Many examples of contradictory results in the literature, often based on quite old and anecdotal reports, are found in the reviews of Kaiser (1984a, 1984b). From his table 1, for example, one concludes that rheumatism can be treated with red, orange, blue, or violet light. A similarly wide set of colors is given for the treatment of blood pressure, which has been reported to be both raised and lowered by blue light; and so on (Kaiser 1984a). Results were contradictory even in the more recent careful studies (Gerrard 1958; Yglesias et al. 1993) on the effect of color on blood pressure described in Section 9.6. Several factors lead to such inconsistencies.

First, there are the specific psychological effects of colors, briefly referred to in Section 9.6 above as well as in Chapter 6, which are difficult to eliminate and which so easily confound physiological reactions. Second, there is the question of the control of the color itself. As described in Chapter 1 in connection with fig. 1.6, orange light, that is light evoking the reaction 'orange' in the eye–brain system (also note the disclaimer as to the use of terms such as 'colored light' or 'orange light' given in Section 1.4 of Chapter 1), can consist of a wide variety of spectral energy distributions. Some of these may contain no spectral orange at all, that is no significant amount of light at about 600 nm. It is rare that spectral energy distributions have been controlled adequately or even discussed in reports on color therapy.

Third, since both the subject and observers are aware of the color of the light being used, neither single nor double blind conditions (Feinstein 1977; Zolman 1993) seem ever to have been achieved. Next, the control of intensity is problematic, involving both photopic vs. scotopic (see fig. 9.3) and narrow-band vs. broad-band possibilities. Then there is the placebo effect (Feinstein 1977; Zolman 1993; Berkow 1992) (also see Section 9.10), which can be stimulated by the expectations of both subject and observer. And so on.

Yet the additive color mixing concepts of Section 1.6.1 of Chapter 1 can be utilized to lead to a set of conditions where the subjects and observers are not aware of the presence or absence of the specific spectral color being tested and where full double blind control can therefore be attained. Various aspects of fig. 9.5 lead to spectral distributions which can serve as test and control light sources to answer specific questions as discussed in Section 9.12.1. There is also the possibility of survey light sources as discussed in Section 9.12.2.

9.12.1. Test and control light sources

Consider the possible illumination arrangements for an experimental design to decide whether, say, narrow-band green light at 520 nm has a specific effect on some physiological process or medical condition. For the present discussion we consider this green light to encompase the fairly narrow range of 520 ± 20 nm as at A in fig. 9.5. (This range of 500 nm to 540 nm could equally well be specified as $18\,519$ cm^{-1} to $20\,000$ cm^{-1}, 2.30 eV to 2.48 eV, and so on, converted by using the data of tables 1.1 and 1.2 of Chapter 1.) Such a spectral distribution could be achieved, for example, by the use of an incandescent light source equipped with sharp cut-off filters. A wider range might be more appropriate for actual studies since the cut-off does not necessarily have to be as sharp as this.

The control illumination for A is provided by an energy distribution such as that of B of fig. 9.5. This contains two bands of light in the blue and yellow-green regions of the spectrum, centered at 480 nm and 570 nm, respectively. By using appropriate relative intensities, this combination gives exactly the same visual color appearance as the spectrally pure 520 nm green at A, but it contains no 520 nm green at all. (These light sources could actually be distinguished by a careful examination of some metameric material described in Section 1.7.6 of Chapter 1, or with a hand spectroscope, items not readily available to either subjects or observers, however.)

A second set of two test and one control light sources is given at C, D, and E in fig. 9.5. All of these appear equally 'white' to the eyes of the subjects and observers (spectroscope and subtle metamerism again excepted), yet C and E contain 520 nm green, while D does not. Note that the incandescent light at E in this figure does contain 520 nm green as indicated by the dashed outline. (Daylight is too variable and fluorescent tube lamps have unsuitable near-discontinuous spectra, see fig. 1.16 of Chapter 1.) The light source D appears white because it consists of an appropriate ratio of the complementary colors 480 nm blue and 580 nm yellow. The relative intensities of the three bands present in C need to be adjusted so that the 'center of mass' of their triangle on the chromaticity diagram of fig. 1.7 of Chapter 1 falls on that white point which corresponds to the specific incandescent light source E.

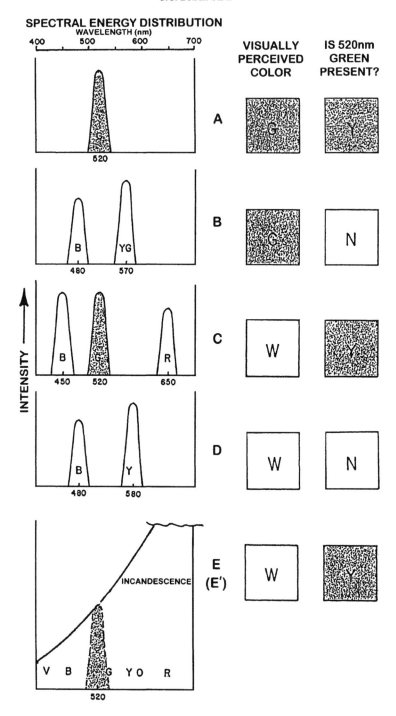

Fig. 9.5. Light sources for double blind testing of a possible effect of 520 ± 20 nm green.

The control of the overall visual light source intensity between A and B in fig. 9.5 is straightforward: both should appear equally bright to the eye. Note, however, that reducing the voltage in incandescent lighting is not always an acceptable option, since this can change the spectral energy distribution. The amount of 520 nm green in C should be the same as that in A if both sets are used together.

These intensity considerations change, however, when light sources C and E are compared. If the overall visual intensity of E is made equal to that of C, the amount of 520 nm green then present in E will be less than that present in C. For equal 520 nm content for comparison with C, the intensity of E needs to be higher, as is indeed shown for E in fig. 9.5. We designate as E' that broad spectral white, similar in energy distribution to E, but reduced in intensity to match C and D in overall visual intensity; E' then contains less 520 nm green than does C or E.

There are then six possible illuminating conditions in fig. 9.5 having these characteristics:

(1) The two light sources A and B appear identically green to the eye, both in color and intensity. Only one of these contains the 520 nm green being tested.
(2) The four light sources, C, D, E, and E', appear white to the eye in color. Three of these, C, D, and E', are intensity matched to the eye; only two of these, C and E', contain the 520 nm green being tested, but not in equal amounts. Finally E is not intensity matched to C, D, or E' but does contain the same amount of 520 nm green as does C.

We can also group these six light sources as follows:

(3) Light sources containing equal amounts of 520 nm green:
    A – appears green to the eye;
    C – appears white to the eye;
    E – appears white, but is brighter than C or E'.
(4) Light source containing less 520 nm green than does (3):
    E' – visual color and intensity matched to C.
(5) Light sources containing no 520 nm green:
    B – visual color and intensity matched to A;
    D – visual color and intensity matched to C and E'.

Although A and B can be dispensed with, there is a definite advantage to including the E' light source. If other spectral wavelengths are also physiologically or therapeutically active (but not necessarily as strongly as the 520 nm green) then a comparison of the strength of the response to E versus E' and a comparison of these with C and D should reveal this. By omitting A and B, it is even possible to disguise the true nature of the experiment from the subject, since all light sources now appear white to the eye.

Similar sets of six light sources, two giving colored illumination and four white, could be used for all regions of the spectrum with the exception of the reds and violet/blues at either extreme end of the visible spectrum. Here the arrangement B of fig. 9.5 is not possible, as an examination of fig. 1.7 of Chapter 1 indicates. One is then limited to the perfectly adequate set of four white sources equivalent to C, D, E, and E'. Once again, three of these include the red (or the violet/blue) being tested, while one does not contain the test color.

In a direct comparison of two different colors (e.g., green vs red, or violet vs orange), designated with subscripts 1 and 2, one again is restricted to whites, now with six light sources:

(α) Two light sources, C_1 and C_2 (equivalent to C), each containing one of the test colors;

(β) Light source D, containing neither test color (if either of the colors used in D in fig. 9.1 occurs in the pair being tested, then a different set of complementary colors would be used);

(γ) Two light sources, E_1 and E_2 (equivalent to E), both containing both test colors, but with only the indicated test-color content intensity matched to that of C_1 and C_2, respectively;

(δ) Light source E_3 (equivalent to E'), with visual intensity matched to C_1 and C_2, containing both test colors, with both at reduced intensities compared to C_1 and C_2, respectively.

9.12.2. Survey light sources

In the absence of any clues as to which part of the spectrum might be physiologically or therapeutically active, a double blind survey experiment could be performed. A number

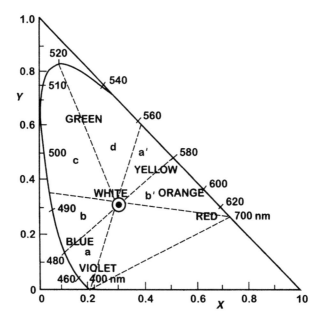

Fig. 9.6. A sectored version of the chromaticity diagram showing color distributions for survey testing.

of possibilities exist. The following provides a reasonably comprehensive survey by using three white-appearing light sources based on dividing up the spectrum as in the sectored chromaticity diagram of fig. 9.6, also shown in different forms in fig. 9.7:

(I)   A light source containing the complementary regions a and a′ in figs 9.6 and 9.7, containing 400 nm to 480 nm violet/blue and 565 nm to 580 nm green/yellow, respectively, in appropriate amounts to produce white;
(II) A light source containing the complementary regions b and b′ in figs 9.6 and 9.7, containing 480 nm to 495 nm blue/green and 580 nm to 700 nm yellow/orange/red, respectively, in appropriate amounts to produce white;
(III) A light source containing the whole spectrum, including all the areas in figs 9.6 and 9.7 in appropriate amounts to produce white, i.e., the same as E in fig. 9.5.

These three light sources are adjusted to appear of equal whiteness and intensity to the eye. All the wavelength limits are only approximate, being dependent on the exact type of incandescent lamps used in III.

If all three light sources, I, II, and III give essentially equal responses, there is probably no specific chromatic effect. A maximum response with light source III would indicate a

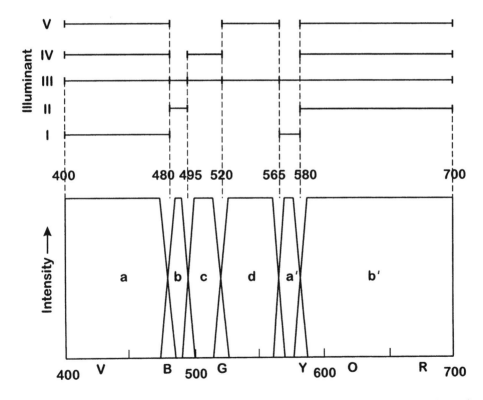

Fig. 9.7. Below: schematic band diagram analogous to fig. 2.6 (intensities, are not to scale); above: the composition of the light sources I to V for survey testing.

specific effect within the green region c plus d in figs 9.6 and 9.7. A maximum response from either light sources I or II would indicate a specific effect within either of the pairs of relevant areas. In each of the last three instances a secondary test is now required to locate the active area precisely. This uses one or two more light sources, each of which contains three bands, of which two are the violet/blue a and the yellow/orange/red b'. The third band is blue/green c (495 nm to 520 nm) for light source IV or green/yellow d (520 nm to 565 nm) for light source V, as shown in fig. 9.7.

For a maximum response from I, this is now compared with IV and/or V to indicate whether the positive response derives from violet/blue a (activity in I and IV and/or V) or from green/yellow a' (activity only in I).

Similarly, for a maximum response from II, this is now compared with IV and/or V to indicate whether the positive response derives from blue/green b (activity only in II) or from yellow/orange/red b' (activity in II and IV and/or V).

For a maximum response from III, a comparison with IV and/or V will distinguish between the two possibly active green regions c and d.

9.12.3. In conclusion

The system of light sources proposed in this addendum can provide experimental designs to locate any specific spectral effects on physiological processes and/or medical conditions. For the first time it permits double blind experimentation, since neither patient nor observer is aware of the nature of the light sources. Thereby both placebo and psychological effects can be completely avoided.

References

American Psychiatric Association, 1994, Diagnostic and Statistical Manual of Mental Disorders, 4th edn (Americal Psychiatric Press, Washington, DC).

Aschoff, J. (ed.), 1981, Handbook of Behavioral Neurobiology, Vol. 4, Biological Rhythms (Plenum Press, New York) pp. 3–545.

Avery, D.H., M.A. Bolte, S. Cohen and M.S. Millet, 1992, Gradual versus rapid dawn simulation treatment of winter depression. J. Clinical Psychiatry 53(10), 359–363.

Avery, D.H., M.A. Bolte, S.R. Dager, L.G. Wilson, M. Weyer, G.B. Cox and D.L. Dunner, 1993, Dawn simulation treatment of winter depression: A controlled study. American J. Psychiatry 150(1), 113–117.

Badia, P., B. Myers, M. Boecker and J. Culpepper, 1991, Bright light effects on body temperature, alertness, EEG, and behavior. Physiology and Behavior 50, 583–588.

Benshoff, H.M., G.C. Brainard, M.D. Rollag and G.R. Lynch, 1987, Suppression of pineal melatonin in Peromyscus leucopus by different monochromatic wavelengths of visible and near-ultraviolet light (UV-A). Brain Research 420(2), 397–402.

Berkow, R. (ed.), 1992, The Merk Manual of Diagnosis and Treatment, 16th edn (Merck and Co., Rahway, NJ) pp. 2647–2649.

Berman, K., R.W. Lam and E.M. Goldner, 1993, Eating attitudes in seasonal affective disorder and bulimia nervosa. J. Affective Disorders 29(4), 219–225.

Binkley, S., 1990, The Clockwork Sparrow (Prentice-Hall, Englewood Cliffs, NJ) pp. 1–246.

Birren, F., 1961, Color Psychology and Color Therapy (Citadel Press, Secaucus, NJ) pp. 3–281.

Birren, F., 1978, Color and Human Response (Van Nostrand Reinhold, New York), pp. 1–126.

Bojkowski, C.J., M.E. Aldhous, J. English, C. Franey, A.L. Poulton, D.J. Skene and J. Arendt, 1987, Suppression of nocturnal plasma melatonin and 6-sulphatoxymelatonin by bright and dim light in man. Hormone and Metabolic Research 19(9), 437–440.

Boulos, Z., S.S. Campbell, A.J. Lewy, M. Terman, D.-J. Dijk and C.I. Eastman, 1995, Light treatment for sleep disorders: Consensus report. VII. Jet Lag. J. Biological Rhythms 10(2), 167–176.

Brainard, G.C., 1998, The healing light: Interface of physics and biology, in: Beyond Seasonal Affective Disorder: Light Treatment for SAD and Non-SAD Disorders, ed. R.W. Lam (American Psychiatric Press, Washington, DC).

Brainard, G.C., B.A. Richardson, L.J. Petterborg and R.J. Reiter, 1982, The effect of different light intensities on pineal melatonin content. Brain Research 233(1), 75–81.

Brainard, G.C., B.A. Richardson, T.S. King, S.A. Matthews and R.J. Reiter, 1983, The suppression of pineal melatonin content and N-acetyltransferase activity by different light irradiances in the Syrian hamster: A dose-response relationship. Endocrinology 113(1), 293–296.

Brainard, G.C., B.A. Richardson, T.S. King and R.J. Reiter, 1984a, The influence of different light spectra on the suppression of pineal melatonin content in the Syrian hamster. Brain Research 294(2), 333–339.

Brainard, G.C., B.A. Richardson, E.C. Hurlbut, S. Steinlecher, S.A. Matthews and R.J. Reiter, 1984b, The influence of various irradiances of artificial light, twilight, and moonlight on the suppression of pineal melatonin content in the Syrian hamster. J. Pineal Research 1(2), 105–119.

Brainard, G.C., P.L. Podolin, S.W. Leivy, M.D. Rollag, C. Cole and F.M. Barker, 1986a, Near ultraviolet radiation (UV-A) suppresses pineal melatonin content. Endocrinology 119(5), 2201–2205.

Brainard, G.C., M.K. Vaughan and R.J. Reiter, 1986b, Effect of light irradiance and wavelength on the Syrian hamster reproductive system. Endocrinology 119(2), 648–654.

Brainard, G.C., A.J. Lewy, M. Menaker, L.S. Miller, R.H. Fredrickson, R.G. Weleber, V. Cassone and D. Hudson, 1988, Dose-response relationship between light irradiance and the suppression of melatonin in human volunteers. Brain Research 454(1–2), 212–218.

Brainard, G., J. French, P. Hannon, M. Rollag, J. Hanifin and W. Storm, 1991, The influence of bright illumination on plasma melatonin, prolactin, and cortisol rhythms in normal subjects during sustained wakefulness. Sleep Research 20, 444.

Brainard, G.C., N.E. Rosenthal, D. Sherry, R.G. Skwerer, M. Waxler and D. Kelly, 1990, Effects of different wavelengths in seasonal affective disorder. J. Affective Disorders 20(4), 209–216.

Brainard, G.C., S. Beacham, J.P. Hanifin, D. Sliney and L. Streletz, 1992, Ultraviolet regulation of neuroendocrine and circadian physiology in rodents and the visual evoked response in children, in: Biological Responses to Ultraviolet a Radiation, ed. F. Urbach (Valdenmar, Overland Park, KS) pp. 261–271.

Brainard, G.C., J.R. Gaddy, F.M. Barker, J.P. Hanifin and M.D. Rollag, 1993, Mechanisms in the eye that mediate the biological and therapeutic effects of light in humans, in: Light and Biological Rhythms in Man, ed. L. Wetterberg (Pergamon Press, Stockholm) pp. 29–53.

Brainard, G.C., F.M. Barker, R.J. Hoffman, M.H. Stetson, J.P. Hanifin, P.L. Podolin and M.D. Rollag, 1994a, Ultraviolet regulation of neuroendocrine and circadian physiology in rodents. Vision Research 34(11), 1521–1533.

Brainard, G.C., J.R. Gaddy, F.L. Ruberg, F.M. Barker, J.P Hanifin M.D. and Rollag, 1994b, Ocular mechanisms that regulate the human pineal gland, in: Advances in Pineal Research, eds M. Møller and P. Pevet (John Libby, London) pp. 415–432.

Bronstein, D.M., G.H. Jacobs, K.A. Haak, J. Neitz and L.D. Lytle, 1987, Action spectrum of the retinal mechanism mediating nocturnal light-induced suppression of rat pineal gland N-acetyltransferase. Brain Research 406(1–2), 352–356.

Campbell, S.S., and D. Dawson, 1990, Enhancement of nighttime alertness and performance with bright ambient light. Physiology and Behavior 48(2), 317–320.

Campbell, S.S., D.-J. Dijk, Z. Boulos, C.I. Eastman, A.J. Lewy and M. Terman, 1995, Light treatment for sleep disorders: Consensus report. III. Alerting and activating effects. J. Biological Rhythms 10(2), 129–132.

Cardinali, D.P., F. Larin and R.J. Wurtman, 1972a, Action spectra for effects of light on hydroxyindole-o-methyltransferases in rat pineal, retina and Harderian gland. Endocrinology 91(4), 877–886.

Cardinali, D.P., F. Larin and R.J. Wurtman, 1972b, Control of the rat pineal gland by light spectra. Proc. Nat. Acad. Sci. USA 69(8), 2003–2005.

CIE, 1987, CIE International Lighting Vocabulary, CIE publication number 17.4 (CIE, Vienna).

Coohill, T.P., 1991, Action spectra again? Photochemistry and Photobiology 54(5), 859–870.

Czeisler, C.A., J.S. Allan, S.H. Strogatz, J.M. Ronda, R. Sanchez, C.D. Rios, W.O. Freitag, G.S. Richardson and R.E. Kronauer, 1986, Bright light resets the human circadian pacemaker independent of the timing of the sleep-wake cycle. Science 233(4764), 667–671.

Czeisler, C.A., M.P. Johnson, J.F. Duffy, E.N. Brown, J.M. Ronda and R.E. Kronauer, 1990, Exposure to bright light and darkness to treat physiologic maladaptation to night work. New England J. Medicine 322(18), 1253–1259.

Czeisler, C.A., T.L. Shanahan, E.B. Klerman, H. Martens, D.J. Brotman, J.S. Emens, T. Klein and J.F. Rizzo III, 1995, Suppression of melatonin secretion in some blind patients by exposure to bright light. New England J. Medicine 332, 6–11.

Daan, S., and A.J. Lewy, 1984, Scheduled exposure to daylight: a potential strategy to reduce "jet lag" following transmeridian flight. Psychopharmacology Bulletin 20(3), 566–568.

Dawson, D., and S.S. Campbell, 1991, Timed exposure to bright light improves sleep and alertness during simulated night shifts. Sleep 14(6), 511–516.

Eastman, C.I., 1990a, What the placebo literature can tell us about light therapy for SAD. Psychopharmacology Bulletin 26(4), 495–504.

Eastman, C.I., 1990b, Circadian rhythms and bright light: Recommendations for shift work. Work and Stress 4, 245–260.

Eastman, C.I., 1991, Squashing versus nudging circadian rhythms with artificial bright light: solutions for shift work? Perspectives in Biology and Medicine 34(2), 181–195.

Eastman, C.I., M.A. Young and L.F. Fogg, 1993, A comparison of two different placebo-controlled SAD light treatment studies, in: Light and Biological Rhythms in Man, ed. L. Wetterberg (Pergamon Press, Stockholm) pp. 371–393.

Feinstein, A.A., 1977, Clinical Biostatistics (Mosby, Saint Louis).

French, J., P. Hannon and G.C. Brainard, 1990, Effects of bright illuminance on body temperature and human performance. Ann. Rev. Chronopharmacology 7, 37–40.

Gaddy, J.R., M.D. Rollag and G.C. Brainard, 1993, Pupil size regulation of threshold of light- induced melatonin suppression. J. Clinical Endocrinology and Metabolism 77(5), 1398–1401.

Gaddy, J.R., F.L. Ruberg, G.C. Brainard and M.D. Rollag, 1994, Pupillary modulation of light-induced melatonin suppression, in: The Biologic Effects of Light, eds M.F. Holick and E.G. Jung (Walter de Gruyter, Berlin) pp. 159–168.

Gerrard, R.M., 1958, Differential effects of colored lights on psychophysiological functions, Doctoral Dissertation (University of California, Los Angeles) pp. 1–301.

Gimbal, T., 1988, Healing Through Colors (C.W. Daniels, Essex, England) pp. 7–175.

Häder, D.-P., and M. Tevini, 1987, General Photobiology (Pergamon Press, Oxford) pp. 3–309.

Hamilton, M., 1967, Development of a rating scale for primary depressive illness. British J. Social and Clinical Psychology 6(4), 278–296.

Hannon, P.R., G.C. Brainard, W. Gibson, J. French, D. Arnall, L. Brugh, C. Littleman-Crank, S. Fleming, J. Hanifin and B. Howell, 1991, Effects of bright illumination on sublingual temperature, cortisol and cognitive performance in humans during nighttime hours. Photochemistry and Photobiology 53, 15S.

Henning, W., 1936, The Fundamentals of Chrome-Orthoptics (Actino Laboratories, Chicago).

Henning, W., 1939, The Practice of Modern Photometry (The American College of Optomotrists, Chicago).

Henning, W., 1940, Procedures in Refraction and Functional Disorders of Vision (Buckeye, Columbus, OH).

Holick, M.F., 1989, Vitamin D: biosynthesis, metabolism, and mode of action, in: Endocrinology, Vol. 2, eds L.J. DeGroot et al. (W.B. Saunders, Philadelphia) pp. 902–926.

Holtz, M.M., J.J. Milette, J.S. Takahashi and W.F. Turek, 1990, Spectral senstivity of the circadian clock's response to light in Djungarian hamsters, in: 2nd Annual Meeting of the Society for Research on Biological Rhythms, Amelia Island, FL, May 10, 1990.

Hutchings, J.B., 1997, Color in Anthropology and Folklore, in: Color for Science, Art, and Technology, ed. K. Nassau (Elsevier, Amsterdam).

Jacobs, G.H., 1992, Ultraviolet vision in vertebrates. American Zoology 32, 544–554.

Jacobs, G.H., and J.F. Deegan, 1994, Sensitivity to ultraviolet light in the gerbil (Meriones unguiculatus): Characteristics and mechanisms. Vision Research 34(11), 1433–1441.

Jagger, J., 1985, Solar-UV Actions on Living Cells (Praeger Publishers, New York) pp. 1–195.

Jimerson, D.C., H.J. Lynch, R.M. Post, R.J. Wurtman and W.E. Bunney, 1977, Urinary melatonin rhythms during sleep deprivation in depressed patients and normals. Life Sciences 20(9), 1501–1508.

Kaiser, P.K., 1984a, Physiological response to color: A critical review. Color Research and Application 9(1), 29–35.

Kaiser, P.K., 1984b, Phototherapy using chromatic, white, and ultraviolet light. Color Research and Application 9(4), 195–205.

Kaplan, R., 1983, Changes in form visual fields in reading disabled children produced by syntonic stimulation. Int. J. Biosocial Research 5(1), 20–33.

Klein, D.C., and J.L. Weller, 1972, Rapid light-induced decrease in pineal serotonin N-acetyltransferase activity. Science 177(48), 532–533.

Klein, D.C., R.Y. Moore and S.M. Reppert, 1991, Suprachiasmatic Nucleus: The Mind's Clock (Oxford University Press, Oxford) pp. 5–456.

Krauskopf, J., 1997, Color vision, in: Color for Science, Art, and Technology, ed. K. Nassau (Elsevier, Amsterdam).

Kripke, D.F., D.J. Mullaney, T.J. Savides and J.C. Gillin, 1989, Phototherapy for nonseasonal major depressive disorders, in: Seasonal Affective Disorders and Phototherapy, eds N.E. Rosenthal and M.C. Blehar (The Guilford Press, New York) pp. 342–356.

Kripke, D.F., 1993, Light regulation of the menstrual cycle, in: Light and Biological Rythms in Man, ed. L. Wetterberg (Pergamon Press, Stockholm) pp. 305–312.

Lam, R.W., E.M. Goldner, L. Solyom and R.A. Remick, 1994, A controlled study of light therapy for bulimia nervosa. Americal J. Psychiatry 151(5), 744–750.

Lewy, A.J., T.A. Wehr, F.K. Goodwin, D.A. Newsome and S.F. Markey, 1980, Light suppresses melatonin secretion in humans. Science 210(4475), 1267–1269.

Lewy, A.J., H.A. Kern, N.E. Rosenthal and T.A. Wehr, 1982, Bright artificial light treatment of a manic-depressive patient with a seasonal mood cycle. American J. Psychiatry 139(11), 1496–1498.

Lewy, A.J., R.L. Sack, L.S. Miller and T.M. Hoban, 1987, Antidepressant and circadian phase-shifting effects of light. Science 235(4786), 352–354.

Lieberman, J., 1986, The effect of syntonic (colored light) stimulation of certain visual and cognitive functions. J. Optometric Vision Development 17.

Lieberman, J., 1991, Light: Medicine of the Future (Bear, Santa Fe, NM) pp. 3–238.

Lin, M.C., D.F. Kripke, B.L. Parry and S.L. Berga, 1990, Night light alters menstrual cycles. Psychiatry Research 33(2), 135–138.

Lynch, H.J., D.C. Jimmerson, Y. Ozaki, R.M. Post, W.E. Bunney and R.J. Wurtman, 1978, Entrainment of rhythmic melatonin secretion in man to a 12-hour phase shift in the light/dark cycle. Life Sciences 23(15), 1557–1563.

Lynch, H.J., M.H. Deng and R.J. Wurtman, 1984, Light intensities required to suppress nocturnal melatonin secretion in albino and pigmented rats. Life Sciences 35(8), 841–847.

MacLaughlin, J.A., R.R. Anderson and M.F. Holick, 1982, Spectral character of sunlight modulates the photosynthesis of previtamin D_3 and its photoisomers in human skin. Science 216(4549), 1001–1003.

McCormack, C.E., and C.R. Sontag, 1980, Entrainment by red light of running activity and ovulation rhythms of rats. American J. Physiology 239(5), R450–R453.

McIntyre, I.M., T.R. Norman, G.D. Burrows and S.M. Armstrong, 1989, Human melatonin suppression by light is intensity dependent. J. Pineal Research 6(2), 149–156.

Milette, J.J., M.M. Hotz, J.S. Takahashi and F.W. Turek, 1987, Characterization of the wavelength of light necessary for initiation of neuroendocrine-gonadal activity in male Djungarian hamsters, in: 20th Annual Meeting of the Society for Study of Reproduction, Urbana, IL, July 20–23, 1987, p. 110.

Moore, R.Y., 1983, Organization and function of a central nervous system circadian oscillator: The suprachiasmatic hypothalamic nucleus. Federation Proceedings 42(11), 2783–2789.

Moore-Ede, M.C., 1993, The Twenty-Four Hour Society: Understanding a World That Never Stops (Addison-Wesley, New York) pp. 3–213.

Moore-Ede, M.C., F.M. Sulzman and C.A. Fuller, 1982, The Clocks That Time Us (Harvard Univ. Press, Cambridge, MA) pp. 1–278.

Moore-Ede, M.C., C.A. Czeisler and G.S. Richardson, 1983, Circadian timekeeping in health and disease. Part 2. Clinical implications of circadian rhythmicity. New England J. Medicine 309(9), 530–536.

Morgane, P.J. and J. Panksepp (eds), 1979, Handbook of the Hypothalamus, Vols 1, 2 and 3 (Marcel Dekker, New York, 1979, 1980).

Nelson, D.E., and J.S. Takahashi, 1991, Comparison of visual sensitivity for suppression of pineal melatonin and circadian phase-shifting in the golden hamster. Brain Research 554(1–2), 272–277.

Nguyen, D.C., J.P. Hanifin, M.D. Rollag, M.H. Stetson and G.C. Brainard, 1990, The influence of different photon densities of 620 nm light on pineal melatonin in Syrian hamsters. The Anatomical Record 226, 72A.

Oren, D.A., G.C. Brainard, J.R. Joseph-Vanderpool, S.H. Johnston, E. Sorek and N.E. Rosenthal, 1991, Treatment of seasonal affective disorder with green light versus red light. American J. Psychiatry **148**(4), 509–511.

Ott, J.N., 1973, Health and Night: The Effects of Natural and Artificial Light on Man and Other Living Things (Devin-Adair, Old Greenwich, CT) pp. 3–208.

Otto, J.L., and K.S. Bly, 1984, Effectiveness of syntonic (colored light) therapy for treating visual disorders in private practice. A Doctoral Dissertation.

Parry, B.L., N.E. Rosenthal, L. Tamarkin and T.A. Wehr, 1987, Treatment of a patient with seasonal premenstrual syndrome. American J. Psychiatry **144**(6), 762–766.

Parry, B.L., S.L. Berga, N. Mostofi, P.A. Sependa, D.F. Kripke and J.C. Gillin, 1989, Morning versus evening bright light treatment of late luteal phase dysphoric disorder. American J. Psychiatry **146**(9), 1215–1217.

Pickard, G.E., and A.J. Silverman, 1981, Direct retinal projections to the hypothalamus, piriform cortex, and accessory optic nuclei in the golden hamster as demonstrated by a sensitive anterograde horseradish peroxidase technique. J. Comparative Neurology **196**(1), 155–172.

Podolin, P.C., M.D. Rollag and G.C. Brainard, 1987, The suppression of nocturnal pineal melatonin in the Syrian hamster: Dose-response curves at 500 nm and 360 nm. Endocrinology **121**(1), 266–270.

Provencio, I., S. Wong, A.B. Lederman, S.M. Argamaso and R.G. Foster, 1994, Visual and circadian responses to light in aged retinally degenerate mice. Vision Research **34**(14), 1799–1806.

Reiter, R.J., 1981, The mammalian pineal gland: structure and function. American J. Anatomy **162**(4), 287–313.

Reiter, R.J., 1991, Pineal gland: interface between the photoperiodic environment and the endocrine system. Trends in Endocrinology and Metabolism **2**, 13–19.

Rey, M.S. (ed.), 1993, IES Lighting Handbook, Chapter 5: Non-visual effects of radiant energy (Illuminating Engineering Society of North America, New York) pp. 135–176.

Rosenthal, N.E., 1993, Diagnosis and treatment of seasonal affective disorder. J. American Medical Association **270**(22), 2717–2720.

Rosenthal, N.E., D.A. Sack, J.C. Gillin, A.J. Lewy, F.K. Goodwin, Y. Davenport, P.S. Mueller, D.A. Newsome and T.A. Wehr, 1984, Seasonal affective disorder. A description of the syndrome and preliminary findings with light therapy. Archives of General Psychiatry **41**(1), 72–80.

Rosenthal, N.E., D.A. Sack, R.G. Skwerer, F.M. Jacobsen and T.A. Wehr, 1988, Phototherapy for seasonal affective disorder. J. Biological Rhythms **3**(2), 101–120.

Rosenthal, N.E., J.R. Joseph-Vanderpool, A.A. Levendosky, S.H. Johnston, R. Allen, K.A. Kelly, E. Souetre, P.M. Schultz and K.E. Starz, 1990, Phase-shifting effects of bright morning light as treatment for delayed sleep phase syndrome Sleep **13**(4), 354–361.

Rosenthal, N.E., D.E. Moul, C.J. Hellekson, D.A. Oren, A. Frank, G.C. Brainard, M.G. Murray and T.A. Wehr, 1993, A multicenter study of the light visor for Seasonal Affective Disorder: No difference in efficacy found between two different intensities. Neuropsychopharmacology **8**(2), 151–160.

Ross, M., and J.M. Olson, 1981, An expectancy-attribution model of the effects of placebos. Psychological Review **88**(5), 408–437.

Ruberg, F.L., D.J. Skene, J.P. Hanifin, M.D. Rollag, J. English, J. Arendt and G.C. Brainard, 1996, Melatonin regulation in humans with color vision deficiencies. J. Clinical Endocrinology and Metabolism **81**(8), 2980–2985.

Sanford, B., S. Beacham, J.P. Hanifin, P. Hannon, L. Streletz, D. Sliney and G.C. Brainard, 1996, The effects of ultraviolet-A radiation on visual evoked potential in the young human eye. Acta Opththalmologica Scandivavica **74**, 553–557.

Schivelbusch, W., 1988, Disenchanted Night: The Industrialization of Light in the Nineteenth century(The University of California Press, Berkeley) pp. 3–221.

Smith, K.C. (ed.) 1989, The Science of Photobiology, 2nd edn (Plenum Press, New York) pp. 1–426.

Society for Light Treatment and Biological Rhythms, 1991, Consensus statements on the safety and effectiveness of light therapy of depression and disorders of biological rhythms. Society for Light Treatment and Biological Rhythms Abstracts, pp. 45–50.

Society for Light Treatment and Biological Rhythms, 1993, Membership directory (Society for Light Treatment and Biological Rhythms, Wilsonville, OR) pp. 1–66.

Spitler, H.R., 1941, The Syntonic Principle: Its Relation to Health and Ocular Problems (The College of Syntonic Optometry, Eaton, OH) pp. 3–210.

Stewart, K.T., J.R. Gaddy, D.M. Benson, B. Byrne, K. Doghramji and G.C. Brainard, 1990, Treatment of winter depression with a portable, head-mounted phototherapy device. Progr. Neuro-Psychopharmacology and Biological Psychiatry 14(4), 569–578.

Stewart, K.T., J.R. Gaddy, B. Byrne, S. Miller and G.C. Brainard, 1991, Effects of green or white light for treatment of seasonal depression. Psychiatry Research 38(3), 261–270.

Takahashi, J.S., P.J. DeCoursey, L. Bauman and M. Menaker, 1984, Spectral sensitivity of a novel photoreceptive system mediating entrainment of mammalian circadian rhythms. Nature 308(5955), 186–188.

Terman, M., and J.S. Terman, 1992, Light therapy for winter depression, in: Biologic Effects of Light, eds M.F. Holick and A.M. Kligman (Walter de Gruyter, New York) pp. 133–154.

Terman, M., J.S. Terman, F.M. Quitkin, P.J. McGrath, J.W. Stewart and B. Rafferty, 1989, Light therapy for seasonal affective disorder. A review of efficacy. Neuropsychopharmacology 2(1), 1–22.

Terman, J.S., M. Terman, D. Schlager, B. Rafferty, M. Rosofsky, M.J. Link, P.F. Gallin and F.M. Quitkin, 1990, Efficacy of brief, intense light exposure for treatment of winter depression. Psychopharmacology Bulletin 26(1), 3–11.

Terman, M., A.J. Lewy, D.-J. Dijk, Z. Boulos, C.I. Eastman and S.S. Campbell, 1995, Light treatment for sleep disorders: consensus report. IV. Sleep phase and duration disturbances. J. Biological Rhythms 10(2), 135–147.

Thiele, G., and H. Meissl, 1987, Action spectra of the lateral eyes recorded from mammalian pineal glands. Brain Research 424(1), 10–16.

Thorington, L., 1985, Spectral, irradiance, and temporal aspects of natural and artificial light, in: The Medical and Biological Effects of Light, eds R.J. Wurtman, M.J. Baum, J.T Potts, (The New York Acad. Sci., New York) pp. 28–54.

Turner, J.A., R.A. Deyo, J.D. Loeser, M. Von Korff and W. Fordyce, 1994, The importance of placebo effects in pain treatment and research. J. American Medical Association 271(20), 1609–1614.

Urbach, F. (ed.), 1992, Biological Responses to Ultraviolet a Radiation (Valdenmar Publishing, Overland Park, KS) pp. 1–418.

US Congress, 1991, Biological rhythms: Implications for the worker (US Government Printing Office, Washington, DC) pp. 3–270.

Vanecek, J., and H. Illnerova, 1982, Night pineal N-acetyltransferase activity in rats exposed to white or red light pulses of various intensities and duration. Experientia 38(11), 1318–1320.

Vaughan, G.M., R.W. Pelham, S.F. Pang, L.L. Loughlin, K.M. Wilson, K.L. Sandock, M.K. Vaughan, S.H. Koslow and R.J. Reiter, 1976, Nocturnal elevation of plasma melatonin and urinary 5-hydroxyindoleacetic acid in young men: Attempts at modification by brief changes in environmental lighting and sleep by autonomic drugs. J. Clinical Endocrinology and Metabolism 42(4), 752–764.

Vision Research, 1994, Special Issue: The Biology of Ultraviolet Reception. Vision Research 34(11), 1359–1539.

Wehr, T.A., D.A. Sack and N.E. Rosenthal, 1987, Seasonal affective disorder with summer depression and winter hypomania. American J. Psychiatry 144(12), 1602–1603.

Wetterberg, L., 1978, Melatonin in humans: physiological and clinical studies. J. Neural Transmission – Supplement 13, 289–310.

Wetterberg, L. (ed.), 1993, Light and Biological Rhythms in Man (Pergamon Press, Stockholm) pp. 1–448.

Wever, R.A., 1985, Use of light to treat jet lag: Differential effects of normal and bright artificial light on human circadian rhythms. Ann. New York Acad. Sci. 453, 282–304.

Winget, C.M., C.W. DeRoshia, C.L. Markley and D.C. Holley, 1984, A review of human physiological and performance changes associated with desynchronosis of biological rhythms. Aviation Space and Environmental Medicine 55(12), 1085–1096.

Wurtman, R.J., M.J. Baum and J.T. Potts (eds), 1985, The Medical and Biological Effects of Light (The New York Academy of Sciences, New York) pp. 1–408.

Yerevanian, B.I., J.L. Anderson, L.J. Grota and M. Bray, 1986, Effects of bright incandescent light on seasonal and nonseasonal major depressive disorder. Psychiatry Research 18(4), 355–364.

Yglesias, M., K.T. Stewart, J.R. Gaddy, G. Zivin, K. Doghramji, W. Thornton and G.C. Brainard, 1993, Does color influence blood pressure and heart rate? in: The 7th Congress Association Internationale de la Couleur (Hungary, Budapest).

Zolman, J.F., 1993, Biostatistics (Oxford Univ. Press, New York).

chapter 10

COLORANTS:
ORGANIC AND INORGANIC
PIGMENTS

PETER A. LEWIS

Sun Chemical Corp.
Colors Group
Cincinnati, OH 45232, USA

Color for Science, Art and Technology
K. Nassau (Editor)
© 1998 Elsevier Science B.V. All rights reserved.

CONTENTS

10.1. Introduction 285

    10.1.1. Color and tint strength 285

    10.1.2. Fastness 286

    10.1.3. Exposure or weatherability 286

10.2. International nomenclature – the Colour Index system 287

10.3. Classification of organic pigments by chemistry 287

10.4. Classification of organic pigments by color 288

    10.4.1. Reds 288

    10.4.2. Blues 297

    10.4.3. Yellows 298

    10.4.4. Oranges 301

    10.4.5. Greens 302

10.5. Classification of inorganic pigments by chemistry 304

10.6. Classification of inorganic pigments by color 304

    10.6.1. Reds 304

    10.6.2. Violets 305

    10.6.3. Blues 305

    10.6.4. Yellows 306

    10.6.5. Oranges 307

    10.6.6. Greens 308

    10.6.7. Browns 308

10.7. Metals as pigments 309

10.8. Pigment dispersion and end use application 309

10.9. Sources of additional information 311

10.1. Introduction

Before entering into any discussion relating to pigments it is first necessary to clearly define what is meant by a *pigment* as opposed to a *dyestuff* (or dye) since in many earlier texts on color the terms "pigment" and "dyestuff" are used almost interchangeably.

A definition of a pigment has been proposed by the Color Pigments Manufacturers Association (CPMA), formerly known as the Dry Color Manufacturers Association (DCMA) in response to a request from the Toxic Substances Interagency Testing Committee. This definition was developed specifically to enable differentiation between a dyestuff and a pigment and as such it is worthwhile reproducing this definition in its entirety:

> "Pigments are colored, black, white or fluorescent particulate organic and inorganic solids which usually are *insoluble* in, and essentially physically and chemically *unaffected* by, the vehicle or substrate in which they are incorporated. They alter appearance by selective absorption and/or by scattering of light.
>
> Pigments are usually *dispersed* in vehicles or substrates for application, as for instance in inks, paints, plastics or other polymeric materials. Pigments *retain* a crystal or particulate structure throughout the coloration process.
>
> As a result of the physical and chemical characteristics of pigments, pigments and dyes differ in their application; when a dye is applied, it penetrates the substrate in a *soluble* form after which it may or may not become insoluble. When a pigment is used to color or opacify a substrate, the finely divided *insoluble* solid remains throughout the coloration process."

A most confusing European term, the use of which should be discouraged, is "pigment dyestuff". This term is meant to refer to insoluble organic pigments devoid of salt forming groups, for example, benzimidazolone orange (PO 36).

10.1.1. Color and tint strength

The most obvious property of a pigment is its hue, that is its color as being distinctly blue, yellow, green, or red and the finer detail that distinguishes a green shade (i.e., greenish) yellow from a red shade yellow. The causes of color in pigments is reviewed in Chapter 4.

Evaluation of any pigment must include a test of full color or masstone that requires inspection of the pigment, undiluted with white, but fully dispersed in a medium that has relevance to the colored formulation. Inspection of this full color shows the hue, intensity, transparency, cleanliness and jetness of the pigment. Comparison of this using side by side evaluation with a standard or specification full color will show how close the test pigment comes to the standard or specification. The full color can then be tinted with a white base such as titanium dioxide to enable the pigment's tinting strength to be assessed. Such a tint is known as a bleached, tinted or reduced draw down or display.

Since a pigment, by definition, is insoluble in the medium in which it is being used, the use of a pigment as a colorant for a system relies upon the pigment being dispersed throughout the chosen system. The system being colored may be a paint, an ink or a polymer such as PVC, polyethylene, polystyrene or similar material. Nevertheless, the color strength developed within the system depends upon the pigment particles being optimally dispersed in a stable, homogeneous manner with minimal aggregates remaining. Pigments are basically colloidal in nature with typical particles being from 0.03 to 0.15 micrometer in diameter and with surface areas of the order of 70 to 90 m^2/g. Special opaque grades are likely to have surface areas considerably smaller; values of from 18 to 24 m^2/g are not unusual. Since physical properties are often a function of exposed surface, the smaller the pigment particles, the higher the surface area and the more susceptible is the pigment to degradation by sunlight, chemicals and other environmental factors that relate to exposure of surfaces.

10.1.2. Fastness

The term fastness is used to relate to how susceptible or durable a pigment is to light, heat, solvents, acidic or alkaline environments, etc. Each of these is quantified using a "Fastness Scale" which rates a pigment from 1 (poor) to 5 (excellent) in all cases except lightfastness, which is sub-divided into 1 through 8, 1 being total failure to 8 being outstanding.

The term *"lightfastness"* refers to a pigment's ability to withstand exposure to light, both direct and indirect, natural and artificial, without suffering any visible change in appearance. The most damaging components of light lie in the Infra Red and Ultra Violet regions of the spectrum and, as such, a rapid evaluation of a pigment's likely reaction to long term exposure to light can be assessed by using equipment that maximizes the exposure of the finished or printed article to UV light. Artificial sources such as the Xenon or carbon arc fade-ometer (ASTM D 822–80) may be used to assess the pigment's behavior in accelerated exposure using the "Blue Wool Scale" to calibrate the measuring equipment (B.S. 1006).

Pigments can deteriorate and lose or change color by reacting with each other, with the medium in which they are suspended, or with the components or impurities in the atmosphere. These topics are covered in Chapter 12, which includes the consideration of typical pigment-containing artists' oil paints.

10.1.3. Exposure or weatherability

It is now generally accepted on a worldwide basis that the true test of a high performance pigment is through prolonged outdoor exposure at specifically chosen sites in the state of Florida. To this end many commercial establishments exist in the state that will provide a controlled service to expose sprayed panels of pigmented coatings or plastics angled in a pre-determined way towards the sun for periods up to five years with assessments at intervals during this period. Typically, many duplicate displays will be exposed per test with one such display being returned to the formulator every six months. Pigments

TABLE 10.1
Colour Index Names and accepted abbreviations.

Colour Index Name	Accepted abbreviation
Pigment Blue	PB
Pigment Green	PG
Pigment Orange	PO
Pigment White	PW
Pigment Black	PBk
Pigment Metal	PM
Pigment Violet	PV
Pigment Yellow	PY
Pigment Brown	PBr
Pigment Red	PR

used in the printing ink industry and similar pigments that cannot be described as "high performance" products do not have to qualify through such rigorous outdoor exposure testing.

10.2. International nomenclature – the Colour Index system

In any modern publication discussing pigments of any description, it is likely that the author will make use of the coding system published as a joint undertaking by the Bradford Society of Dyers and Colourists (BSDC) in the United Kingdom and the Association of Textile Chemists and Colorists (AATCC) in the United States. This system is known as the "Colour Index" and as such is a recognized Trade Mark. The Colour Index (CI) identifies each pigment by giving the compound a unique "Colour Index Name" and a "Colour Index Number". This description is valuable to pigment consumers for documents such as Material Safety or Hazard Data Sheets, and for chemical composition identification acceptable to most government bodies. For example DNA Orange has the Colour Index name of Pigment Orange 5 (PO 5) and the Colour Index number of 12075 and as such is unique in this respect since no other pigment will carry this identification. A table of the accepted abbreviations is presented as table 10.1.

10.3. Classification of organic pigments by chemistry

Pigments may be broadly divided into opaque or hiding whites, blacks and colored toners. All opaque whites are inorganic compounds and, since they are not colored as such, they fall outside the contents of this chapter. Such opaque whites include the following:

Lithophone (co-precipitate of barium sulfate, $BaSO_4$, and zinc sulfide, ZnS).
Zinc oxide, ZnO.
White lead (basic lead carbonate), $2Pb(CO_3)_2$, $Pb(OH)_2$.
Antimony oxide, Sb_2O_3.

Titanium white (mixture of titanium dioxide, TiO_2 and blanc fixe, $BaSO_4$).

Titanium dioxide (rutile form), TiO_2.

The major black pigment used in today's market is pure carbon black, Pigment Black 7, also known as Furnace Black or Channel Black. Lamp or Vegetable Black, Pigment Black 6 and Pigment Black 9 or Bone Black, from charred animal bones, are used only for specialized applications such as artists' colors and crayons. Aniline Black, Pigment Black 1, one of the few organic blacks, suffers from low tint strength and tint fastness when compared to the more economical carbon blacks. Other inorganic blacks available include black iron oxide, $FeO \cdot Fe_2O_3$, Pigment Black 11 and the spinel blacks such as iron copper chromite black spinel, the solid solution $CuO \cdot Fe_2O_3 \cdot Cr_2O_3$.

Colored organic pigments contain a characteristic grouping or arrangement of atoms known as a "*chromophore*" which imparts color to the molecule. A characteristic of most organic colors is the occurrence of alternating double bonds throughout the molecule, that is the system is a conjugated system, in conjunction with a benzenoid skeleton that results in the availability of a diffuse electron cloud that allows the phenomenon of color by transition of these electrons among the molecular orbitals of the molecule. In addition, the organic pigment is likely to feature a number of modifying groups called "auxochromes" that alter the primary color of the pigment in a more subtle way, such that a red is shifted to a more yellow shade or a blue to a more red shade whilst still maintaining the primary hue of red or blue rather than pushing the hue over to an orange or a violet (also see Section 4.5.1).

Perhaps the most important of the chromophores is the azo chromophore (–N=N–). The naphthol reds, monoarylide and diarylide yellows, benzimidazolones, pyrazolones and azo condensation pigments are all examples of organic pigments that feature the azo chromophore. Of equal importance is the phthalocyanine structure based upon tetrabenzotetra-azaporphin. Organic pigments are also derived from heterocyclic structures such as trans-linear quinacridone (PV 19) and carbazole dioxazine violet (PV 23). Finally there are pigments that result from vat dyestuffs and miscellaneous metal complexes. The causes of color in organic pigments is reviewed in Section 4.5.1.

10.4. Classification of organic pigments by color

10.4.1. Reds

Many reds finding widespread use within the color consuming industries contain the azo chromophore (–N=N–) and as such are termed 'azo reds'. Their use centers around their economics, since these colors are tinctorially strong and of relatively minor expense when compared with reds such as those derived from quinacridone or vat dyestuffs. A further subdivision is possible into acid, monoazo metallized pigments such as Manganese Red 2B (PR 48:4) and non-metallized azo reds such as the 'Naphthol Reds'. Typically each of the metallized type contain an anionic grouping such as sulfonic (–SO$_3$H) or carboxylic acid (–COOH) which will ionize and react with a metal cation such as calcium, manganese or barium to form an insoluble metallized azo pigment. Conversely non-metallized azo reds do not contain an anionic group in their structure and therefore cannot complex with a metal cation.

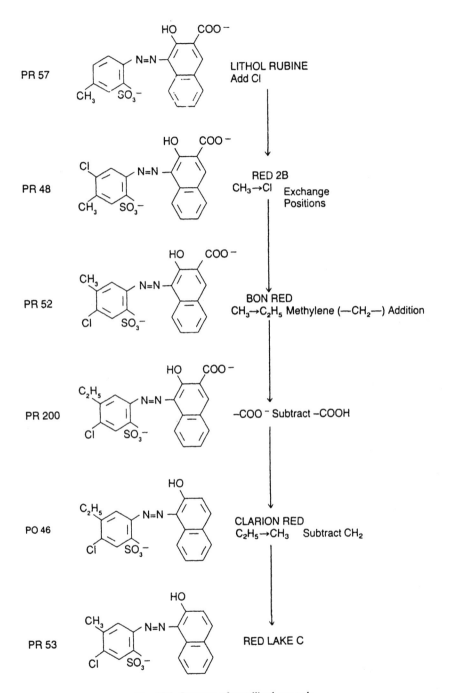

Fig. 10.1. Structure of metallized azo reds.

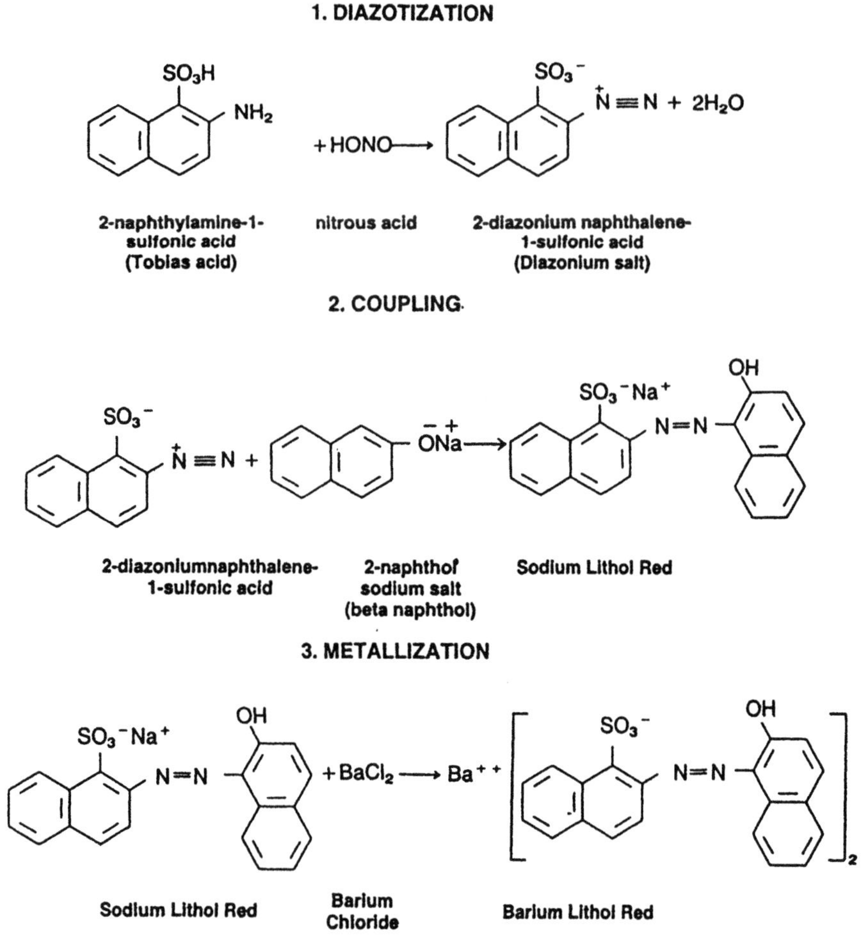

1. DIAZOTIZATION

2-naphthylamine-1-   nitrous acid   2-diazonium naphthalene-
sulfonic acid   1-sulfonic acid
(Tobias acid)   (Diazonium salt)

2. COUPLING

2-diazoniumnaphthalene-   2-naphthol   Sodium Lithol Red
1-sulfonic acid   sodium salt
                            (beta naphthol)

3. METALLIZATION

Sodium Lithol Red   Barium   Barium Lithol Red
                           Chloride

Fig. 10.2. Lithol red manufacture.

Figure 10.1 illustrates the structure of a series of metallized azo reds that are of considerable commercial importance and that find use within the printing ink and plastics industry with limited use in the coatings industry.

10.4.1.1. *Lithol Red*

The reaction sequence used to manufacture Lithol Red is shown in fig. 10.2. This pigment's major use is in the printing ink industry, particularly Barium Lithol for publication gravure printing as the red for four color printing. Lithol Reds will hydrolyze under strongly acidic or alkaline conditions to give weaker, yellower shade products.

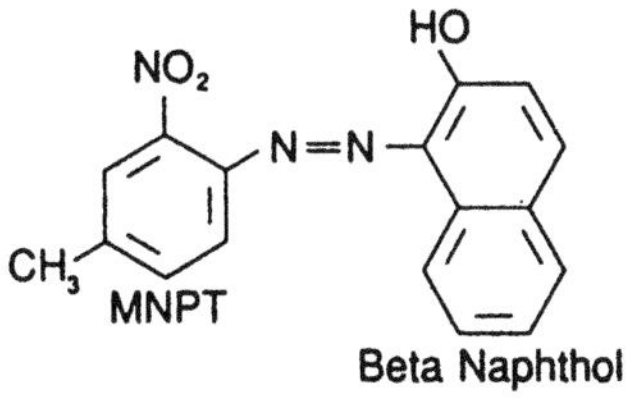

Fig. 10.3. Toluidine red, Pigment Red 3.

10.4.1.2. Permanent Red 2B

Use in highly alkaline systems is again severely restricted. The barium salt is characterized by a clean, yellow hue as compared to the bluer calcium salt. The barium salt has poorer lightfastness and weaker tinting strength but a slightly better bake stability as compared to the calcium salt. The Manganese Red 2B has sufficiently improved lightfastness to allow its use in implement finishes and aerosol spray cans for touch-up paints. This salt is bluer, dirtier and less intense as compared to the calcium salt.

10.4.1.3. Lithol Rubine Red

This blue shade red was discovered in 1903 and has found widespread use in the printing ink industry, both offset and solvent as the process "magenta" of four color printing.

10.4.1.4. BON Reds

Characterized by outstanding cleanliness, brightness and color purity the manganese salt offers a very blue shade with improved lightfastness as compared to the calcium salt. As such the manganese salt is suitable for outdoor applications and, as with the Manganese Red 2B, can be used in blends with Molybdate Orange (PR 104) to give bright, economical reds.

10.4.1.5. Toluidine Red

This pigment, shown in fig. 10.3, is a non-metallized azo red, chemically the reaction product from coupling the diazonium salt of 2-nitro-4-toluidine (MNPT) onto 2-naphthol (beta naphthol). Almost the entire US production of Toluidine Red, an amount in excess of 0.75 million kilos, is consumed by the coatings industry. The pigment provides a bright, economical red of acceptable lightfastness, when used in full shade, coupled with a high degree of color intensity and good hiding power.

10.4.1.6. Naphthol Reds

Naphthol Reds are a group of pigments that exhibit good tinctorial properties combined with moderate fastness to heat, light and solvents. They are extremely acid, alkali, and soap resistant pigments, properties which lead to their use in masonry paints, latex emulsions and for printing applications needing alkali resistance. Figure 10.4 illustrates the generic structure of the naphthol molecule and gives the chemistry of several of the types commercially available.

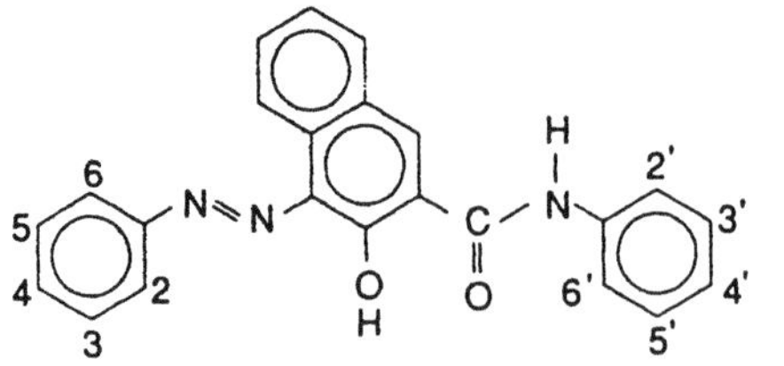

Colour Index Name	Colour Index Number	Substituents					
		2	4	5	2'	4'	5'
PR 2	12310	Cl	H	Cl	H	H	H
PR 7	12420	CH₃	Cl	H	CH₃	Cl	H
PR 9	12460	Cl	H	Cl	OCH₃	H	H
PR 10	12440	Cl	H	Cl	H	CH₃	H
PR 14	12380	NO₂	Cl	H	CH₃	H	H
PR 17	12390	CH₃	H	NO₂	CH₃	H	H
PR 22	12315	CH₃	H	NO₂	H	H	H
PR 23	12355	OCH₃	H	NO₂	H	H	NO₂
PR 112	12370	Cl	Cl	Cl	CH₃	H	H

Fig. 10.4. Generic structure of the naphthol reds showing details of some of the older naphthol reds.

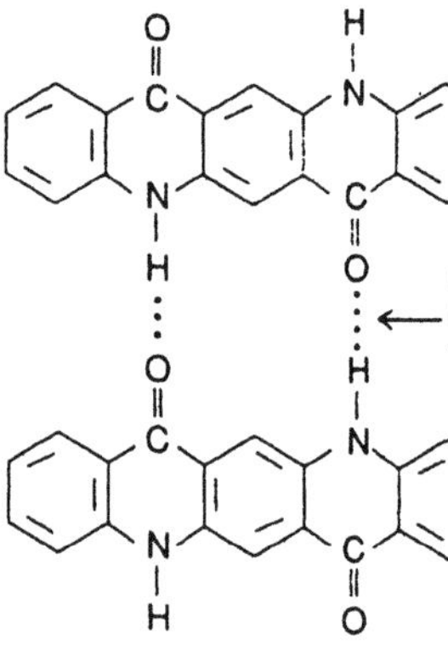

Fig. 10.5. Trans linear Quinacridone, Pigment Violet 19, showing proposed inter-molecular hydrogen bonding thought to confer additional stability to the molecule.

10.4.1.7. High Performance Reds

The high performance reds, considered here, fall into four basic classes: quinacridone reds and violets, vat dyestuff-based reds such as perylenes, benzimidazolone reds and disazo condensation reds.

Quinacridone Reds are heterocyclic in nature as shown in fig. 10.5. Addition of differing auxochromic groups such as methyl (–CH₃) and chlorine (–Cl) gives Pigment Red 122 and Pigment Red 202 respectively, both described as magentas. As a group, the quinacridones find their primary use in plastics, automotive, industrial and exterior finishes. The pigments combine excellent tinctorial properties with outstanding durability,

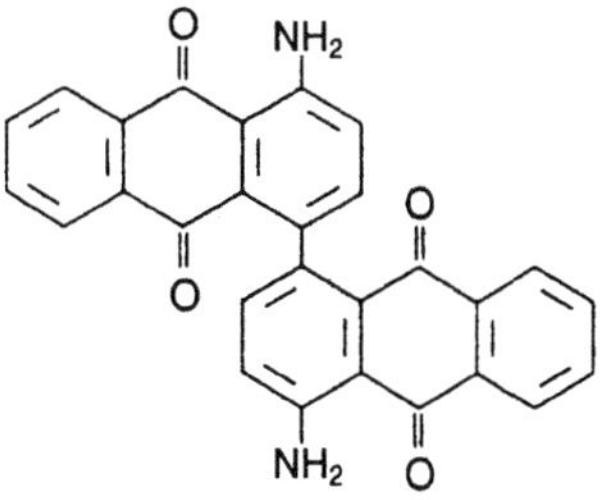

Anthraquinone Red (PR 177)

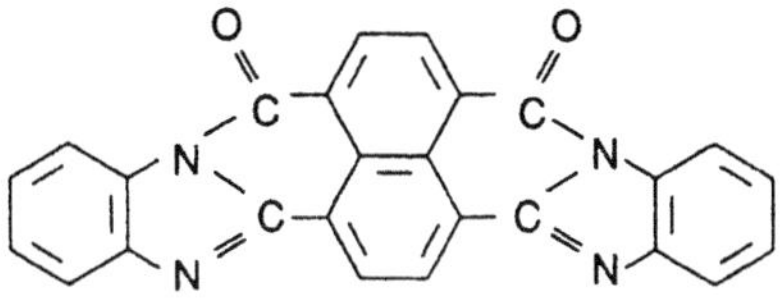

Perinone Red (PR 194)

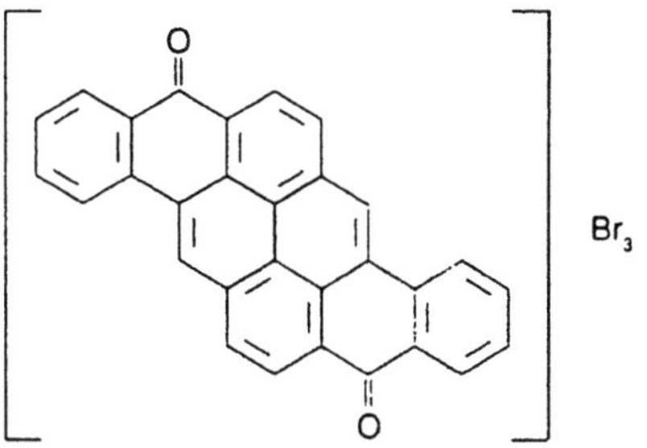

Brominated Pyranthrone Red (PR 216)

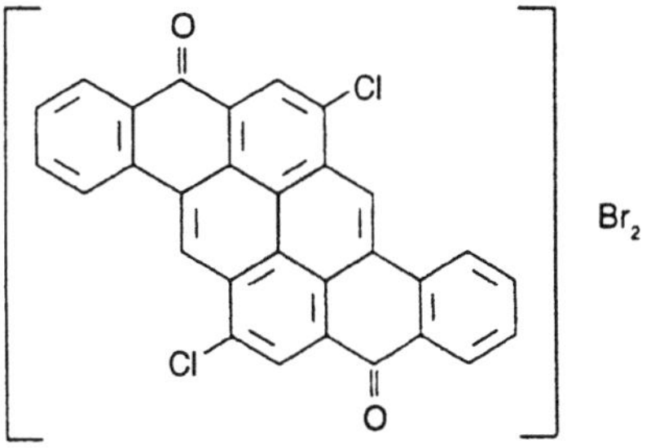

Pyranthrone Red (PR 226)

Fig. 10.6. Typical heterocyclic vat pigment reds.

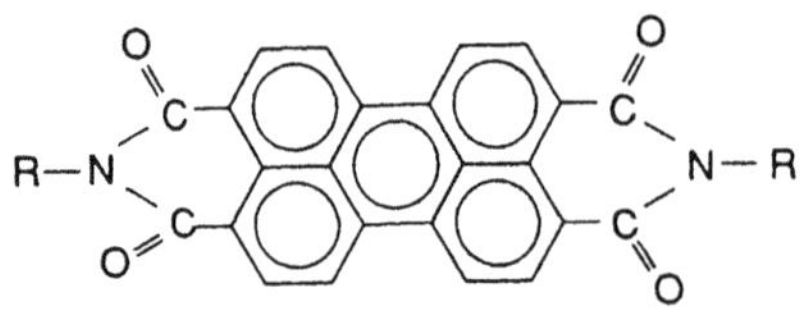

R = -C₆H₄OC₂H₅	PR 123	Vermillion
R = -C₆H₃(CH₃)₂	PR 149	Scarlet
R = -CH₃	PR 179	Maroon
R = -C₆H₅OCH₃	PR 190	Red
R = -H	PV 29	Bordeaux
R = -C₆H₅Cl	PR 189	Yellow Shade Red
R = -⟨⟩-N=N-⟨⟩	PR 178	Red

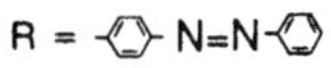

Fig. 10.7. Generic structure depicting perylene based reds.

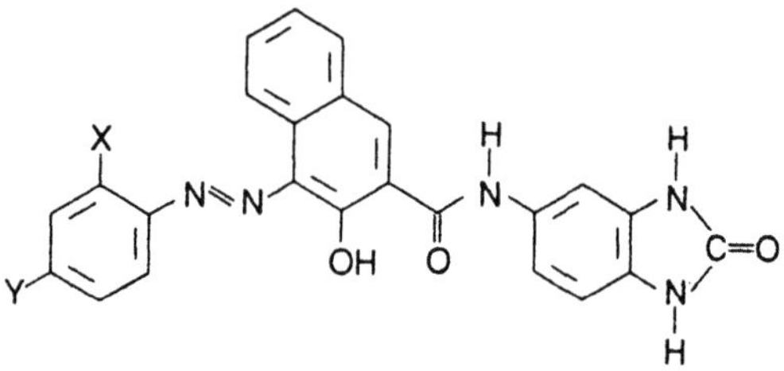

Colour Index Name	X	Y
PR 171	OCH₃	NO₂
PR 175	COOCH₃	H
PR 176	OCH₃	CONHC₆H₅
PR 185	OCH₃	SO₂NHCH₃
PR 208	COOC₄H₉	H

Fig. 10.8. Benzimidazolone based reds.

solvent fastness, lightfastness, heat fastness and chemical resistance. Examples of five quinacridone magentas are shown on color plate 28, both in masstone as well as in two dilutions. The differences among these five pigments are accentuated as they become more extended with white.

Vat Red pigments are based upon anthraquinone as shown in fig. 10.6. Their major use is in the automotive marketplace for clean, bright red finishes.

Perylene Red pigments, shown in fig. 10.7, provide pure, transparent shades and novel styling effects when used in metallic aluminum and mica finishes. In addition to high performance coating finishes these colors also find a major outlet in the fibers industry.

Benzimidazolone Based Reds are illustrated in fig. 10.8. Benzimidazolone reds show excellent fastness to light at all depths of shade, good weatherability, and excellent fastness to overspraying at elevated temperatures.

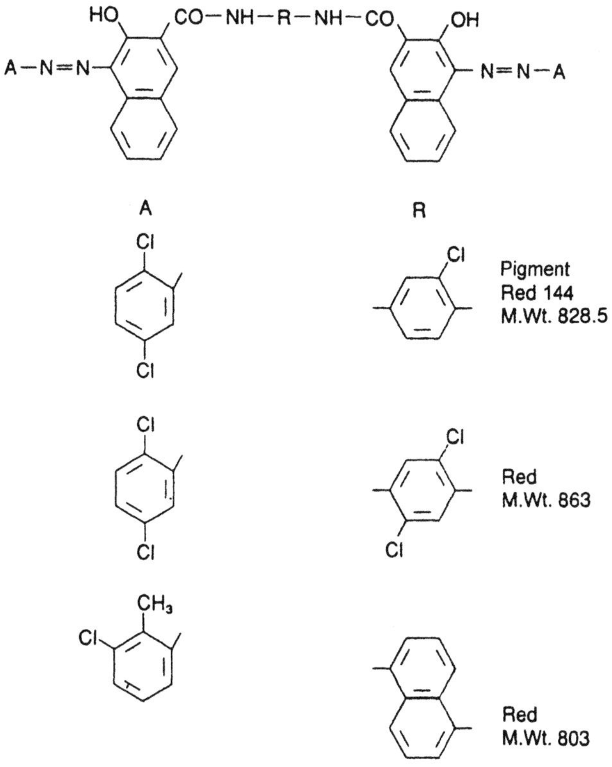

Fig. 10.9. Typical disazo condensation reds.

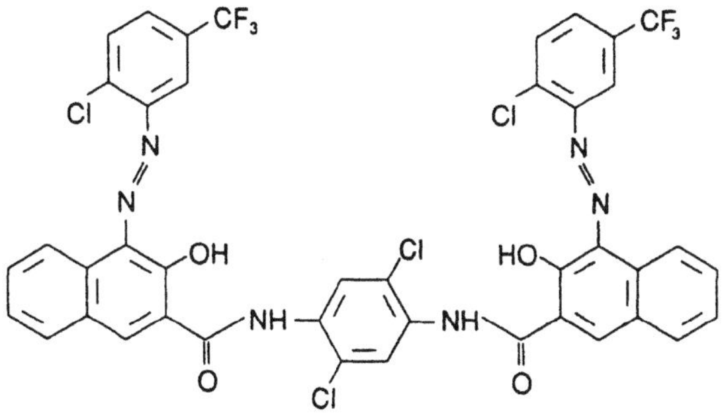

Fig. 10.10. Structure of Pigment Red 242, a disazo condensation pigment.

Disazo Condensation Reds have been available commercially in Europe since 1957 and in the United States since 1960. Their outstanding fastness properties have resulted in their use in high quality industrial finishes in addition to plastics and fibers. Figure 10.9 illustrates typical structures of the many disazo condensation reds. Pigment Red 242, fig. 10.10, is a bright yellow shade disazo condensation pigment with excellent fastness properties that is finding increased use in high quality industrial finishes.

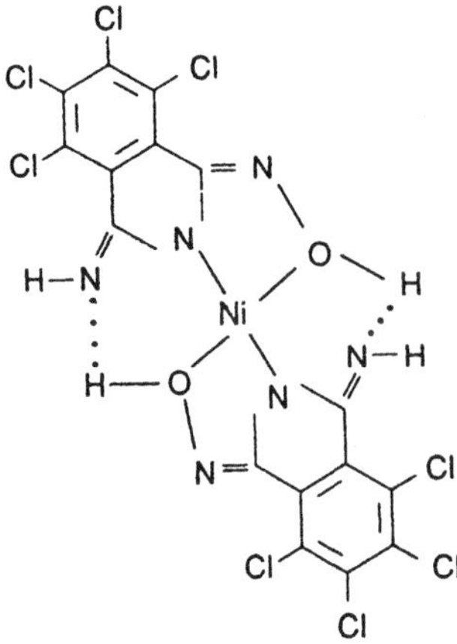

Fig. 10.11. Structure of Pigment Red 257, a nickel complex.

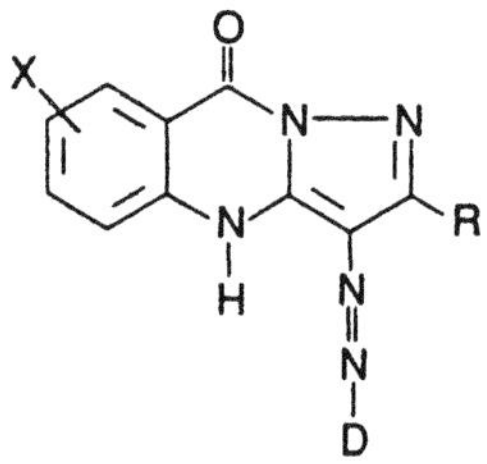

Fig. 10.12. Structure of typical pyrazoloquinazolone.

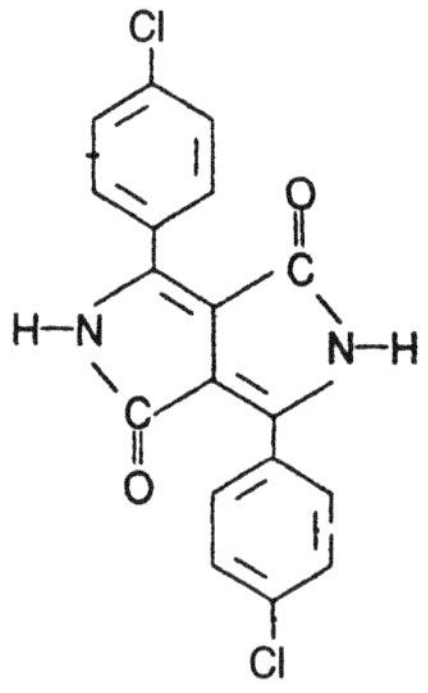

Fig. 10.13. Structure of pyrrolo-pyrrole red, Pigment Red 254.

10.4.1.8. Novel High Performance Reds

In recent years several novel, high performance, organic reds have been commercialized
and targeted directly at the requirements of the coatings and plastics marketplace. Pigment Red 257, fig. 10.11, is a nickel complex with a red-violet masstone and a magenta

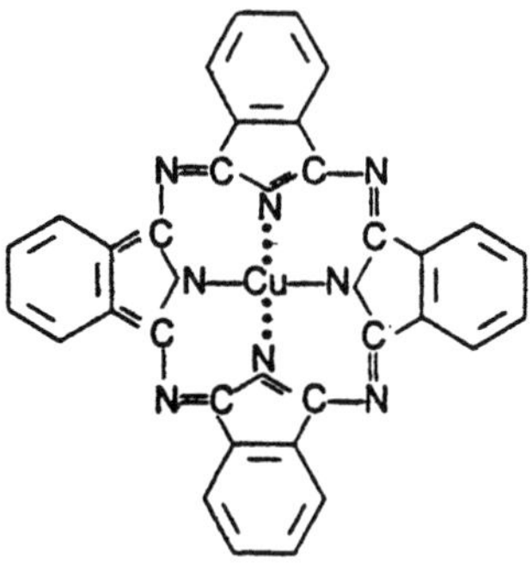

Fig. 10.14. Structure of Copper Phthalocyanine Blue.

undertone. The pigment is particularly useful in the formulation of high quality industrial and automotive coatings. Pigment Reds 251 and 252 are both based on the pyrazolo-quinazolone structure as shown in fig. 10.12. Each pigment exhibits excellent brightness of hue at full shades, good gloss retention and high scattering power combined with good light and weather fastness. Recently a series of novel reds based upon the pyrrolo-pyrrole structure, fig. 10.13, have been marketed into the automotive coatings industry.

10.4.2. Blues

10.4.2.1. Copper Phthalocyanine Blue

The most important and most widely used blue throughout all applications of the pigment consuming industry is copper phthalocyanine blue, Pigment Blue 15, fig. 10.14. First described in 1928 by chemists working for the Scottish Dye Works (now part of I.C.I.) this pigment has steadily increased in importance to become a product with worldwide significance. Copper phthalocyanine is commercially available in two crystal forms, the alpha and the beta form. The alpha crystal is described as Pigment Blue 15, 15:1 and 15:2 and is a clean, bright red shade blue. The beta crystal is described as Pigment Blue 15:3 and 15:4 and is a clean green or peacock shade. The beta form is the most stable crystal form and readily resists recrystallization. The alpha form, conversely, is the least stable or meta form, which readily converts to the more stable, green shade, beta crystal. As such the crystal requires special, proprietary treatments to produce a red shade product that is stable to both crystallization and flocculation. Copper phthalocyanine gives excellent performance in most applications but there is considerable variation between both the chemical and crystal types available. Use of any of the unstabilized grades in strong solvents or in systems that experience heat during dispersion or application will result in a shift in shade to the greener side and a loss of strength as recrystallization takes place within the unstabilized crystal. Copper phthalocyanine blue is the choice of ink makers, paint makers and plastic color formulators throughout the world when a blue color is called for at any tint level. Several forms are shown in color plate 31; also see Section 10.8.

10.4.2.2. Miscellaneous Blues

Indanthrone Blue, Pigment Blue 60, belonging to the class of pigments described as "vat pigments", is a very red shade, non-bronzing, flocculation resistant, pigment with outstanding fastness properties.

Carbazole Violet, Pigment Violet 23, a complex heterocyclic, is an intense red shade blue pigment that possesses excellent fastness properties. Its relatively high cost and hard nature limit widespread use. It is used at very low levels to produce "brighter whites" by imparting a bluer hue to the undertone of the white.

10.4.3. Yellows

10.4.3.1. Monoarylide yellows

These are azo pigments as represented in fig. 10.15.

Hansa Yellow G, Pigment Yellow 1, is a bright yellow pigment that has a major use in trade sales, emulsion and masonry paints. Its major disadvantages are its poor bleed resistance in most popular solvents, poor lightfastness in tint shades and very inferior bake resistance due to its tendency to sublime.

Hansa Yellow 10G, Pigment Yellow 3, greener in shade than Pigment Yellow 1, is used in the same types of applications and suffers from the same deficiencies as Pigment Yellow 1 with the exception that Pigment Yellow 3 is suitable for use in exterior applications at high tint levels.

Miscellaneous Monoarylide Yellows, Pigment Yellow 65, offering a redder shade than the previous two yellows discussed, is used in trade sales, latex, and masonry paints. A

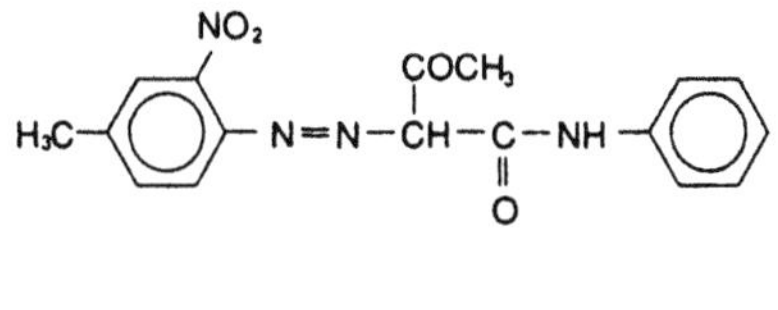

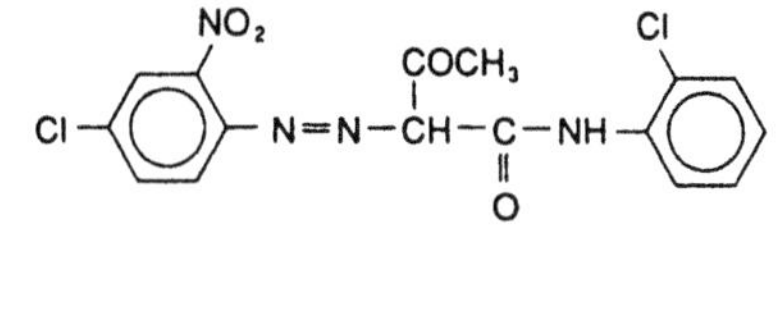

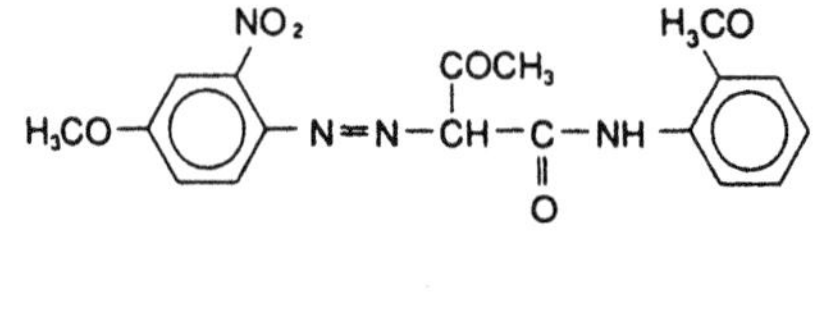

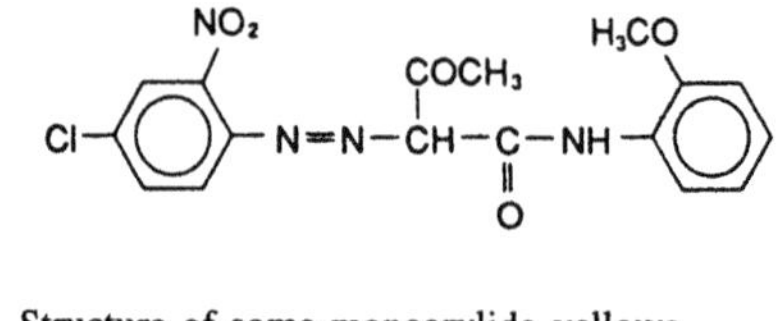

Fig. 10.15. Structure of some monoarylide yellows.

more recent application is for road traffic marking paints which are specified as being lead free (see color plate 29). The bleed resistance and baking stability are little improved over Pigment Yellows 1 and 3. Pigment Yellow 74, a pigment that is considerably stronger and somewhat greener than Pigment Yellow 1, is suitable for outdoor applications. Major outlets, as with all the monoarylide yellows, are in latex, trade sales and masonry paints. Pigment Yellow 75, a red shade yellow that has only recently found considerable application in the color industry as a replacement for lead containing medium chrome yellow, is used in road traffic marking paints.

10.4.3.2. Diarylide Yellows

The commercially important yellows are shown in fig. 10.16. Table 10.2 gives a summary of the properties of the major diarylide yellow pigments of commercial significance.

Each of the diarylide yellows offers low cost, reasonable heat stability and moderate chemical resistance. The major worldwide market for this class of yellows is the printing ink industry followed closely by the plastics industry. These yellows are approximately twice as strong as the monoarylide yellows dealt with previously, furthermore they offer improved bleed resistance and heat fastness. Nevertheless, none of the diarylide yellows have durability properties that would allow for their use in outdoor exposure situations.

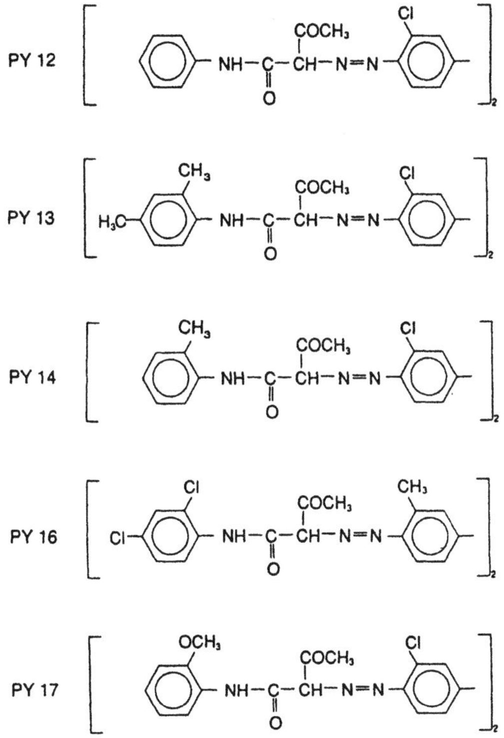

Fig. 10.16. Structure of some diarylide yellows.

TABLE 10.2
Properties of the Diarylide Yellows.

Colour Index Name	Common Name	Comments
PY 12	AAA Yellow	Poor lightfastness and bleed resistance. Printing inks major use.
PY 13	MX Yellow	Improved heat stability and solvent fastness vs. PY 12. Major use in printing inks
PY 14	OT Yellow	Green shade. Poor tint lightfastness. Used in packaging inks
PY 16	Yellow NCG	Bright green shade. Improved heat and solvent fastness
PY 17	OA Yellow	Green shade. Poor lightfastness. Very transparent
PY 55	PT Yellow	Red shade. Poor lightfastness. Isomer of PY 14
PY 81	Yellow H10G	Bright green shade
PY 83	Yellow HR	Very red shade. Improved transparency and lightfastness vs. PY12
PY 106	Yellow GGR	Green shade. Poor lightfastness. Major use in packaging inks
PY 113	Yellow H10GL	Very green shade. Better heat and solvent fastness vs. PY12
PY 114	Yellow G3R	Red shade. Some use in offset inks
PY 126	Yellow DGR	Improved mixed coupling vs. PY 12. Used in printing inks
PY 127	Yellow GRL	Bright red shade. Poor lightfastness. Offset inks principal use
PY 152	Yellow YR	Very opaque, red shade. Poor lightfastness

TABLE 10.3
Properties of the Benzimidazolone Yellows.

Colour Index Name	Common Name	Comments
PY 120	Yellow H2G	Medium shade. Good solvent fastness. Excellent lightfastness. Used in industrial finishes
PY 151	Yellow H4G	Greener shade. Good solvent fastness. Excellent lightfastness. Used in industrial and refinish paints
PY 154	Yellow H3G	Redder than PY 151. Good solvent fastness. Excellent lightfastness. Automotive grade
PY 156	Yellow HLR	Redder shade. Transparent, good exterior durability in full shade and tint. Used in exterior paints
PY 175	Yellow H6G	Very green shade. Good solvent fastness. Excellent lightfastness. Exterior applications

10.4.3.3. Benzimidazolone Yellows

Illustrated in fig. 10.17, these yellows have exceptional fastness to heat, light and over-striping. Used initially for the coloring of plastics, these pigments are now finding increased use in the coatings industry where their excellent fastness properties are demanded. Table 10.3 gives a summary of the properties of this class of pigments.

10.4.3.4. Heterocyclic Yellows

Each of these yellow pigments contain a heterocyclic molecule within their structure as shown by the examples presented in fig. 10.18. In spite of their apparent complexity

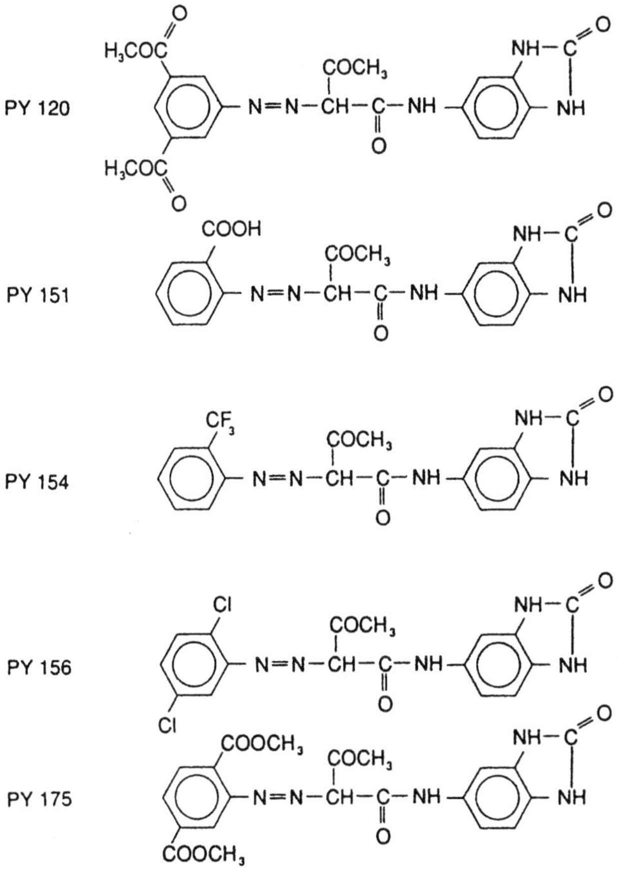

Fig. 10.17. Structure of the benzimidazolone yellows.

these new high performance yellows continue to be introduced to satisfy the exacting demands of the coatings and plastics industry.

10.4.4. Oranges

10.4.4.1. Azo Based Oranges

These oranges show considerable variation in structure as can be seen from fig. 10.19.

Orthonitroaniline Orange, Pigment Orange 2, is used in printing inks and is not recommended in coatings due to it's poor solvent fastness and lightfastness.

Dinitroaniline Orange, Pigment Orange 5, offers good lightfastness in full tone and moderate solvent fastness. As such it finds outlets in printing inks, latex paints and air dry finishes. Its poor baking stability rules out its use in high bake enamels.

Benzimidazolone Orange, Pigment Orange 36, has a bright red shade with high tint strength. In its opacified form this pigment offers excellent fastness to both heat and solvents and a hue similar to Molybdate Orange (PO 104). As such it is used in automotive

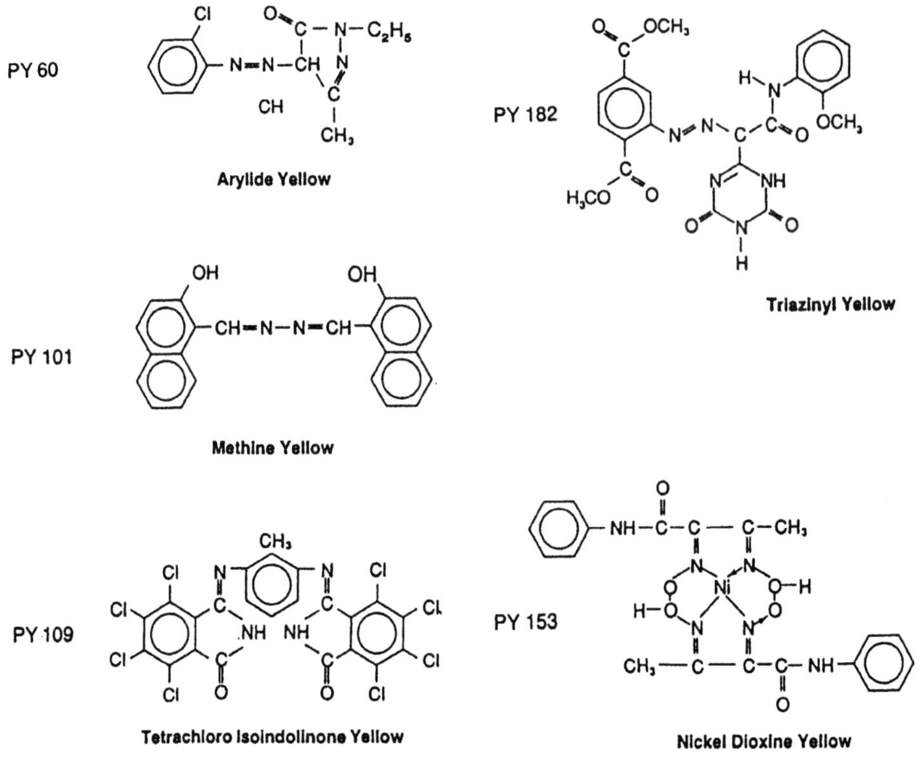

Fig. 10.18. Structure of some heterocyclic yellows.

and high quality industrial formulations which must be lead free and which formerly used the lead based Molybdate Orange.

10.4.5. Greens

10.4.5.1. Copper Phthalocyanine Green

When a self shade green is required, rather than a green produced by mixing blue and yellow, then copper phthalocyanine is the pigment of choice. Copper phthalocyanine green is based upon halogenated copper phthalocyanine blue, using either chlorine or a mixture of chlorine and bromine to replace the hydrogen atoms on the molecule. Pigment Green 7 is a blue shade green made by introducing 13–15 chlorine atoms into the copper phthalocyanine molecule, as in fig. 10.20, whereas Pigment Green 36 is a yellow shade green based upon a structure that involves progressive replacement of chlorine in the phthalocyanine structure with bromine. The most highly brominated product, Pigment Green 36, has an extreme yellow shade and contains upwards of twelve bromine atoms. Both greens are shown in color plate 31. They exhibit outstanding fastness properties to solvents, heat, light and outdoor exposure. They can be used equally effectively in both masstone and tints down to the very palest depth of shades. Metal-

PO 2

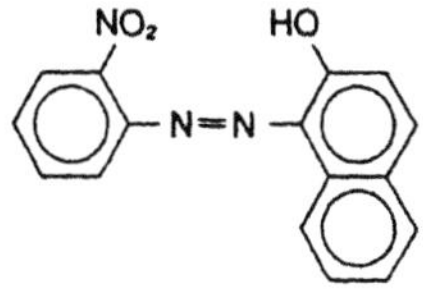

Orthonitroaniline Orange

PO 5

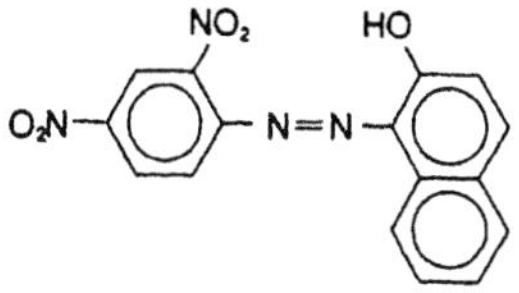

Dinitroaniline Orange

PO 13

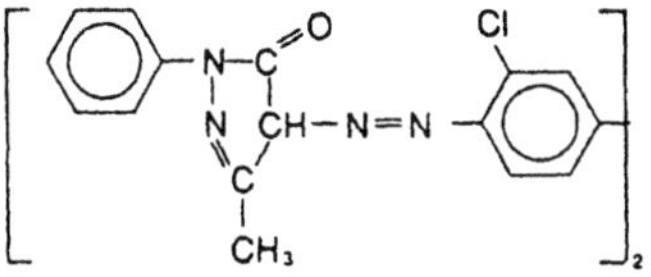

Pyrazolone Orange

Fig. 10.19. Structure of some oranges based upon the azo chromophore.

Pigment Green 7

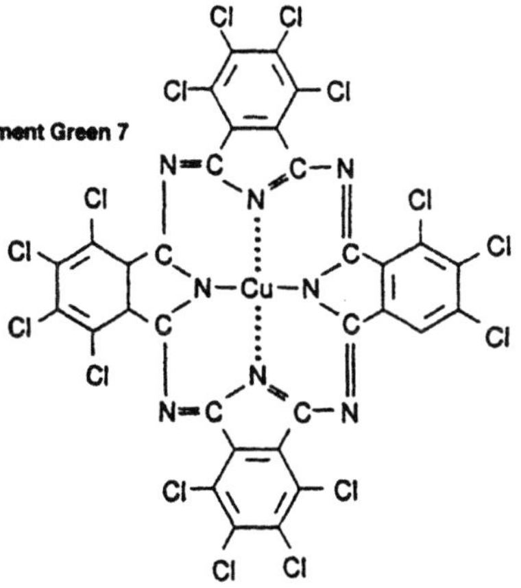

Fig. 10.20. Suggested structure of phthalocyanine green (PG 7).

lic automotive paints may feature phthalocyanine green at all shade depths. Approximately 50% of the worldwide production of copper phthalocyanine green is consumed by the coatings industry, the remainder going into the plastics and printing ink marketplace.

10.4.5.2. Miscellaneous Greens
There are other commercially available organic green pigments that may find some use
in coloring printing inks, plastics and paint. Pigment Greens 1, 2 and 4 are triphenyl
methane based dyes complexed with phospho tungsto molybdic acid (PTMA) to allow
their use as pigments. Their fastness properties are adequate for most ink applications
but they find little use in paint or plastics. Pigment Green 10, Nickel Azo Yellow or
Green Gold is the most lightfast azo pigment currently in commercial production, being
lightfast at all range of shade from deep tones to pale tints. The pigments does, however,
show poor overstripe fastness when used in baking enamels.

10.5. Classification of inorganic pigments by chemistry

Broadly speaking the colored inorganic pigments are either lead chromates, metal ox-
ides, sulfides or sulfoselenides with a few miscellaneous pigments such as cobalt blue,
Ultramarine blue, iron blue and bismuth vanadate yellow. In addition to the inorganic
pigments there also exist a series of pigments classed as "mixed metal oxides" such as
Zinc iron chromite brown (PB 33), Cobalt chromite green (PG 26), Cobalt titanate green
(PG 50), Manganese antimony titanate brown (PY 164), Cobalt aluminate blue (PB 28
& PB 36), and Chrome antimony titanate buff (PB 24). Both Pigment Yellow 53 and
Pigment Yellow 184 typify the pigments falling into the classification of mixed metal
oxides. Several inorganic pigments are shown in color plate 30.

Organic pigments have a higher oil adsorption, leading to higher rheology dispersions,
a greater tinctorial strength and a higher cost on a pure toner basis when compared to
inorganic pigments. Properties such as lightfastness, bleed, durability, chemical resistance,
etc., vary with the specific comparison being made; it is not always true to say that
inorganics are superior in all these aspects.

The unique use of bismuth oxychloride as a Nacreous or Pearlescent pigment goes
back to the early 1960s when the chemical was used as a substitute for natural pearl
essence in finger nail enamels. These nacreous pigment particles are thin, translucent
platelets of high refractive index that partially transmit and partially reflect incident
light. Further research resulted in synthetic nacreous pigments, derived for the most part,
from titanium dioxide coated and ferric oxide coated micas. More complex synthetic
inorganic nacreous pigments, in which absorption colorants are deposited on top of the
titanium dioxide layer, and which can display two distinct colors, dependant upon the
viewing angle, have been available since the mid 1970s.

Numbers in parentheses, e.g., {7}, indicate the mechanism that causes the color in
these inorganic pigments, as discussed in Sections 4.3 to 4.7 of Chapter 4.

10.6. Classification of inorganic pigments by color

10.6.1. Reds

10.6.1.1. Iron Oxide Reds
Pigment Red 101 (Synthetic) and Pigment Red 102 (Natural), $Fe_2O \cdot xH_2O$, also carry
such names as Haematite, Mars Red, Ferrite Red, Rouge, Turkey Red and Persian Gulf

Oxide {4}. The wide range of red iron oxide shades available, in addition to their acid and alkali resistance and their economy, accounts for the large volumes of these pigments used in today's paint and furniture finishes marketplace.

10.6.1.2. Molybdate Orange

Pigment Red 104, a very yellow shade red with the empirical formula of $PbCrO_4 \cdot xPbMoO_4 \cdot yPbSO_4$ is an opaque pigment with high solvent fastness, moderate heat fastness and good economy {7}. On the negative side it has poor alkali and acid resistance. Nontreated grades also tend to darken markedly on prolonged exposure to the environment.

10.6.1.3. Cadmium Reds

Pigment Red 108, cadmium sulfoselenide red is cadmium sulfoselenide, a solid solution, $CdS \cdot xCdSe$ {9}. The pigment's hue is determined by the amount of cadmium sulfoselenide incorporated into the solid solution and, to a lesser extent, the temperature of processing. Cadmium Red features excellent stability to heat, alkali, solvents and light when used at high tint and masstone levels (see color plate 15).

10.6.1.4. Mercury Cadmium Red

Pigment Red 113, is yet another solid solution, $CdS \cdot xHgS$ {9}. It offers good hiding properties and good solvent resistance with excellent brightness, but its inferior heat and acid resistance when compared to Pigment Red 108 has limited its use to plastics, artists colors and opaque printing inks.

10.6.2. Violets

Technically blue shade reds, the two most important inorganic violet pigments are Pigment Violet 15, Ultramarine Violet {4}, prepared by the oxidation of Ultramarine Blue, Pigment Blue 29 (see below) and Pigment Violet 16, Manganese Violet, $MnNH_4P_2O_7$. Ultramarine Violet has good heat and light fastness and a brilliant red shade. It will react with metals to form sulfides. It finds use in cosmetic applications, artists colors and specialty acrylic poster paint. Manganese Violet does not possess the same brightness of hue as Pigment Violet 15, has only moderate opacity and poor alkali resistance; two grades are shown in color plate 30. On the plus side the pigment has high lightfastness and superior fastness to solvents and overstripe bleed. Major uses are in the plastics and cosmetics industry.

10.6.3. Blues

10.6.3.1. Iron Blue

Pigment Blue 27, also known as Iron Blue, Chinese Blue, Bronze Blue, Prussian and Milori Blue, is ferric ammonium ferrocyanide, $FeNH_4Fe(CN)_6 \cdot xH_2O$, shown in color plate 30 {7}. It offers good resistance to weak acids but markedly poor resistance to even mild alkali. Furthermore it has a tendency to "bleach out" on storage, losing almost all its color, when incorporated into a paint formulation that contains oxidizable vehicles such as linseed oil. The pigment has only acceptable lightfastness properties when used at masstone levels.

10.6.3.2. Cobalt Blue

Pigment Blue 28, $CoAl_2O$, is chemically very inert, offers excellent heat stability and lightfastness and high opacity {4}. Major uses include coatings and plastics with particular emphasis on the vinyl sidings marketplace.

10.6.3.3. Ultramarine Blue

Pigment Blue 29, going by such varied common names as Laundry Blue, Dolly Blue and lapis lazuli, is $Na_6Al_6Si_6O_{24}S_4$ and its major use is as a component of laundry powders and detergent soaps. It is shown in color plates 12 and 30 {7}.

10.6.3.4. Cobalt Chromite Spinel

Pigment Blue 36, $Co(Al,Cr)_2O_4$, another mixed metal oxide, known as Cerulean Blue {4}. It offers high chemical resistance and outstanding outdoor durability. As such it is used in high quality coatings applications, artists colors, camouflage coatings, ceramics and vinyl siding.

10.6.4. Yellows

10.6.4.1. Strontium Yellow

Pigment Yellow 32, $SrCrO_4$, finds a primary use in corrosion inhibiting coatings. This pigment has poor tint strength, low opacity and unsatisfactory alkali and acid resistance which limits its more widespread use {7}.

10.6.4.2. Chrome Yellow

Pigment Yellow 34, has the empirical formula $PbCrO_4 \cdot xPbSO_4$. Various types exist that differ in the ratio of the lead sulfate to the lead chromate and as such are described as Medium chrome, Primrose and Lemon chrome yellows {7}. A typical Primrose chrome will contain 23–30% lead sulfate in the solid solution whereas a Medium chrome will contain 0–6% lead sulfate. Primrose chrome exhibits a very green shade and offers good lightfastness, high opacity and low rheology, coupled with economy of use. The coatings industry, closely followed by the ink and plastics industry, is the largest consumer of Primrose chrome. Medium chrome is seen widely used in road marking paints in the United States where the law requires a yellow line.

10.6.4.3. Cadmium Zinc Yellow

Pigment Yellow 35, yet another solid solution, is $CdS \cdot xZnS$ {9}. The hue is readily altered by varying the ratio of the two components. Levels of zinc sulfide of 14–21% give a green or primrose shade whilst 1–7% gives a redder shade achieving a "golden" hue. Incorporation of barium sulfate during manufacture produces a lithophone version, Pigment Yellow 35:1. Cadmium zinc yellows offer bright, clean, opaque pigments with excellent resistance to heat, light and strong solvents.

10.6.4.4. Zinc Chromate

Pigment Yellow 36, $4ZnO \cdot K_2O \cdot 4CrO_3 \cdot 3H_2O$ a bright, green shade yellow is used primarily in corrosion inhibiting coatings {7}. Its poor tinctorial strength and poor resistance to acid and alkali severely limits this pigments use elsewhere.

10.6.4.5. Cadmium Sulfide Yellow

Pigment Yellow 37, CdS, can have hues ranging from a green shade to a very red shade by simply varying the calcination conditions {9}. Offering excellent stability to heat, light, acids and alkali, this pigment's only major drawback is its tendency to fade in the presence of moisture. Major use is in plastic applications with minor use in artists' colorants, coated fabrics, opaque printing inks and leather goods (see color plate 15).

10.6.4.6. Iron Oxide Yellows

Pigment Yellow 42 (synthetic), and Pigment Yellow 43 (natural) are both $FeO \cdot xH_2O$ {4}. The natural oxides also contain clay and various other minor minerals. Available under several names, often related to the country of origin or the pigments history, the natural yellow oxide is also called Indian Ochre, Ocher, Sienna and limonite. Iron oxide yellows are economical pigments with excellent lightfastness, weatherability, opacity and flow properties. On the downside they are dull in masstone and only exhibit fair tinctorial strength and moderate baking stability at best. It is their value in use that has resulted in their widespread acceptance throughout the pigment consuming marketplace.

10.6.4.7. Mixed Metal Oxide Yellows

Various mixed metal oxide yellows exist under the classification of miscellaneous yellows. All are solid solutions of oxides containing two or more metals in their structure {4, 7}. These oxides offer excellent outdoor durability and heat fastness, their major use now being in the market for colored vinyl siding.

10.6.4.8. Bismuth Vanadate/Molybdate Yellow

Pigment Yellow 184, was introduced into the marketplace as recently as 1985, is $4BiVO_4 \cdot 3Bi_2MoO_6$ {7}. A green shade yellow used principally for brilliant solid shade in both automotive and industrial coatings, it has excellent weatherfastness coupled with good hiding power and gloss retention.

10.6.5. Oranges

10.6.5.1. Cadmium Orange

Pigment Orange 20, cadmium sulfoselenide orange, is $CdS \cdot xCdSe$ {9}. A change in ratio of the solid solution components gives pigments that are bright yellow (PY 35) to bright red (PR 108). These pigments' major use is in plastics and in industrial coatings, and for color coding applications where chemical and heat resistance are principal requirements (see color plate 15).

10.6.5.2. Chrome Orange

Pigment Orange 21, is $PbCrO_4 \cdot xPbO$ {7}. Shades varying from a yellow shade to a red shade are produced, dependent upon the alkalinity maintained during the reaction sequence. As with all lead containing pigments, the product will darken on exposure to the atmosphere, the rate dependant upon the sulfur content.

10.6.5.3. Cadmium Mercury Orange

Pigment Orange 23, is $CdS \cdot xHgS$ {9}. Again, various hues can be obtained by controlling the formation of the mixed crystal. An extremely heat stable pigment with excellent chemical resistance, weatherability and solvent fastness.

10.6.6. Greens

10.6.6.1. Chrome Green
Pigment Green 15, is merely a mixture of a green shade chrome yellow (PY 34) and iron blue (PB 27). As such, Chrome Greens offer a range of hues with a light yellow shade to a deep dark shade, providing good hiding, high tint strength and a moderate chemical resistance at an economical price.

10.6.6.2. Chromium Oxide Green
Pigment Green 17, Cr_2O_3 has a unique use in camouflage paints because of its ability to reflect infra red light. Otherwise it finds use in roofing granules, ceramics and security inks. It is shown in color plate 10 {4}.

10.6.6.3. Hydrated Chromium Oxide Green
Pigment Green 18, also known as Viridian Green or Guignets Green, is $Cr_2O_3 \cdot 2H_2O$. It is a bright, blue shade green with high chroma and outstanding fastness properties and is used in security inks, exterior paints and artists colors, exterior paints and artists colors {4}.

10.6.6.4. Mixed Metal Oxide Greens
Pigment Green 26, cobalt chromite green, $CoCrO_4$ is a blue green shade with excellent weather ability, light and heat fastness, and is used in ceramics, roofing granules, plastics and high temperature cured systems. Pigment Green 50, cobalt titanate green, Co_2TiO_4, offers high infra red reflectivity coupled with outstanding chemical and heat fastness and outdoor durability and as such finds use in camouflage paints {4 and 7}.

10.6.7. Browns

Natural Iron Oxides are mined from either "iron oxide" mines, operating principally to supply ore as feedstock for blast furnaces, with a small offtake directed to the pigment industry, or from "pigment" mines which operate solely to supply pigmentary grade ore. The color's major outlet is as a colorant for furniture finishes, caulk and paints {4 and 7}.

Synthetic Brown Oxide is also known as brown magnetite iron oxide, Pigment Brown 6, produced by controlled oxidation of Pigment Black 11 and may be represented as $Fe_2O_3 \cdot xFeO \cdot yH_2O$. Pigment Brown 11, magnesium ferrite, is $MgO \cdot Fe_2O_3$. The volume of all types of brown oxides used in colorants is generally low since most browns are achieved by mixing yellow, red and black pigments. As a class these pigments have good chemical resistance and high tint strength and as such do find some use in wood stains and furniture finishes. Ten iron oxide pigments are shown in color plate 30 {4 and 7}.

Mixed Metal Oxide Browns of variable composition find use in the pigments marketplace and are identified as Pigment Browns 24, 33 and 35 {4 and 7}. They are used in plastics, rubber, fibers and for tarnish proof bronze colored printing inks.

10.7. Metals as pigments

There exist a series of pure metals that are used as pigments in their own right to achieve coloring effects specific to their particle size, shape and distribution. Such "Pigment Metals" are derived from metals such as aluminum, nickel and zinc. The cause of their colors {8} in covered in Section 4.6.1.

Aluminum Flake. Manufactured from the pure metal and possessing many of the characteristics of the aluminum itself, this pigment, Pigment Metal 1, has excellent corrosion resistance properties, is highly reflective to visible light and is very stable as a result of the formation of a very thin, clinging oxide coating that forms spontaneously over the surface of the finely divided, flaked pigment. Aluminum pigments have a distinct lamellar morphology that allows the pigment, when incorporated into a thin film or coating, to align as a multilayer parallel to the substrate and surface of the film. Such an aligned film gives the pleasing "metallic" effect associated with high quality finishes seen on today's automobiles. Flakes can have a thickness ranging from 0.1 to 2.0 micrometers and a diameter varying from a low of 0.5 micrometers to a high of 200 micrometers.

Zinc Pigment in the form of both zinc dust and zinc powder, Pigment Metal 6, is used in zinc-rich paints applied to bare steel to minimize rusting. The three most common grades of zinc dust, regular through ultrafine, have a particle size ranging from 8 to 3 micrometers.

Nickel Powders are used as electrically conductive pigments in various polymer and resin systems. The use of reflective nickel flakes in transparent coatings has been researched and developed since 1973 for use in automotive and appliance finishes.

10.8. Pigment dispersion and end use application

Dry pigments comprise a mixture of primary particles, aggregates, and agglomerates that must be wetted before dispersion forces can take full effect and enable the production of a stable, pigmentary dispersion in the medium of choice. The particle size distribution of a pigment, while having little effect on the actual hue of the color, plays a considerable role in the transparency and gloss achieved by incorporating the pigment into a system. Large particle sizes, of the order of 20 m^2/g surface area, are likely to give opaque dispersions with good hiding characteristics and better light fastness as compared to the same pigment type with a small particle size and high surface area of the order of 80 m^2/g. Thus, for example, Quinacridone Red, Pigment Violet 19, is available as an opaque, clean, high hiding, pigment with a surface area of 18–21 m^2/g and with a low oil adsorption or as a transparent, clean, red with a surface area of 75–80 m^2/g that is ideally suited to systems incorporating metallic or mica flakes where transparent pigments are vital to the system. A pigment with a small particle size distribution will have a high surface area as a consequence and the resultant dispersion may have too high a rheology such that the system gels and will not flow. Particle size and surface area are always a trade off between durability, transparency, opacity and rheology.

The ideal dispersion consists of a homogeneous suspension of primary particles. The forces which hold these primary particles together as either aggregates or agglomerates are quite large and are termed Van der Waals attractive forces. The process of dispersion

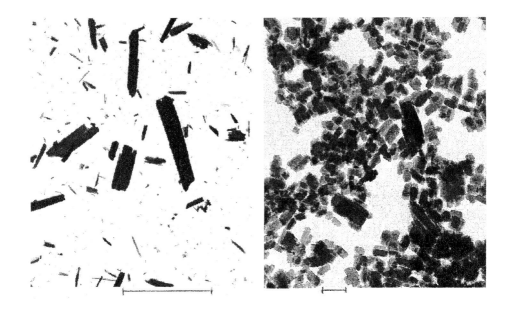

Fig. 10.21. Copper phthalocyanine blue pigment PB 15:3; left: crude crystals, scale = 10 micrometers; right: finished pigment, scale = 0.1 micrometer.

consists of first excluding air from around the particles to fully wet the particles and then breaking down any aggregates and agglomerates to their primary particles. The particles are then dispersed in the resin/solvent system to reduce the distance between the particles and prevent re-aggregation. The ease with which this wetting, de-aggregation and dispersion sequence takes place is referred to as the dispersibility of the pigment and varies with the pigment type and vehicle system.

The wetting or deaeration process is physical in nature as it requires that the vehicle or solvent displace the air on the surface of the pigment particles. This process is often easier with organic solvent systems since the pigment is basically hydrophobic, having a greater affinity for organic solvents and oleoresinous materials than for aqueous systems. Alkyd, acrylic, polyester and oleoresinous systems, etc. are good pigment wetters and are, therefore, suitable by themselves for this process stage. Water-based systems are poor wetters of dry pigments and require the presence of a surfactant to aid in the wetting process.

The dispersion stage involves the input of large amounts of energy to fully disperse the pigment throughout the vehicle system. Many types of milling equipment are available that will impart the necessary energy to the system and achieve optimum dispersion in the process.

A typical crude copper phthalocyanine blue pigment is shown at the left in fig. 10.21. A wide range of particle sizes can be seen, up to 10 micrometer in length, with a surface area near 20 m^2/g. After mechanical attrition, the resulting PB 15:3 finished pigment is shown at the right in this figure. Particles are much more uniform in size, with the largest about 0.1 micrometer and a surface area near 100 m^2/g. An alternative diminution process to achieve pigmentary size involves the acid swelling process, where the pigment is dissolved in concentrated sulfuric acid and precipitated on dilution with ice water. However, this process has ecological problems in the disposal of the large quantities of waste acid.

The end use applications of colored pigments are as varied as the colors themselves, whatever the application, however, the pigment has to exist as a dispersion and not a solution. One of the reasons quinacridones cannot be used to color nylon is because the nylon acts as a solvent, destroying the structure of the pigment. Without doubt the largest consumer is the printing ink industry, from lithographic offset, both heatset and sheetfed, flexo gravure, publication and packaging gravure, gift wrap, tag and label through posters, can printing and specialty books. The four-color printing process, based upon printing "yellow, magenta, cyan and black" to produce any color of the spectrum uses millions of pounds annually of diarylide yellow (PY 12 or PY 13), rubine (PR 49:1 or PR 57:1), phthalocyanine blue (PB 15:3) and carbon black (PBk 7). The next largest consumers of pigments are the plastics and coatings industry. Both are alike in the respect that color concentrates or intermix bases are prepared of a selection of colors which are then mixed together to give a custom colored end product. Pigments are dispersed into rubber, both synthetic and natural, and plastics such as polypropylene, vinyl, high and low density polyethylene, and polystyrene and coatings resins ranging from polyesters, alkyds, acrylics, and latex emulsions polyurethanes. These three end uses consume in excess of ninety percent of the pigments produced worldwide. Other notable applications include textiles, cosmetics, non-woven fibers, artists colors, fiber glass, roofing granules and candle wax.

10.9. Sources of additional information

1. American Association of Textile Chemists and Colorists, P.O. Box 12215, Research Triangle Park, NC 27709, USA.

2. Dry Color Manufacturers Association, North 19th St, Arlington, VA, 22209.

3. Ehrich, F.F., 1968, Pigments, in: Kirk-Othmer Encyclopedia of Chemical Technology, Vol.15 (Wiley, New York).

4. Fytelman, M., 1978, Pigments, in: Kirk-Othmer Encyclopedia of Chemical Technology, 3rd edn, Vol. 15 (Wiley, New York).

5. Geissler, G., 1981, Polymer Paint Colour J., Sept. 30, 614–623.

6. Hopmeir, A.P., 1969, Pigments, in: Encyclopedia of Polymer Science and Technology, Vol. 10 (Interscience, New York) pp. 157–193.

7. Lewis, P.A., 1987, Pigment Handbook, Vol. 1, 2nd edn (Wiley, New York).

8. Lewis, P.A., 1988, Organic Pigments, FSCT Monograph Series (Philadelphia, PA 19107).

9. Lewis, P.A., 1991, Organic Pigments, in: Coatings Technology Handbook (Marcel Dekker, New York).

10. Lubs, H.A., 1955, The Chemistry of Synthetic Dyes and Pigments, ACS Monograph No. 127 (American Chemical Society, Reinhold, New York).

11. Mills, W.G.B., 1962, Paint Chemists Handbook (Scott Greenwood, London).

12. Moser, F.H. and A.L. Thomas, 1963, Phthalocyanine Compounds, ACS Monograph No. 157 (American Chemical Society, Reinhold, New York).

13. Muzall, J.M. and W.L. Cook, 1979, Mutagenicity Res. **67**, 1–8.

14. NPCA Raw Materials Index, Pigments Section (National Paint and Coatings Association, Washington, DC 20005).

15. NPIRI Raw Material Data Handbook, Vol. 4, Pigments (National Printing Ink Research Institute, Lehigh University, Bethlehem, PA 18015).

16. Parfitt, G.D., 1981, Dispersion of Powders in Liquid, 3rd edn (Applied Science, London).

17. Patterson, D., 1967, Pigments, An Introduction to Their Physical Chemistry (Elsevier, Amsterdam).

18. Patton, T.C., 1973, Pigment Handbook, 3 vols (Wiley, New York).

19. Patton, T.C., 1964, Paint Flow and Pigment Dispersion (Interscience, New York).

20. Remington, J.S. and W. Francis, 1954, Pigments; Their Manufacture, Properties and Use (Leonard Hill, London).

21. Venkataraman, K., 1952 (Vol. 1) to 1978 (Vol. 8), The Chemistry of Synthetic Dyes (Academic Press, New York).

chapter 11

COLORANTS: DYES

J. RICHARD ASPLAND

University of Clemson
Clemson, SC, USA

Color for Science, Art and Technology
K. Nassau (Editor)

CONTENTS

11.1.	Introduction	316
	11.1.1. Dyes and Pigments	316
	11.1.2. Historical background	316
	11.1.3. Textile uses of dyes	318
	11.1.4. Other uses for dyes	318
11.2.	Textile substrates	318
	11.2.1. Natural and man-made fibrous polymers	318
	11.2.2. Physical and chemical fiber characteristics	319
	11.2.3. Fiber-blends	320
11.3.	Color application categories	320
	11.3.1. Anionic dyes for cellulosic fibers	320
	11.3.2. Ionic dyes for ionic fibers	324
	11.3.3. Nonionic dyes for man-made fibers	326
	11.3.4. Azoic combinations	327
11.4.	Color fastness	328
11.5.	The dyeing process	328
	11.5.1. Dyes in solution	329
	11.5.2. Dye sorption	329
	11.5.3. Dye diffusion	329
	11.5.4. Batch dyeing	329
	11.5.5. Continuous dyeing	330
11.6.	Sorption isotherms	330
	11.6.1. Freundlich isotherms	330
	11.6.2. Langmuir isotherms	331
	11.6.3. Nernst isotherms	331
	11.6.4. Hybrid sorption isotherms	332
	11.6.5. Dye–fiber combinations – blends	332
11.7.	Dyeing variables – machinery	333
	11.7.1. Batch dyeing machinery	333
	11.7.2. Continuous dyeing machinery	337

11.8. Dyeing variables – auxiliaries 341

    11.8.1. Buffers 341

    11.8.2. Carriers 341

    11.8.3. Compatibilizers 341

    11.8.4. Levelers 342

    11.8.5. Fixing agents and stainblockers 342

    11.8.6. Miscellaneous 342

11.9. Textile pigmentation processes 343

11.10. Summary 343

References 343

11.1. Introduction

11.1.1. Dyes and Pigments

The superficial distinction between dyes and pigments is quite clear (Lewis 1997). Pigments are invariably insoluble particles, and any process in which the particles become embedded or trapped within a polymer is properly referred to as a pigmenting or pigmentation process (Aspland 1993a).

In the fiber industry, pigments can be incorporated in the polymer dope, melt or solution, prior to its extrusion as fibers. In the textile industry, pigments can be mechanically applied to the surface of fabrics in the presence of an acrylic binder which, on curing, leaves a pigmented, acrylic polymer layer coating the fiber surfaces (Schwindt and Faulhaber 1984). However, both these processes are frequently and erroneously referred to as dyeings. The terms: dope-dyeing, solution-dyeing and (worse) pigment-dyeing, are all in common use. Unlike dyes, pigments by themselves have no intrinsic attraction for fibers.

The semantic problem arises primarily from a misunderstanding of what a dyeing process really is. All dyeing processes involve the transport of dyes, colorant derivatives or color precursors, in the form of molecules or molecular ions, from an external phase to an internal phase within the substrate, i.e., from a dyebath or from a fiber surface to a location inside the fiber. This external phase is usually water, but may be high temperature air (Ingamells 1989) and, occasionally, is a non-aqueous solvent (AATCC 1993).

Pigment particles, although small, are still far too large to be transportable across the phase boundary and into the internal phase of polymeric substrates (see Section 11.5). But, there are a number of pigments (notably *vat pigments* (Lewis 1997)) which, by means of reversible chemical changes, can be dissolved and the soluble derivatives can participate in dyeing processes (see Section 11.3). The question is, should these pigments be referred to as dyes? If in doubt, there is nothing wrong with referring to them as colorants, for this term implies nothing about their method of application.

11.1.2. Historical background

Archeologists have found dyed fabrics and felts more than 45 centuries old, and the history of dyeing no doubt could be extended back even further were it not for the susceptibility of natural fibers to degradation by exposure to air, sunlight, water, enzymes and micro-organisms. The known history of decorative pigmentation dates back at least as far as the Aurignacian Period (ca. 28 000 to ca. 22 000 B.C.) and is exemplified by the cave paintings at Lascaux, France (see Chapter 6 and color plate 26). During the intervening centuries, and by a method belatedly ascribed to Edison, man has found a variety of satisfactory colorants and developed many diverse means of coloration.

11.1.2.1. Natural dyes

During the past millennia, all manner of vegetable materials have been used as sources of natural dyes and these include leaves, berries, flowers, roots, and bark from plants grown in many parts of the world. Lichens have been widely used. By contrast, only a few natural dyes, such as lac, have been derived from insects. The royal purple, tyrian purple, formerly used to dye the robes of middle-eastern kings and Roman emperors, was derived from the mollusc, Murex brandaris, found on the eastern shores of the Mediterranean. The use of the *Colour Index* (Colour Index 1971a), where many such dyes are listed, is discussed in Chapter 10, Section 10.2.

Natural dyes, with few exceptions, do not have the resistance to color change (known as color-fastness) on exposure to light, washing and other agencies, of the synthetic dyes available today. The only blue among them is indigo; there are no greens, and even the brightest among them cannot approach the brightness of the selection of synthetic dyes in the world marketplace at this time. It is curious that synthetic indigo, currently the most widely used blue dye of all, is prized to a large extent for its color-fastness deficiencies. These include its ability to fade on washing, particularly in the presence of hypochlorite bleaches, and its poor resistance to removal by abrasion, notably with lava rocks in the stone washing and acid washing processes. These properties permit indigo dyed denim to make a unique fashion statement.

Natural dyes fall mainly into the *acid* and *acid-mordant* dye application categories, see Section 11.3, although there are several dyes which have an attraction (substantivity) for cellulosic fibers. There are at least two *vat* dyes (indigo and tyrian purple) and there is one *basic* dye. It is a serious error to suppose that a return to the use of natural dyes would be environmentally desirable. Because of their poor colorfastness, most natural dyes are applied along with metallic mordants such as copper, iron, tin, aluminum, chromium and cobalt salts, to promote the formation of dye-metal-fiber complexes of different color and better fastness. When metal salts are used as mordants, to react with dyes on the textile goods, an excess of metal ions is required. The metal ions end up in the plant effluent. If all dye houses were to switch to natural (mordant) dyes, the environment would be endangered.

11.1.2.2. Synthetic dyes

The Englishman, W.H. Perkin, synthesized and purified the first synthetic dye (Mauveine) in 1856, during an abortive attempt to synthesize quinine. Since then, the dyestuff industry, particularly in Europe, has grown apace and spearheaded the growth of the whole organic chemical industry.

The explosive growth in synthetic dyes, was at least partially due to the fact that the products had a level of brightness (saturation, chroma) and color strength previously unattainable, to say nothing of hues unavailable from extracts of plants and animals.

By the early part of the twentieth century, many new dyes had been developed for the natural fibers available. These fell into the following dye application categories: direct, vat, sulfur and *azoic* combinations for the cellulosic fibers, principally cotton; acid, acid-mordant and *pre-metallized* dyes for the animal fibers, primarily wool and silk; and basic dyes, which could be used for silk and also for cellulosic fibers with the use of a mordant. During this same period, various regenerated cellulosic fibers, e.g., viscose and cuprammonium rayons, were found to dye with the same dyes and in a similar manner to cotton, and a new class of dyes, *disperse* dyes, was developed for secondary cellulose acetate.

The big thrust into man-made fibers began to gather steam in the late 1940s and early 1950s, with nylon, polyester and acrylic fibers, but no new dye application categories were required in order to dye them. The most important subsequent development has been the *reactive* dyes for cellulosic fibers, introduced in 1956, coinciding with the centennial of Perkin's discovery of mauveine. There are currently several thousand synthetic dyes available to dyers (Buyer's Guide (annual)).

11.1.3. Textile uses of dyes

As will be seen later in more detail, Sections 11.3, 11.4 and 11.7, dyes can be selected to dye all but a few fibrous substrates to almost any shade desired, with fastness properties which meet all but the most stringent of requirements. Nevertheless, it is important to note that not all dyes are applicable to all fibers.

Textiles fibers can be dyed as staple fibers in the form of raw-stock, tow, sliver, or as staple or continuous filament yarns, knit-goods, woven goods, carpets or garments, i.e., at virtually any stage of the textile manufacturing process. The later in the manufacturing process, the greater the economic advantages.

11.1.4. Other uses for dyes

An important use for some dyes is given in Chapter 10, where soluble dye anions are precipitated by various metallic cations to produce metallized pigments (Lewis 1997). Basic dyes may be precipitated by phosphotungstomolybdic acids.

Selected dyes can be used for coloring paper, for textile printing, for coloring food, drinks and cosmetics, for hair-dyeing, for leather-goods dyeing, woodstaining and for dyeing non-woven products. There are small amounts of specialized products used as laboratory indicators and indicators for medical testing. Dyes are also used in photographic and xerographic applications, and some are used in printing inks for special purposes. They may even be used for anodic coloration of metal surfaces.

11.2. Textile substrates

11.2.1. Natural and man-made fibrous polymers

It used to be conventional to sub-divide fibers into natural, regenerated and synthetic fibers, with regenerated fibers being those in which the backbone of the fibrous polymer had not been synthesized, but had pre-existed in a natural product, e.g., cellulose in cotton-linters or wood pulp, which could be converted into viscose-rayon, or secondary cellulose acetate. Now, it is normal to subdivide fibers into only the two groups: natural or man-made. But this division is no help in deciding how and with what types of dyes the individual fibers might be dyed. A fiber classification for the purpose of dyers must combine a selection of chemical and physical fiber characteristics.

11.2.2. Physical and chemical fiber characteristics

Most dyeings take place in water and at temperatures which range from 30°–130°C. It is important that the goods remain strong enough when wet to be mechanically handled during dyeing. Another important characteristic of fibers is their ability to permit the entry of water, bearing dissolved dye, into the fiber phase. The moisture absorbing characteristics of fibers can be equated with the moisture regain. This is the tendency of fibers to absorb water from the air and is generally measured at 20°C and 65% relative humidity. Values for the major chemical fiber types may be found (Morton and Hearle 1975). Most natural fibers have a high moisture regain and are known as hydrophilic, while most man-made fibers have a low regain and are relatively hydrophobic. The absence of moisture regain explains the problems which face those who would dye glass or polypropylene fibers. However, both may be surface pigmented.

The molecular fine-structure within the fibers plays a critical role when comparing two or more fibers of the same chemical characteristics. This fine-structure is known as the fiber morphology, and is often crudely characterized in terms of the relative contributions of crystalline and amorphous regions within the fibers.

Dyeing morphologically different fibers in the same bath and observing the color contrasts obtained is several orders of magnitude better than any other known analytical technique for detecting (although not intimately characterizing) morphological differences between fibers (Kobsa 1982). This often leads to a condition known as *barré*, or stripiness. Morphological differences in man-made fibers can easily be introduced in drawing, heat-setting or texturing, or any other processes involving heat or tension (Hallada and Holfeld 1975).

The chemical repeat groups present in fibrous polymers can help to explain the hydrophobic or hydrophilic characteristics of the fiber, but they are also important in determing the influence of acidic or basic solutions on hydrolytic stability of the fiber. For example: *secondary cellulose acetate* looses its acetyl groups in water at temperatures of about 100°C, becoming more similar to viscose rayon; *wool*, hydrolyses readily in strongly alkaline solutions, to yield amino acids. Most people know not to wash wool in alkaline detergents. Fewer people are aware that wool can be boiled for hours, without damage, in quite strongly acidic solutions. *Polyester* fibers, not readily accessible to water (hydrophobic) at temperatures below ca. 75°C, can be surface-eroded with caustic alkalis, resulting in fibers of lower diameter and a very pleasant soft hand.

But, for dyeing several important fibers, see Section 11.6, it is the influence of mild alkalis and acids on end-groups and pendant groups on the polymer molecules which determines the ionic character of the fibers in the dye bath, and which dictates the uptake of molecular dye ions, in turn.

Of these groups, the most prevalent in fibers are the amino and carboxylic acid groups in wool, silk and nylon, which in water, and depending on the pH, can become ammonium cations or carboxylate anions respectively, and the sulfate and sulfonate anions associated with some acrylic and modacrylic fibers as well as cationic dyeable polyester and nylon, see Sections 11.5 and 11.6 (Beckmann 1979).

The variety of physical forms in which fibers may be dyed dictates the kind of machinery necessary for processing them. This topic will be introduced and briefly discussed in Section 11.7.

11.2.3. Fiber-blends

For aesthetic, economic and performance reasons two or more fibrous polymers can be blended in innumerable ways. Not only that, but blends can be dyed to achieve a variety of effects which are not possible when dyeing substrates containing only one fiber type. For example: all the fibers in the blend can be dyed to the same color, called a union, or union shade; the different fibers may be dyed to different depths, called tone-in-tone; the individual fibers may be dyed to different depths and hues, called cross-dyeing, or one or more of the fibers may be left white, called a reserve (Aspland 1993b; Shore 1979). Polyester/cotton blends are easily the most prevalent. A means of categorizing dye-fiber combinations, for the purpose of simplifying the dyeing of blends, is given in Section 11.6.5.

11.3. Color application categories

All dyes are organic or organometallic chemicals. There are no inorganic dyes. Dyes, like organic pigments (Lewis 1997), are aromatic compounds containing chromophores and auxochromes (the causes of their colors covered in Section 4.5.1 of Chapter 4), but they also contain groups which will either ensure their solubility in a dyeing medium or will provide the means for their being made soluble.

Organic chemists would certainly prefer to classify dyes by the chemical nature of the chromophores. But these are an amazingly diverse collection of chemical types, some of which have only found utility in two or three commercial dye products. By far the most important of these chemical categories of dyes are those based on AZO, anthraquinonoid (AQ), triphenylmethane (TPM) and phthalocyanine (PC) chromophores. There are literally dozens of others (Colour Index 1971b).

Classification by color application categories is much simpler, having fewer subdivisions. It is also noteworthy that some chemical (chromophoric) types, such as the AZO derivatives, may be found in many different dye application categories. In the next sections, the general nature of dyes from the different application categories will be examined (Colour Index 1971c).

11.3.1. Anionic dyes for cellulosic fibers

11.3.1.1. Direct dyes

These dyes are known as 'direct' because of the directness and simplicity of their application to the cellulosic fibers, which include cotton, viscose rayon, flax, sisal, hemp and other bast fibers. Cotton is by far the most important of these fibers, with about 6 billion pounds per annum being produced in the USA alone.

Direct dyes are often relatively large and linear molecules which are soluble in water by virtue of a plurality of sodium sulfonate substituents, $-SO_3Na$. These groups dissociate in water to give colored sulfonate anions and the corresponding number of sodium cations. The anions are substantive (attracted) to cellulose when applied from a bath containing electrolyte (Colour Index 1971d). Most direct dyes can be described by means of the general formula

Dye $-(SO_3^-)_n \cdot n Na^+$, where n is almost always 1–5.

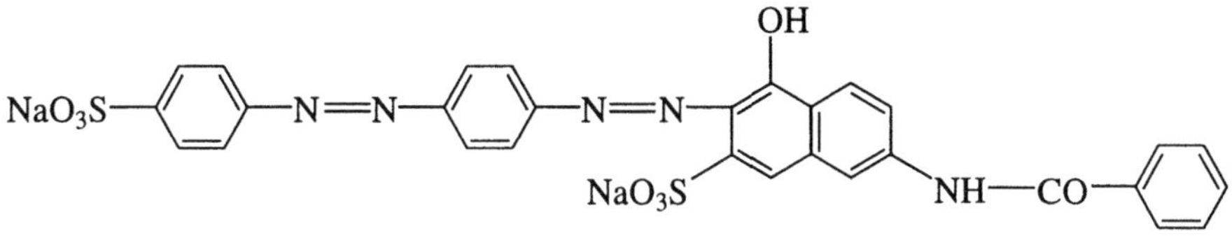

C.I. Direct Red 81

Fig. 11.1. A typical AZO direct dye structure.

Dyeings of direct dyes, as a group, have relatively poor fastness (resistance) to washing, moderate light fastness (resistance to fading on exposure to light) and a good color gamut; but they are not characterized by particular brightness. Direct dyes are moderate in cost and are easy to apply. A typical direct dye structure is shown in fig. 11.1, where the sub-structure, –N=N–, is the azo group.

11.3.1.2. Vat dyes

The vat dyes (colors) are sold in very finely divided pigmentary form either as pastes or powders. They may only be legitimately called dyes because the reduced form of the dye is a water soluble anion, known as a leuco ("white") compound, which will dye cellulose in the presence of electrolyte, in the same manner as do direct dyes. Akaline reduction requires sodium hydroxide and sodium dithionate ($Na_2S_2O_4$).

After the leuco vat anion has dyed the cellulose, the reducing agent and alkali are washed away and the dyeing is exposed to an oxidizing solution – frequently acetic acid and hydrogen peroxide. Such an oxidizing bath converts the leuco dye, within the fiber, to its original pigmentary form. But now the insoluble pigment is entrapped by the fiber.

Vat colors are costly and complex chemicals, generally based on derivatives of anthraquinone. Nonetheless, indigo – which is not an anthraquinone derivative – is the best known, oldest and most widely used of the vat colors. But indigo is not a good example for the behaviour of vat colors, see Section 11.1.3. Vat dyeings generally have excellent wet, light and abrasion fastness. Vat dyeings are more limited in their color gamut than are direct dyeings, since they lack a good solid red, and none of them is particularly bright. However, as a group, vat dyeings have outstanding fastness to bleaching. This makes vat colors extremely useful for cellulosic goods which must be exposed to industrial laundering, e.g., uniform fabrics.

The reversible, redox chemical reactions common to all vat colors are shown in fig. 11.2. Note, that the number of carbonyl groups undergoing these reactions are always present in multiples of two. Typical vat structures are shown in fig. 11.3.

11.3.1.3. Sulfur dyes

In many ways sulfur dyes (colors) are analogous to vat dyes (colors). They are pigmentary, but reducible to water-soluble anionic forms which dye cellulosic fibers readily. However, their dyeing functional groups are di-sulfides which reduce to sodium mercaptides, see

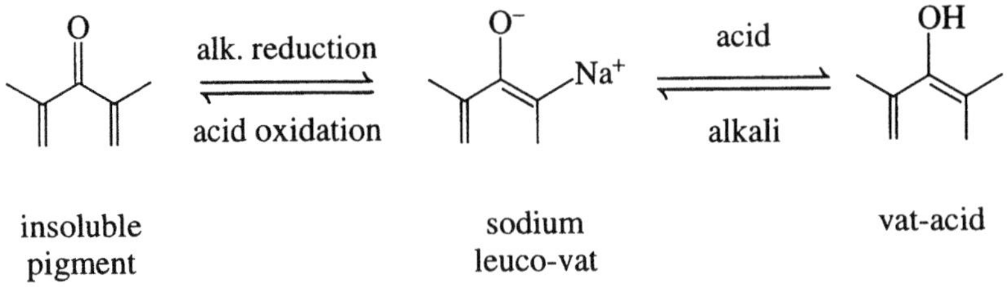

Fig. 11.2. Vat dye functional groups and their dyeing reactions.

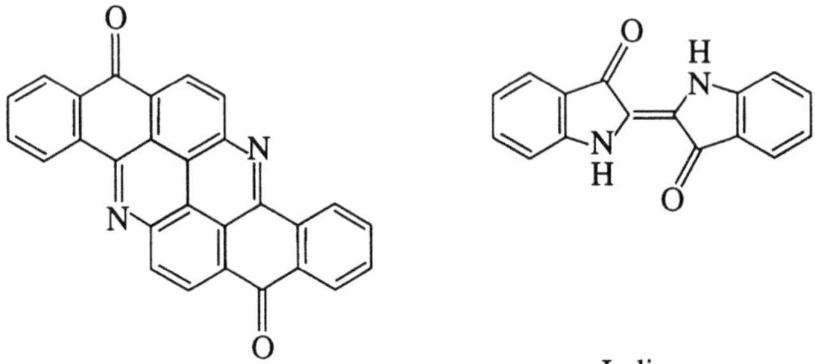

C.I. Vat Yellow 1   C.I. Vat Blue 1

Fig. 11.3. Two vat dye structures.

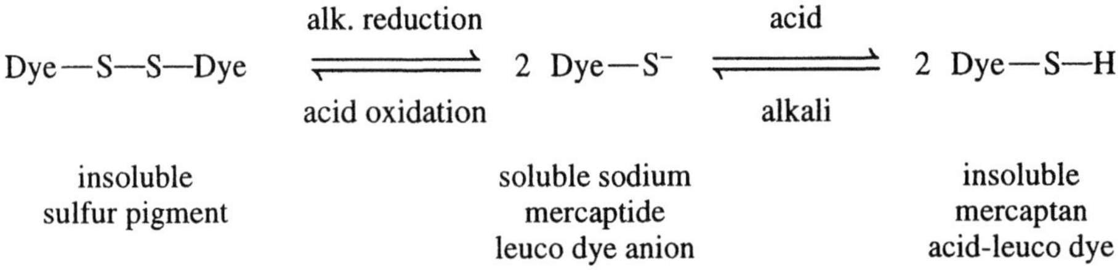

$$\text{Dye—S—S—Dye} \underset{\text{acid oxidation}}{\overset{\text{alk. reduction}}{\rightleftharpoons}} 2\ \text{Dye—S}^- \underset{\text{alkali}}{\overset{\text{acid}}{\rightleftharpoons}} 2\ \text{Dye—S—H}$$

| insoluble sulfur pigment | soluble sodium mercaptide leuco dye anion | insoluble mercaptan acid-leuco dye |

Fig. 11.4. Sulfur dye functional groups and their dyeing reactions.

fig. 11.4. The structures of their chromophores are for the most part unknown, and even the Colour Index (Colour Index 1971e) must be content merely to name the intermediate chemicals from which they are produced. Sulfur colors reduce more readily than vat colors and require only sodium sulfides to dissolve them. They are more difficult than vat colors to oxidize, requiring acidic sodium bromate plus a vanadate catalyst. This combination of properties enables manufacturers to sell them as relatively stable, pre-reduced liquids, i.e., as the concentrated solutions of the sodium salt of the leuco dye anion.

Sulfur colors are the most economical of colors, but have, as a group, the narrowest color gamut of all. Their dyeings lack true reds, have only dull yellows and are most

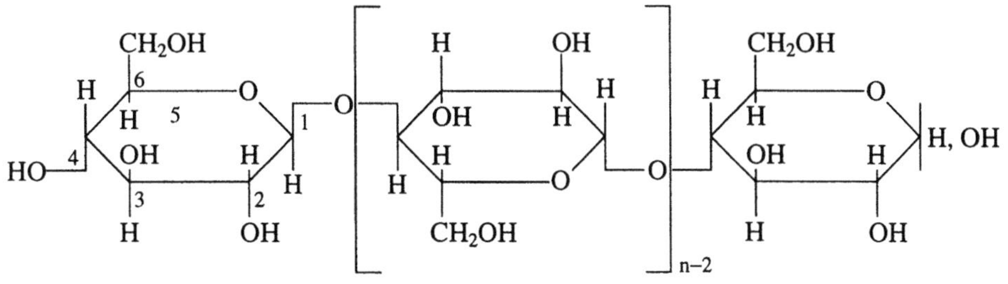

Fig. 11.5. Structural representation of cellulose.

valuable for heavy blacks, navy blues, browns and dark greens. There is no black to compare with a full shade of sulfur black. The light fastness of their dyeings is moderate to good in heavy shades and their wash fastness is good. However, they do have relatively poor fastness to abrasion and remarkably poor resistance to chlorine bleaching. In many ways the fastness properties parallel those of indigo dyeings, and sulfur colors are often used for fashionable denim goods.

11.3.1.4. Reactive dyes

The first fiber reactive dyes were water soluble products designed specifically to react with cellulose – particularly with the hydroxy groups so prevalent in cellulosic fibers – and to form covalent dye-fiber bonds. Cellulose is technically a 1,4-B-D-glucan, in which each repeat unit has the atomic formula $C_6H_{10}O_5$, in which are included three hydroxy groups – one primary and two secondary, see fig. 11.5. For simplicity, the cellulosic polymers may be represented by the simplified general formula, Cell-OH.

Reactive dyes, like direct dyes, are usually solubilized by the inclusion of sodium sulfonate groups, $-SO_3Na$, and their anions have some substantivity (but not usually so much as direct dyes) for cellulose, in the presence of electrolytes. But, unlike direct dyes, they have additional groups whose purpose is simply to facilitate reaction with the cellulose to produce new, colored cellulose derivatives. The benefit to be derived from such colored co-valent derivatives is extremely high resistance to color removal by washing. These dyes can be represented by the general formula dye-X, where -X is a leaving group. The leaving group is displaced from the original dye molecule by reactive nucleophiles and becomes the X^- anion.

The important overall reactions which take place when reactive dyes are introduced to cellulose in the presence of alkali, can now be written as follows:

$$\text{Cell-OH} + \text{OH}^- \leftrightarrow \text{Cell-O}^- + H_2O, \tag{11.1}$$

$$\text{Cell-O}^- + \text{Dye-X} \rightarrow \text{Cell-O} - \text{Dye} + X^-, \tag{11.2}$$

$$\text{OH}^- + \text{Dye} - X \rightarrow \text{Dye-OH} + X^-. \tag{11.3}$$

The primary acid-base equilibrium reaction (11.1) generates some of the highly reactive cellulosate ion by means of another powerful nucleophile – the hydroxide ion. The

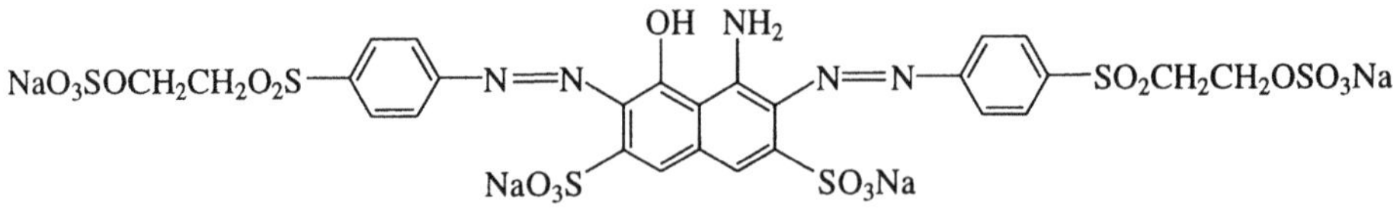

Fig. 11.6. An important AZO reactive dye – C.I. Reactive Black 5.

position of the equilibrium was found to be further to the right than most chemists had dared to expect, but both nucleophiles are always in evidence, cf. equations (11.2) and (11.3). The dye–fiber reaction (11.2) gives the desired product, but the influence of the hydrolysis reaction (11.3) results in wasted dye, and gives an undesirable, colored by-product. Nevertheless, the overall success of the reactive dyes in the market place indicates that the early promise is being fulfilled.

Suitable chemical groups to activate the leaving group, fall into two categories which require different reaction mechanisms (Rys and Zollinger 1993). However, the *overall* reaction is usually the expulsion of an anion, X^-, and the formation of a cellulose – dye bond. Today, the primary activating groups are still the original s-triazinyl ring and the masked vinyl sulfone side chain, $-SO_2-CH_2CH_2-OSO_3Na$, as exemplified by the reactive dye structure given in fig. 11.6. Here, the hydroxide ion first unmasks the reactive vinyl group.

Reactive dyeings have excellent wash fastness, a wide range of light fastness from moderate to excellent, and an incomparably brilliant range of colors for cellulosic fibers. However, they are expensive, their applications is not easy, they require large amounts of salt, and they can produce an excessive amount of colored effluent. They cannot compete with vat dyeings for fastness to bleaching or commercial laundering.

11.3.2. Ionic dyes for ionic fibers

11.3.2.1. Acid dyes
Acid dyes are a very extensive group whose origins and name pre-date synthetic dye chemistry. Many naturally occurring dyes have long been known to have higher substantivity for animal fibers when acids are present in the dyebath. A large majority of synthetic acid dyes can be given the same general formula as direct dyes: Dye-$(SO_3^-)_n \cdot n$Na$^+$. Their molecules tend to be smaller than those of direct dyes, and the number of sulfonate groups is correspondingly lower. The value n can be 1–4, but is often only 1, particularly for those acid dyes being selected to dye nylon.

Acid dyes are presently used to dye nylon, silk and wool. Nylon, which is produced at the rate of about 4 billion pounds per annum in the USA, is easily the major substrate for these dyes, and tufted carpet is the major textile use for nylon. Because the colored part of the acid dye molecule is anionic, it should not be surprising that such dyes would be attracted to cationic groups, primarily the ammonium groups, $-NH_3^+$, within the wool, silk and nylon fibers. Typical acid dye structures are shown in fig. 11.7.

Chrome dyes are a special sub-group of acid dyes which can form resonance-stabilized dye:metal complexes with metals such as chromium. In these complexes the valency and

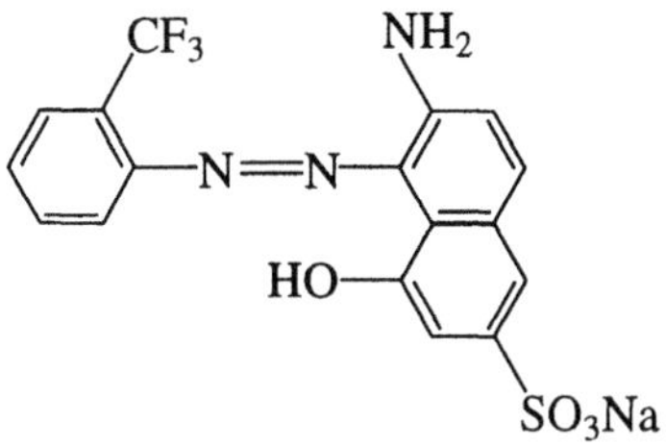

C.I. Acid Red 337 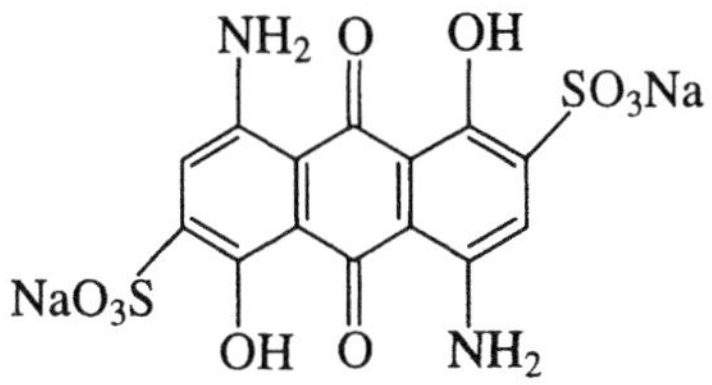   C.I. Acid Blue 45

Fig. 11.7. AZO and anthraquinonoid (AQ) acid dye structures.

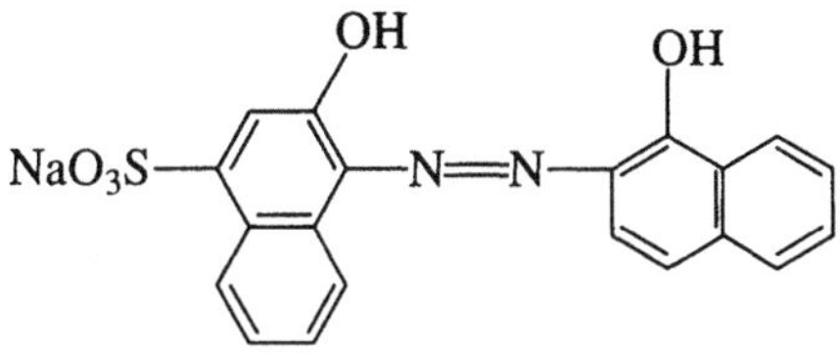

C.I. Mordant Black 3

Fig. 11.8. Mordant (chrome) acid dye structure.

the coordination number of the chromium are both satisfied, and the resulting colors are larger, more stable molecules, with better fastness properties and generally with a different color than those of the original acid dyes from which they are formed. Before complexing with chromium these dyes often contain o,o'-dihydroxy azo groups, fig. 11.8.

The process of dyeing such dyes, which requires treatment with chromium derivatives before, during or after dyeing, is still widely used for dyeings of outstanding fastness, but generally dull shades on wool. Worldwide, approximately 35% of all wool is dyed with these dyes. The unfixed chromium presents a serious environmental problem in the effluent and the dyed goods may ultimately be a problem in landfills.

11.3.2.2. Premetallized dyes
A partial answer to the environmental problems of chrome dyes is the use of anionic dyes in which a metal ion has already been coordinated during manufacture. These are the premetallized (acid) dyes, which are available as 1:1 and 2:1 dye:metal complexes. As the use of 1:1 dye:metal complexes is somewhat limited, only the 2:1 complexes will be represented here, fig. 11.9. The metal ion is usually either a trivalent chromium or cobalt ion, and the product is anionic without the need for a sulfonate group.

11.3.2.3. Basic (cationic) dyes
The earliest synthetic dyes, including mauveine, methylene blue and malachite green, were basic dyes. These are referred to quite interchangeably as cationic dyes, and they

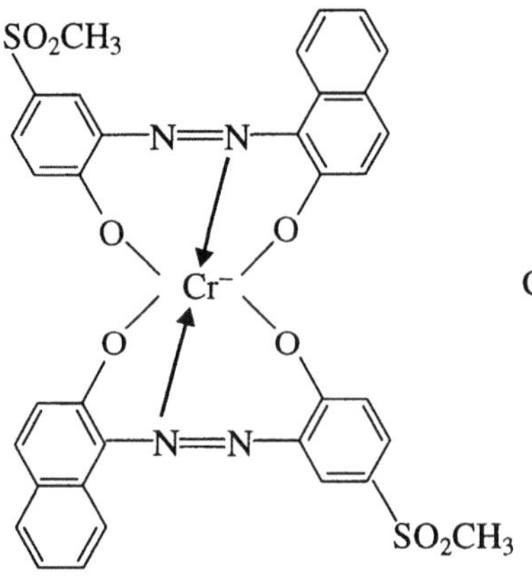

SO$_2$CH$_3$

Cr$^-$

SO$_2$CH$_3$

C.I. Acid Black 60

Fig. 11.9. 1 : 2 Premetallized acid dye structure.

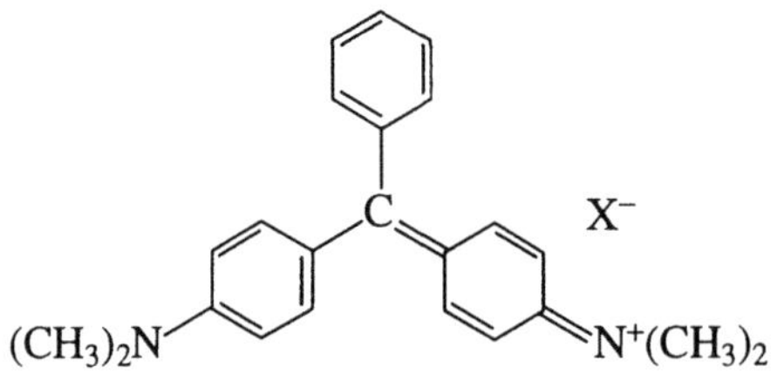

Malachite Green
C.I. Basic Green 4

(CH$_3$)$_2$N

C

X$^-$

N$^+$(CH$_3$)$_2$

Fig. 11.10. An early basic dye.

do indeed consist of a single highly colored cation, with a simple inert counter-ion: Dye$^+$X$^-$ where, X$^-$ may be a chloride ion. Their biggest utility today is for dyeing hydrophobic man-made fibers with anions attached to the polymer molecules (Beckmann 1979).

The basic dyes developed a reputation for poor light fastness on the natural fibers but, when dyed onto hydrophobic fibers, their light-fastness is good. This can be explained readily once it is understood that the presence of water always has a material effect in accelerating the fading of dyeings exposed to light. The basic dyes have an enormously wide gamut of intense, brilliant shades, several of which are fluorescent and as a group, on man-made fibers they have moderate to good fastness to both light and washing. Figure 11.10 shows an older basic dye structure, which is a derivative of triphenylmethane (TPM).

11.3.3. Nonionic dyes for man-made fibers

Here there is only one application category of dyes. Disperse dyes are hydrophobic, non-ionic products and are sold as fine ground powders or dye pastes. They are most suited to dyeing polyester to a wide gamut of shades with a broad range of light, wash, chlorine and other fastness properties. Disperse dyeings are capable of satisfying the fastness

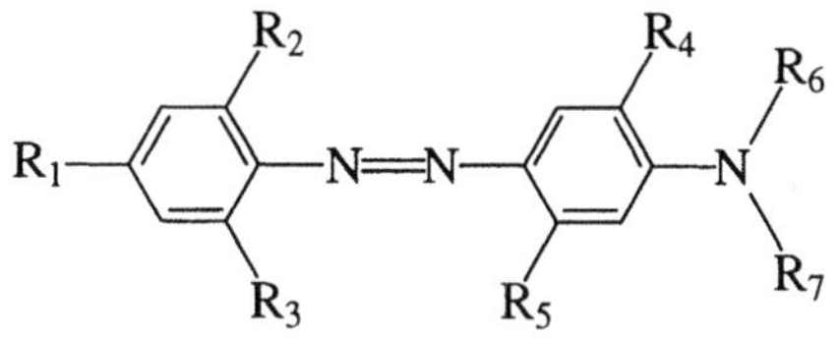

C.I. Disperse Orange 3
has: $R_1 = NO_2$;
$R_2-R_7 = H$.

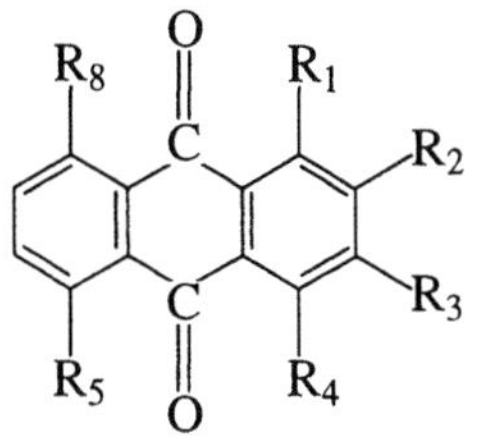

C.I. Disperse Red 9
has: $R_1 = NHCH_3$;
$R_2-R_8 = H$.

Fig. 11.11. Structural bases of most AZO and AQ disperse dyes.

requirements of all but the most particular of customers. There is one such customer – the automotive industry.

Disperse dyes have found some utility on many hydrophobic fibers. They are often applied to nylon, on which they have generally poorer fastness properties than acid dyes. They are the only generally satisfactory dye category for dyeing secondary cellulose acetate.

Disperse dyes are derived largely from simple monoazo and anthraquinonoid chromophores. Despite their hydrophobicity, disperse dyes do indeed dye from their solutions (single molecules) in the application medium, see Section 11.6.3. The structural skeletons of most azo and anthraquinonoid (AQ) dyes are shown in fig. 11.11. Disperse dyes have the useful property of subliming at ca. 200°C. This forms the basis of continuously dyeing polyester/cotton blends, Section 11.7.2.

11.3.4. Azoic combinations

These are not really dyes at all, but are essentially colorless intermediates which are the precursors of azo pigments. The coupling components are frequently derivatives of 2-hydroxy, 3-naphthoic acid arylamides. These, when dissolved in alkali, ionize to give anionic naphtholates, which have some substantivity for cellulose (just like other anionic dyes) in the presence of electrolytes. The other component of the combination – the diazo component – is a stabilized diazonium salt (or its amine precursor). When a diazonium salt is applied to cellulosic fibers already impregnated with a naptholate, coupling takes place and azo pigments are formed in the fiber.

Starting with basic intermediates, which are inexpensive, this cost advantage is largely sacrificed to the complexity of a 2-step dyeing process. However, azoic combinations are still the only way the very deepest shades of orange, scarlet, red and bordeaux can be economically achieved on cotton and cotton/polyester blends.

11.4. Color fastness

Color-fastness is the resistance of colored materials to color change or loss as the result of exposure to different agencies. These agencies are very diverse (Technical Manual (annual)), and include light, water, dry-cleaning solvents, washing, burnt-gas fumes, ozone, industrial laundering (with hypochlorites), and abrasion. Several of these processes and the techniques used for the preservation of color are discussed in Chapter 12.

The whole business of color-fastness testing is complicated. In the attempt to set both quality control and end-use performance standards for colored materials, fastness test methods must try to accommodate the variety of end-uses and performance expectations of many different customers for the materials. The expectations are legion, and frequently ill-defined. Color-fastness testing is further complicated by the number of organizations which have devoted huge amounts of time and money in the development of test methods. For example in the USA there are the *American Association of Textile Chemists and Colorists* (AATCC)[1], the *International Organization for Standardization* (ISO), and the *Society of Automotive Engineers* (SAE), who have a common interest in the light fastness of automotive fabrics. For insight into the difficulties encountered by those wishing to establish a relationship between product performance and test methods, see (Soll 1990).

Why are there so many different test methods? Light fastness testing would be a good example to look at briefly. To what kind of light is the colored material likely to be exposed? If the answer is daylight, then further questions arise immediately: at what latitude; is the light direct or indirect; at what season; at what time of day; how many hours per day; how long is the material in the dark between exposures; is the material exposed to the weather or covered; at what surface temperature; behind glass or not; how great is the contribution of UV light; what is the relative humidity; are industrial gas fumes present, and if they are, at what concentrations... This list goes on and on.

The principal purpose when conducting accelerated light-fastness testing is only to obtain an idea of the color-fastness properties of a dyeing relative to that of others, determined under the same conditions. Is it surprising that an ordinary purchaser of a domestic carpet has not the slightest feel for what constitutes a normal expectancy of light-fastness for the carpet which was installed in the living room?

Fastness scales have been discussed (Lewis 1997). The light fastness of dyeings improves with the depth of shade. The wet fastness, and most other fastness properties, gets worse as the depth of shade increases. As might have been anticipated, the cost of dyes usually increases with the expectancy of good fastness.

11.5. The dyeing process

There are three separate steps in the dyeing process: dye sorption by the fiber (substrate); dye diffusion into the fiber, and dye fixation on the fiber. In batch dyeing (i.e., in dyeing a fixed weight of goods in a fixed weight of dyebath), the first step, dye sorption, is achieved by circulating the dyebath through the goods, or moving the goods through the

---

[1] American Association of Textile Chemists and Colorists, P.O. Box 12215, Research Triangle Park, NC 27709, USA.

bath, or both. In continuous dyeing, the first step is the mechanical placement of dyes at or near to the fiber surfaces. It is the second step, dye diffusion, which requires the dye to be in a monomolecular form. Once the monomolecular dye has diffused into the fiber, fixation within the fiber can be due to three causes: physical bonding between dye and fiber; chemical bonding between dye and fiber, and conversion of the soluble coloring matter into an insoluble pigmentary form.

11.5.1. Dyes in solution

Members of the following dye application categories are intrinsically water-soluble salts: acid, basic, direct, reactive. Basic dyes are cationic; the others are anionic. Sulfur and vat colors are reduced to give the anionic leuco-compounds.

The chemical potential of ionized dyes in solution is related to the ionic product: [color ion] $\times$ [counter ion]^n. Here n is the number of ionic charges on the color-ion, which for anionic colors can be 1–6 or more, but for cationic colors is normally equal to one; the square brackets denote concentrations in moles/l. Since the counter-ion concentration is electrolyte dependent, so is the chemical potential. For the non-ionic disperse dyes the chemical potential is proportional to the dye solution concentration. The chemical potential of the dissolved dye in the dyebath is one of the factors which drives the dyeing process.

11.5.2. Dye sorption

Before dissolved dye can diffuse into the fibers it must first pass the negatively charged barrier surrounding all fibers immersed in water (Giles 1993). Basic dyes, being cationic, are attracted to the surface of all fibers by electrostatic attraction to the barrier layer. But anionic compounds would be repelled were it not for the addition of high concentrations of ions to the dyebath, usually in the form of common salt or sodium sulfate. These ions effectively swamp out, or mask, the surface negative potential of the fibers by raising the background ionic strength. Once the barrier is passed, the different dyes are sorbed due to some of a whole range of physical forces including van der Waal's forces, hydrogen-bonding, Heitler–London dispersion forces, hydrophobic-bonding and coulombic (electrostatic) attraction (Giles 1993).

11.5.3. Dye diffusion

The higher chemical potential in the bath drives the dyes from the surface and into the fiber, where they are initially held by the same forces of attraction which caused them to be sorbed at the fiber surface. It is the wide range of molecular sizes, configurations, and ionic charges, plus the hydrophilic/hydrophobic character of the dyes and the fibers which ensures that some colors diffuse (dye) more rapidly than others, and that some are attracted and held more strongly than others.

11.5.4. Batch dyeing

Liquor ratios are the relationships between the weight of the dyebath and the weight of the goods being dyed. Sometimes this is expressed as a ratio, e.g., 20:1 liquor: goods, but sometimes it is convenient to refer to it as a single number, e.g., $L = 20$. In both these cases there would be 20 times as much water as goods (by weight). Practically, L can be in the 2–50 range, see Section 11.7.

Exhaustion is the percentage of the dye added initially which has transferred from the dyebath to the fiber after a given dyeing time. When the dyeing is taken to completion it is economically desirable that the exhaustion (%E) should be as high as possible, to avoid having excess dye in the waste water. Values of 85–95 for E would not be unusual.

When the chemical potential of dye in the bath is high, the rate of dyeing exceeds that rate at which color leaves the fiber. But, if a colored fiber (containing a soluble dye) is placed in bath without dye, the chemical potential of dye in the fiber exceeds that in the bath and color loss from the fiber into the solution takes place. This process may be called leveling or stripping. At the end of a dyeing, the rates of color loss from fiber to bath and bath to fiber are equal, i.e., a dynamic equilibrium is established. At this stage, the dye is partitioned between the concentration in the solution (dyebath), written $[D]_s$, and that in the fiber, $[D]_f$ (both expressed as grams/gram). Here the partition coefficient, K, has no dimensions, and can be written: $K = [D]_f/[D]_s$, and it can be shown that: %E $= 100K/(K + L)$. The amount of dye to be used in a batch dyeing is usually expressed as a percentage on the weight of the goods (o.w.g.).

11.5.5. Continuous dyeing

The amount of liquor applied to fabric, moving continuously through a dye range, is generally expressed as the % wet-pick-up, or the increase in the weight of the wet fabric over that of 100 units of the same fabric when dry. This is written: % w.p.u. If the added weight is water containing a known concentration of dye, the % dye o.w.g. can easily be calculated.

11.6. Sorption isotherms

At dyeing equilibrium there are concentrations of residual dye in the dyebath, $[D]_s$ and dye on the fiber, $[D]_f$. If a series of dyeings are conducted under the same conditions but at different initial concentrations of dye in the bath, a series of values of $[D]_s$ and $[D]_f$ can be obtained. When these are plotted against one another, a sorption isotherm is obtained. There are only 3 principle types of sorption isotherms, indicating that basically there are only 3 broad types of dyeing behavior. These isotherms are known as: Freundlich, Langmuir and Nernst types, and they are illustrated in fig. 11.12.

11.6.1. Freundlich isotherms

These are smooth continuous curves of the form: $y = kx^n$, where n has no physical significance, and $[D]_s = x$; $[D]_f = y$. The relationship is quite empirical. There is no limiting value (saturation value) for dye on the fiber, $[D]_f$. In other words, as one keeps increasing the concentration dye in the dyebath, more and more dye goes onto the fiber. All anionic dyes for cellulosic fibers follow similar isotherms and may be given the designation C. They are indicative of non-ionic, or weak bonding forces between the dye and the fiber.

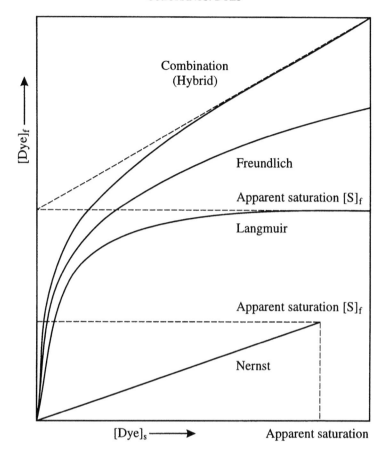

Fig. 11.12. Sorption isotherms for dyes.

11.6.2. Langmuir isotherms

These isotherms show apparent saturation values $[S]_f$ for the concentration of dye in the fiber, and are of the form: $y = kx[S]_f/(1 + kx)$. When x is very large, $y = [S]_f$. These are the isotherms applicable to dyeing ionic dyes on ionic fibers with the opposite charge, e.g., acid, premetallized and chrome dye anions on wool, silk and nylon fibers (cationic); and, basic dye cations on acrylic, modacrylic, cationic dyeable nylon and cationic dyeable polyester fibers (anionic). The former dye-fiber attractions may be designated A, and the latter B.

11.6.3. Nernst isotherms

These are straight lines up to an apparent fiber saturation value: $y = Kx$. However, the x-axis in this isotherm is not the concentration of dye in solution $[D]_s$, but the concentration of dye in the bath. Some of this dye may not be dissolved. Beyond the saturated solubility of the dye in water, no further dye can dissolve, and therefore no further dye can be partitioned between the bath and the fiber. This is why there is an apparent fiber saturation value. The Nernst isotherm is applicable to the sparingly soluble disperse dyes which form hydrophobic bonds with hydrophobic fibers. These dye fibers interactions may be called class D. They are of greatest importance when dyeing polyester or cellulose acetates with disperse dyes and of less importance when dyeing nylon with disperse dyes because nylon is dyed with better fastness using acid dyes.

11.6.4. Hybrid sorption isotherms

When strongly hydrophobic acid dyes with large molecules and a single negative charge dye hydrophobic fibers such as nylon, there is the possibility of hybrid Nernst and Langmuir isotherms. Such dyes can be dyed to $[D]_f$ values which exceed the apparent saturation value of the fiber, as with acid dyes which are strongly hydrophilic, have several sulfonate groups and have small molecules. For graphic clarity, the slope of the Nernst isotherm used to create the hybrid (combination) isotherm is steeper than the one used in the lower part of fig. 11.12.

11.6.5. Dye–fiber combinations – blends

The number of possible fiber combinations which might be used to give economic, aesthetic or physical property benefits, is legion. Designating each fiber with the appropriate letter to indicate the dyeing possibilities (A–D) as in Table 11.1 helps to systematize the general dyeing characteristics of a multitude of potential blended fabrics. These letters may be referred to as substantivity (dye–fiber attractiveness) groups. Using this nomenclature, cotton/polyester blends would be CD blends and wool/nylon carpets would be AA blends (Aspland 1993a). It is easy to envisage that dyeing fibers from different substantivity groups (e.g., CD blends) might be complex. It should also be clear that fibers from the same substantivity group (e.g., AA blends) can have different levels of attraction for the dyes. This would make dyeing both fibers simultaneously to the same depth (union) very challenging (see Section 11.2.3).

TABLE 11.1

Substantivity groups of major textile fibers.

Major textile fibers	Substantivity group (Section 11.6)
Cotton	C
Viscose rayon	C
Nylons	A (D)
Wool	A
Silk	A
Acrylic, modacrylic	B
Polyester	D
Secondary acetate	D
Glass, polypropylene	–

11.7. Dyeing variables – machinery

Commercially acceptable dyeings require that all areas of the goods being dyed are given substantially equal exposure to the dye being applied. Clearly, different types of dyeing machinery are required for the different physical forms of fibers.

11.7.1. Batch dyeing machinery

There are three general categories of batch dyeing machine.

11.7.1.1. Circulating dyebath: stationary goods

The classical machine in the category is a *package dyeing* machine, figs 11.13 to 11.15. Here the yarn is wound onto special metal spring or perforated plastic forms which can be mounted, one on top of the other, onto perforated stainless steel posts in the machine. The fit between adjacent packages must be tight, and the packages of uniform density. The dyebath is pumped up through the posts, out through the packages (inside-out) and back to the pump, or, by reversing the flow, from outside the packages into the post (outside-in). In dyeing, inside-out and outside-in are alternated.

The same principles are used for dyeing raw-stock in the form of staple fibers. However, here the raw stock is loaded and packed into a perforated container until it resembles a large single package of yarn. Similarly, hose can be packed into baskets for dyeing, where each basket is like a single oversized yarn package. In both cases, uniformity of packing density is the key to a uniform dyeing.

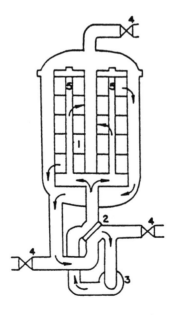

Fig. 11.13. Package dyeing machine: 1 – packages; 2 – flow-reversing valve; 3 – pump; 4 – valves; 5 – perforated posts.

Fig. 11.14. Small package dyeing machine. Courtesy Gaston county dyeing machine Co.

Fig. 11.15. Large package dyeing machine. Courtesy Gaston country dyeing machine Co.

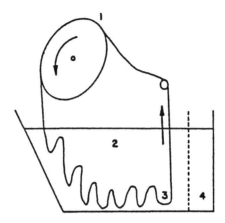

Fig. 11.16. Dye beck: 1 – elliptical reel; 2 – dyebath; 3 – fabric; 4 – compartment for adding dyes and chemicals.

Warp-knit fabrics are frequently dyed by first rolling them in full width on a perforated post, known as a beam. The resultant roll is then mounted in a *beam dyeing machine* and the whole roll is dyed as if it were a single package. These machines are usually laid on their side for the convenience of loading the beam horizontally.

All the above equipment can be run at low liquor ratios, $L = 5$–10 and, being enclosed, can be designed to run at elevated temperatures (100–130°C) under pressure.

11.7.1.2. Circulating goods: stationary dyebath

The machines in this category are usually used to dye fabrics. The machines include beck, jig, and garment dyeing machines.

Beck (or *winch*) *dyeing* machines are the most commonly used worldwide. The essential features are shown in fig. 11.14. The fabric, in rope form, is pulled round and round by the circulating elliptical reel. The process normally uses relatively long liquor ratios, $L = 15$–30 and is only of value for goods insensitive to creasing and crack marks, e.g., cotton single or double knit goods, in tubular form. Becks are usually enclosed, but rarely pressurized.

If goods are sensitive to creasing, they must be processed in open (full) width, and this could be achieved on a *jig dyeing* machine or *jigger*. Here cloth on a roll is threaded under the surface of a dyebath, out again, and wound onto another roll. As one roll unwinds, the other rolls up the cloth, wet with dyebath liquor. When the first roll is empty and the second one full, the direction is reversed. The fabric passes through the dyebath several times, forwards and backwards. These machines can run using very low liquor ratios, $L = 5$–10, but problems arise with dyebaths containing mixtures of dyes which may not have the same attraction to the fibers. This differential attraction (substantivity) is called differential strike, and leads to a problem known as tailing, in which the two ends of the fabric are not the same shade. Careful dye selection is required.

Nowadays, most *garment dyeing machines* are simply glorified front-loading domestic or commercial laundry machines. Of course, there are many refinements to ensure

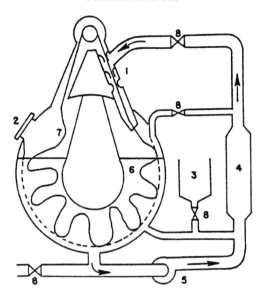

Fig. 11.17. Jet-dyeing machine: 1 – venturi; 2 – inspection port; 3 – mixing tank; 4 – heat exchanger; 5 – pump; 6 – dyebath; 7 – fabric; 8 – valves.

robustness, higher temperatures, easy loading and unloading. Liquor ratios are generally higher than those used in the other machines mentioned, $L = 30\text{--}50$.

11.7.1.3. Circulating dyebath and goods

Perhaps the most important dyeing machine development in the last 40 years is the *jet dyeing machine*, fig. 11.17. There are many different configurations of jet dyeing machines, but all operate on the principle that the cloth, in rope form, is moved by the tangential impact of the dyebath on the goods at the nozzle of the jet (venturi). These machines can be pressurized, run at low liquor ratios, $L = 5\text{--}15$, and are particularly useful for dyeing polyester and polyester/cotton goods without the need for carriers to assist in dyeing the polyester, see Section 11.8. Jets are at the same time more versatile, but more expensive, than becks.

11.7.2. Continuous dyeing machinery

The majority of continuously dyed goods is dyed as open width fabric on a conventional full continuous dye range. Designed for dyeing polyester/cotton blended fabrics, the first half of the range is primarily to dye the polyester portion of the goods by the thermofixation (sublimation) of disperse dyes, see below. The ranges generally operate at ca. 100 meters/min. and there can be 400–600 meters of fabric threaded through them. For dyeing 100% cotton or for nylon carpet dyeing, the first half of the range may be eliminated.

The mechanical range components, in order, are: A-frame; accumulator; padder I (dye-pad); pre-dryer; dryer; Thermosol oven; padder II (chem-pad); steamer; wash-boxes (at least three or four sets); dryer, and A-frame. Goods arrive at one end of the

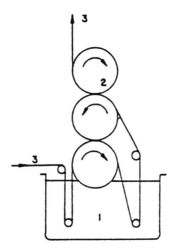

Fig. 11.18. Pad mangle: 1 – dye liquor; 2 – squeeze rolls; 3 – fabric.

range on a cloth holder such as an A-frame, which is simply a means of supporting a roll of fabric. The fabric passes to an accumulator, which holds enough piled-up cloth to allow an empty A-frame to be replaced by a full one and for the two pieces of goods to be seamed together without stopping the range. Next the goods enter the padder.

11.7.2.1. Padder
This piece of equipment is critical to the overall result. The goods enters the liquid in the pad-trough (which contain water plus dissolved or dispersed dyes and chemicals), is drawn through the liquid and picks up the pad-liquor along with dyes and chemicals. The goods enters the nip and is squeezed uniformly across its full width by the pad rolls (bowls). The pressure at the nip may be adjustable over a wide range. Excess liquid returns to the pad trough.

A padder is shown, fig. 11.16, in which there are three rolls, two dips and two nips (also see fig. 11.19). However, there is a multitude of other possibilities. The rolls can be steel or coated with rubber of different degrees of hardness. It is important to note that the best padders have their roll profiles controlled from within by hydraulic pressure. This ensures nip uniformity across rolls which could be as much as 4 meters in width. The wet-pick-up at this stage can range from 60–400% o.w.g. depending on the goods, but 70–80% is more normal.

11.7.2.2. Pre-dryer: dryer
The wet goods next enters the pre-dryer, which heats the water extremely rapidly with I.R. radiant heaters, reducing the water content of the goods to a low level so rapidly that (with the use of an antimigrant) the dissolved and dispersed chemicals cannot migrate to the drying surfaces of the goods. The antimigrant is a polymer whose concentrated solutions in water are extremely viscous and cannot flow rapidly.

Fig. 11.19. A dye padder.

11.7.2.3. Thermosol oven (thermofixation)
Here the temperature is raised to 200–215°C for 30–90 seconds. Under these conditions, non-ionic disperse dyes, which may have been dried onto the fiber surfaces, sublime as individual dye molecules. These molecules have substantivity for polyester fibers and polyester dyeing takes place. This marks the halfway point in the process. The dry goods (with its polyester dyed) is either collected at an A-frame, or proceeds to a second padder from which more dye or chemicals may be applied.

11.7.2.4. Steamer
After chemicals or dyes have been applied at the second padder, the goods again have a water content of 60–400% w.p.u., or the equivalent of a liquor ratio from 0.6–4. The goods then normally proceeds to a roller steamer containing saturated steam at 100°C, fig. 11.20. Here the low liquor-ratio dyebath on the fiber is heated up to the boil and dye diffusion processes are completed in 2–3 minutes.

11.7.2.5. Wash-boxes
These have to perform a number of important 'post-dye-diffusion' functions, which differ with the classes of dyes applied: vat and sulphur dyeings require removal of reducing

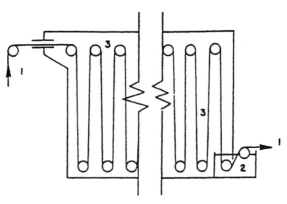

Fig. 11.20. Roller steamer: 1 – fabric; 2 – water seal; 3 – steam chamber.

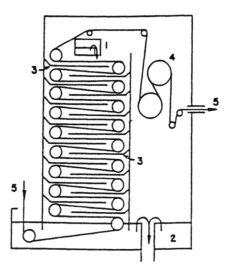

Fig. 11.21. Counterflow washer: 1 – wash-water inlet; 2 – water drain; 3 – baffles; 4 – tensioning rolls; 5 – fabric.

agents, exposure to oxidants and removal of oxidants; vats also require "soaping" to develop their greatest fastness properties; reactive dyeings need to be repeatedly scoured to remove hydrolysed by-products; disperse dyes require dispersing agents and other residuals to be scoured free from fiber surfaces. Modern wash boxes are built on counterflow principles, by which the cloth and the treatment liquid flow in opposite directions for greater energy efficiency and lower water use. An example is shown schematically, fig. 11.21. Drying and take-up of the goods onto A-frames completes the process. Continuous processing requires long lengths of goods to be efficient and cost-effective, and less than 2000 meters of a particular color is considered inadequate. Clean-up is a problem, and the order in which different colors are run should be selected with care.

11.8. Dyeing variables – auxiliaries

There are many different kinds of chemicals which influence dyeing processes and the dyed products.

11.8.1. Buffers

In all dyeing processes where pH control is important, buffer systems may be used. These include four different types.

Acid dyeing for wool, silk and nylon. Here the pH of the bath determines the overall charge on the fiber molecule and thereby its attraction for anionic dyes, so buffering the bath is important. Here the pH can range from 2.5 to 8, depending on the dye–fiber combination.

Basic dyeing and disperse dyeing. Here it is important to keep the pH slightly on the acid side, not because it effects the dyeing processes, but because many of the dyes themselves are sensitive to irreversible changes at a pHs of 7 and higher.

Reactive dyeing. Primarily for cellulosic fibers, the rates of dye–fiber reaction and hydrolysis here are functions of the pH. Because some reactive groups are more reactive than others, it is imperative that dyeings should be under strict pH and temperature control.

Azoic combinations. When diazotizing aromatic amines but, more particularly, when coupling their diazonium salts with sodium naphtolates, the pH must be carefully controlled.

11.8.2. Carriers

In the early days of polyester batch dyeing, it was found that the rate of dyeing at the boil with disperse dyes was barely adequate, except for a very few dyes more suitable for dyeing secondary acetate. This is because polyesters have second order glass transition temperatures (T_g, ca. 75°C) only just below the atmospheric boiling point of water, 100°C. No dyeing takes place below T_g and at 100°C the rate is still not sufficiently great for most practical purposes.

It was found that a variety of miscellaneous, water immiscible, volatile, organic, colorless solvents, if dispersed in the dyebaths, would materially accelerate the dyeing and, in effect, carry the dyes into the fibers. Hence, the name carrier. One by one, these materials have been found to be environmentally undesirable and have been withdrawn. This leaves, as the only acceptable alternative, dyeing polyester at high temperatures (HT) from 110–130°C, under pressure.

11.8.3. Compatibilizers

Sometimes anionic dyes and cationic dyes must be used in the same bath during the dyeing of particular fiber blends, such as acrylic/wool blends. Such dyes would normally co-precipitate to form insoluble complex salts. However, this problem can be averted by means of compatibilizers. These are often colorless polyethanoxy derivatives of cationic or anionic character. For example, if an anionic compatibilizer is added to a cationic dye solution, a complex salt is formed which is dispersed or suspended by the polyethanoxy group. Now, the anionic dye can be safely added to the mix.

11.8.4. Levelers

Dyers require a number of different tools to control the color uniformity of the dyed goods while they attempt to complete dyeing in the shortest time possible. These tools include temperature controls, control of the circulation of the bath relative to the goods, and the use of leveling agents. Leveling agents are particularly useful in ionic dyeing systems where the substantivity and the rate of dyeing depend on the attraction of oppositely charged dye-fiber species. They are normally molecules with non ionic solubilizing groups coupled with cationic or anionic groups. Some levelers are amphoteric with the ionicity changing with pH. They parallel compatibilizers closely.

Anionic levelers are anions which control the rate of acid-dyeings by competing with the colored dye anions for the fiber cations (wool, silk, nylon). The more substantive the leveler, and the higher its concentration, the slower the acid dye anion will be able to dye. However, these same compounds may be levelers for basic dyes. They complex with the cationic dyes in solution, lowering the effective concentration of the cations, and slowing the rate of dyeing. Dyeing can only take place due to the progressive breakdown of the complex and the liberation of free dye cations, which may then interact with an anionic fiber (acrylics, modacrylics, cationic dyeable nylon and polyester).

Cationic levelers are the reverse of the anionic levelers. They compete with basic dyes for anionic fibers, thereby slowing down the rate of dyeing. They complex with anionic dyes in solution, lowering the effective concentration of the dyes, and slow down the rate at which these can dye cationic fibers.

11.8.5. Fixing agents and stainblockers

The same principles, of coulombic attraction between oppositely charged species, are effective here also. Only the use of the principles is different.

Direct dye fixatives are cationic compounds which are used to aftertreat direct dyeings. They form less-soluble complex salts of the anionic dyes, and cause the wet-fastness of the dyeings to be increased.

Acid dye fixatives involve a different mechanism. Complex anionic compounds (originally tannic acid salts, now, synthetic tanning agents – syntans) are applied to the cationic wool, silk or nylon fibers after dyeing. However, they are of such high molecular weights that they become entangled at the fiber surfaces and form a richly anionic surface layer. This layer repels dye anions already in the fiber and prevents them from escaping, which leads to improved wet fastness. It also repels dye anions trying to get into the fiber. This is the technology on which stain-blocker carpets are based and is the reason they resist staining by artificial fruit drinks, which are colored with acid dyes.

11.8.6. Miscellaneous

The general effect of salts on raising the chemical potential of ionic dyes in solution has already been noted. Surfactants are added to dye baths for a variety of reasons, but they may all result in foaming. They may be added as *penetrants or wetters* to ensure that all the fiber surfaces are prepared for the dyeing process. *Dispersants* may be added as security against dispersed compounds aggregating or flocculating, e.g., disperse dyes. In any case, *defoamers* are added to counteract the foam, which can be a distinct problem in dyeing. *Lubricants and softeners* are often added to ensure that the fibers slip readily, relative to one another, obviating defects such as creases and crack marks, quite apart from giving the dyed fabric a more desirable hand (feel).

11.9. Textile pigmentation processes

Mass pigmentation of fibers has been mentioned earlier, as has the process of padding pigment dispersions onto fabrics in the presence of a binder (Ingamells 1989). Most textile printing in the USA is pigment printing, also in the presence of the ubiquitous binder.

However, in a batch process known erroneously as pigment dyeing, garments may be surface pigmented by an exhaust process, better referred to as *exhaust pigmentation*. Here the cotton or polyester/cotton garments are treated with a substantive, cationic acrylic pre-polymer, which is attracted to the surface negative charges generated by all fibers in water. The garments are then rinsed. Next, they are treated with a dispersion of pigments, which are conventionally dispersed using anionic dispersants. These anionic dispersions are attracted to the cationic pre-polymer. Finally the garments, with pigments adhering to all the fiber surfaces, are wet with an acrylic binder dispersion, followed by drying and curing. Although the process imitates batch dyeing, it is a true pigmentation, since all the color is bound on the fiber surfaces.

11.10. Summary

Many aspects of dyes and dyeing have not been covered. However, the coverage runs the danger of seeming long-winded to the uninitiated and perfunctory to the cognoscenti. It is the author's hope that somewhere in the middle there are curious readers who will find something here of interest. Let us not forget this represents a brief account of a huge industry, in which ca. 10 billion pounds of fiber is dyed each year in the USA alone, using ca. $1 billion of dyes.

References

(Note: for AATCC see footnote on page 328).

AATCC, 1993, Textile Solvent Technology – Update' 73, AATCC, Atlanta, GA, 1993 (AATCC, Research Triangle Park, NC).

Aspland, J.R., 1993a, in: Textile Chemist and Colorist 25(10), 31.

Aspland, J.R., 1993b, Textile Chemist and Colorist 25(9), 79.

Beckmann, W., 1979, in: The Dyeing of Synthetic Polymer and Acetate Fibers, ed. D.M. Nunn (Society of Dyers and Colourists, Bradford, UK) Ch. 5.

Buyer's Guide (annual), Buyer's Guide for the Textile Wet Processing Industry, published annually as the July issue of the Textile Chemist and Colorist.

Colour Index, 1971a, 3rd edn, Vol. 3 (Society of Dyers and Colourists, Bradford, UK, and AATCC, Research Triangle Park, NC).

Colour Index, 1971b, 3rd edn (Society of Dyers and Colourists, Bradford, UK, and AATCC, Research Triangle Park, NC) Vol. 4, 6 et. seq.

Colour Index, 1971c, 3rd edn (Society of Dyers and Colourists, Bradford, UK, and AATCC, Research Triangle Park, NC) Vol. 2, 3.

Colour Index, 1971d, 3rd edn, (Society of Dyers and Colourists, Bradford, UK, and AATCC, Research Triangle Park, NC) Vol. 2, p. 2005.

Colour Index, 1971e, 3rd edn (Society of Dyers and Colourists, Bradford, UK, and AATCC, Research Triangle Park, NC) Vol. 3.

Giles, C.H., 1993, in: Textile Chemist and Colorist, Vol. 25-10, Ch. 2.

Hallada, D.P. and W.T. Holfeld, 1975, Canadian Textile Journal **92**, 37.

Ingamells, W.C., 1989, The Theory of Colouration of Textiles, ed. A. Johnson (Soc. Dyers and Colourists, Bradford, UK) Ch. 4.

Kobsa, H., 1982, Book of Papers, AATCC National Technical Conference, Charlotte (AATCC, Research Triangle Park, NC) p. 85.

Lewis, P.A., 1997, in: Color for Science, Art and Technology (Elsevier, Amsterdam) Ch. 10.

Morton, W.E. and J.W.S. Hearle, 1975, Physical Properties of Textile Fibers, 2nd edn (Heinemann, London) p. 170.

Rys, P. and H. Zollinger, 1993, in: Textile Chemist and Colorist, Vol. 25-10, Ch. 4.

Schwindt, W. and G. Faulhaber, 1984, Review of Progress in Colouration **14**, 166.

Shore, J., 1979, in: The Dyeing of Synthetic Polymer and Acetate Fibers, ed. D.M. Nunn (Society of Dyers and Colourists, Bradford, UK) Ch. 7.

Soll, M., 1990, Textile Praxis International, August, p. 799.

Technical Manual (annual), published annually by the AATCC, Research Triangle Park, NC.

chapter 12

COLOR PRESERVATION

KURT NASSAU

Lebanon, NJ, USA
(AT&T Bell Telephone Laboratories
Murray Hill, NJ, USA, retired)

Color for Science, Art and Technology
K. Nassau (Editor)

CONTENTS

12.1. Introduction 347

12.2. Chemical interactions between colorants 348

12.3. Chemical interaction effects on colorants 348

12.4. Atmospheric interactions with colorants 348

12.5. Photochemical deterioration 349

    12.5.1. Photodegradation in organic colorants 350

    12.5.2. Photodegradation in inorganic colorants 351

12.6. Conclusions and further reading 351

References 352

12.1. Introduction

This topic is most frequently the concern of the museum curator faced with the preservation of color. The objects involved include paintings; fabrics; positive and negative photographic images; drawings and documents on paper, parchment, and other substrates; furniture; buildings; and a wide variety of other objects of historical and/or artistic importance.

Preservation is made difficult by the complexity of the materials involved. Consider just water- and oil-based paints. A paint contains one to several colorants that selectively absorb and reflect light, usually pigments (see Chapter 10), although dyes (see Chapter 11) could also be present. The pigments are dispersed by extended grinding in a fluid, the vehicle or binder; a solvent or thinner is also frequently present. The vehicles used for water colors have included solutions or dispersions of gums, glues, egg white, and even boiled rice. Natural resins and drying oils, such as rosin, copal, lac, and linseed oil were the vehicles for earlier paints (including lacquers and varnishes) frequently combined with turpentine and/or other thinners. For modern paints a variety of synthetic vehicles dominate, such as alkyd resins used for oil paints, and polyvinyl and acrylic resins used for water-based latex paints. In addition to pigment, vehicle, and thinner, many other ingredients may be present: plasticizers to improve solubility and flexibility; extenders or fillers to reduce cost and improve mechanical properties; driers or catalysts to speed up setting or hardening; fungicides; surfactants; and so on.

Undesirable changes in color can have four origins, using paint as a representative color medium. First, there may be chemical interactions within the paint itself, covered in Section 12.2. Second, there may be chemical interactions between the paint and another material such as the substrate below the paint or inappropriate cleaning agents, covered in Section 12.3. Next, there can be chemical interactions between the paint and the atmosphere itself or with pollutants in the atmosphere, covered in Section 12.4. And, last, there can be various effects of light, frequently simultaneously involving the atmosphere as well, covered in Section 12.5.

Color fastness is the ability of a material to withstand fading in several forms such as light fastness and wash fastness; it is covered to some extent in Chapters 10 and 11. Wash fastness includes several components, derived from the solubility of the dye, from deterioration of the dye itself, or from interaction with a variety of chemicals present in the wash, including soap or detergents, bleaches, and alkalies (or sometimes acids), all intensified by the elevated temperature usually used in laundering.

One might think that the colors of inorganic colorants are much more stable than those of organic colorants. Yet, as will be seen in Section 12.5, both types are subject to a number of degradation processes.

12.2. Chemical interactions between colorants

Within paints, interactions between pigments themselves can produce color changes. Thus lead and copper pigments (e.g., lead chromate yellows to reds and azurite blue, respectively) can react with sulfur-containing pigments (such as the cadmium yellows to reds), or with sulfur present in the egg-yolk used in egg-tempera painting, to give black lead and copper sulfides.

Bitumen-based paints, such as Van Dyke brown and asphaltum, can impede the hardening of the paint vehicle itself, resulting in wrinkling and cracking, and thus can lead to more ready access of the atmosphere to the pigments and an acceleration of atmospheric-induced deteriorations.

All such processes may be initiated and are usually accelerated by heat, moisture, and energetic photons, as discussed below in Section 12.5.

12.3. Chemical interaction effects on colorants

Alkalies present in the plaster behind fresco paintings or in inappropriate cleaning agents can convert the pigment white lead into the brown lead oxide PbO_2, blue azurite into the black copper oxide CuO, Prussian blue into the brown iron hydroxide $Fe(OH)_3$, the green verdigris in copper patina to blue, and can also change the colors of "lake" pigments.

Even traces of sulfur-containing fixers not adequately washed out of photographic prints or negatives will convert silver grains to silver sulfide, with fading of the image; contact with sulfur-containing paper will produce the same reaction.

All of these processes may be initiated and are usually accelerated by heat, moisture, and energetic photons, as discussed below in Section 12.5.

12.4. Atmospheric interactions with colorants

The oxygen present in air can produce color changes by oxidizing copper pigments to produce the black copper oxide CuO. Sulfide pigments, such as the cadmium yellows to reds, can oxidize in the presence of moisture to produce the white cadmium oxide CdO plus sulfuric acid H_2SO_4. This released acid is itself a very destructive agent. Sulfuric acid (as well as other acids) can lighten the color of blue ultramarine, even bleaching it completely to white, and can change the colors of "lake" pigments. Acid can also destroy the substrates canvas, paper, and other organic substances on which the integrity of the paint depends. Sulfuric acid can change the pigment white lead to lead sulfate without changing the color. The lead sulfate, however, has a lower refractive index than the white lead and so the hiding power is reduced, converting opacity into translucency, and sometimes revealing overpainted pentimenti (second thoughts). Several other carbonate and hydroxide-containing pigments can react similarly.

Ozone O_3 is a very powerful oxidant, even in minute amounts. In addition to the oxygen-induced changes listed above, it can decolorize essentially all organic dyes and pigments as well as destroy paper, fabrics, rubber, and so on. Hydrogen peroxide H_2O_2 and other peroxides are less active oxidants than ozone, but do cause the same types of damage more slowly.

An atmosphere polluted with hydrogen sulfide H_2S (from the burning of sulfur-containing fuels) or with other volatile sulfides, can cause all of the sulfur-produced color changes described in Section 12.2. Air poluted by industrial emissions can contain a wide variety of organic solvents and other substances with the potential for causing a broad range of damage.

Here again, all these processes may be initiated and usually are accelerated by heat, moisture, and the energetic photons of Section 12.5.

12.5. Photochemical deterioration

In general it can be said that only absorbed light causes changes. It is however not always true that the amount of damage is proportional to the amount of light absorbed, as is usually assumed, since a chemical chain reaction may produce many molecular changes from the absorption of just a single photon. At times, however, the photochemically induced change can prevent further deterioration, as when a light-darkened varnish layer absorbs additional light and thus protects the pigments underneath.

The more energetic the photons, the more likely is it that degradation will result. Of the visible and adjacent parts of the spectrum, ultraviolet is the most deleterious, with violet and blue also frequently active but at a slower rate and lesser level. At the same time, even infrared can be dangerous when intense, as in cinema projectors where it must be filtered out to prevent excessive heating which can damage the colors in film.

As seen in table 12.1, incandescent lighting contains about one third of the amount of ultraviolet of fluorescent lighting, which itself contains one tenth as much ultraviolet as does unfiltered daylight. Window glass reduces the ultraviolet content of daylight by about one third, and special colorless UV-filtering 'Plexiglass' reduces it to about one tenth. For display of the most precious items, such as the US Declaration of Independence in Washington DC, the special yellow 'Plexiglass UF3' removes both ultraviolet as well as the blue end of the visible spectrum with significant, but acceptable interference with viewing. Sealing such items in an atmosphere of a chemically inert gas such as nitrogen, argon, or helium also aids in their preservation.

TABLE 12.1

The relative amounts of damaging ultraviolet in equal quantities of light.

Illumination	Relative damage
Vertical skylight, open	100
Vertical skylight, window glass	34
Vertical skylight, colorless UV-filtering plexiglass	9
Fluorescent lighting	9
Incandescent lighting	3

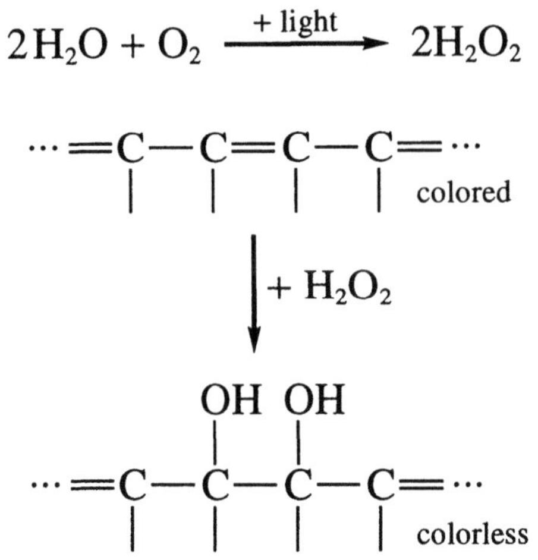

$$2\,H_2O + O_2 \xrightarrow{\;+\;light\;} 2\,H_2O_2$$

$$\cdots = C - C = C - C = \cdots \quad \text{colored}$$

$$\Big\downarrow + H_2O_2$$

$$\cdots = \overset{OH}{\underset{|}{C}} - \overset{OH}{\underset{|}{C}} - C - C = \cdots \quad \text{colorless}$$

Fig. 12.1. Photodegradation by the oxidation and break-up of a conjugated double bond colorant system via hydrogen peroxide H_2O_2.

12.5.1. Photodegradation in organic colorants

Photodegradation of organic colorants results when an energetic photon, usually in the ultraviolet or blue region of the spectrum, is absorbed by a molecule and elevates the molecule or some part of it into an excited state. This activation process may be direct, then called photo-oxidation. Here an excited state is induced by absorption of the energy of the photon into a double bond in the conjugated system of a colorant molecule. This excited double bond can now react with oxygen in the atmosphere, resulting in its conversion to a single bond. This occurs by a process of oxidation, e.g., by the addition of an OH to each end of the double bond. The process breaks up the conjugated system, and the result can be a loss of color. A darkening can also occur, as in the organic varnishes used to protect paintings, pastels, etc.

In an indirect activation, the energetic photon interacts with oxygen and moisture in the atmosphere to form hydrogen peroxide, which then bleaches the dye, again by destroying double bonds by oxidation as in fig. 12.1. Essentially any organic substances, including colorants, art media and substrates such as paper, fabric, parchment, wood, etc. can have their color altered by these processes. In some materials, such as wood and amber, either a lightening or a darkening can result.

Another indirect change is photoreduction, when an energetic photon excites the hydrogen-containing part of a molecule, for example in the canvas substrate of a paint. Excited hydrogen atoms can then add to each end of a double bond of an adjacent organic pigment, thus destroying the conjugated system and the color as in fig. 12.2. The substrate itself also deteriorates from this process, then called phototendering or substrate activation; this is particularly deleterious in silk.

Physical changes affecting the integrity of materials can also accompany these color changes, including the turning brittle or total disintegration of paper and fabrics.

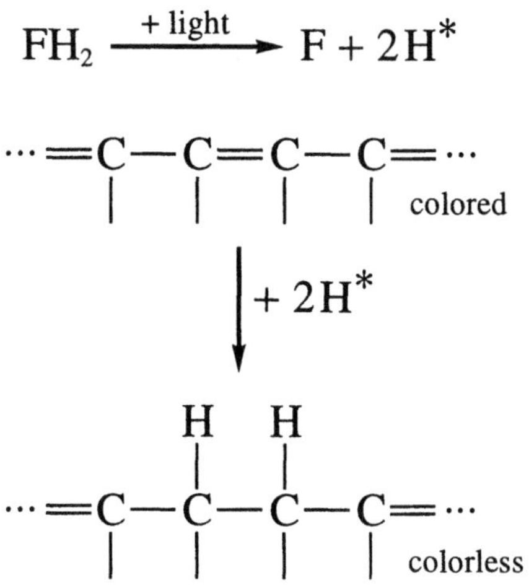

$$FH_2 \xrightarrow{+ \text{ light}} F + 2H^*$$

$$\cdots=C-C=C-C=\cdots \quad \text{colored}$$

$$\Big\downarrow + 2H^*$$

$$\cdots=C-\overset{\overset{\displaystyle H}{|}}{C}-\overset{\overset{\displaystyle H}{|}}{C}-C=\cdots \quad \text{colorless}$$

Fig. 12.2. Photoreduction involving a fiber FH_2 and the reduction and break-up of a conjugated double bond colorant system via active hydrogen H^*. An asterisk indicates excess energy.

12.5.2. Photodegradation in inorganic colorants

As with organic colorants, photo-oxidation can be either direct by oxidation of part of the colorant molecule or indirect via hydrogen peroxide. Darkening can be produced by photo-oxidation in chromium-containing pigments, such as chrome yellow and molybdate orange, with the formation of the chromium oxide chrome green Cr_2O_3 and in lead-containing pigments, such as yellow massicot and red minium, with the formation of the brown lead oxide PbO_2. The white pigments lithopone and zinc white can also turn yellow as well as produce flaking or chalking; in the case of lithopone it is the oxidation of the ZnS to the bulkier $ZnSO_4$ that produces the expansion that leads to the chalking.

A particularly serious direct photochemical change is the conversion of the red pigment vermillion, the mercury sulfide cinnabar HgS, to the same composition but different structure of the black compound metacinnabar (see also Chapter 4, Section 4.6.2 and table 4.2). In a poorly formulated paint this change can begin in as little as five years.

12.6. Conclusions and further reading

In general, storage in a controlled low humidity atmosphere free of pollution, at a constant (preferably low) temperature, and in the absence of light will permit colors to be retained over millenia, as in sealed Egyptian tombs. However, such conditions do not permit the viewing essential to the functioning of a museum! It is inevitable that the light which permits viewing necessarily produces some deterioration. A knowledge of the deterioration processes is required for the adjustment of the museum environment to provide maximum survivability consistent with public viewing.

While some of the processes involved in colorant deterioration are well understood, the majority have not been studied in detail and much work remains to be done. Work in progress is reported in journals such as: *Journal of the American Institute for Conservation; Museum Management and Curatorship* (Butterworth-Heinemann); and *Studies in Conservation* (International Institute for Conservation).

Some of the scientific and technical aspects involved in colorant deterioration are covered in Chapter 4. A useful source is Brill's *Light, Its Interaction with Art and Antiques* (Brill 1980), best read in conjunction with Nassau's *The Physics and Chemistry of Color* (Nassau 1983), an abbreviated treatment of which appears here as Chapter 4. Invaluable are the more than 20 volumes in the Butterworth-Heinemann (Boston, London, etc.) *Series in Conservation and Museology.* Topics covered include the conservation of artists' pigments, textiles, antiquities, including manuscripts, paintings, wall paintings, glass, buildings, and geological materials, including minerals, rocks, and meteorites (e.g., Thomson 1986; Nassau 1992).

References

Brill, T.B., 1980, Light; Its Interactions with Art and Antiques (Plenum Press, New York).

Nassau, K., 1983, The Physics and Chemistry of Color; The Fifteen Causes of Color (Wiley, New York).

Nassau, K., 1992, Conserving light-sensitive minerals and gems, in: The Care and Conservation of Geological Materials: Minerals, Rocks, Meteorites and Lunar Finds, ed. F.M. Howie (Butterworth-Heinemann, Boston and London).

Thomson, G., 1986, The Museum Environment, 2nd edn (Butterworth-Heinemann, Boston and London).

chapter 13

COLOR IMAGING:
PRINTING AND PHOTOGRAPHY

GARY G. FIELD

California Polytechnic State University
San Luis Obispo, CA, USA

Color for Science, Art and Technology
K. Nassau (Editor)
1998 Elsevier Science B.V.

CONTENTS

13.1.   Introduction   356

13.2.   Subtractive color reproduction fundamentals   356

13.3.   Imaging processes   358

      13.3.1.   Photographic film imaging systems   358

      13.3.2.   Photographic CCD imaging systems   358

      13.3.3.   Electronic scanning for printing   359

      13.3.4.   Lithographic printing   359

      13.3.5.   Intaglio and relief printing   359

      13.3.6.   Screen or stencil printing   360

      13.3.7.   Direct imaging printing   360

      13.3.8.   Halftone and other imaging processes   361

      13.3.9.   Mathematical and computational principles   362

13.4.   Color separation   364

      13.4.1.   Color separation for photography   364

      13.4.2.   Color separation for printing   366

13.5.   Evaluation conditions   367

13.6.   Color and image constraints   368

      13.6.1.   Color gamut   368

      13.6.2.   Image structure   372

13.7.   Color reproduction quality objectives   374

      13.7.1.   Tone reproduction or lightness adjustment   374

      13.7.2.   Color balance or hue shifts   376

      13.7.3.   Color correction or saturation adjustment   377

      13.7.4.   Color reproduction models   378

      13.7.5.   Image definition requirements   379

      13.7.6.   Interference patterns   380

      13.7.7.   Surface characteristics   380

13.8.   Research directions   381

      13.8.1.   Process developments   382

      13.8.2.   Quality developments   382

13.9.   Bibliography   383

13.10. Appendixes 384

    13.10.1. The Demichel equations 384

    13.10.2. The Neugebauer equations 385

    13.10.3. The masking equations 386

    13.10.4. The Clapper transformation 386

References 387

13.1. Introduction

This chapter covers color imaging with particular focus upon photographic imaging and photomechanical reproduction processes. Television and video display imaging are not covered in this chapter as these topics are addressed in Chapter 15. Motion picture photography is not covered to any significant degree, but many of the still photography concepts apply with equal validity to motion picture photography. The primary emphasis in the photomechanical reproduction part of this chapter is on process color printing; i.e., pictorial printing in natural color, rather than flat or solid color printing for such non-pictorial products as type-only labels or posters.

13.2. Subtractive color reproduction fundamentals

All color printing and virtually all color photography is based on the subtractive color reproduction principle (see Section 1.6.2 of Chapter 1). The primary colorants are yellow, magenta and cyan pigments or dyes. Each colorant should transmit two thirds of the visible spectrum and absorb one third. A white substrate is used as a base upon which the colorants are superimposed. In the case of color transparencies (color slides, sheet film, or motion picture film) the base that supports the colorant layers is clear film.

Color printing processes typically use a black image in addition to the subtractive primaries. Figure 13.1 illustrates the difference between one version of "ideal" and typical actual properties of a substrate and three subtractive primaries. Sometimes, for special color quality printing requirements, such additional colors as red, green, blue, orange, pink, or pale blue may also be used to supplement the basic colors.

Theoretically, the subtractive reproduction process works by selectively subtracting specific portions of the incident white light. Magenta, for example, subtracts the green wavelengths (about 500–600 nm) from white, thus allowing the red (about 600–700 nm) and blue (about 400–500 nm) portions to be transmitted as in fig. 13.1. The subsequent color is a bluish red that is called magenta. Yellow absorbs blue and transmits red and green, and cyan absorbs red and transmits blue and green. If each colorant absorbs one third of the visible spectrum then all light is absorbed and the image area in question appears black when they are all combined. Similarly, when any two primary colors are combined, a secondary color is formed. If, for example, yellow (which absorbs blue) and cyan (which absorbs red) are combined, the resulting color is green because neither yellow nor cyan absorbs green.

The white substrate initially reflects all color; therefore, as more colorant is applied, less color is perceived, i.e., we approach black. For surface colors (images viewed by

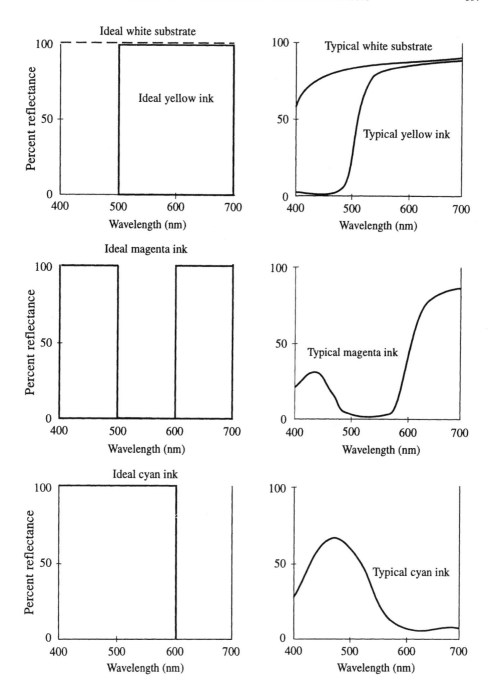

Fig. 13.1. One version of "ideal" substrate, yellow, magenta, and cyan ink reflectance curves, left, together with typical actual substrate and printing ink absorption curves (right).

reflected light), the color perception will be influenced by the print's gloss and texture. The viewing conditions will also exert a powerful influence on the perception of color images in either transparent or reflective form.

13.3. Imaging processes

13.3.1. Photographic film imaging systems

Photographic imaging systems (Langford 1982) employ an optical system, a photosensitive receptor, and the appropriate mechanical, chemical and electronic support devices and systems. In some cases, a finished photographic image is produced directly in the camera, whereas in other cases, subsequent processing is required.

The basic requirements of a camera include: a light-tight compartment, a flat film plane that is perpendicular to the optical axis, and a means of controlling the amount of light that reaches the film during exposure. The camera lens should form a high quality image over the area of the film plane, and should be adjustable to accommodate focus and lighting variables.

The most common photosensitive material used in cameras consists of a clear film base upon which are coated three emulsion layers: one layer is sensitive to blue light and forms the yellow image, another layer is sensitive to green light and forms the magenta image, and the third layer is sensitive to red light and forms the cyan image. The nature of the image on the processed film depends upon the emulsion type chosen by the photographer. In some cases, the film is processed to form a transparent positive color image. In other cases, the image is in negative form and must be used to subsequently expose the ultimate photosensitive layers which are coated upon a white reflective base. The resulting photographic print may be produced at a larger or smaller size than the original negative through the use of appropriate optical–mechanical devices.

Conventional photographic materials are processed in a series of chemical solutions in order to render the latent image visible. The so-called "instant" photography systems commonly produce reflective color prints in the camera via the activation of integral chemical pods after exposure has been made. These systems usually produce a positive.

13.3.2. Photographic CCD imaging systems

Another photosensitive system that is now being employed in camera systems is the charge coupled device (CCD) which replaces film. The CCD converts the incident light into a series of electrical signals that represent the red, green, and blue content of the image. The signals are processed in a computer before being output via one or more visual display technologies. Color monitor image displays ("soft" images) are used for routine inspection of electronic images prior to generating a physical ("hard" image) output. A variety of technologies may be used to produce hard copy output, including: thermal dye transfer, electrostatic imaging, ink jet printing, and conventional silver-based photographic materials. Hard copy output systems are based on the subtractive color primaries yellow, magenta, cyan, sometimes supplemented by black. Soft copy output is based on the additive color primaries blue, green and red.

13.3.3. Electronic scanning for printing

The imaging processes used in the printing industry for the photomechanical reproduction process have been based upon electronic scanning devices as far back as the early 1950s. Such systems began to handle the majority of color imaging work during the 1970s. Today, virtually all color imaging tasks in the printing industry are handled by electronic scanning and related systems.

The color scanners used by the printing industry (Field 1990a) have two basic configurations: (1) the rotary drum system employs photomultiplier tubes (PMT) and microscope optics to scan the original photograph or drawing which is mounted on the scanning drum; (2) the flat bed system employs one or more linear CCDs in conjunction with an optical system to traverse the original, which is mounted on the flat bed.

The scanning process captures the information content of the original, separates the image, via filters, into red, green and blue records which are, in turn, processed through the scanner's computer system in order to generate film (usually) images representing the yellow, magenta, cyan and black (YMCK) content of the reproduction as illustrated in color plate 38. The output images are scaled, corrected, enhanced, and assembled to suit the layout and quality requirements of the printed product. The images will also incorporate a periodic or random image structure that is suited to the printing process being used.

13.3.4. Lithographic printing

The most common printing process is the lithographic process. Lithography uses a flat-surface plate with an image that will selectively attract or repel ink. The plate is wrapped around the appropriate cylinder of a lithographic printing machine. Typically, the inked image is transferred to a thin sheet of rubber, the "blanket", on another cylinder before being transferred to the substrate. The process is often called "offset" lithography because of the two-stage transfer process. In practice, however, the offset principle may be used with other processes, but such usage is rare.

The substrates available for use in the offset lithographic process include metal, plastic, and (more commonly) paper and paperboard. Substrates are fed into the printing machine in either sheet or roll ("web" printing) configurations. Printing machines are usually designed so that all colors are sequentially applied during one pass through the press before being dried or "set". Perfecting presses print both sides of the sheet or web during one pass through the press. Such presses that are used for color work are usually web presses because most web machines are equipped with drying systems whereas most sheet fed presses are not. (See Field (1988, pp. 99–131) and Bruno (1986) for more details of the conventional printing processes.)

13.3.5. Intaglio and relief printing

The gravure process works on the intaglio principle: the image is recessed in a metal cylinder. The cylinder rotates in a trough of ink and is scraped clean by a "doctor" blade before the cylinder is pressed into contact with the substrate, which is commonly in web form. Image transfer is effected through capillary action or through electrostatic assist technologies.

The gravure process uses a fluid liquid ink system and relies on a cellular image structure to contain the ink in the image areas of a fast-rotating cylinder. The wet image on the substrate is dried prior to the application of subsequent colors in contrast to the web offset lithography process, where all images are printed wet-on-wet before being dried or set. Better ink transfer is achieved if inks are dried before the application of following images.

Most gravure cylinders are produced by direct electro-mechanical engraving techniques; i.e., no intermediate films are used between the electronic scanning and engraving machines. In recent years, a number of direct exposure plate systems are being used by lithographic printers. In most cases, a laser source is used to expose a plate on the line-by-line basis that is used for exposing film. In order to speed the plate exposure process, a series of laser exposure heads are used.

The relief printing processes employ a raised image area which is preferentially inked prior to pressing the image carrier on to the substrate and transferring the inked image. There are two types of relief process: letterpress and flexography. Letterpress uses a non-resilient plate, made of metal or hard plastic, and (like lithography) a paste-type ink. Printing may be effected via flat or rotary type machines. The process has been in decline for a number of years. The flexographic process uses a soft rubber or plastic plate, a liquid ink system, and web fed substrates.

The image resolution of relief and intaglio systems is lower than that of the lithographic process because of the physical limitations of the etching processes relative to the image carrier requirements. The use of one process over another is, however, usually determined by economic factors. The relatively expensive gravure process is generally confined to long-run work. Better than 95 percent of all printing is handled by lithography, intaglio and relief processes. In fact, such processes handle virtually all of the long-run work. The prominent exception is the use of the screen printing process for printing glass bottle or jars.

13.3.6. Screen or stencil printing

The screen or stencil printing process employs a stencil attached to a thin nylon or steel mesh. A squeegee is used to force ink through the mesh on to the substrate in those areas not protected by the stencil. This screen printing process uses a higher viscosity ink than either the gravure or flexographic processes and can print on virtually any substrate. The screen printing process lays down a thicker ink film than any other printing process and, therefore, produces high densities and good color saturation, but necessitates the use of extensive drying systems.

The comparatively low resolution and slow speed of the screen printing process usually confines its use to such short run products as showcards, posters, ceramics printing and T-shirts. The flexibility of the process and its short-run low-cost structure make it a significant, albeit minor, imaging process.

13.3.7. Direct imaging printing

The newer printing processes do not use a conventional image carrier (plate, cylinder, or stencil); rather, a fresh image is generated for each "impression". A computer drives an ink jet printer, a laser imaging driven electrophotographic system, or a similar direct imaging technology. Such systems can now produce high quality images at low relative cost for short run requirements. Economic factors favor other processes for larger images and long run work.

13.3.8. Halftone and other imaging processes

Printed images are generally produced via the technique known as the halftone process. This process resolves the image into a series of elements that, when expressed as a percentage, cover from zero to 100 percent of the image element. In theory, there are an infinite number of steps between the extreme values, but in practice, digital halftone images rarely have more than 145 steps. A halftone process is required in order to create the illusion of tonal variation. For most processes, it is not possible to vary the density of image elements or dots; therefore, the tonal perception is created by varying the area of the dot. In order for the tonal illusion to work, the halftone dots must be fine enough that they are not resolved by the eye at normal viewing distance.

The color computer section of the scanner computes the required halftone dot sizes while the screening computer determines the dot shape, screen frequency and screen angle. Screen frequency is usually dependent upon the printing process and the substrate. Newspapers may use 85 halftone lines per inch in contrast to high quality art or photography books that may use 300 lines per inch.

The colors will be placed on angles which are chosen to minimize moiré interference patterns and to provide a misregister allowance. These limitations restrict the number of high density colors to four, and even then there may be moiré problems. Recent developments in stochastic or random screening have eliminated these problems because the image no longer forms a regular dot but, rather, consists of a number of randomly distributed elements that are equivalent in area to the dot they replace. Halftone screen or subject moiré does not occur with random screens.

Apart from stochastic dot structures, the screening computer can vary the shape of regular dots to take round, square, elliptical, or other shapes. Elliptical dots, for example, will help to smooth grainy tonal values at the expense of some loss of image resolution. Square dot screens, by contrast, will more faithfully reproduce fine detail and graininess.

Gravure halftone dot cells are usually formed by an engraving stylus that cuts deeper and larger cells for shadow tones, and shallower and smaller cells for highlight or lighter tones. This process is one where both area and density are variable. The thinner highlight ink films help to produce cleaner colors in those areas.

In contrast to the photomechanical processes, photographic tonal variations are due to the concentration of microscopic silver particles (or their dyed equivalents). The dye or silver concentration is related to the brightness values of the original scene. The particles are so small that the images are said to be "continuous tone".

The hard copy thermal dye transfer images that are produced from digital camera input or as photomechanical proofs may have 128 tonal levels per color, and 300 lines per inch of image elements. Such images are usually accepted as equivalent to continuous tone prints by most observers. Ink jet, electrostatic, or other hard copy imaging systems generally do not have the same image resolution or tonal variation as the thermal dye transfer technology.

13.3.9. Mathematical and computational principles

The transformation from original colors to reproduced colors may be handled by several distinct mathematical or computational techniques. The "original" may mean the actual scene or object, or it could mean an intermediate reproduction, such as a photograph. The "reproduction" is the pictorial representation of the original in its final form, relative to purpose. A photograph, for example, would be classed as a reproduction if it is mounted in an album or is framed for display. The same photograph would be classed as an original if it is used in the photomechanical reproduction process to produce printed versions in books or magazines. In either case, however, the transformations from one colorant system to another may be described in mathematical terms. These approaches have been reviewed by Field (1988, pp. 223–269, 321–329), and are summarized here.

The Neugebauer equations, building on the work of Demichel, consider a given area of a photomechanical reproduction to consist of eight distinct fractional areas: unprinted white paper, the three areas covered by each of the yellow, magenta and cyan primaries, the three areas covered by each of the blue, green and red secondary colors (i.e., the areas where two primaries overlap), and the area covered by black (i.e., the overlap of all three primary colors). The Demichel equations are given in Appendix 13.10.1. The eight fractional areas are fused or blended by the eye in order to produce the sensation of a single color as illustrated in color plate 33.

The Neugebauer equations express the R, G and B tristimulus response in terms of fractional area reflectances of the eight distinct color areas of the reproduction, and are given in Appendix 13.10.2. In practice, these equations are modified to compensate for the effects of light scattering related to the substrate and the colorant layers. Such modifications include the Yule–Nielsen n factor for light loss in halftone images, and Warner's over color modification (OCM) factor to compensate for additivity and proportionality failure. These correction factors are usually developed by empirical methods for a given combination of ink and substrate.

A further deviation from the original Neugebauer equations concerns the use of black as a fourth color. Under such conditions, the symmetry of the original versions is upset, and more than one solution is possible.

The practical application of the Neugebauer equations has been confined to the generation of look up tables (LUT) consisting of pre-solved values for yellow, magenta, cyan and black that are linked to given input values of red, green and blue (commonly tristimulus values but, theoretically, any R, G, B values). The real-time computation of high resolution image color requirements is too slow for practical applications.

A second approach to the quantification of the relationships between original and reproduced colors is the use of second or higher order versions of the masking equations. A linear first order matrix conversion of RGB to CMY values, as given in Appendix 13.10.3, will produce inaccuracies due to such non-linear effects as proportionality failure and additivity failure. Inaccuracies may be reduced through the use of second order transformation (Clapper 1961) of the masking equation that is given in Appendix 13.10.4. Higher order equations may be used to reduce the transformation errors even further. A recent comparative study (Heuberger et al. 1992) has shown that fourth order equations provide accuracy approaching that obtained by a modified Neugebauer equation applied on a "cellular" basis.

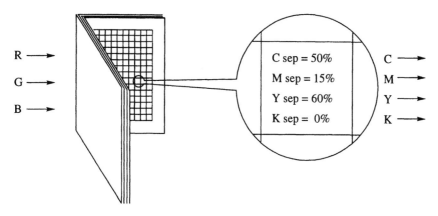

Fig. 13.2. The principle of a look up table: Input red, green, blue (R, G, B) signals correspond to the page, row, and column values. Output cyan, magenta, yellow, black $(C, M, Y, K$, respectively) values are given by the table entry.

A final method of quantifying the relationships between original and reproduced colors involves the use of a printed color chart. The chart is produced on the required ink-substrate system and displays a large number of colors made from known combinations of halftone dot values. The printed colors are measured in terms of their RGB values and stored together with the corresponding YMC values. The previously cited study by Hueberger et al. (1992) suggests that the chart should have 5800 colors if linear interpolation will be used, and 132 600 colors if interpolation is not used. These specifications will produce halftone values within two percent of the ideal values.

The calibration or input values for either the Neugebauer or the masking equations are also generated from printed samples that represent the printing conditions. The differences between the approaches revolve around the size of the initial test target and the subsequent complexity of the measurement and computation processes.

The Neugebauer equations use a relatively simple calibration target, but involve complex and lengthy calculations. Fourth order matrices involve the use of moderately complex calibration targets and fairly quick computations. The empirical approach requires an elaborate calibration target and a large number of measurements. An extensive empirical LUT will not involve any calculations, as illustrated in fig. 13.2, but an abridged table will require the real-time computations of interpolated values.

The choice of the method used for generating a LUT is based on market considerations (quality requirements and variety of color output options) and on equipment manufacturing concerns (computing speed, memory and cost constraints). Theoretically, the resulting image quality should be virtually the same regardless of the method used for generating the LUT.

Today, it is common practice to separate the input scanning (image capture) and output recording aspects of image reproduction. The scanned image is stored in digital form pending optional image processing steps prior to the recording stage. Older scanning systems, some of which are still in use, recorded the image almost simultaneously as it was scanned or captured. The non linearities of the reproduction system require the use of supplementary empirical controls in order to achieve a satisfactory solution of the masking equations, upon which such scanners are based.

The black printer computation is usually based on proprietary systems that incorporate particular strategies regarding the replacement of appropriate 3-color combinations with black. These techniques are known as Gray Component Replacement (GCR) or Under Color Removal (UCR), the latter method being confined to near neutrals only. In practice, the black printer is used to extend the density range (or lightness values) in darker tonal values; consequently, there are many approaches that will produce visual results that are virtually identical.

Non-halftone photographic image processing systems tend to be simpler than the halftone systems because of greater substrate and colorant standardization and because of fewer proportionality and additivity failure problems. Such systems are generally based on matrix transformation approaches.

13.4. Color separation

13.4.1. Color separation for photography

The separation of original colors into their equivalent blue, green and red absorption values prior to their reproduction via a yellow, magenta and cyan colorant system may be accomplished in a number of ways. Color separation in the photographic process occurs at the instant the photograph is taken, in contrast to the photomechanical process, where distinct images are produced sequentially prior to their ultimate superimposition.

The early subtractive color photography processes used "one shot" cameras to produce three separation negatives simultaneously. A system of mirrors, beam splitters, and filters were employed in these cameras. The Technicolor motion picture process was the most common commercial application of this process. The subsequent images were produced from dyed positive conversions of the separation negatives. For many years, however, the common methods of color photography utilize the integral tripack system whereby all three photosensitive emulsions are coated on to a common base.

The emulsion sensitivities are adjusted in manufacturing or are restricted through the use of a yellow filter layer within the emulsion pack, in order to produce independent records of the blue, green, and red absorptions of the original subject. When the latent images are processed, a dye coupling (or similar) process is used to produce the yellow, magenta and cyan dye layers. The mechanism involves the reaction with a coupler of byproducts from the developing process in the immediate vicinity of the exposed silver grains. The developed and undeveloped silver grains are dissolved after the dyes are formed.

The sensitivity of the emulsion layers varies from manufacturer to manufacturer and from product to product within a manufacturer's range. The sensitivities are chosen to satisfy the color, speed, granularity, processing technology and other design parameters. Similarly, the absorption characteristics of the dye layers will also differ from product to product. Typical color film sensitivities and dye layer absorptions are illustrated in fig. 13.3.

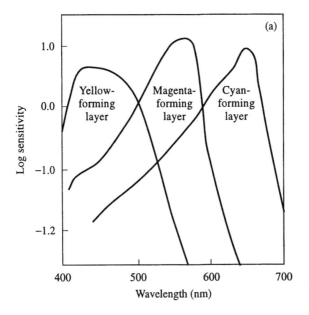

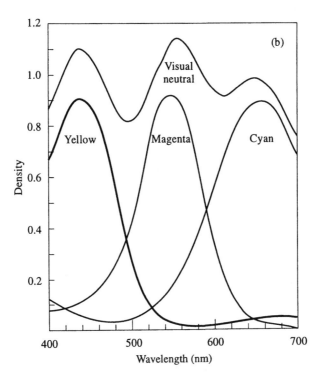

Fig. 13.3. Typical color film sensitivity curves (a) and dye absorption characteristics (b).

One color photograph of a given scene may differ from another color photograph of a different type exposed at the same time and perspective. Apart from differences in proprietary technology, consumers often prefer the appearance of one photograph over another for a given subject. Such preferences may be reversed for different subjects. In other words, there is evidence to suggest that individuals have color preferences that represent distortions of reality. Caucasian skin tones, for example, are usually reproduced slightly tanner than the reality. A number of complex psychophysical factors influence these preferences which are ultimately manifest in a broad range of photographic products with differing color responses and colorant systems.

13.4.2. Color separation for printing

Until the late 1960s, most photomechanical color separations were produced on cameras by making sequential exposures through filters from the original. The subsequent separation negatives were the first stage in a number of steps that resulted in a set of image carriers (plates or cylinders, usually) that were suitable ink receptors. The camera-based color separation processes have now been replaced by electronic color scanning systems. Such systems were first patented in the late 1930s but were not used commercially until the 1950s. The advent of inexpensive digital computers accelerated the use of scanning systems: by the mid-1980s, probably more than 95 percent of color separations were made on scanners.

The photomultiplier tube (PMT) based scanners separate the colors simultaneously. The original is wrapped around a drum that rotates at high speed while a scanning head traverses the drum. The scanning head consists of a microscope lens, a series of mirrors, beam splitters, filters, lenses and the PMTs. The photomultipliers convert light into proportional electrical energy that, in turn, is used as the input signal to the scanner's color computer or intermediate storage media. Some PMT-based scanners can handle a high dynamic range (optical density 4.00+) at resolutions greater than 18 000 lines per inch.

The charge coupled device (CCD) based scanners separate the colors sequentially. The original is mounted on a flat table and is traversed by an optical system that utilizes a system of lenses, mirrors and filters. Some scanners use three CCDs to traverse the original one after the other, while other scanners use one CCD and require three separate passes, with a different filter for each pass. The CCD scanners are not as expensive as the PMT scanners; however, the CCD scanners cannot match the dynamic range or resolution performance of the best PMT scanners. CCD system images are also subject to slight quality degradation because of the effects of optical flare.

The sensitivities of PMT and CCD scanner systems are based on proprietary combinations of optical devices and electronic photoreceptors. In practice, it is possible to configure the scanning system to produce a colorimetric response, but there may be a number of computational reasons why different responses may be preferred in practice. A key complication, however, can occur when the scanner interprets a given color differently from the response of the human eye to that color. Such differences are caused by different and non-standard light sources as well as by different color responses.

The color correction and transformation processes within individual scanning systems will incorporate the appropriate controls required for making the preferred color adjustments. In practice, the color response differences tend to be corrected within the overall color adjustment process rather than as a discrete activity. In certain cases, however, particular original colors will cause problems of a metameric nature (see Section 1.7.6 of Chapter 1 and Section 2.2.7 of Chapter 2) that will necessitate the use of user-defined local color adjustment strategies.

The subsequent image processing in the scanner and imaging system addresses the scaling, halftoning, and assembly requirements. The output may be recorded on film or directly on a plate, cylinder or proof. Cylinder recording is accomplished via an electromechanical engraving machine that cuts cells into the cylinder's surface.

13.5. Evaluation conditions

The viewing conditions for pictorial images exert a greater influence on color perception than in those cases where flat or non-pictorial colors are being evaluated. The light source's color temperature, intensity, spectral distribution; the color, size, and intensity of the surround; and the presence of nearby non-neutral reflecting surfaces can all influence the perception of the color image.

Fortunately, color viewing conditions for photographic and photomechanical processes are the subject of standards. The US standard is ANSI PH 2.30-1989, which is the general foundation for an ISO standard currently under revision. The ANSI standard specifies a 5000 K color temperature, a color rendering index of at least 90, surface luminance for the transparency illuminator of 1400 cd/m^2, and an illuminance on the viewing plane of 2200 lux for reflection materials. The geometry of illuminations and viewing, the environmental conditions, diffusion characteristics, and the borders around a transparency are all subjects of the standard. Lower intensity conditions are specified in part 2 of the standard for routine inspection, display, or judging.

Standard viewing conditions help reduce color communication problems during the color reproduction process. A number of barriers to effective communication still exist, including: physiological differences in the visual processes of observers, psychological differences or preferences between observers, and the buyer–seller dynamics of the color image purchasing process. The lack of effective communication tools exacerbates color communication difficulties. Verbal instructions tend to be emotion-laden and very imprecise, while quantitative instructions tend to be too abstract and too complex for making rapid evaluations of images in a production setting.

In practice, a physical color sample represents the best practical way of specifying image color adjustment objectives. A printed color chart, for example, that contains known color combinations produced under controlled conditions, may be used to select desired colors that are identified by specific production coordinates. Reference color charts (as opposed to the stable Munsell and similar color order systems) must be replaced periodically because of substrate discoloration or pigment fading.

Color charts and other physical references must be used with care in image color specification tasks because of the influence of surround colors (see Section 3.4.5 of Chapter 3 and color plate 8). The colorimetric values of a reference color and an image color may be identical, but may appear to be different because of the color areas adjacent to the image area in question. The perception of the image color will be influenced by the visual interaction of the size, shape and color of the surround color(s) with the size, shape and color of the image color in question. Viewing distance will also exert an influence.

13.6. Color and image constraints

The quality of photographic images is constrained by color gamut and image structure limitations. The two-dimensional representation of three-dimensional scenes imposes further restrictions on the perceived naturalness of the reproduction that have not been solved in a satisfactory manner. It is certainly true that, in some cases, stereoscopic color transparency images produce results of startling and impressive realism; however, the special viewing condition severely restricts the use of this process. The use of lenticular screens on special photographic prints produces a 3-dimensional effect at the expense of image disturbances that are due to the screen; therefore, its use is generally confined to the 3D photography enthusiast's market. The following discussion is limited to two dimensional or monoscopic representations of images. Yule (1967) has published an extensive review of the color and image constraints that follow.

13.6.1. Color gamut

The characteristics of the dyes or pigments are the key factors in determining the color gamut of an imaging system. The ideal absorption characteristics of the yellow, magenta, and cyan colorants were stated earlier in connection with fig. 13.1. In practice, yellow approaches the ideal: the absorption of red and green light is low. Magenta colorants, however, absorb substantially more blue light than the ideal, while cyan colorants absorb more blue and green light than the ideal. The high red light absorption of cyan, the high green light absorption of magenta, the high blue light absorption of yellow, and the low red light absorption of magenta are all generally satisfactory.

The color gamut restrictions due to these colorant characteristics include low saturation ("dirty") blues and greens, and low saturation magentas and cyans. Indeed, the red-to-yellow range is the only part of the color space where the reproduction may match high saturation colors. In practice, however, excellent reproductions are possible in the vast majority of cases because most reproductions (or scenes) do not contain highly saturated colors. In those rare cases where such strongly saturated colors are found, as in flowers on seed packets, for instance, it may be necessary to use supplementary colors to help enhance the gamut in a given area.

The lightness (or density) values of printed reproductions are usually limited by the physical requirements of printed ink films. The thick films associated with screen printing require extensive drying, whereas, the thickness of ink films printed by offset lithography are restricted by the image transfer requirements of wet films. In practice, therefore, the use of a black printer is required in order to raise the maximum density of printed reproductions to a level that produces acceptable image contrast and sharpness.

Photographic mechanisms do not rely on the superimposition of wet images; consequently, the reproduced 3-color lightness values are usually quite satisfactory. Color transparencies can, for example, approach optical density 4.0 in deep shadow tones. Photographic reflection prints and photomechanical reproductions rarely exceed maximum optical density of about 2.2.

The substrate's whiteness, brightness, smoothness and absorptivity can all influence the color gamut. In this sense, it is more accurate to refer to the color gamut of a colorant–substrate system rather than to the gamut of a colorant system.

TABLE 13.1
Additivity failure (marked *) of densities in a typical printing example.

Printed colors	Blue filter	Green filter	Red filter
Yellow	0.94	0.08	0.04
Magenta	0.59	1.07	0.13
Cyan	0.10	0.40	1.17
Yellow + Magenta	1.53	1.15	0.17
Red overlap	1.32*	1.07*	0.14*
Yellow + Cyan	1.04	0.48	1.21
Green overlap	0.93*	0.47*	1.20*
Magenta + Cyan	0.69	1.47	1.30
Blue overlap	0.61*	1.33*	1.16*

The whiteness of a substrate refers to its neutrality. Ideally, substrates should be perfectly neutral, but in practice most paper substrates have a slight yellowish hue. The human eye will, however, readily realign color space with reference to a slightly non-neutral substrate in the absence of a true white reference. In those cases when comparisons are being made between images on substrates with different whiteness values, the adaptation adjustment will not happen and the visual process will perceive the images as being different.

The brightness of a substrate, from an image reproduction point of view, refers to how much light is being reflected from the substrate. Ideally, a substrate should reflect 100 percent of the incident light, but in practice, the total reflectance value rarely exceeds 90 percent (see fig. 13.1). Low brightness levels will lower image contrast and darken light tonal values.

Smoothness and absorptivity influences on gamut apply more to photomechanical systems than to photographic systems. Surface reflections, interimage reflections, and intrasubstrate reflections associated with the superimposition of wet inks on an absorbent substrate combine to influence the color gamut of the systems: cyan ink films become less saturated or grayer and magenta ink films become redder as absorption increases and gloss decreases. The lowest levels of printed ink film color distortion are associated with substrates having 100 percent gloss and zero absorptivity. Substrates are chosen to reflect cost and product requirements, so it is inevitable that some reproduction systems will use substrates that reduce the gamut (e.g., newsprint).

Additivity failure occurs when the sum of the densities of individual primaries does not equal the sum of the appropriate secondary color densities. The blue, green and red filter densities of a solid green, for example, should equal the sum of the blue, green and red densities of the individual yellow and cyan colorants that were superimposed to produce the green. Typically, the sum of the individual primary color densities exceeds the densities of the appropriate secondaries. This effect reduces the saturation of the secondary colors to levels below what is theoretically possible from the colorant system as illustrated in table 13.1.

Additivity failure is caused in photographic and photomechanical reproduction systems by interimage light scattering and loss, first surface reflections, and because the commonly used pigments and dyes are not perfectly transparent. Ink trap problems, where a wet ink film will not transfer to a previously printed wet ink film as well as it will transfer to unprinted substrate, is another reason for additivity failure in the photomechanical printing processes. The spectral absorptions of the filters being used when making the measurements will influence the additivity computations.

Proportionality failure is primarily a problem in photomechanical reproduction systems. This refers to the failure of halftone values to retain the proportional saturation values of the solid or 100 percent value of the ink film. Typically, light halftone tint values reproduce at lower saturation or "dirtier" than corresponding thin continuous films of the same ink printed at the same primary density as a given halftone tint, as indicated in fig. 13.4.

A primary cause of proportionality failure is the influence of the halftone screening process. The blending or fusion by the eye of a color halftone dot and a white unprinted area distorts the subsequent color. The influence is less pronounced for yellow pigments because of the low absorption in the red and green segments of that color. In other words, proportionality failure diminishes as the measured "wanted" colors, red and green for the yellow pigment, approach the corresponding densities of the substrate.

Proportionality failure is lower for finer screen halftones and for halftones printed on uncoated paper because of the influence of light penetrating into the paper. Light passes through the ink film, is diffused within the substrate, and emerges, in part, within the white areas that surround the halftone dots. The relative contribution of this optical "fringe" to the perceived tone is greater for finer screens because the fringe effect occupies proportionally more of the unprinted white space between halftone dots than with coarser screen rulings.

First surface light scattering and multiple internal reflection effects will also influence proportionality failure. The gravure printing process exhibits very little proportionality failure because the fluid ink tends to spread over the substrate in a thin film and, therefore, exhibits less of a defined halftone structure than do other processes.

The surface characteristics of a substrate can exert a significant influence on the appearance of the reproduction. Textured substrates, for example, scatter reflected light from the surface which will, when combined with the light reflected from the image, produce a disruption in the perceived smoothness of large even color areas. The tactile qualities of the substrate may, however, outweigh this effect in the mind of the consumer. Wedding photographs, for example, are often printed on a "silk finish" paper because the texture is perceived to offer a higher quality result for that kind of image. Such preferences preclude the specification of a single ideal texture.

The gloss of a substrate influences the lightness and saturation values of a given reproduction. Low gloss, rough substrates exhibit considerable first surface diffuse light scattering effects. The diffuse scattered white light combines with the underlying image reflection to, in effect, "dilute" the saturation and lightness values. The image sharpness and resolution will also be adversely affected. Glossy, smooth substrates will maximize the saturation, density (lightness), sharpness and resolution of photographic images. These outcomes, however, may not be desired in all situations. The photomechanical reproduction of a water color or pastel drawing will retain the integrity and feel of the original

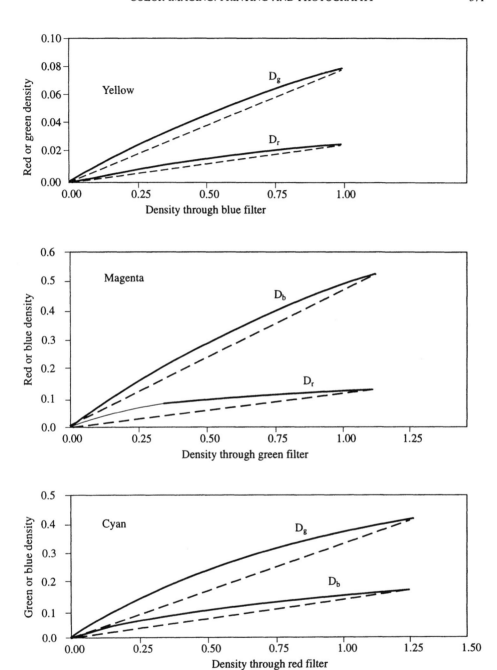

Fig. 13.4. Graphical representation of proportionality failure; perfect proportionality is represented by the dashed lines.

only if it is printed on an uncoated, low gloss substrate. Ideal gloss, therefore, also depends on the nature of the original, the intended use of the reproduction, and intangible artistic or creative factors.

13.6.2. Image structure

The resolution of a photographic imaging system is dependent upon the optical system and the resolving power of the light-sensitive emulsion. The latter is usually dependent upon the fineness of the silver halide grains. The required resolution usually depends upon the ultimate purpose of the image. If great enlargements are required, then high resolution, large format film is used. If modest enlargement "snapshot" photographs define the objectives, then relatively small format, coarser grain (and higher speed) films may be preferred.

The resolution of a photomechanical system is influenced, to a large degree, by the smoothness of the substrate and the resolving power of the particular printing process (the lithographic process generally has the highest resolution). The halftone screen frequency, the "screen ruling", is selected to suit the process-substrate combination and may range from 85 lines per inch for newspaper printing up to 300 lines per inch for high quality photography or art books. A halftone "line" could be made up of anywhere from 12 to 24 recording lines; therefore a 300 line per inch halftone screen may have a recording resolution of 7200 lines per inch. As shown in fig. 13.5, the scanning pitch is one half that of the screening pitch, i.e. one half of a row of dots is recorded every revolution of the recording drum. Halftone prints have been made with screen rulings as fine as 600 lines per inch, but about 300 halftone lines per inch represents the maximum ability of the human eye to resolve fine detail under optimal lighting conditions. The stochastic or random dot screens are capable of delivering higher resolution per unit frequency than conventional halftone images.

Image sharpness of photographs is influenced by the lens, camera focus and the silver halide emulsion. Sharpness is a visual sensation related to the abruptness of change in tonal value at the edge of an object. This edge quality effect may be quantified by a measure called acutance. In general, sharp images are preferred over unsharp images, but in some cases (portrait photography) a deliberate soft-focus effect may be desired. Sharpness within a photograph is related to the distance of objects from the camera and the aperture of the lens. Smaller lens openings produce greater depth of field, but the accompanying higher diffraction causes some loss of resolution. In general, the optimal aperture for a lens is two or three f stops down from maximum aperture.

The perceived sharpness of a photomechanical reproduction may be increased by the use of the edge enhancement controls of the scanner. Such adjustments may help to partially compensate for the sharpness-reducing effects of newsprint and other absorbent substrates. If the edge enhancement controls are set at too high levels, however, image graininess may increase and the reproduction may appear unnatural.

The ideal level of image sharpness is not known with any certainty. Photographs of people and natural scenes probably do not benefit from high levels of artificial sharpening. Photographs of such products as fabrics, machinery, jewelry and other manufactured items may benefit from high sharpness values. The scanner sharpness settings are also adjusted

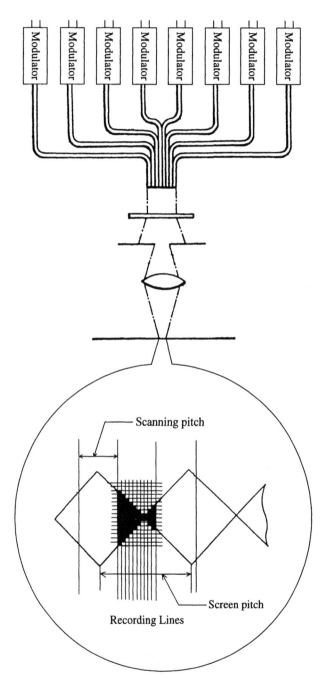

Fig. 13.5. Halftone dot formation in digital imaging systems.

according to the substrate: higher levels are used for highly absorbent uncoated substrates, and lower levels are used for low absorptivity coated substrates.

Graininess or mottle is a random disturbance to the perceived evenness or smoothness of what should be an even tonal value or area. Granularity is the quantitative measure of graininess. Photographic grain effects are dependent on the characteristics of the emulsion, the processing techniques and the degree of enlargement. There are some tradeoffs to be made between graininess and sharpness (Bartleson 1982).

Photomechanically induced graininess effects are almost always due to excessive scanner sharpness settings. Mottle is a printing process-based effect and is influenced by the characteristics of the inks, substrates, and the press operating techniques and tradeoffs.

Ideally, graininess and mottle should be nonexistent or at low levels. In practice, some graininess and mottle may be inevitable because their elimination sometimes produces some undesirable secondary effect (Field 1990b) that may be more objectionable than a slight grain or mottle. Early versions of stochastic screens sometimes caused a graininess-like effect in smooth tones.

Interference patterns are the periodic or regular moiré patterns that result from overlapping halftone screen images in the photomechanical reproduction process. The nature of the angling requirements means that some visible moiré is inevitable, but this effect is successfully minimized with fine screen rulings and appropriate screen angle selection. The multi-angle selection destroys some resolution but same-angle printing would be too difficult to control in most cases. Such tradeoffs are part of the dynamic balancing process that must be employed when producing an optimal reproduction of a given original under a given set of manufacturing conditions.

13.7. Color reproduction quality objectives

The conditions which define optimal color reproduction quality are difficult to specify with any certainty. This state of affairs exists because the definitions of optimality will vary according to the image and its purpose. Quality enhancements are commonly applied on a situational basis: hair texture is emphasized for a shampoo package illustration, subtle tones and details are enhanced in a bride's gown, or the facial details are reproduced to advantage. All three of these elements could be present in the same image, but would require different treatments in order to optimize the reproduction of a given element.

13.7.1. Tone reproduction or lightness adjustment

Tone reproduction refers to the relationship between original tonal values and corresponding reproduced tones. Given that the tonal values of the original generally exceed the reproduced tonal values, some kind of tonal compression is inevitable. The reproduction scale may also influence the optimal tonal compression. The work by Bartleson (1968) applying brightness constancy concepts to the study of tone reproduction initiated an important direction for tone reproduction research; specifically, support for the notion that proportionate brightnesses through the tone scale are not equally important in determining quality. Jorgensen (1977), working with black and white halftone reproductions, developed preferred tone reproduction curves for high key (mainly light tones)

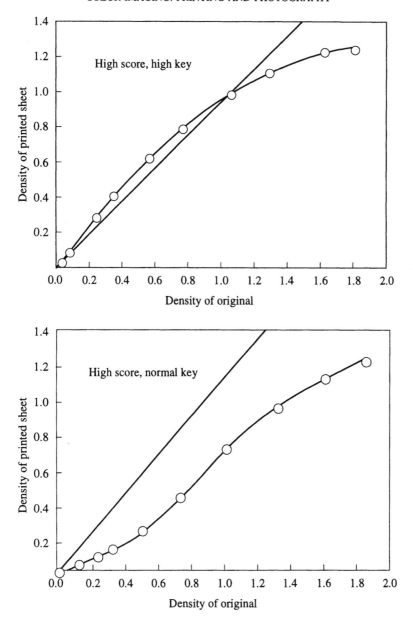

Fig. 13.6. Tone reproduction curves of "high score" prints.

and normal reproductions. He found that when image tone quality decisions had to be made, most observers tended to trade off the less important picture details in favor of the "interest area". The interest area was reproduced at a slope of about 1.0 while other areas were compressed. This concept has been applied in empirical form to color image processing systems. The widespread use of color monitors for display of images in process has made the operator's tone reproduction tasks easier: the image may be adjusted on the monitor until a pleasing reproduction is achieved by empirical methods. The accuracy of such adjustments depends upon how well the monitor has been calibrated to the reproduction media. Figure 13.6 shows Jorgensen's tone reproduction curves for high key and normal originals that were derived from high score rankings of sample prints by a panel of observers.

The tone reproduction of photographic transparencies is determined by the sensitometric characteristics of the light sensitive system and on the lighting strategies employed by the photographer. Different parts of the tone scale may be emphasized by appropriate placement of light sources or reflectors. In the case of color prints the same strategies apply, but may be supplemented through the use of appropriate exposure and processing techniques when the prints are made in the darkroom.

Tone reproduction adjustments of photomechanical reproductions are made to reflect the interest area requirements of the original and the dot gain characteristics of the printing system (ink-paper-press). In practice, the effects of dot gain may be incorporated into the lookup table for a given printing system. If so, the tone reproduction adjustments made by the scanner operator are aimed at enhancing the tone reproduction of the original in light of customer specifications and feedback via proof markups.

Poor quality originals, such as those that appear over or under exposed, may be improved in the reproductions via the application of correction curves. Reproductions of faded color originals may also be adjusted by the application of correction curves for the dye layer in question.

The starting point for white is also an important value when optimizing tone reproduction quality. Such unimportant tonal areas as specular reflections are reproduced without tonal information. In those cases where the original was photographed on a foggy morning, there may not be a true white in the original. The scanner or system operator's task is to set the just printable white point to represent the lightest tone value that contains detail. In those cases where there is no white reference, the operator relies on an equivalent target white (usually optical density about 0.35 density for a color transparency) for setting the minimum printable tones.

The shadow values may be manipulated to some degree by the GCR/UCR controls together with contrast adjustments of the black printer. Higher levels of GCR/UCR will lower the maximum optimal density of the reproduction and limit contrast in the shadow or darker tones. A steep-slope black separation together with low or no use of GCR/UCR, and near solid or 100 percent tonal values in all four colors, will produce conditions that will result in satisfactory tonal separation and density in dark tones. The scanner or system operator must be alert to those situations where there are no true shadow tones (the foggy morning photograph) and ensure that the dark tones are not artificially darkened. In other words, the operator must exercise judgment when classifying original images as poorly exposed as opposed to those that were merely exposed under abnormal conditions.

13.7.2. Color balance or hue shifts

The yellow, magenta and cyan image layers are color balanced to produce visually neutral grays. If the gray scale is not neutral, the entire reproduction will exhibit a color cast.

Overall casts in photographs may be due to the use of film that was not balanced to the lighting conditions used when the photograph was made. Improper processing or selective dye layer fading may also produce a color cast.

Color casts of photographs may be excluded from the photomechanical reproduction by the use of appropriate techniques. A complementary color–compensating filter of the right strength may be laid over a color transparency to neutralize the cast, or the tone scale adjustment for a particular separation may be manipulated (made darker or lighter) in order to compensate for the unbalanced layer.

The yellow, magenta and cyan halftone images must also be balanced to produce a neutral gray. A special test image, the gray balance chart, is initially printed with the ink-substrate-press conditions to be employed. The resulting chart is examined to find the dot percentage combinations that will result in a 3-color neutral gray at the target gray levels. The selected gray balance points represent the point of correct color balance. Color cast removal and gray balance adjustment may be set during the same scanner setup step, but the gray balance requirements are usually incorporated into the scanner's or system's look up table.

The ideal requirement for color balance is realized when the neutral colors are free of any unintended color cast. This objective is relatively simple to achieve.

13.7.3. Color correction or saturation adjustment

The unwanted absorptions of the yellow, magenta and cyan colorants are the primary reasons why color correction processes (fig. 13.3) are built into the image reproduction processes. The most visible evidence of the correction processes are the orange mask layers that may be observed on processed color negative film. The strength and nature of these correction dyes are designed to compensate for the deficiencies of the subsequent color print making process.

Look up table (LUT) values incorporate the color correction adjustments that are necessary for optimum color rendition from that set of colorants. The LUT values also incorporate the corrections required to compensate for any color distortions related to the color scanning or color separation process. ANSI-defined standard reference targets are available for this purpose: IT8.7/1 and IT8.7/2 for input characterization, and IT8.7/3 for output characterization.

Color adjustment problems arise when the saturation component of the color gamut exceeds that available from the colorant system at hand. The saturation of non gamut colors will be compressed in the final reproduction. What is not clear, however, is how the colors adjacent to the non gamut color should be treated in the compression process (Field 1989).

One saturation compression strategy is simply to ignore the colors that exceed the system's gamut. The subsequent reproduction will place the non-gamut color on the nearest gamut boundary. In other words, there will be no difference in the reproduction between original colors that did lie on the gamut boundary, and non-gamut colors that are now reproduced as a boundary gamut color.

An alternate saturation compression strategy is to compress the saturation of all colors equally. In this strategy, the non-gamut color is again reproduced at the nearest gamut boundary point, but the original colors that lay at the gamut boundary are reproduced at a lower saturation. In other words, some of the saturation difference between the original non-gamut color and the gamut boundary color is retained in the reproduction because the original boundary gamut color is deliberately reproduced at lower saturation.

The common saturation compression strategy is the second one described above, i.e., the "compress all colors equally" strategy. There are circumstances, however, where the former strategy would be preferred. If a gamut boundary original color represents a critical color in, say, a clothing catalog, and the non-gamut color is in a relatively unimportant area, then the gamut boundary color should be reproduced as accurately as possible, while the non-gamut color is ignored.

The optimal saturation compression strategy will depend upon the circumstances and may not become obvious until trial proofs have been produced. Customers will mark corrections on proofs to compensate for saturation compression that may even require some slight hue shifts of reproduced colors that have the correct hue. The exact nature of the corrections or adjustments is not intuitively obvious and it may take several iterations before the reproduced color represents the best possible compromise.

13.7.4. Color reproduction models

A number of attempts have been made to combine the hue, saturation and lightness values into a single color reproduction model or index that may be used to predict the optimal reproduction of a given original via a given reproduction system. Such models probably have greater applicability to photographic reproductions than to photomechanical reproductions, but certainly have considerable applications to both systems.

The color reproduction models developed by Pointer (1986) and Johnson (1995) upon the foundation of Hunt's (1991, 1994) color appearance model, have produced some promising results. In practice, and this is especially true for photomechanical reproductions, the reproduction requirements of a particular tonal interest area or the saturation of a given gamut boundary color will often require a deviation from the predictions of a model. A successful model will do much to help raise the overall level of color reproduction quality, but it will still require the use of artistic judgment and color separation modification skills in order that the best results are obtained from certain originals with a particular reproduction requirements.

Rhodes (1978) proposed a more empirical approach to color optimization that was based on the use of color monitors in research studies. The subsequent widespread use of color monitors for production tasks has done much to assist an operator in making judgments about the optimal hue, saturation and lightness requirements for a given reproduction. Such judgments are valid only if the monitor is calibrated to the particular reproduction process being used. Even then, the resolution, gloss, texture and other characteristics of the reproduction cannot be captured on the monitor. These characteristics may exert an influence on the preferred reproduction of a given color; therefore, neither color reproduction models nor color monitors are capable of handling the entire range of factors that together produce the desired overall quality. Both approaches, however, have done much to raise the general level of color reproduction quality and to reduce quality variability.

13.7.5. Image definition requirements

Leaving aside the limitations imposed by substrate and process, the resolution requirements are determined by the nature of the original image and the viewing conditions. A billboard poster may, for example, appear to have low resolution when it is inspected at arm's length, but will appear to be quite satisfactory when viewed at a distance.

Some original images contain much information that should be retained in the reproduction in order to inform the viewer of features or details. A photograph of a stereo system or a short wave radio, for example, may contain much detail about the controls that is of interest to the consumer. The catalog image of the device in question must either be reproduced with a fine enough halftone screen to retain the important detail, or be reproduced at a size where the detail is retained for the screen ruling being used. In other words, if the image is reproduced large enough, a relatively coarse screen will retain the required detail. If the image is small, then the screen ruling should be correspondingly finer in order to retain important detail.

Sometimes the product form dictates the resolution requirements. Photography books, especially monographs, are often reproduced with very fine screen rulings (about 300 lines per inch) in order to more closely simulate the detail of the original photograph.

Photographs also have varying resolution requirements. Motion picture images will appear to be of low resolution and quite grainy at close viewing distances or if the motion is stopped, but will be quite satisfactory at normal theater viewing distances and in motion. Aerial reconnaissance photographs, on the other hand, are often viewed through hand held magnifiers and, therefore, have very high resolution requirements.

Sharpness requirements will also vary according to circumstances. Portrait photographs are sometimes made with a deliberate soft focus effect, and photographs of objects or people in motion often present the illusion of motion by having the background or the object appear unsharp or blurred. In most cases, however, photographic images have a high sharpness requirement that is consistent with the lens, the exposure conditions, the photosensitive emulsion, and the processing conditions.

The sharpness of printed color reproductions is dependent, in part, upon the accuracy of register between the yellow, magenta, cyan and black images. The required accuracy depends on the sharpness of the image: soft focus images have a greater tolerance for misregister. For sharp images, it is generally accepted that a register variation of up to one half of a row of dots is acceptable (Jorgensen 1982). Finer screen rulings, therefore, have smaller physical misregister allowances, than do coarser screen rulings.

Electronic enhancement of an image may help to increase apparent sharpness. Such adjustments, if overdone, can emphasize graininess and produce an unnatural appearance. To some extent, the subject matter and the printing conditions (especially substrate) will determine the optimal enhancement level.

Image sharpness may be improved by increasing ink film thickness (IFT) on press. An increase in IFT will, however, lead to a loss of resolution and a possible increase in mottle or ink film graininess. Changes in IFT will also influence saturation and lightness, a fact which serves to demonstrate the interrelated nature of quality attributes. It is not possible, in many cases, to maximize one attribute without suffering an accompanying degradation of another attribute. Optimal quality requires the right balance between conflicting attributes. Attempts have been made (Engeldrum 1995) to integrate sharpness, color quality, and defects measures into a broad image quality model through the application of linear and nonlinear regression techniques. Unfortunately, the model coefficients cannot be interpreted as guides to the relative importance of each image quality component, because the scales are interval scales, rather than fundamental measures. The number of contributors to pictorial image quality are such that it may be impossible to derive a single satisfactory index.

13.7.6. Interference patterns

The quality objectives regarding interference patterns are relatively simple: there should not be any interference patterns. While it is true that some photographers make effective use of film grain as part of the creative process, the absence of artifacts in images is generally a universal objective.

It was stated in Section 13.7.5 that the optimization of certain image properties is in conflict with optimization of other properties. Graininess and mottle, therefore, should be minimized to the extent that other image structure and color attributes are not compromised.

The photomechanical reproduction process, by nature, has interference patterns that are extremely difficult to eliminate. Overlapping halftone screens, even if they are angled to avoid objectionable moiré, will create a noticeable pattern with all but the finest screen rulings. The use of stochastic screens minimizes moiré problems and improves resolution as shown in fig. 13.7. The objectionability of any given effect is largely a function of the subject. Areas of busy detail, for example, rarely appear grainy.

The actual printing process may introduce a number of unwanted spots or patterns. Some of these are regular or periodic while others are random in nature. In most cases, these spots or patterns may be eliminated during the production process; but, in some cases, unwanted spots or patterns may be part of a more general systemic characteristic.

13.7.7. Surface characteristics

The surface characteristics of an image intended for viewing by reflected light can exert significant influence on the appearance. Hunter (1975) has published a substantial review of appearance factors that includes, but goes well beyond, the standard colorimetric considerations. The "ideal" surface characteristics for a given reproduction is open to some interpretation. A textured substrate, for example, will influence the perceived image structure and color appearance while introducing a tactile element. The texture may superimpose a regular physical pattern over the image and introduce a micro variation in surface reflection that may vary with changes in lighting, viewing angle and viewing distance. In this context, there is no "good" or "bad", rather, a combination of optical and physical qualities best suited to the image and its purpose.

Gloss is somewhat easier to discuss. The gloss of a reproduction should approximately match that of the original when artist-generated originals are being considered. In general, the same can be said for photographic originals, but higher gloss will produce higher color saturation, density, and sharpness. Glare will influence perception at high gloss levels; therefore, an intermediate gloss value may be preferred in most cases (Swanton and Dalal 1995).

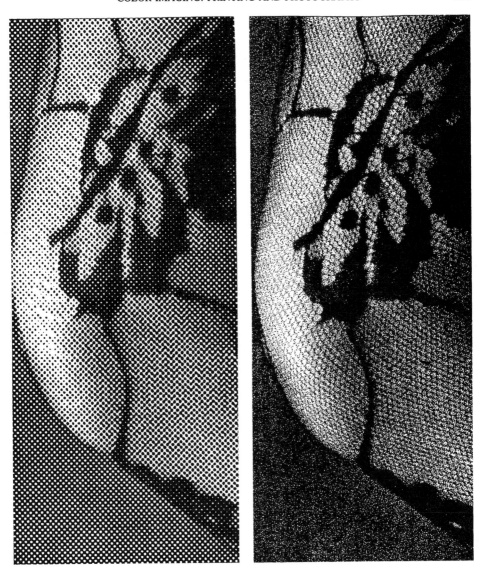

Fig. 13.7. Enlarged views of a conventional halftone image structure, left, and a stochastic or random-dot halftone image structure, right. Should be viewed from a distance.

13.8. Research directions

The developments discussed to date have focused upon color reproduction process adjustments for improved quality. Color reproduction process research has always taken this approach: what is the best quality we can achieve from the system and materials at hand and how can we achieve this quality level on a consistent basis?

The related research focuses upon the individual technologies and materials within the color reproduction cycle. Improvements in the performance of a particular element may lead to overall improvement of the color reproduction system and lead to the reexamination of established practices.

13.8.1. Process developments

A number of technologies have been developed in recent years to enable the direct transfer of digital image data from the imaging system to the printing press. The direct transfer systems eliminate the use of film (and plates in some cases) and shorten the elapsed time between the color separation and printing processes. Such systems have been promoted for the short run color printing markets that fall in the gap between those served by machine generated photographic prints and conventional printing processes. The main attraction of direct-to-press imaging technologies appear to be time savings and an economically attractive production cost for particular short run color printing markets. The image resolution of such systems, however, do not yet match the best conventional imaging systems.

Some direct-to-plate lithographic systems will function without the water (fountain solution) application that is common in conventional lithography. Non-direct waterless plate systems have also been marketed in recent years. The absence of water allows the printer to use thinner ink films (in conventional systems, water is emulsified in the ink) and thus achieve higher resolution and reduced mottle. The color saturation and density achievable by such systems may exceed conventional systems. A drawback of waterless systems is that a cooling system must be used to stabilize ink temperature throughout the run.

The major equipment developments in color separation involve the ongoing application of automated set up routines to color scanners. Some of these approaches are claimed by vendors to be based on artificial intelligence concepts. These scanners are focused on maximizing productivity and, while not necessarily optimizing quality, they also set the scanner to produce a very acceptable result.

Photographic color print machines also make use of computer-assisted set up and operation. The most recent development, the APS advanced photosystem, involves the encoding of the negative with the exposure in a form that is readable by the print machine. Exposure data are used within the machine to help optimize the subsequent exposure of the print.

13.8.2. Quality developments

One of the main barriers against using more than the standard four colors for process color printing has been the difficulty with moiré patterns when halftone images are overlapped. If the extra colors have low density, pale blues or pinks, for example, the low image contrast, coupled with the subsequent reduced light tonal values in cyan and magenta, will minimize moiré problems. Greeting card manufacturers have been using this technique for many years to help expand color gamut at the lighter tonal values because proportionality failure of magenta and cyan halftone images reduces the saturation of associated light tones.

It was not until the frequency modulated, or stochastic or random dot, halftone screening technique was introduced that it became feasible to use high density colors to expand the color gamut in darker tonal areas. The almost random nature of the halftone screen elements has meant that moiré patterns are no longer a significant problem when printing more than four high density colors. Until this development, supplementary reds or greens, for example, had to be printed as solids in selected areas in order to avoid moiré problems.

The most common high density supplementary colors are those closest to the subtractive secondary colors: blue, green, and red. These colors are those that most often appear weak or desaturated in reproductions produced by the lithographic process. Additivity failure, which includes the lack of perfect ink transfer ("trapping") to previously-printed wet ink films is the primary reason for less-than-satisfactory secondary color areas. Six and, in some cases, eight color presses are available to accommodate the needs of the "high fidelity" (as it is often called) color printing market. The use, for example, of "duotone black" (two black impressions) techniques for high quality black and white photography publications is now quite routine. The similar use of supplementary colors for expanded color gamuts is likely to become more common for specialist high quality market niches. The optimal selection of supplementary colors depends on the relative importance of proportionality and additivity failure effects to a particular class of images or printed products.

The random dot or stochastic screening technology that makes six or seven color moiré-free printing possible, also helps to improve certain measures of image definition. The random and small nature of the image elements result in reproductions of higher resolution by comparison with conventional halftone reproductions of the same nominal screen ruling. This improvement in resolution occurs because the smaller image elements have less of an effect in breaking up the image detail than does the conventional halftone screen. These and similar developments offer the promise of extracting more quality from what is an increasingly well-optimized system.

Color photographic materials continue to exhibit gains in resolution and color saturation. These developments further expand the range of color reproduction possibilities that are available. The fact that such products find a market niche is evidence that the concept of optimal color reproduction has a number of meanings. Indeed, a process with elements of art, commerce and science cannot help but be multi faceted and sometimes intangible.

13.9. Bibliography

1. Evans, R.M., W.T. Hanson Jr. and W.L. Brewer, 1953, Principles of Color Photography (Wiley, New York, NY).

2. Eynard, R.A. (ed.), 1973, Color: Theory and Imaging Systems (Society of Photographic Scientists and Engineers, Washington, DC).

3. Field, G.G., 1988, Color and Its Reproduction (Graphic Arts Technical Foundation, Pittsburgh, PA).

4. Field, G.G., 1990, Color Scanning and Imaging Systems (Graphic Arts Technical Foundation, Pittsburgh, PA).

5. Hunt, R.W.G., 1995, The Reproduction of Colour, 5th edn (Fountain Press, Kingston-upon-Thames, England).

6. Hunter, R.S., 1975, The Measurement of Appearance (Wiley, New York, NY).

7. Pearson, M. (ed.), 1971, Proceedings of the ISCC Conference on the Optimum Reproduction of Color (Inter Society Color Council and Rochester Institute of Technology, Rochester, NY).

8. Pearson, M. (ed.), 1992, Proceedings of the Conference on Comparison of Color Images Presented in Different Media (Technical Association of the Graphic Arts and the Inter Society Color Council, Rochester, NY).

9. Sturge, J.M., V.K. Walworth and A. Shepp (eds.), 1989, Imaging Processes and Materials (Van Nostrand Reinhold, New York, NY).

10. Yule, J.A.C., 1967, Principles of Color Reproduction (Wiley, New York, NY).

13.10. Appendixes

13.10.1. The Demichel equations

Demichel's equations are based on the following assumptions:

Let:

fractional dot area of yellow $= y$,

fractional dot area of magenta $= m$,

fractional dot area of cyan $= c$,

therefore:

area not covered by yellow $= 1 - y$,

area not covered by magenta $= 1 - m$,

area not covered by cyan $= 1 - c$,

and:

area not covered by yellow or magenta $= (1 - y)(1 - m)$,

area not covered by cyan or yellow $= (1 - c)(1 - y)$,

area not covered by magenta or cyan $= (1 - m)(1 - c)$,

area not covered by yellow, magenta or cyan $= (1 - y)(1 - m)(1 - c) =$ white.

Similarly:

area covered by yellow and magenta, but not cyan $= ym(1 - c) =$ red,

area covered by cyan and yellow, but not magenta $= cy(1 - m) =$ green,

area covered by magenta and cyan, but not yellow $= mc(1 - y) =$ blue,

area covered by yellow, magenta, and cyan $= ymc =$ black,

and:

area that is solely yellow (i.e., not magenta and not cyan) = $y(1-m)(1-c)$ = yellow,

area that is solely magenta (i.e., not yellow and not cyan) = $m(1-y)(1-c)$ = magenta,

area that is solely cyan (i.e., not yellow and not magenta = $c(1-y)(1-m)$ = cyan.

The following expression, therefore, is the mathematical representation of the yellow, magenta, and cyan dot areas in a given color: $(1-y)(1-m)(1-c) + y(1-m)(1-c) + m(1-y)(1-c) + c(1-y)(1-m) + ym(1-c) + cy(1-m) + mc(1-y) + ymc$.

13.10.2. The Neugebauer equations

The Neugebauer equations are based on the following conditions:

Let:

R_y = red-light reflectance of yellow ink,

R_m = red-light reflectance of magenta ink,

R_c = red-light reflectance of cyan ink,

R_{ym} = red-light reflectance of the overlap of yellow and magenta,

R_{cy} = red-light reflectance of the overlap of cyan and yellow,

R_{mc} = red-light reflectance of the overlap of magenta and cyan,

R_{ymc} = red-light reflectance of the overlap of yellow, magenta, and cyan.

These definitions apply similarly to the green-light and blue-light reflectances.

Assume red-light reflectance from white paper = 1; then, by incorporating the expressions for the reflectance properties of the inks into the Demichel equations that represent the halftone dot areas, we have:

$$R = (1-y)(1-m)(1-c) + y(1-m)(1-c)R_y$$
$$+ m(1-y)(1-c)R_m + c(1-y)(1-m)R_c + ym(1-c)R_{ym}$$
$$+ yc(1-m)R_{yc} + mc(1-y)R_{mc} + ymcR_{ymc},$$
$$G = (1-y)(1-m)(1-c) + y(1-m)(1-c)G_y$$
$$+ m(1-y)(1-c)G_m + c(1-y)(1-m)G_c + ym(1-c)G_{ym}$$
$$+ yc(1-m)G_{yc} + mc(1-y)G_{mc} + ymcG_{ymc},$$
$$B = (1-y)(1-m)(1-c) + y(1-m)(1-c)B_y$$
$$+ m(1-y)(1-c)B_m + c(1-y)(1-m)B_c + ym(1-c)B_{ym}$$
$$+ yc(1-m)B_{yc} + mc(1-y)B_{mc} + ymcB_{ymc},$$

where R = the total red-light reflectance, G = the total green-light reflectance, and B = the total blue-light reflectance.

13.10.3. The masking equations

The masking equations may be solved by a first order RGB to CMY conversion.

Let:

D_r, D_g, D_b represent the red, green, and blue densities of the three-color combination;

C, M, Y are principal densities of the individual subtractive primary inks;

c_b, c_g, for example, are the ratios of the blue and green densities of the cyan ink to its red density. The corresponding ratios for the magenta and yellow inks are m_r, m_b, y_r, and y_g.

Thus, according to the additivity rule, we have:

$$D_r = C + m_r M + y_r Y,$$
$$D_g = c_g C + M + y_g Y,$$
$$D_b = c_b C + m_b M + Y.$$

The masking requirements can be expressed in this form:

$$C = a_{11} D_r + a_{12} D_g + a_{13} D_b,$$
$$M = a_{21} D_r + a_{22} D_g + a_{23} D_b,$$
$$Y = a_{31} D_r + a_{32} D_g + a_{33} D_b.$$

Where the a_{11} to a_{33} values are the unknown coefficients. These coefficients represent the percentage masks required for correction of the unwanted densities of the inks.

These equations can be rearranged into a more useful form:

$$C = a_{11}\left(D_r + \frac{a_{12}}{a_{11}} D_g + \frac{a_{13}}{a_{11}} D_b\right),$$
$$M = a_{22}\left(\frac{a_{21}}{a_{22}} D_r + D_g + \frac{a_{23}}{a_{22}} D_b\right),$$
$$Y = a_{33}\left(\frac{a_{31}}{a_{33}} D_r + D_g + \frac{a_{32}}{a_{33}} D_b\right).$$

13.10.4. The Clapper transformation

A second order transformation suggested by Clapper uses:

$$c = a_{11} D_r + a_{12} D_g + a_{13} D_b + a_{14} D_r^2 + a_{15} D_g^2 + a_{16} D_b^2$$
$$+ a_{17} D_r D_g + a_{18} D_r D_b + a_{19} D_g D_b,$$

and similarly for m and y.

The values D_r, D_b, and D_g are the densities of the printed color patches and c, m, and y are ink amounts needed to match the original color. The a_{11} to a_{19} values are the unknown coefficients that are calculated by regression analysis. Clapper suggested that this equation could be used for simplified calibration of analog color scanners.

References

Bartleson, C.J., 1968, Criterion for tone reproduction. Journal of the Optical Society of America **58**(7), 992–995.

Bartleson, C.J., 1982, The combined influence of sharpness and graininess on the quality of colour prints. The Journal of Photographic Science **30**(2), 33–38.

Bruno, M.H., 1986, Principles of Color Proofing (GAMA Communications, Salem, NH) pp. 13–18, 36–46, 65–114.

Clapper, F.R., 1961, An empirical determination of halftone color-reproduction requirements, in: Proceedings of the Technical Association of the Graphic Arts (Rochester, NY) pp. 31–41.

Engeldrum, P.G., 1995, A framework for image quality models. Journal of Imaging Science and Technology **39**(4), 1–6.

Evans, R.M., W.T. Hanson Jr. and W.L. Brewer, 1953, Principles of Color Photography (Wiley, New York, NY).

Eynard, R.A. (ed.), 1973, Color: Theory and Imaging Systems (Society of Photographic Scientists and Engineers, Washington, DC).

Field, G.G., 1988, Color and Its Reproduction (Graphic Arts Technical Foundation, Pittsburgh, PA).

Field, G.G., 1989, Color correction objectives and strategies, in: Proceedings of the Technical Association of the Graphic Arts (Rochester, NY) pp. 330–349.

Field, G.G., 1990a, Color Scanning and Imaging Systems (Graphic Arts Technical Foundation, Pittsburgh, PA) pp. 23–40.

Field, G.G., 1990b, Image structure aspects of printed image quality. The Journal of Photographic Science **38**(5/6), 197–200.

Heuberger, K.J., Z.M. Jing and S. Persiev, 1992, Color transformations and lookup tables, in: Proceedings of the Technical Association of the Graphic Arts, Vol. 2, ed. M. Pearson (TAGA/ISCC, Rochester, NY) pp. 863–881.

Hunt, R.W.G., 1991, Revised colour-appearance model for related and unrelated colours. Color Research and Application **16**, 146–165.

Hunt, R.W.G., 1994, An improved predictor of colourfulness in a model of colour vision. Color Research and Application **19**, 23–26.

Hunt, R.W.G., 1995, The Reproduction of Colour, 5th edn (Fountain Press, Kingston-upon-Thames, England).

Hunter, R.S., 1975, The Measurement of Appearance (Wiley, New York, NY) pp. 4–43, 65–80.

Johnson, T., 1995, A complete colour reproduction model for graphic arts, in: Proceedings of the Technical Association of the Graphic Arts, Vol. 2 (Rochester, NY) pp. 1061–1076.

Jorgensen, G.W., 1977, Preferred tone reproduction for black and white halftones, in: Advances in Printing Science and Technology, ed. W.H. Banks (Pentech Press, London) pp. 109–142.

Jorgensen, G.W., 1982, Control of color register, Research progress report No. 114 (Graphic Arts Technical Foundation, Pittsburgh, PA).

Langford, M.J., 1982, The Master Guide to Photography (Alfred A. Knopf, New York, NY) pp. 96–159.

Pearson, M. (ed.), 1971, Proceedings of the ISCC Conference on the Optimum Reproduction of Color (Inter Society Color Council and Rochester Institute of Technology, Rochester, NY).

Pearson, M. (ed.), 1992, Proceedings of the Conference on Comparison of Color Images Presented in Different Media (Technical Association of the Graphic Arts and the Inter Society Color Council, Rochester, NY).

Pointer, M.R., 1986, Measuring colour reproduction. The Journal of Photographic Science **34**(3), 81–90.

Rhodes, W.L., 1978, Proposal for an empirical approach to color reproduction. Color Research and Application **3**(4), 197–201.

Sturge, J.M., V.K. Walworth and A. Shepp (eds.), 1989, Imaging Processes and Materials (Van Nostrand Reinhold, New York, NY).

Swanton, P.C. and E.N. Dalal, 1995, Gloss preferences for color xerographic prints. Journal of Imaging Science and Technology **40**(2), 158–163.

Yule, J.A.C., 1967, Principles of Color Reproduction (Wiley, New York, NY) pp. 151–232, 328–346.

chapter 14

COLOR ENCODING
IN THE PHOTO CD SYSTEM

EDWARD J. GIORGIANNI

and

THOMAS E. MADDEN
Eastman Kodak Co.
Rochester, NY, USA

Color for Science, Art and Technology
K. Nassau (Editor)

CONTENTS

14.1. Introduction 392

    14.1.1. The *Kodak Photo* CD System 392

    14.1.2. Hybrid imaging systems 393

    14.1.3. Digital color encoding 394

14.2. Color-imaging systems 394

    14.2.1. Stage 1: Image capture and color separation 394

    14.2.2. Stage 2: Signal processing 395

    14.2.3. Stage 3: Color reconstruction and image display 395

    14.2.4. Multiple input/output systems 396

14.3. Input signal processing 396

    14.3.1. Input compatibility 398

    14.3.2. Creating input compatibility 398

    14.3.3. A single input medium and a single input scanner 399

    14.3.4. A single input medium and multiple input scanners 400

    14.3.5. Multiple input reflection print media 401

    14.3.6. Multiple reflection print and transparency film inputs 401

    14.3.7. Multiple reflection print, transparency, and negative film inputs 402

    14.3.8. The *Photo CD* System solution to input compatibility 403

14.4. Data metrics for color encoding 405

    14.4.1. The data metric for the *Photo CD* System – *Kodak PhotoYCC* Color Interchange Space . 406

    14.4.2. Video requirements 406

    14.4.3. Color gamut 406

    14.4.4. Luminance dynamic range 409

    14.4.5. Digital quantization 410

    14.4.6. Image compression 410

    14.4.7. Formation of *PhotoYCC* Space digital values 412

    14.4.8. Color encoding implementation 413

14.5. Output signal processing 414

    14.5.1. Colorimetric transformation 414

    14.5.2. Gamut adjustment 415

    14.5.3. Output drive signals 417

14.6. Discussion 417

    14.6.1. Device-independent color 418

    14.6.2. Color-appearance colorimetry and encoding 419

    14.6.3. *Photo CD* Format images and color management 419

    14.6.4. Other applications of *Kodak PhotoYCC* Color Interchange Space 420

    14.6.5. Summary 420

14.7. Appendixes 420

    14.7.1. Calculation of reference image-capturing device tristimulus values 420

    14.7.2. CCIR-709 primaries 421

    14.7.3. RGB nonlinear transformations 421

References 422

14.1. Introduction

The *Kodak Photo CD* System creates compact discs of images by scanning and digitally encoding images from photographic media. *Photo CD* Discs can be used to produce video and hardcopy images on a variety of devices and media.

The *Photo CD* System is an example of hybrid color imaging, which combines photographic and electronic imaging technologies. Because hybrid systems may include many sources of input and multiple modes of output and display, the control of color in such systems can become quite complex.

One of the keys to the successful management of color in a hybrid imaging system is the use of an appropriate method for image color encoding. The encoding must meet three objectives: it must enable input from all of the image sources supported by the system; it must provide for efficient storage and interchange of image data; and it must allow for the production of images on all of the output devices and media supported by the system.

In this chapter, the inventors of the color-encoding method used in the *Photo CD* System describe how each of these objectives can be achieved in practical hybrid imaging systems. Device-dependent and device-independent color, colorimetric and noncolorimetric encoding, luminance dynamic range, color gamut, quantization, spatial compression, output optimization, and other topics related to digital color encoding are discussed. The color encoding used in *Kodak PhotoYCC* Color Interchange Space is described in detail.

The concepts of Chapters 1, 13, and 15 are relevant to various parts of this discussion. While references to specific sections of those chapters are not included, it is suggested that they be read prior to reading this chapter.

Photo CD and *PhotoYCC* are trademarks of Eastman Kodak Company.

14.1.1. The Kodak Photo CD System

In 1992, Eastman Kodak Company introduced the *Kodak Photo CD* System (Eastman Kodak 1991) (fig 14.1). This sophisticated color-imaging system creates compact discs of digital images by scanning and digitally encoding photographic images from media such as negatives, transparencies, and reflection prints. *Photo CD* Discs can be used to produce video images on television receivers and monitors using special disc players. Video images can also be displayed on computer systems using CD-ROM drives and appropriate software. Hardcopy images, such as 35 mm slides, larger-format transparencies, reflection prints, and color separations for graphic arts applications can be produced from *Photo CD* Discs on systems equipped with film writers, thermal printers, and other types of digital hardcopy devices (Giorgianni et al. 1993).

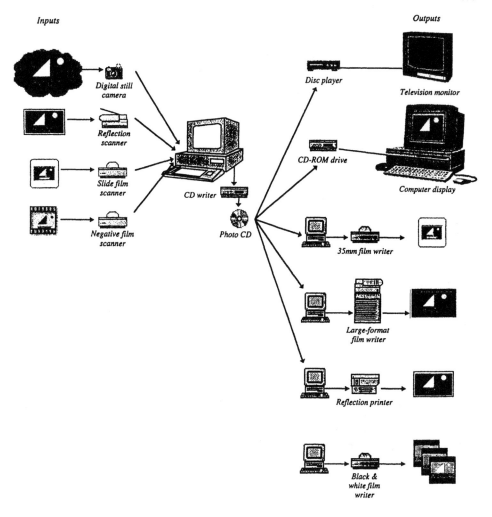

Fig. 14.1. *Photo CD* system image creation and use.

14.1.2. Hybrid imaging systems

The *Photo CD* System is a *hybrid* color-imaging system. Hybrid imaging systems incorporate both photographic and electronic imaging technologies, and well-designed systems can enhance the advantages inherent in each of these component technologies. For example, a powerful photocomposition system can be produced by combining the superior imaging characteristics of chemically based photographic media with the flexibility of electronic image editing. The image manipulation capabilities provided by hybrid imaging have found wide application in the motion picture and television industries, particularly for special effects and other post-production work. Hybrid imaging technology is the basis of the color electronic prepress industry, and it has made possible the rapid development and expansion of desktop color imaging. In the *Photo CD* System, advanced hybrid technology allows digital images to be created from disparate types of photographic media and used in applications ranging from simple video playback devices to complex commercial imaging systems.

14.1.3. Digital color encoding

One of the keys to the successful performance of the *Photo CD* System, and to that of any other hybrid color-imaging system, is the use of an appropriate method for numerically representing and digitally encoding color image information. The color-encoding method must meet three fundamental objectives: it must permit input from all of the image sources supported by the system; it must provide for efficient storage and interchange of image data; and it must allow for the production of images on all of the output devices and media supported by the system (DeMarsh and Giorgianni 1989).

This chapter shows that while various methods for digitally encoding color images are available, the invention of a new method (Giorgianni and Madden 1997) – based on a fundamentally different encoding concept – was required to meet the comprehensive objectives of the *Photo CD* System. This method and its associated *PhotoYCC* Space data metric will be explained in detail as part of a broader discussion of color-encoding techniques for color-imaging systems.

14.2. Color-imaging systems

All complete color-imaging systems – whether chemical, electronic, or hybrid – must perform five functions: image capture, color separation, signal processing, color reconstruction, and image display. These functions are generally grouped in three stages.

14.2.1. Stage 1:
Image capture and color separation

An original scene is comprised of individual color stimuli, emanating from each discrete spatial location within the scene. The stimulus at a given location consists of a continuous spectral distribution of radiant flux, formed when light from the source illuminating the scene is spectrally modulated by the reflectance, transmittance, and absorptance characteristics of the object at that location (Billmeyer and Saltzman 1981) (fig. 14.2). In the first stage of an imaging system, a fraction of the light from each original stimulus is captured, and three (or more) separate color records or signals are formed. This process, called image capture and color separation, can be accomplished in a photographic system by three overlaid emulsion layers that are spectrally sensitive to red, green, and blue light. Electronic image capture and color separation can be achieved using a single solid-state sensor comprised of a mosaic of individual light-sensitive elements overlaid with red, green, or blue filters, or by the use of three sensors preceded by an appropriate optical arrangement of beam splitters and color filters.

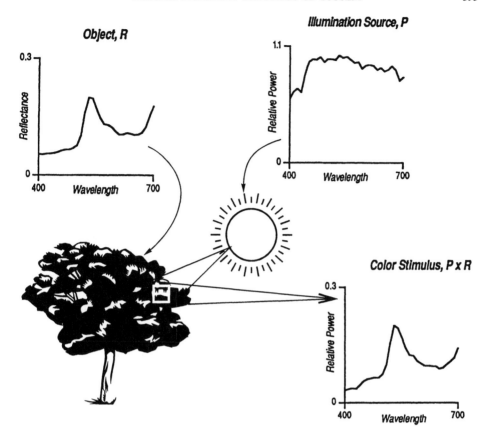

Fig. 14.2. Formation of an original-scene color stimulus.

14.2.2. Stage 2:
Signal processing

In the second stage of a color-imaging system, image signals produced in the image-capture/color-separation stage are made suitable for output by the use of appropriate signal processing. Signal processing typically includes linear and nonlinear transformations of individual color signals, and linear and nonlinear transformations involving interactions among these color signals. The processing may be chemical in the case of conventional photography or a combination of analog and digital processing in the case of electronic or hybrid color-imaging systems.

14.2.3. Stage 3:
Color reconstruction and image display

In color reconstruction, the processed signals from the second stage are used to control the color-forming elements of the output medium or device. For self-luminous displays, this color reconstruction process directly forms a displayed output image. A three-beam video projector, for example, produces a displayed image by additively combining modulated intensities of red, green, and blue light. For the display of hardcopy images, a source of illumination is required. Photographic media, for example, typically use cyan, magenta, and yellow image-forming dyes to modulate, by absorption or subtraction, the transmitted or reflected light of a viewing illuminant to produce a displayed output image (Hunt 1995).

14.2.4. Multiple input/output systems

Hybrid imaging systems often must support multiple sources of input images and multiple types of output devices and media. This can best be accomplished by dividing the digital signal processing stage into two substages – input signal processing and output signal processing. The approach is illustrated in fig. 14.3, which depicts a multiple input/output system where each input or output has its own associated digital signal processing, which will be referred to in this chapter as a *transform*. This arrangement of parallel inputs and parallel outputs makes possible – and requires – a *color-encoding specification* into which each input can record and from which each output can receive image data. The overall success of the imaging system will depend, in large part, on the appropriate design of the color-encoding specification.

A color-encoding specification includes both a color-encoding method and a color-encoding data metric. The color-encoding *method* defines the actual meaning of the encoded data, while the color-encoding *data metric* defines the numerical units in which encoded data are expressed. A complete color-encoding specification may also include other attributes, such as data compression method and data file format. While not strictly part of the color encoding, these attributes must be completely defined so as to permit interchange of images, encoded according to the encoding specification, among imaging systems.

In the following sections, the functions of input and output signal processing, the role of the color-encoding specification, and the factors that influence the selection of color-encoding methods and data metrics in multiple input/output hybrid color-imaging systems will be discussed.

14.3. Input signal processing

The principle function of input signal processing is to enable the system to successfully accept input images from all media and devices supported by the system. Input signal processing translates data measured from each picture element (pixel) of an input image to encoded values expressed in terms of the particular color-encoding specification defined for the system (refer again to fig. 14.3). In order for the color-encoding specification to effectively define the encoded image values, input signal processing must be based on an encoding method that creates what the authors have termed "input compatibility" among all input image sources (Giorgianni et al. 1993).

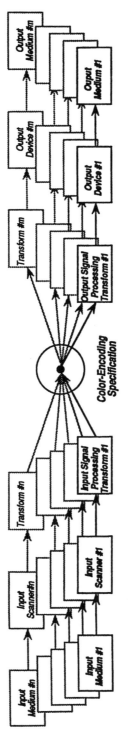

Fig. 14.3. Use of a color-encoding specification in a multiple input/output color-imaging system.

14.3.1. Input compatibility

Input compatibility means that image data from all input sources are encoded on a common basis. Moreover, input compatibility means that input image data are expressed in terms of the color-encoding specification such that encoded values completely and unambiguously specify the color of each pixel. Interpretation of input-compatible encoded color data does not require knowledge of the particular input image medium that was the source of that data, nor does it require knowledge of the device that was used to measure the image. When input compatibility is achieved, a commonality is created in the data encoded from all sources of input images, and an independence of all system outputs from all system inputs is established. As a result, input compatibility provides a system with several important features:

(1) Each output device can produce images from any image encoded in the color-encoding specification, regardless of the type of input medium or device that was used to produce the encoded image data.
(2) Image data encoded from one medium can be intermixed with image data encoded from another. For example, image data encoded from two different types of media can be seamlessly merged to produce a homogeneous-appearing composite image.
(3) The number of signal-processing transforms required by the system is minimized. Without the input/output independence established by input compatibility, each output device would require a different transform to produce images encoded from each type of input. The number of transforms for the system therefore would equal the product of the number of inputs and outputs (32 transforms for 4 inputs and 8 outputs, for example). With input compatibility, however, the number of transforms is simply the sum of the number of inputs and outputs (12 transforms instead of 32 in this example).
(4) New input sources can be added to the system without requiring the addition of new corresponding output transforms.

Input compatibility is easily achieved in imaging systems that are limited to particular subsets of input types. However, in systems where the inputs are numerous and disparate, producing input compatibility can become quite difficult. Consider, for example, a color-imaging system that must utilize highly dissimilar sources of input color images, such as photographic negatives and positive transparencies. These two types of input media differ in that the optical densities of negatives increase, while the densities of positive transparencies decrease, as a function of increasing exposure. As a result, image data measured directly from these media is fundamentally different and incompatible. The meaning of measured values alone is ambiguous, i.e., the interpretation of the color associated with a given set of values depends on whether the values were measured from a negative or a positive medium. Appropriate input signal processing is required to create compatibility where it does not inherently exist.

14.3.2. Creating input compatibility

In order to create input compatibility, all fundamental disparities among the various types of inputs must be eliminated by the color-encoding process. To achieve this, it is necessary to identify some color property – a particular aspect of color – that all of the inputs have in common. It is that aspect of color that must then be measured and digitally encoded in order to specify completely and unambiguously the color of each encoded pixel. Some measurement and encoding options include:

- *Densitometry* – measurement and encoding according to a defined set of spectral responsivities.
- *Standard CIE colorimetry* – measurement and encoding according to the spectral responsivities of the CIE Standard Observer, using standard colorimetric techniques (CIE 1986).
- *Advanced colorimetry or color appearance* – colorimetric measurement with encoding that includes colorimetric adjustments for certain physical and perceptual factors determined according to perceptual experiments and/or models of the human visual system (Naytani et al. 1990; Hunt 1991b).

In the following sections, these measurement and encoding methods will be examined, and a series of increasingly complex imaging systems will be analyzed. In each example, a determination will be made as to which aspect of color can be measured and encoded to satisfactorily produce input compatibility among the system input sources in the simplest way.

This analysis will provide an opportunity to discuss several important methods commonly used for color encoding images in practical hybrid imaging systems. A knowledge of each of these encoding methods is essential in order to fully understand *Photo CD* System color encoding. In particular, it is necessary to understand the capabilities and limitations of each alternative method in order to understand why the requirement to support input from a highly disparate array of media – including photographic negatives, transparencies, and reflection prints – required the invention of a fundamentally different method of color encoding for the *Photo CD* System.

14.3.3. A single input medium and a single input scanner

In this first example, the system will be restricted to having just one input medium and a single scanner. The system might consist of a particular reflection scanner being used exclusively to scan images from a particular photographic paper. Almost any form of densitometric measurement, such as red, green, and blue (RGB) scanner values, is sufficient here for compatibility because all scanned input data in this system are inherently "input compatible". There are no fundamental disparities in the scanned image data because all data are produced from the same scanner measuring images on the same medium. The meaning of the scanned values therefore is completely unambiguous.

This simple system illustrates an important distinction between the specific aspect of color that is being encoded and the numerical encoding data metric that subsequently will be used to store encoded data. In this example, the color aspect being encoded for compatibility is a particular scanner's densitometric measurements of images on a particular medium. These measured values can be numerically encoded using various data metrics (i.e., numerical units), which can be almost any function of the measured values. The data metric could, for example, include the logarithm or the cube root of the measured values, or the scanned values might instead be converted to CIE colorimetric values. Such conversions may be desirable for other reasons, as will be discussed later; but it is critical to realize that data metric selections and color-space conversions alone can neither create nor destroy input compatibility. If input compatibility does not inherently exist among the inputs, it must be created by the encoding method. *Compatibility can never be created simply by the use of a common encoding data metric because data metric conversions do not create anything new; they simply convert data from one set of numerical units to another.*

14.3.4. A single input medium and multiple input scanners

In this somewhat more complex system, while there is still only one input medium, there are now multiple input scanners having different densitometric characteristics. (This example also would apply to a system having a single scanner with densitometric characteristics that change over time.) Because there again is only one input medium, almost any form of densitometric measurement can be used in this system. Unlike in the previous example, however, input compatibility will not result from such measurements alone. The measured values in this system are not inherently input compatible because they are ambiguous, i.e., the meaning of a set of scanned values would depend on which scanner was used to measure the values (or on when the values were measured by a scanner whose densitometric characteristics change over time). Therefore, input compatibility must be created because it does not inherently exist.

In order to create input compatibility in this example, the input signal processing of each scanner in the system must include an appropriate correction transform, based on a calibration of the particular device. The correction transform primarily would compensate for differences in the spectral responsivities of the particular scanner from those of a defined set of responsivities. The transform would also correct for other measurement differences due to analog circuitry, analog-to-digital conversion, spatial nonuniformities, and other properties of the individual scanner. When each scanner is used with its associated correction transform, each will produce identical values for corresponding pixels of the same input image.

In one practical implementation of this approach, all of the scanners at a major motion picture special effects studio were calibrated to produce values equal to those measured according to a particular ISO densitometric standard. Many color electronic prepress systems also successfully use color encoding based solely on calibrated densitometry. Scanner calibration is all that is required to create input compatibility in these systems because a single input medium is always used. Calibrated densitometric values therefore have a consistent and unambiguous meaning.

In the *Photo CD* System, and in other hybrid color-imaging systems that include multiple types of input media, input transforms based on scanner calibration alone are not sufficient to create input compatibility. However, the *Photo CD* System, as well as all other successful hybrid imaging systems, does incorporate scanner correction transforms as part of its input signal processing. These corrections, which essentially make all system scanners behave identically, will be assumed in all of the examples that follow.

14.3.5. Multiple input reflection print media

In this third example, the system is no longer limited to a single input medium – it now supports input from an assortment of reflection–print media that may have been produced using various printing inks, conventional photographic dyes, thermal-transfer dyes, or other types of colorants. Because the spectral absorptions of the image-forming colorants differ from medium to medium, visually matched colors on the different media are metameric – not spectral – matches. Therefore, encoding based on the densitometric measurement techniques used in the previous examples would not achieve input compatibility for these media. The meaning of a set of calibrated RGB densitometric values alone would be ambiguous in this system because the color associated with those values would depend on which reflection medium was measured.

The most straightforward solution to input compatibility for these input media is based on the one color-related property they have in common: they are all designed to be directly viewed by a human observer. Input compatibility would be achieved, by definition, if visually matched colors on the different input media produced identical encoded values. Therefore, if the illumination conditions for metameric matching of the input reflection print media of this system are specified, encoding based on standard CIE colorimetric measurements would be sufficient to achieve input compatibility among the media for those particular conditions.

Standard colorimetric values can be measured directly by colorimetric scanners, i.e., by scanners having spectral responsivities equivalent to the defined spectral responsivities of the CIE Standard Observer (or to any other set of color-matching functions). Alternatively, measurements from noncolorimetric (RGB) scanners can be transformed to colorimetric values using appropriate input signal processing transforms. Note, however, that a noncolorimetric scanner would require a different transform for each type of input medium having a different set of image-forming colorants.

14.3.6. Multiple reflection print and transparency film inputs

Although the colorants varied from medium to medium in the previous example, all of the input media were of the same basic type. In this fourth example, the system supports two different types of inputs – photographic transparency films and reflection prints. Since both types of media are designed to be viewed directly by a human observer, it might seem logical to again encode according to CIE colorimetric measurements. However, color encoding based on a standard CIE colorimetric specification alone – such as CIE XYZ, CIELAB, or CIELUV (CIE 1986) – will not achieve input compatibility of images input from a combination of reflection prints and transparencies.

The reason for this is that the two media types are designed to produce images intended to be viewed in specific, and very different, environments. Reflection prints are designed to be viewed in an environment in which the illumination of the image is similar in luminance level and chromaticity to the illumination of the rest of the environment. Most transparency media, such as typical 35 mm slide films, are designed to be projected and viewed in a darkened room. Because an observer's perception of color will differ in these two environments (Bartleson and Breneman 1967; Breneman 1977, 1987; Bartleson 1980), reflection print and transparency film media are designed to produce different colorimetric reproductions. The reproduction characteristics of each medium include colorimetric compensations that account for any perceptual effects that will be induced by the medium's respective viewing environment.

In the case of a 35 mm slide film, these colorimetric compensations include: adjustments in reproduced densities in accordance with the observer's general brightness adaptation to the dark-projection environment; adjustments in reproduced luminance contrast in accordance with the observer's lateral brightness adaptation induced by the dark areas surrounding the projected image; and adjustments in reproduced color balance in accordance with the observer's incomplete chromatic adaptation to the chromaticity of the projector illuminant.

A measuring device, such as a colorimetric input scanner, is of course not subject to perceptual effects. As a consequence, its measurements of images on disparate types of positive media, having different compensations in their colorimetric reproductions, will not correspond to the appearances of those images. For example, a projected 35 mm slide image that visually matches a reflection print image would colorimetrically measure as being darker, higher in luminance contrast, and more cyan-blue in color balance than the reflection print. These characteristics would be readily apparent in a reflection print made according to the transparency's measured colorimetry, rather than to its color appearance (see color plate 34).

This example demonstrates that a color-encoding method limited to a standard CIE colorimetric specification is not sufficient to achieve input compatibility among media that have been designed to be viewed in different environments. The media of this example could, however, be encoded in terms of their color appearance by the use of advanced forms of colorimetry that include considerations for perceptual effects. This form of color encoding is discussed further in Section 14.6.2.

14.3.7. Multiple reflection print, transparency, and negative film inputs

While a color-appearance approach to image encoding may be used for reflection prints, photographic transparencies, and other forms of positive images, it is problematic for systems that are required to accept input from both positive and negative imaging media. Because negative media are not designed to be viewed directly by a human observer, their own color appearance is inherently incompatible with the appearance of positive images.

One approach that can be used to achieve compatibility among positive and negative imaging media is to create, from scanned negative data, "virtual" positive images, i.e., computed images having the colorimetry of optical prints that might be made from the input negative. These computed positive images, and images scanned from actual positive media, can then be encoded in terms of color appearance. A problem with this approach is that it generally results in a loss of at least some information that was recorded on the negative. While this loss may be acceptable for certain applications, it would be inappropriate in the *Photo CD* System where negative media are often used expressly for their capability to record an extensive range of color information.

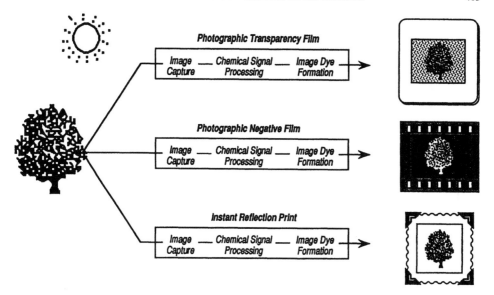

Fig. 14.4. Capture of color stimuli on multiple input media.

14.3.8. The Photo CD System solution to input compatibility

Although each color-encoding method discussed so far may be useful for certain imaging applications, none meets the *Photo CD* System objectives for supporting input from negatives, transparencies, and reflection prints, while preserving the full range of original-scene information recorded on these media. In order to fully meet these objectives, the authors developed a fundamentally different concept of color encoding for the *Photo CD* System (Giorgianni and Madden 1993).

As in each of the previous examples, the approach is based on creating input compatibility by encoding in terms of a particular color-related property that all of the inputs have in common. *What is fundamentally different about Photo CD System encoding, however, is that the color property that provides the basis for this compatibility can be independent of the particular color reproduction characteristics of the input media.*

To understand this concept, consider the photographic transparency, negative, and instant reflection-print input media shown in fig. 14.4. By design, the spectral sensitivities, chemical signal processing, and image-forming-dye spectra are different for these media. So if each were used to photograph an identical color stimulus (or a scene comprised of a collection of individual color stimuli), the resulting reproductions would be significantly different in spectral composition, colorimetric measurement, and color appearance.

From fig. 14.4, however, it can be seen that there is one color-related property that all of the inputs have in common: the spectral power distribution of the stimuli being photographed by each of the media. This point of commonality, then, is the key to the *Photo CD* System method of color encoding. *Photo CD System encoding is based on the colors of original stimuli, not on the colors reproduced from those stimuli on the imaging media being scanned.*

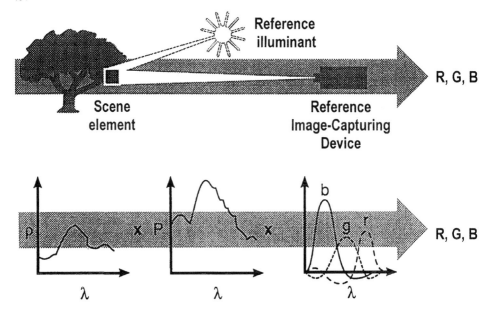

Fig. 14.5. Capture of a color stimulus using a reference image-capturing device.

Encoding according to original color stimuli enables the system to support input from any source. Moreover, it does so by creating virtually complete input compatibility, even from the most disparate types of input media and devices. In addition, this form of encoding allows the color reproduction of the entire imaging chain, from an original scene to the final displayed image, to be optimized for each output device, medium, and viewing environment.

While encoding based on original color stimuli provides a basis for achieving input compatibility and optimum color reproduction, a method is required for efficiently representing the continuous spectral power distributions of those stimuli with just three channels of data. For that purpose, the authors created the concept of a reference image-capturing device (Giorgianni et al. 1993; Eastman Kodak 1991).

Figure 14.5. shows this hypothetical reference image-capturing device (with defined spectral responsivities) and an individual color stimulus (comprised of a scene object having a spectral reflectance or transmittance and a scene illuminant having a spectral power distribution). When the reference image-capturing device captures light from the color stimulus, it produces tristimulus values R, G, and B (Appendix 14.7.1). More will be said later about the characteristics of this reference device, but it is important to note here that its spectral responsivities correspond to a set of color-matching functions. *Thus, reference image-capturing device RGB tristimulus values correspond directly to CIE colorimetry of original-scene color stimuli.*

Figure 14.6 shows the relationship of the reference image-capturing device to actual *Photo CD* System inputs. Color-image signals from the reference device would be directly encoded without additional input signal processing. Signals from any other source of

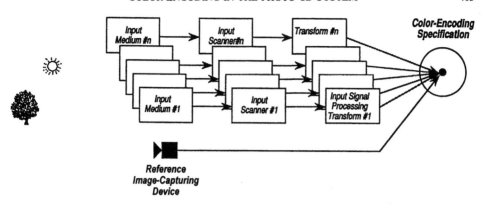

Fig. 14.6. Encoding from multiple input media/scanners and a reference image-capturing device.

input, such as a scanned film, are transformed by the input signal processing to produce the same encoded values as those that would have been produced by the reference device had it captured the same original-scene stimuli. The encoded values for each stimulus of a given scene therefore would be the same regardless of which device or medium was actually used to capture that scene. The signal processing required to achieve this result will be described later in this chapter.

Compatibility is achieved in the *Photo CD* System, then, by the transformation of scanned image data from each input medium to a unique definition and a unique tristimulus encoding. This definition and its numerical description are the foundation of *Photo CD* System color encoding:

- The definition of each *Photo CD* Format image is: *The colorimetry of the original scene that caused the image to form on the input imaging medium being scanned.*
- The tristimulus values that are used to encode this original-scene colorimetry are: *Those values that would have been produced by the reference image-capturing device had it captured the same original scene.*

14.4. Data metrics for color encoding

In the preceding discussion of color-encoding methods, it was shown that the choice of data metric (i.e., numerical units) is unrelated to the creation of input compatibility. Nevertheless, it is often conceptually and operationally convenient to define a single data metric for a system once input compatibility has been achieved.

The particular requirements of a specific color-imaging system often will dictate the most appropriate data metric for that system. For example, in a desktop imaging system that allows the user to directly adjust image data values, a data metric that incorporates familiar perceptual color attributes, such as hue, saturation, and lightness, might be useful. Other systems might have applications that would suggest a different type of data metric. Systems vary widely in their principal uses, productivity and quality requirements, and computational capabilities; so there is no single "best" data metric for all systems.

The *PhotoYCC* Space data metric, which will be described in the following section, was developed by the authors to meet the specific color-encoding and engineering requirements of the *Photo CD* System. A description of its properties will provide an opportunity to discuss several important issues concerning data metrics.

14.4.1. The data metric for the Photo CD System – Kodak PhotoYCC Color Interchange Space

In the *Photo CD* System, there are four requirements directly related to the specification of its data metric:

(1) The metric must be capable of encoding a wide gamut of colors and an extensive dynamic range of luminance information.
(2) The metric must provide for image compression incorporating spatial subsampling.
(3) The metric must encode images such that the effects of digital quantization are virtually undetectable.
(4) The metric must allow the system to produce excellent quality video images on television receivers and computer workstations using minimal digital signal processing.

The specific influence that each of these requirements had on the development of the *PhotoYCC* Space data metric is described in the following sections.

14.4.2. Video requirements

In order for the *Photo CD* System to produce high quality images on existing video systems without requiring any special monitor adjustments, the output video signals produced from *Photo CD* Discs must closely conform to industry standards for video-output devices. Because it is highly desirable to minimize output video signal processing requirements, it would seem logical to encode image data on *Photo CD* Discs directly according to existing video standards. There is a problem with this approach, however, in that it is inconsistent with the color gamut requirements for the *Photo CD* System.

14.4.3. Color gamut

The chromaticity limits of a color gamut associated with an encoding data metric based directly on existing video standards are shown in the CIE u', v' chromaticity diagram of fig. 14.7 (also see color plate 5). The figure shows the chromaticity locations of a typical set of video RGB primaries, as defined by the chromaticities of the red, green, and blue phosphors of a video display. To a first approximation, the gamut of colors that can be displayed using these primaries is indicated by the area of the chromaticity diagram included in the triangle formed by connecting the points corresponding to the chromaticity coordinates of each of the three primaries. (This is only an approximation because the actual gamut is three-dimensional, and the chromaticity limits of the gamut will differ at different relative luminance levels.)

Figure 14.7 also shows the chromaticity limits for colors that can be produced using various sets of photographic image-forming dyes. As shown in the figure, some of the photographic colors are outside the displayable color gamut of the video system. In

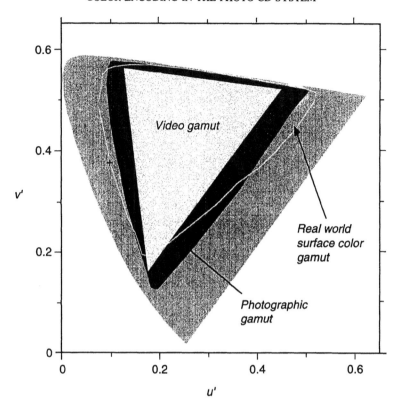

Fig. 14.7. Chromaticity gamut boundaries for CRT phosphors, photographic dyes, and real-world surface colors on CIELUV; also see color plate 5.

particular, the video display gamut does not include a wide range of cyan and cyan-blue colors that are within the photographic color gamut. Hardcopy output images of full photographic quality therefore could not be produced from digital images encoded in terms of a data metric that is restricted to the video color gamut. This is illustrated in color plate 35, where the right image represents the full color gamut of the original photographic image and the left image represents a restriction to the video color gamut. Figure 14.7 also shows chromaticity limits for a collection of colors representing real-world surface colors (Pointer 1980). This collection includes natural colors as well as manufactured colors, such as those produced by printing inks, pigments, dyes, and other forms of colorants. Because many of these colors lie well outside the video display gamut, they also could not be accurately encoded in, or reproduced from, a data metric restricted to the video color gamut. This was an important consideration, because the *Photo CD* Format was intended to be capable of supporting future devices and media having color gamuts greater than those of currently available products.

So there was a conflict. If *Photo CD* Format images were color encoded using a data metric defined strictly according to existing video standards, no further digital signal processing would be required to produce video signals for display. But, as has been

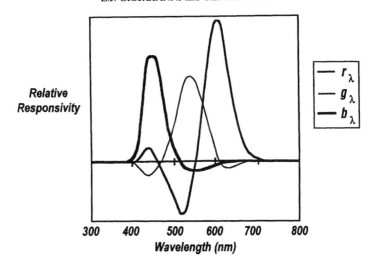

Fig. 14.8. Reference image-capturing device spectral responsivities.

shown, the color gamut of that metric would not include all photographic and real-world colors. On the other hand, if the data metric were defined in terms of a different set of primaries whose color gamut included all colors of interest, a significant amount of additional signal processing would be required in order to transform encoded values to corresponding values appropriate for video displays.

This conflict between output signal processing requirements and color gamut was resolved in the *Photo CD* System by the use of a data metric that is based on the tristimulus values formed by the reference image-capturing device (introduced in Section 14.3.3). In particular, the successful resolution of this conflict resulted from an appropriate specification of the red, green, and blue spectral responsivities for this device. These responsivities, shown in fig. 14.8, were defined to correspond to the color-matching functions for the video primaries specified in CCIR Recommendation 709, "The HDTV Standard for the Studio and for International Programme Exchange" (now ITU-R BT.709) (CCIR 0001). The chromaticities of these primaries are given in Appendix 14.7.2. The advantage of using these particular responsivities is that positive tristimulus values of the reference device correspond to signal values that would be produced by a typical video camera. Therefore, only minimal signal processing is required to produce standard video signals from these positive values.

Unlike an actual video camera, however, the hypothetical reference image-capturing device is also capable of forming negative tristimulus values to represent colors outside the displayable color gamut of a video system. This capability exists because the reference device is a mathematical concept, not an actual instrument. As such, its color channels can have imaginary spectral responsivities with negative sensitivity to light at certain wavelengths (as shown in fig. 14.8). Colors outside the standard video chromaticity gamut therefore produce reference tristimulus values where at least one of the values is negative. For example, the cyan color indicated by the + mark in fig. 14.7 would be represented by values that are positive for green and blue, but negative for red.

Because *Photo CD* System encoding is based on a reference device whose spectral responsivities are derived from video primaries, the *Photo CD* System data metric provides encoded values that can be easily transformed to values appropriate for all video applications. At the same time, because the reference device is capable of forming both positive and negative tristimulus values, the chromaticity limits of its data metric are unrestricted. A remaining consideration influencing an encoding data metric's color-gamut encoding capabilities is its encodable luminance dynamic range.

14.4.4. Luminance dynamic range

Most original scenes contain information that covers a broad range of relative luminance levels. In many cases, a considerable amount of that information occurs at high relative luminance levels, including levels that are above the relative luminance of a perfect white (i.e., a nonfluorescent, diffuse, 100% white reflector). Specular highlights, such as those produced by sunlight reflecting from water or polished surfaces, are one source of such information. Diffuse highlights, such as those produced by some areas of a wedding dress illuminated from above the camera angle, represent another important source of scene information above perfect white. In addition, areas outside the principal area of a scene may be highly illuminated and thus may produce luminance levels above that of a perfect white within the principal area of the scene. For example, a cloudy sky may contain areas having luminance levels well above the luminance of a white object in the principal area of a scene. Fluorescent colors also may produce, at certain wavelengths, light levels greater than those of a perfect white.

Because of the visual importance of this above-white information, photographic materials are designed to record an extensive range of luminance information. For example, when normally exposed, a typical photographic transparency film has the ability to record (and discriminate) RGB exposures up to two times greater than those that would be produced by a perfect white. Photographic negatives have an even greater light-capturing range. In order to produce output images of photographic quality from scanned photographic input media, then, it is necessary to encode values corresponding to an extensive dynamic range of original-scene luminance information. Referring to color plate 36, note the loss of highlight information in the image on the left, which was produced from encoded values limited to existing video specifications for luminance dynamic range compared to the image on the right, which was produced using the greater dynamic range of luminance values of the original photographic image.

The tristimulus values formed by the hypothetical reference image-capturing device were specified to have an unlimited luminance dynamic range in order to retain the extensive range of luminance information recorded in photographic media. When this capability is combined with the capability of forming both positive and negative tristimulus values, the color gamut of the reference device – and thus of the *Photo CD* System color space – becomes unlimited. In practice, the luminance dynamic range that is encodable in digital form depends on the number of bits that are used for encoding. For example, in applications that use 24-bit encoding, the luminance dynamic range extends to twice the relative luminance of a perfect white (Madden and Giorgianni 1993), and the overall encodable dynamic range corresponds to a ratio of luminance values of approximately 350:1.

14.4.5. Digital quantization

Because it was recognized that many *Photo CD* System applications would be limited to 24-bit color (8 bits per channel), digital quantization was a concern. If linear RGB tristimulus values were digitized at eight bits per color channel, visually detectable quantization artifacts would be produced. This is illustrated in fig. 14.9 where the tail of each vector arrow represents the CIELAB $a*$, $b*$ values of a given original color, and the head represents the closest reproduction that can be made from digitally quantized color signals representing that color. The lengths of the vectors are an indication of the magnitude of the color errors resulting from the quantization. The quantization effects are also demonstrated in the left image of color plate 37. (In order to illustrate the effect clearly, the quantization shown in the figure and the color plate was performed at five bits, rather than eight bits, per color channel.)

To minimize quantization artifacts in the *Photo CD* System, linear RGB tristimulus values are first transformed to nonlinear R'G'B' values (see fig. 14.10 and Appendix 14.7.3). This process results in a more perceptually uniform distribution of quantization errors, as shown in fig. 14.11 and in the right image of color plate 37. (Again, for clarity, the quantization was performed at five bits per color channel rather than eight.) Although other non-linear transformations could have been used, this particular set of equations was chosen because the transformation for positive RGB values corresponds to the video camera opto-electronic transfer characteristic defined in CCIR Recommendation 709 (CCIR 0001). The use of this nonlinear transformation enhances the correspondence of *PhotoYCC* Space encoded images to existing video standards and further simplifies the signal processing required to produce video signals. Note that in addition to the CCIR equations, corresponding equations have been provided for negative tristimulus values in order to fully preserve the color gamut capabilities of the data metric.

14.4.6. Image compression

Before images are written to standard *Photo CD* Discs, they are decomposed into a sequence of image components in a hierarchy of resolutions, ranging from 128 lines by 192 pixels to 2048 lines by 3072 pixels (Eastman Kodak 1991). Higher resolution images, such as 4096 lines by 6144 pixels and greater, can be stored on *Photo CD* Discs intended for professional applications. In order to store a practical number of images on a disc (more than 100 images on a standard disc, for example), images are stored in a compressed form.

The compression process includes a technique called *chroma subsampling* (Rabbani 1991). In this technique, which takes advantage of certain perceptual characteristics of the human visual system, luma (achromatic) information is stored at full resolution while chroma (non-achromatic) information is stored at a lower resolution. In general, images produced from full-resolution luma and lower-resolution chroma data will be perceived to have the visual quality of full-resolution color images.

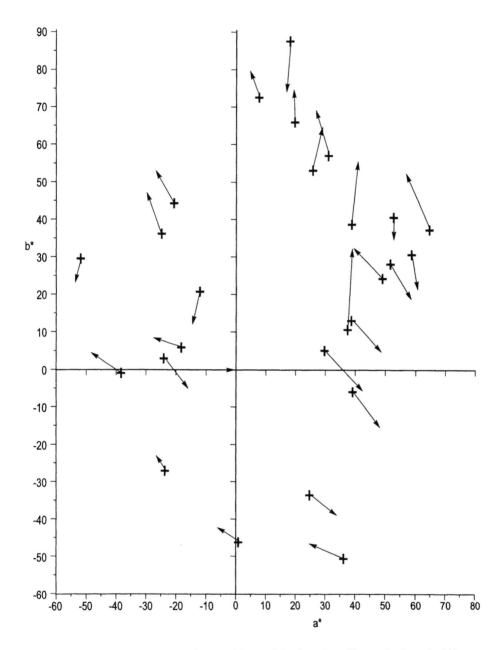

Fig. 14.9. Color errors due to quantization of *linear* tristimulus values (illustrated using only 5 bits per color channel).

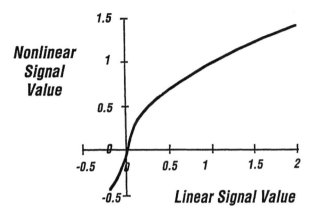

Fig. 14.10. Nonlinear transformation function for reference image-capturing device tristimulus values.

In the *Photo CD* System, the compression process begins with a transformation of the nonlinear R'G'B' values to form separate luma and chroma channels. This transformation is accomplished by the following equations:

$$Luma = 0.299\,R' + 0.587\,G' + 0.114\,B',$$
$$Chroma_1 = -0.299\,R' - 0.587\,G' + 0.886\,B', \tag{14.1}$$
$$Chroma_2 = 0.701\,R' - 0.587\,G' - 0.114\,B'.$$

These equations correspond to those incorporated in video-signal encoder and decoder circuits, and their use also enhances the correspondence of *PhotoYCC* Space encoded images to video standards.

As part of the image compression process, most image components are stored with a non-subsampled luma channel and two chroma channels that have been spatially sub-sampled by a factor of two in the horizontal and vertical directions. Thus, only one value from each of the chroma channels is required for every four values from the luma channel. The subsampled image data are further compressed using Huffman encoding techniques (Rabbani 1991).

14.4.7. Formation of PhotoYCC Space digital values

The final step in the encoding process is the conversion of the luma/chroma signals to digital values. For 24-bit (8-bits per channel) encoding, *PhotoYCC* Space values are formed according to the following equations:

$$Y = (255/1.402)Luma,$$
$$C_1 = 111.40\,Chroma_1 + 156, \tag{14.2}$$
$$C_2 = 135.64\,Chroma_2 + 137.$$

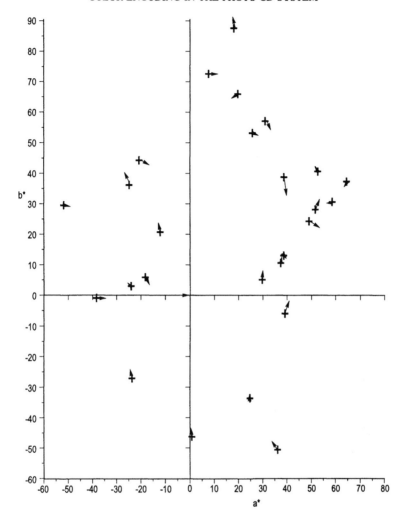

Fig. 14.11. Color errors due to quantization of *nonlinear* tristimulus values (illustrated using only 5 bits per color channel).

The values for the coefficients and constants of these equations were specified such that virtually all surface colors, including those found in nature and those produced by manufactured colorants, can be uniquely encoded. The color-encoding gamut was determined from published work and from additional studies by the authors.

14.4.8. Color encoding implementation

The *PhotoYCC* Space digital color encoding that has been described is implemented in the *Kodak Photo CD* System in the following steps:

(1) A photographic image is scanned, and RGB transmission (or reflection) densitometric values are determined for each pixel;

(2) The scanned densitometric RGB values are corrected, according to individual scanner calibration transforms, to correspond to those values that would have been measured by a defined reference scanner;

(3) The corrected densitometric values are used to determine the RGB exposures recorded by the photographic input medium;

(4) The photographic RGB exposures are (optionally) color corrected for overall level and color balance;

(5) The (color corrected) photographic RGB exposures are transformed to equivalent RGB tristimulus values for the reference image-capturing device;

(6) The reference image-capturing device RGB tristimulus values are transformed to nonlinear $R'G'B'$ values;

(7) The nonlinear $R'G'B'$ values are transformed to $Luma$, $Chroma_1$, and $Chroma_2$ values; and

(8) The $Luma$, $Chroma_1$, and $Chroma_2$ values are scaled and converted to Y, C_1, and C_2 $PhotoYCC$ Space digital values.

14.5. Output signal processing

Color values encoded in terms of a Color-Encoding Specification, such as $PhotoYCC$ Color Interchange Space, must be converted to output-device code values suitable for forming an output image for display. These code values control drive signals that in turn control the amounts of the color-forming elements, such as the dyes of a thermal printer or the luminous phosphors of a CRT, that will make up the final image. The conversion of encoded color values to output code values is accomplished by an output signal processing transform. In multiple-output systems, each output device requires a different output transform (refer again to fig. 14.3).

Output signal processing performs three functions: colorimetric transformation, gamut adjustment, and output code value formation. A general description of these functions and their implementation in $Photo\ CD$ System output signal processing is given in the following sections.

14.5.1. Colorimetric transformation

The first function of output signal processing, required by all methods of color encoding, is to transform encoded colorimetric values to values that are appropriate for the selected output. Output colorimetric values must be adjusted to compensate for alterations in an observer's color perception caused by the differing conditions of the environments in which output images may be viewed, and they must be further adjusted to account for various physical factors that may differ in those environments.

In $Photo\ CD$ System output signal processing, original-scene colorimetric values encoded in $PhotoYCC$ Color Interchange Space are transformed to colorimetric values that are appropriate for reproduced images viewed in the environment associated with the selected output. This transformation compensates for changes in perceived luminance contrast that may be induced by the relative luminances of the areas surrounding the viewed images, for the influence on perceived luminance contrast and chroma due to the absolute level of viewing illumination, and for the observer's state of chromatic adaptation in the environment that will be used for viewing images produced by the selected output. In addition, this transformation accounts for the effects of flare (stray light) that is anticipated to be present in the particular output viewing environment.

In some applications, the colorimetric transformation may include deliberate alterations to the original-scene colorimetric values in order to produce output images having specific color-appearance characteristics. The color-appearance characteristics could, for example, correspond to those of a particular imaging medium, device, or system. Other alterations, such as desired hue rotations, chroma increases or decreases, etc., may also be included in the colorimetric transformation.

14.5.2. Gamut adjustment

The second function of output signal processing is to account for specified output colorimetric values that are not physically reproducible by the selected output device or medium. This function, sometimes called *gamut mapping,* replaces out-of-gamut colorimetric values with substitute values that are attainable by the output. The criteria used in performing these substitutions may vary according to the application for which the output is used. For example, the mapping may be based on the shortest three-dimensional distance in color space from the out-of-gamut color values to the gamut boundary of the output. Alternatively, the mapping may be based on a criterion that places a greater emphasis on maintaining a particular color aspect, such as hue, of the out-of-gamut color.

Color gamut adjustments are particularly important in *Photo CD* System output signal processing because the range of color information that can be encoded in *PhotoYCC* Color Interchange Space is greater than that displayable by any existing output device or medium. It is particularly critical to adjust the extensive luminance dynamic range of *PhotoYCC* Space image data to a range that is appropriate for a particular output. In *Kodak Photo CD* Players, for example, *PhotoYCC* Space values are transformed to produce standard analog video signals consistent with the luminance dynamic range of conventional television receivers.

For *Photo CD* System digital video applications, appropriate digital processing must be used to produce video code values that are consistent with a displayable range of relative luminance values. For example, the following equations, which are based on CCIR Recommendation 601-1 "Encoding Parameters of Digital Television for Studios" (now Recommendation ITU-R BT.601-1) (CCIR 0002), may be used to convert 8-bit-per-channel *PhotoYCC* Space values to RGB video 8-bit code values:

$$\left. \begin{aligned} Y_{video} &= 1.3584\,Y, \\ C_{1\,video} &= 2.2179(C_1 - 156), \\ C_{2\,video} &= 1.8215(C_2 - 137), \end{aligned} \right\} \tag{14.3}$$

$$\left. \begin{aligned} R_{video} &= Y_{video} + C_{2\,video}, \\ G_{video} &= Y_{video} - 0.194\,C_{1\,video} - 0.509\,C_{2\,video}, \\ B_{video} &= Y_{video} + C_{1\,video}. \end{aligned} \right\} \tag{14.4}$$

TABLE 14.1

% scene white	Y	R_{video}	G_{vide}	B_{video}	R'_{video}	G'_{video}	B'_{video}
1	8	11	11	11	13	13	13
2	16	22	22	22	23	23	23
5	34	46	46	46	47	47	47
10	53	72	72	72	71	71	71
15	67	91	91	91	88	88	88
20	79	107	107	107	102	102	102
30	98	134	134	134	126	126	126
40	114	156	156	156	145	145	145
50	128	175	175	175	161	161	161
60	141	192	192	192	176	176	176
70	152	207	207	207	188	188	188
80	163	221	221	221	201	201	201
90	173	235	235	235	213	213	213
100	182	247	247	247	223	223	223
107	188	255	255	255	229	229	229
120	199	(271)	(271)	(271)	240	240	240
140	215	(292)	(292)	(292)	249	249	249
160	229	(311)	(311)	(311)	253	253	253
180	243	(330)	(330)	(330)	254	254	254
200	255	(347)	(347)	(347)	255	255	255

A scale factor of 1.3584 is included in these equations so that a 90% reflectance white in an original scene (encoded in *PhotoYCC* Color Interchange Space as $Y = 173$, $C_1 = 156$, $C_2 = 137$) is transformed to RGB_{video} code values of 235, 235, 235, in accordance with current digital-video specifications. Table 14.1 shows the 8-bit digital values of the Y channel of *PhotoYCC* Color Interchange Space and the corresponding RGB_{video} code values obtained using these equations for a neutral scale (represented by a series of percent-scene-white values).

The RGB_{video} code values produced using these equations correspond to the full dynamic range of luminance values that can be encoded in terms of *PhotoYCC* Color Interchange Space at 8-bits per channel. As can be seen from the table, luminance information above 107% scene white produces code values greater than 255. If these RGB_{video} digital code values were used directly by a 24-bit video display device, any information corresponding to those values would be lost, because all values greater than 255 would be clipped, i.e., set equal to 255. To prevent this, it is necessary to transform either the *PhotoYCC* Space values or the corresponding RGB_{video} code values, such that new $R'G'B'_{video}$ values are formed that are within the range of the display code values. The transformation must be nonlinear in order to prevent an unacceptable shift in the overall brightness of the image. In many software applications, this transformation is implemented using one-dimensional lookup tables. For example, the $R'G'B'_{video}$ code values listed in table 14.1 represent the result of applying the lookup table shown in fig. 14.12 to the original RGB_{video} code values. Note that the full luminance dynamic range of the

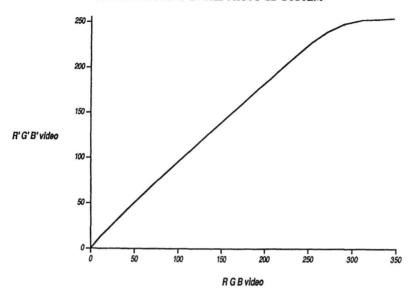

Fig. 14.12. Example lookup table for digital video display of RGB$_{video}$ values derived from *PhotoYCC* space values.

8-bits-per-channel *PhotoYCC* Space encoding, which extends to 200% of scene white, now produces R′G′B′$_{video}$ code values that are no greater than 255.

14.5.3. Output drive signals

The third function of output signal processing is to derive output-device code values that will produce, on the selected output device or medium, output colors having the colorimetric values determined from the colorimetric transformation and gamut-adjustment functions. This final function consists of two steps: output *characterization* and output *calibration*.

In *Photo CD* System applications, a characterization transform is developed from data obtained by measuring the colorimetric values of a large number of color patches, produced from an array of device code signals on a reference device that is representative of the selected output device. The characterization transform is then used to relate desired output colorimetric values to the reference output device code values required to produce those values. To account for individual output device differences from the reference device, calibration transforms are derived using a much smaller set of color patches. In most applications, the characterization transform remains constant while the calibration transform is updated periodically to account for changes that may occur in the output device over time.

14.6. Discussion

This chapter has described how the color-encoding objectives of the *Photo CD* System were achieved by the invention of a unique encoding method, based on the concept of a reference image-capturing device, and by the development of a new data metric, *Kodak PhotoYCC* Color Interchange Space. It is sometimes assumed that such color-encoding objectives can similarly be achieved by methods based on what is termed "device-independent color" and by the use of the metrics of standard colorimetry. These assumptions, and some concluding thoughts on color encoding and color management, are discussed in the following sections.

14.6.1. Device-independent color

"Device-independent color" generally refers to encoding that is not based on device-specific units, such as RGB light intensities or CMYK dye amounts. The term most commonly refers to the use of standard colorimetric methods and standard metrics such as CIE XYZ, CIE u'v', CIELAB, and CIELUV. Perhaps because this form of encoding will work in certain restricted cases, as in the example described in Section 14.3.5, several common misconceptions, or "myths", regarding device-independent color persist:

- *Myth 1 – Device-independent color encoding allows the input of images from multiple types of media and devices.* In reality, the successful input of images from disparate sources requires creating input compatibility by encoding an aspect of color that is common among all inputs. The encoding of standard colorimetric values alone cannot create compatibility among most types of input media and devices, as was explained in Sections 14.3.6 and 14.3.7.
- *Myth 2 – A device-independent color metric is necessary to provide common image data for interchange among imaging systems.* As was shown in this chapter, colorimetric measurements may provide a basis for color encoding in certain applications. However, it is not always desirable, and it is never actually necessary, to express encoded values derived from those measurements in terms of standard colorimetric units. In many applications, it is more practical to express colorimetric values in other terms, such as physically realizable device primaries (rather than CIE primaries), color-signal values for a reference output device, or tristimulus values for a reference input device (as in *PhotoYCC* Space).
- *Myth 3 – CIE standard colorimetric values are device independent and thus can be used to specify colorimetric values for output on multiple media and devices.* As explained in Section 14.5.1, encoded colorimetric values must always be modified for output to different devices, media, and viewing environments. Use of standard CIE colorimetric values does not eliminate this requirement.
- *Myth 4 – Only standard colorimetry metrics are device independent.* In the opinion of the authors, the true test of the device independence of a metric is its encoding capabilities, not its conformance to existing standards. An alternative definition is suggested: *A metric is device independent if it is not restricted to either the luminance dynamic range or the color gamut achievable by physically realizable devices.*
  While this alternative definition includes standard colorimetry metrics, it also includes metrics that may be more practical to implement in actual imaging applications. *PhotoYCC* Color Interchange Space, for example, is based on the primaries of a physically realizable display device. Yet the metric is colorimetrically rigorous and can encode an essentially unlimited color gamut and luminance dynamic range.

At the same time, however, it provides color values that can be used with minimal signal processing on practical output devices.

- *Myth 5 – CIE colorimetric values describe the appearance of color.* CIE colorimetry is designed for quantifying the trichromatic characteristics of color stimuli. Standard CIE colorimetry alone does not, however, specify the color appearance of those stimuli. A specification of color appearance requires advanced forms of colorimetry that additionally account for adaptation and other perceptual factors.

14.6.2. Color-appearance colorimetry and encoding

The previous section, and the example systems described in this chapter, have shown that color encoding based on a standard CIE colorimetric specification alone is not sufficient for use with media that have been designed for different viewing environments. More generally, standard colorimetric encoding is not sufficient for use whenever observer perceptions are altered by adaptation. Adaptive effects may be induced by the environments in which images are viewed, and they may also result from adaptation to the displayed images themselves (Bartleson and Breneman 1967; Breneman 1977, 1987; Bartleson 1980).

There has been considerable research on these forms of visual adaptation. By using controlled conditions of illumination and observation, it has been possible to characterize perceived color shifts due to adaptation effects and to express such shifts in colorimetric terms. Properly applied, this information can be used to determine colorimetric values corresponding to appearance matches between different media and different viewing environments (Breneman 1987; Naytani et al. 1990; Hunt 1991a). Advanced forms of colorimetry that include colorimetric adjustments for perceptual factors therefore can be used to color encode images in terms of their color appearance under defined sets of viewing conditions. While this method of color encoding can meet most of the *Photo CD* System objectives, the transformations involved can be quite complex. Therefore this method of color encoding is best suited for imaging systems equipped with fairly comprehensive color-management support. As described in the next section, *Photo CD* Format images can be used on color-managed systems based on color appearance as well as on other methods of color encoding.

14.6.3. Photo CD Format images and color management

Photo CD Format images can be used in a number of ways in imaging systems that incorporate *color-management*, i.e., systems that employ appropriate hardware and software to adjust and control color. Because *PhotoYCC* Space values represent original-scene colorimetry, *Photo CD* Format images are particularly useful in color-managed applications, such as product advertising, where the accurate reproduction of original-scene colors is a principal objective. However, *PhotoYCC* Space values can also be transformed in color-managed systems to provide other forms of color information. For example, encoded *PhotoYCC* Space data can be used together with documentation information contained in each *Photo CD* Format image file header to determine colorimetric values for each pixel of the scanned input medium. These values can then be transformed to input-color-appearance values for use in color-management systems based on this form of encoding.

Color management can also be used to transform *PhotoYCC* Space values to values corresponding to those from other image sources for use in more specialized color-managed imaging systems. For example, a *Photo CD* Format image can be transformed to the densitometric or colorimetric values of a scanned photographic transparency film for use in a transparency-based electronic prepress system. Note that because of the input compatibility created by the *Photo CD* System encoding method, the transformation that would accomplish this can be applied to any *Photo CD* Format image, including an image that may not have originated on a photographic transparency film.

14.6.4. Other applications of Kodak PhotoYCC Color Interchange Space

Although the *PhotoYCC* space data metric was developed expressly for the *Photo CD* System, its features make it an attractive choice as a data metric for the interchange of images and other forms of color information among imaging systems. Colorimetric values, either measured from an image or otherwise specified, can be interchanged in terms of *PhotoYCC* Space values by first transforming from standard colorimetry to color-appearance colorimetry, and by then further transforming to the original-scene space of *Kodak PhotoYCC* Color Interchange Space. This transformation process is essentially the inverse of the output colorimetric transformation process described in Section 14.5.1. Images and colors encoded for interchange using this technique are encoded in terms of their *own* color appearance. However, these encoded values still are compatible with standard *Photo CD* Format images, and they can be used without special treatment in all *Photo CD* System applications.

14.6.5. Summary

In this chapter, the authors have described various color-encoding techniques for hybrid color-imaging systems. It was shown how the *Photo CD* System encoding method and its *PhotoYCC* Space data metric resulted from an analysis of the specific input, output, and image storage requirements of the *Kodak Photo CD* System. It was also shown that while the *PhotoYCC* Space data metric provides a rigorous colorimetric definition for the *Photo CD* System, the color-encoding method itself is based on encoding concepts that go well beyond a standard colorimetric color specification. The unique properties of this encoding method allow *Photo CD* Format images to be used in a wide variety of applications, ranging from simple video playback devices to complex color-managed systems.

It is hoped that this chapter has served to further the understanding of digital color encoding in general and of the color encoding of the *Photo CD* System in particular.

14.7. Appendixes

14.7.1. Calculation of reference image-capturing device tristimulus values

Reference tristimulus values for an original-scene object are computed using the following equations:

$$R = k_r \sum_{\lambda=380}^{780} S_\lambda R_\lambda \bar{r}_\lambda,$$

$$G = k_g \sum_{\lambda=380}^{780} S_\lambda R_\lambda \bar{g}_\lambda, \tag{14.5}$$

$$B = k_b \sum_{\lambda=380}^{780} S_\lambda R_\lambda \bar{b}_\lambda,$$

where R, G, and B are red, green, and blue tristimulus values; S_λ is the spectral power of the illuminant at each wavelength λ; R_λ is the spectral reflectance or transmittance of an object; and $\bar{r}_\lambda$, $\bar{g}_\lambda$, and $\bar{b}_\lambda$ are the color-matching-function spectral responsivities of the reference image-capturing device. Normalizing factors k_r, k_g, and k_b are determined such that R, G, and B tristimulus values of 1.00 will result for a perfect white object.

14.7.2. CCIR-709 primaries

The CIE chromaticity coordinates of the CCIR Recommendation 709 RGB primaries are:

		R	G	B
x	=	0.64	0.30	0.15
y	=	0.33	0.60	0.06
z	=	0.03	0.10	0.79

14.7.3. RGB nonlinear transformations

To minimize quantization artifacts in the *Photo CD* System, linear RGB tristimulus values are transformed to nonlinear R'G'B' values, prior to the formation of YCC values, according to the following equations:

For R, G, $B \geqslant 0.018$:

$$R' = 1.099\, R^{0.45} - 0.099,$$
$$G' = 1.099\, G^{0.45} - 0.099, \tag{14.6}$$
$$B' = 1.099\, B^{0.45} - 0.099.$$

For R, G, $B \leqslant -0.018$:

$$R' = -1.099|R|^{0.45} + 0.099,$$
$$G' = -1.099|G|^{0.45} + 0.099, \tag{14.7}$$
$$B' = -1.099|B|^{0.45} + 0.099.$$

For $-0.018 < R$, G, $B < 0.018$:

$$R' = 4.50\, R,$$
$$G' = 4.50\, G, \tag{14.8}$$
$$B' = 4.50\, B.$$

References

Bartleson, C.J., 1980, Measures of brightness and lightness. Die Farbe **28**(3/6).

Bartleson, C.J. and E.J. Breneman, 1967, Brightness reproduction in the photographic process. Photogr. Sci. Eng. **11**, 254.

Billmeyer Jr., F.W. and M. Saltzman, 1981, Principles of Color Technology (Wiley, New York).

Breneman, E.J., 1977, Perceived saturation in stimuli viewed in light and dark surrounds. J. Opt. Soc. Am. **67**(5), 657.

Breneman, E.J., 1987, Corresponding chromaticities for different states of adaptation to complex visual fields. J. Opt. Soc. Am. A **4**(6), 1115.

CCIR, 0001, Recommendation 709, Basic parameter values for the HDTV standard for the studio and for international programme exchange (now Recommendation ITU-R BT.709).

CCIR, 0002, Recommendation 601-1, Encoding parameters of digital television for studios (now Recommendation ITU-R BT.601-1).

CIE, 1986, Publication 15.2 "Colorimetry", 2nd edn (CIE, Vienna).

DeMarsh, L.E. and E.J. Giorgianni, 1989, Color science for imaging systems. Physics Today, September 1989.

Eastman Kodak, 1991, *Kodak Photo CD* System – A Planning Guide for Developers (Eastman Kodak Company, Rochester, NY).

Evans, R.M., 1948, An Introduction to Color (Wiley, New York).

Evans, R.M., 1959, Eye, Film, and Camera in Color Photography (Wiley, New York).

Giorgianni, E.J. and T.E. Madden, 1993, U.S. Patent No. 5 267 030, Methods and associated apparatus for forming image data metrics which achieve media compatibility for subsequent imaging applications.

Giorgianni, E.J. and T.E. Madden, 1997, Digital Color Management: Encoding Solutions (Addison-Wesley-Longman, Reading, MA).

Giorgianni, E.J., et al., 1993, Fully Utilizing *Photo CD* Images (Eastman Kodak Company, Rochester, NY) Chs 1, 2, 4.

Hunt, R.W.G., 1991a, Measuring Color, 2nd edn (Ellis Horwood, Chichester, England, and Simon and Schuster, Englewood Cliffs, NJ).

Hunt, R.W.G., 1991b, Revised colour-appearance model for related and unrelated colors. Color Res. Appl. **16**(3), 146–165.

Hunt, R.W.G., 1995, The Reproduction of Colour in Photography, Printing, and Television, 5th edn (Fountain Press, Tolworth, England).

Judd, D.B. and G. Wyszcecki, 1975, Color in Business, Science, and Industry (Wiley, New York).

Madden, T.E. and E.J. Giorgianni, 1993, U.S. Patent No. 5 224 178, Extending dynamic range of stored image database.

Nayatani, Y., et al., 1990, Color-appearance and chromatic-adaptation transform. Color Res. Appl. **15**(4), 210–221.

Pointer, M.R., 1980, The gamut of real surface colors. Color Res. Appl.

Rabbani, M., 1991, Digital Image Compression Techniques (SPIE Optical Engineering Press, Bellingham, Washington).

Stiles, W.S. and G. Wyszcecki, 1967, Color Science (Wiley, New York).

Wright, W.D., 1964, The Measurement of Color, 3rd edn (Hilger and Watts, London).

chapter 15

COLOR DISPLAYS

HEINWIG LANG

Broadcast Television Systems GmBH
Darmstadt, Germany

Color for Science, Art and Technology
K. Nassau (Editor)
© 1998 Elsevier Science B.V. All rights reserved.

CONTENTS

15.1. Introduction 425

15.2. Display systems 425

    15.2.1. Spatio-temporal architecture of the display image 425

    15.2.2. The cathode ray tube color display 429

    15.2.3. Liquid crystal displays 431

    15.2.4. Projection displays 433

    15.2.5. Other display technologies.. 434

15.3. Colorimetric calibration of a display 435

    15.3.1. What is calibration good for? 435

    15.3.2. Determination of display transfer functions 435

    15.3.3. Determination of primary tristimulus values 437

    15.3.4. Stability, uniformity and signal-independence 440

15.4. Colorimetry of displays.. 441

    15.4.1. Device-dependent color spaces for displays 441

    15.4.2. Transformation into device-independent color systems 444

    15.4.3. The color gamut of a display.. 446

    15.4.4. Color discrimination on a display 448

15.5. Applications 449

    15.5.1. Television displays 449

    15.5.2. Displays in design and in reproduction media............................. 450

    15.5.3. Displays in physiological optics and ophthalmology 450

    15.5.4. Displays in color teaching 451

15.6. Appendix

    Derivation of transformation matrices 453

References 455

15.1. Introduction

Color displays are used together with computers in many applications which have nothing to do with color, and where color is used only as an organizing category for the display of information. But color displays are also important tools in all fields of color design, color reproduction, color science and other color technologies. This chapter treats the color display mainly as a tool for the generation of specified colors or color patterns.

In the first section the spatial and temporal structure, pixels, lines and their relation to spatial and temporal resolution are treated. A description of the most important types of color display follows, concluding with a perspective to future display technologies. Section 15.3 is dedicated to the problems connected with display calibration and Section 15.4 to the specification of the colors on a display in different color systems. Section 15.5 finally deals with applications of color displays and describes problems characteristic for these applications (Sproson 1983; Travis 1991; Lang 1995).

15.2. Display systems

15.2.1. Spatio-temporal architecture of the display image

The subject of this chapter is the electronic color display, in which an electronic video signal is transformed into a two-dimensional image. The two main applications of this type of displays are television and computer displays. This article will deal mainly with computer displays, but not exclusively. Together with a keyboard and a mouse, the display represents an important element of the interface between computer and the human user.

Most displays have an *aspect ratio* (ratio of image width to image height) of 4 : 3, modern TV displays have an aspect ratio of 16 : 9.

In computer displays the image is composed of *pixels*. The pixel (derived from 'picture element') is the smallest element with a defined color and/or brightness in a display image. The resolution of a display is defined by the number of pixels in a horizontal line and the number of pixels in a vertical row (or the number of lines). The standard resolution of the VGA (Video Graphic Adapter) graphic standard for personal computers is 640×480 pixels. This means that on a display with an aspect ratio of 4 : 3 the pixels form a square grid, because $640 : 480 = 4 : 3$.

In analogue television systems the number of lines and the bandwidth of the video signal are the system parameters defining the resolution of the image. In digital television systems the number of lines and the number of sampling points along a line are the corresponding parameters.

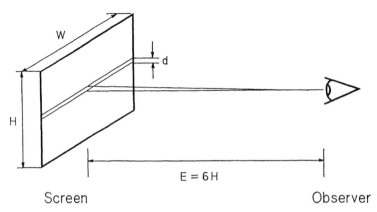

Fig. 15.1. Derivation of the relations between picture height H, picture width W, viewing distance E and resolution; d is the distance between two adjacent lines.

The dimensions of the screen are generally characterized by the length of the diagonal of the screen in inches. In the case of an *aspect ratio* of 4 : 3 the height (H) and width (W) of the screen can be calculated from the diagonal (D) using simple geometry to obtain:

$$H = 3D/5, \qquad W = 4D/5. \tag{15.1}$$

In the case of a screen diagonal with $D = 14$ inch $= 356$ mm we have $H = 8.4$ inch $= 213$ mm and $W = 11.2$ inch $= 284$ mm. The distance between the centers of two adjacent pixels in the case of the VGA standard of 640×480 pixels is 0.44 mm.

Generally, natural images are displayed in a television display and therefore the line structure should not be visible. Human *visual acuity* is defined by the angle between two points or lines, which are just noticeable as being different. For normal daylight illumination the visual acuity is about 1 minute of arc or $1/60$ degree. As the proposed *viewing distance* (distance between television screen and observer) is $6\times$ picture height, the number of lines of the television system necessary to make the line structure invisible can be calculated (fig. 15.1). The height of the screen is seen at an angle of $\mathrm{arctg}(1/6) = 9.5$ deg. If the lines shall appear under $1/60$ deg to the observer, the number of lines per picture height is $9.5/(1/60) = 570$, which corresponds roughly to the number of active lines of the television systems. In the USA NTSC-System (National Television Society Council) the number of active lines is 485, in the EBU-System (European Broadcasting Union) there are 575 active lines.

For a computer display the viewing distance is normally 2 or 3 times the picture height. It is therefore seen at 27 or 18 deg. If the pixel structure is expected to be invisible, the number of pixels per picture height should exceed 1620 in the case of a viewing distance of $2\times$ picture height or 1080 for $3\times$ picture height. In most cases however computer images are not expected to imitate natural images with continuous transitions between different colors. They have to display information and it is therefore not necessary that the pixel structure be invisible. Nevertheless, even in the case of displaying information in the form of diagram or geometrical forms, the visibility of the pixel structure can be disturbing, especially along oblique or curved lines (Chen and Hasegawa 1992).

The *color* on the display is generated by the additive mixture of three primaries (see Section 1.6.1 of Chapter 1). This type of color reproduction is totally different from the subtractive mixture used in photographic reproduction or the autotypic mixture applied in printing (see Section 13.2 of Chapter 13). Whereas in photography and printing the colors cyan (C), magenta (M) and yellow (Y) are used as primary colors (in printing together with black (K) as a fourth primary color), additive mixing requires red (R), green (G) and blue (B) as primaries, if the mixtures of these primaries shall cover a large gamut. The color of a pixel is therefore defined by the intensities of these primaries. These intensity values are called the *Display tristimulus values R, G* and *B*.

Correspondingly, the *video signal* generated in the computer to drive the display must consist of three components E_R, E_G and E_B. The relation between the video signal and the corresponding tristimulus value of the pixel is nonlinear and is described by a function called *transfer function* of the display (see Sections 15.4.1 and 15.4.2 below).

The video signal contains the information for the tristimulus values of the pixels in successive temporal order. The first pixel is at the top left, then the pixels of the first line follow, and the last is at the right end of the bottom line. The *frame period* is the length in milliseconds of the video signal containing the informations of all lines or one total frame. The inverse of the frame period is the *frame rate* and is measured in Hertz, Hz. In the case of progressive scanning (see below) the frame rate is identical with the *vertical frequency*. The *horizontal frequency* is the number of lines displayed in one second, measured in kHz, and its inverse is the line period. In the USA NTSC television system the frame rate is 30 Hz and the horizontal or line frequency is 15 750 Hz (EBU-System: 25 Hz and 15 650 Hz). The number of line periods per frame period is hence 525 (625). The number of active or visible lines is however only 485 (575) as already mentioned. The difference of about 3 msec is needed for the electron beam in the cathode ray tube to jump from the lower right to the top left corner of the screen. For the same reason a certain fraction of the line period is needed for the beam to jump from the left end of one line to the beginning at the right end of the following line. These times are called vertical and horizontal *blanking periods*. These blanking periods are also needed for the synchronization pulses contained in the video signal, which indicate the beginning of a new line or a new frame.

Modern *HDTV systems* (High Definition Television) are designed for large screen displays with an aspect ratio of 16 : 9. HDTV systems based on a frame period of 30 Hz used in the USA and Japan have 1125 lines per frame, whereas the European HDTV system based on a 25 Hz frame period has 1250 lines per frame. For television systems these parameters are fixed and a television monitor can normally only display pictures in one mode. HDTV images cannot be displayed on normal television monitors. The socalled *multisync computer displays* are however able to display images in different scanning modes with different vertical and horizontal frequencies.

The generation of the image on the displays now in use follows either the *flying spot principle* or the *light valve principle*. Examples for flying spot type displays are the cathode ray tube (CRT, see Section 15.2.2 below) and the laser display. In these displays

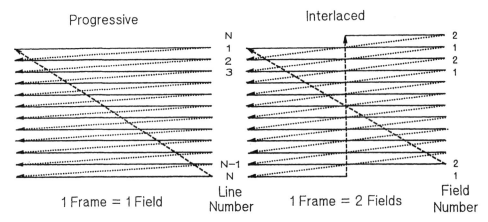

Fig. 15.2. Progressive and interlaced scanning modes.

the pixels are switched on one after the other, but are dark most of the time. In a laser display the laser beam writes pixel after pixel and line after line on the screen and, in a certain moment, only one pixel is bright. On the CRT screen the electron beam stimulates pixel after pixel. As the stimulated phosphor has a certain after-glow, at a given moment not only one but a number of pixels are bright. The liquid crystal (LC) display is an example of the light valve display type (see Section 15.2.3 below). Here all pixels are permanently bright. The pixels are filters (light valves or switches) with switchable or variable transparency. They are illuminated by a light source and their transparency can be changed once every frame period, but is changed only if the color or brightness of this part of the screen has altered during this period.

The brightness of every pixel of a flying spot-type display is modulated with the vertical frequency. Therefore the vertical frequency must be high enough to be averaged by the human visual system and to give the impression of a constant brightness. If the vertical frequency is not high enough to allow sufficient integration, the image is more or less *flickering*. To make the flickering totally invisible, the vertical frequency must lie above the *flicker fusion frequency*, FFF. This FFF is between 20 Hz for very low and about 80 Hz for very bright lights, it is lower in the center of the visual field than at its periphery, and it is slightly different for different individuals. For a normal display brightness of about 100 Candela/m^2 the FFF has values between 50 and 70 Hz.

Frame rates of 25 Hz as they are used in European and 30 Hz as in most other television systems are far too low to deliver flicker-free images. This was the reason for the introduction of the *interlaced scanning mode* (fig. 15.2). In contrast to the *progressive scanning mode*, where line after line from top to bottom is displayed, in interlaced scanning the lines are packed into two fields which contain only every second line and are displayed successively. The first field contains the odd line numbers and the second field contains the even line numbers. As every frame contains two fields, the field rate is twice the frame rate. The vertical frequency, the number of the periods in which the flying light spot moves from top to bottom in one second, is identical to the field rate and

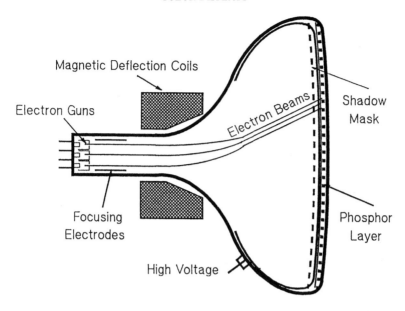

Fig. 15.3. Schematic of a shadow mask picture tube.

the vertical frequency is responsible for the visibility of flicker. Frame rates of 25 and 30 Hz lead in the case of interlaced scanning to vertical frequencies of 50 and 60 Hz, which are above the fusion frequency for moderate brightness values.

A disadvantage of the interlaced scanning mode is the effect of *line flickering*. If the distance between display screen and observer is small enough to render the lines visible, a horizontal border between black and white jumps between two adjacent lines with the frame rate, that is half the vertical frequency. For normal viewing distances in television (6× picture height) line flickering is hardly visible, but for the smaller viewing distances usual for computer displays it is visible and can be very disturbing. Progressive scanning modes are therefore preferred for computer displays.

15.2.2. The cathode ray tube color display

The ancestors of the modern cathode ray color display are the cathode ray tube, CRT which was invented in 1897 by K.F. Braun, and the shadow mask tube, described in principle in a patent of the German engineer, Walter Flechsig in 1938 and technologically developed in the USA between 1950 and 1954 to a mass product serving as the key component of the television receiver. The availability of a low-cost color display was the necessary condition for the successful introduction of color television in 1954.

Figure 15.3 shows a shadow mask tube. The neck of the tube contains three separate sources of electrons, the electron guns. They produce three beams of electrons, which are focused and accelerated by electric fields and deflected by magnetic fields, generated in deflection coils outside the tube neck. The deflecting magnetic fields are modulated in such a way that the point where the electron beams hit the front screen of the tube

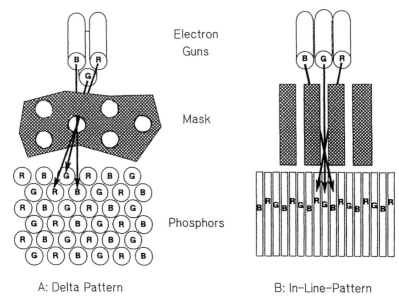

Electron
Guns

Mask

Phosphors

A: Delta Pattern   B: In-Line-Pattern

Fig. 15.4. Arrangement of electron guns, mask and phosphor patches in delta (A) and in-line (B) picture tubes.

describes the line pattern of fig. 15.2. The beams pass through a metal mask before they impinge on a layer of cathodoluminescent materials, the phosphors, covering the inner side of the front screen.

The energy released by the slowing down of the electrons inside the phosphor is changed partly into heat, partly into electromagnetic radiation. This latter effect is called cathodoluminescence. Only a fraction of this radiation is inside the visible part of the spectrum, the rest is emitted as ultraviolet radiation and X-rays. As this radiation is harmful it must be blocked by the front plate of the tube, which contains special materials to absorb this radiation.

The color of the visible radiation created in the phosphor depends on the chemical nature of the phosphor. The phosphor screen of a color display tube consists of a regular pattern of three different phosphors, emitting in the red, green and blue regions of the spectrum (also see Chapter 4, Section 4.8.6, and color plate 3).

The shadow mask is situated a few millimeters in front of the phosphor screen and consists of a regular pattern of holes or slots. These allow only the electrons of one beam to hit the phosphor patches of one color. Figure 15.4(a) shows the effect of the shadow masks in the case of the arrangement of the classical delta gun tube. The electron guns are arranged in the form of a triangle (delta), and the phosphor patches as well. Figure 15.4(b) shows the three electron guns in a linear (in-line) arrangement, and a corresponding mask consisting of stripes.

The spacing between the centers of adjacent slots in the mask is between 0.2 and 0.4 mm and is called *pitch*. The mask slots do not correspond to the pixels, because the electron beams generally pass through more than one slot and stimulate a number of phosphor triplets. Especially in multisync displays, the dot pattern of the mask must be finer than the line and pixel pattern of the graphic mode with the highest resolution to avoid interference (moiré) patterns between both structures.

The alignment between the mask and the phosphor screen is very critical. Misalignment can be caused by the heating effect of the electrons absorbed by the mask during operation. Misalignment leads to *purity errors* caused by electrons of one beam hitting the wrong phosphor patches. Purity errors can also be caused by magnetic build-up on the mask leading to deflection of the beams. This magnetization of the mask can be removed by *degaussing* the mask with an alternating magnetic field. Most high quality CRT displays automatically degauss when the power is on, while some have a separate degauss button.

The phosphors of the shadow mask tube are inorganic crystals in grains of 5–10 microns, forming a layer 10–30 microns thick. The phosphors have to comply with a number of demands. To achieve bright images, phosphors with high luminous efficiency are needed. The emission after stimulation (after-glow) should be not too short to reduce flicker, but short enough to avoid smearing, that means that the after-glow must be sufficiently reduced after one frame period (Compton 1989; Raue et al. 1989). For television screens, the chromaticities of the light emitted by the phosphors is specified (DeMarsh 1974, 1993; SMPTE 1982).

With the *white balance* of the display, the amplifiers driving the three electron guns are adjusted to give electron beams that generate the amounts of red, green and blue adding to a specified white or neutral color in the case of three equal color video signals of maximum voltage (white level). Furthermore the tracking of the three guns has to be controlled to give neutral colors of the same chromatically in the case of equal color video signals of variable levels.

The CRT display is still the first choice for high quality color displays. It has the advantage of being a mass product and the resolution available is as high as with any other display system. Disadvantages of the CRT display are its large volume and weight, especially for large displays, as well as the necessary high voltage (20–30 kV). The shortcomings related to image quality are the sensitivity of the contrast to ambient light, the limitation of the peak luminance (between 100–200 Candela/m^2), the imperfect uniformity of color and brightness, the interdependence of the three electron guns and the flickering of the image at lower vertical frequencies.

15.2.3. Liquid crystal displays

Liquid crystal (LC) displays use materials that have some properties of liquids and some of crystals in a certain temperature domain. They can be cast by pouring and they have a partially ordered molecular structure. The molecular structure influences the optical properties and leads, for example, to double refraction, the property to change the state of polarization of transmitted light. Furthermore the crystal structure and hence the optical properties of liquid crystals can be influenced by electric fields.

For color displays the '*twisted nematic*' (TN) liquid crystals are most common. The molecules in this type of liquid crystals are arranged in layers. In each layer the molecules have a common orientation, and the orientation changes from layer to layer. If a TN crystal is packed between glass plates, the molecules of its top layer are oriented at right

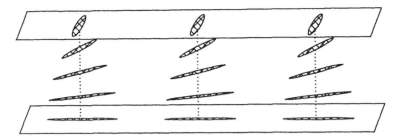

Fig. 15.5. Orientation of molecules in a twisted nematic liquid crystal sandwiched between parallel glass plates.

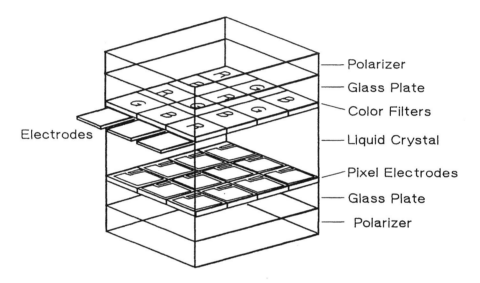

Fig. 15.6. Schematic section through a color LCD display.

angles to the molecules of the bottom layer. The polarization plane of linearly polarized light transmitted through this sandwiched TN-LC is rotated by 90 deg (fig. 15.5). If the glass plates are placed between crossed polarizers the light can pass through this TN-LC cell, while in the case of parallel polarizers it is stopped.

If an electric field is applied to the TN-LC cell, the orientation of the layers and the rotation of the polarization plane of transmitted light is changed. In the case of parallel polarizers the cell becomes transparent, in the case of crossed polarizers it becomes opaque. The transparency of the cell can thus be varied between zero and a maximum value by modulating the electric field (Becker and Neumeier 1992).

A monochrome LC display is composed of a matrix of TN-LC cells, each cell forming one pixel. The electrodes for the generation of the modulating electric fields are transparent and driven by addressable thin film transistors. In the case of a color LC display, one pixel must consist of at least three TN-LD cells with color filters in the primary colors red, green and blue (fig. 15.6). The whole display must be uniformly back-lighted.

An important advantage of the LCD is its flat shape and light weight compared to the CRT, as well as the flicker-free image. Disadvantages are the fact that the contrast of the LCD depends on the viewing angle and optimum contrast is achieved only for zero viewing angle. This is a consequence of the polarizers. Another problem is the slow response of the LC which causes smear in the case of dynamic images. The filters necessary for the color display together with the absorption of the polarizers reduce the luminous efficiency and the maximum luminance of the display.

15.2.4. Projection displays

Projection displays are the only available solution for large area displays, because CRT displays have maximum diagonal of about 1 meter and the largest LCDs are much smaller. In the past a number of different projection devices have been developed, but I shall restrict myself to CRT- and LC-projection-displays, which are most common at present.

The *CRT-projection display* consists of three monochrome CRTs, which have no mask and only one electron gun each, like a black-and-white image tube. Each CRT has a phosphor emitting either red, green or blue light, so that there is one tube for each primary. The three CRTs are driven separately by the signals for the red, green and blue channels. The three images in the three primary colors are projected and superimposed on the screen either by three different projection lenses, or they are superimposed by a system of semi-transparent or dichroic filters and are then projected together by a single projection lens (fig. 15.7).

As the CRTs in a projection display have no mask, they can be much brighter than shadow mask tubes. Typical luminous fluxes coming out of the projection lens are 200–500 Lumen. This results in luminances between 60–150 cd/m^2 on a normal screen, which is similar to values in direct-view CRT displays.

LC displays can be used together with an overhead projector for demonstrations. The light efficiency is not very high in this case because of the necessary mosaic color filter. Therefore *LC projection displays* use three monochrome LC displays for the three colors. They are illuminated by the same light source, with the luminous flux split by dichroic mirrors into three beams of different color (fig. 15.8). For example, the first dichroic mirror reflects the blue part of the spectrum while the green and red parts of the

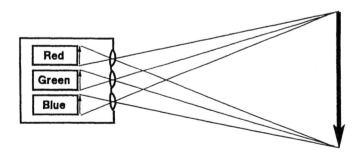

Fig. 15.7. CRT projection system.

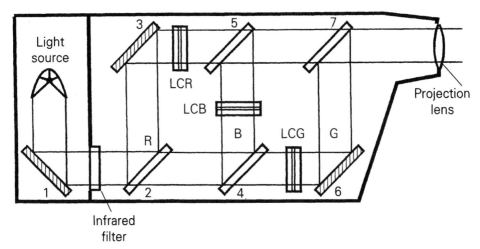

Fig. 15.8. Schematic of a LC light valve projection display; 1, 3, and 6 are fully reflecting mirrors; 2, 4, 5, and 7 are semi-transparent mirrors.

spectrum are transmitted. The second dichroic mirror separates the green and red parts. After the separated beams have passed the three LC cells, they are again superimposed by mirrors and are projected by a single lens system.

The colors of the three primaries of an LC projection display are defined by the spectral transmittances and reflectances of the mirrors and filters used to separate and unite the light beams. They can be modelled by the design of these optical parts. An advantage of the LC projector is that it is flicker-free. A common problem for all projected images is that they are more sensitive to ambient room light than direct-view displays because of the high reflectance of the screen. Less impairment by scattered light can be achieved by rear projection systems.

15.2.5. Other display technologies

Today the color image display market is shared nearly totally by CRT and LC displays. Both technologies have severe disadvantages (large volume of the CRT, insufficient image quality and/or high price of color LCD). The industry is therefore developing alternative technologies for displays. An ideal display should be flat and light weight, allow small as well as large screens, consume less power and need only low voltage, produce pictures with luminance high enough for use in daylight, have highly saturated primaries, emit no radiation that could be harmful, contain only parts and materials which can be recycled and of course should be produced as a low cost mass product.

It seems unrealistic to expect the availability of a display with all these features in the near future. The plasma display (PD) and the electroluminescent (EL) flat panel display (ELD) are two technologies under development, which seem to promise at least important steps in this direction. Both plasma and EL displays are available as monochrome or two color displays, but full color displays with sufficient resolution exist only as prototypes.

The *plasma display* uses a plasma discharge in a gas between transparent electrodes producing UV radiation, which stimulates phosphors emitting red, green and blue light. Flat panel plasma displays with diagonals larger than 75 cm (30 in.) and only a few cm thickness have been reported. Electroluminescent materials, used in the *electroluminescent display*, emit light if they are stimulated by an alternating electric field. A thin film of electroluminescent semiconductor is packed between transparent insulating layers, which are themselves packed between transparent electrodes. Sandwiches of these five layers form the elements of a thin film EL (TFEL) display (Reuber 1993).

15.3. Colorimetric calibration of a display

15.3.1. What is calibration good for?

Calibration of a display establishes a relation between the video signals E_R, E_G, E_B driving the display and the CIE tristimulus values X, Y, Z (see Section 2.2.6 of Chapter 2; also Section 15.4.2) of a corresponding area on the display. It is the necessary prerequisite for a specification of any color on the display in a display-independant color system and for its comparison with colors outside this display.

This relationship can be expressed in the form of a *look-up table*, in *analytical form* or in a combination of both. If each of the three digitized color video signals can take 256 different values in the case of 8 bit/channel quantization, a look-up table would have $256^3 = 16\,777\,216$ entrance addresses, and the same number of color coordinate triplets would have to be stored in the table. It is unrealistic to measure so many colors on a screen, therefore the number of entries in the table must be reduced drastically and values between have to be interpolated. Another disadvantage of the look-up table is that it is valuable only for one state of the display. If for example the brightness control is changed, a different table is needed.

For this reason the relations between color signals and colors on the screen are preferably expressed in analytical form as far as possible. The assessment of these relations consists of two steps: the measurement of the relation between a single color signal and the luminance of the corresponding primary color on the screen, and the colorimetric measurement of the tristimulus values of the primaries. The former relation is called the *transfer function*, TF of that color channel. Transfer functions of CRT as well as of LC displays are essentially nonlinear.

If the transfer functions of the three color channels and the tristimulus values of the primaries are known, the corresponding screen color for each combination of color signals can be calculated using the simple laws of additive mixture, if a number of assumptions about the stability, uniformity and independence of the channels are justified. We will discuss the validity of these assumptions in Section 15.3.3.

15.3.2. Determination of display transfer functions

The physical reason for the nonlinearity of the CRT transfer function is the nonlinear dependence of the electron beam current on the video signal applied to the cathode of the electron gun. In the case of the LC display, it is the nonlinear dependance of the rotation angle of the TNT molecules and the nonlinear relation between this angle and the transmission of the TNT-LC cell between crossed polarizers.

For the CRT display the nonlinearity can be analytically described by an expression of the form

$$L = E^{\gamma}. \tag{15.2}$$

This equation is valid if L is the luminance of the excited primary when no ambient light is reflected from the screen. E is the signal driving this primary if the brightness control is adjusted to give zero luminance $L = 0$ for $E = 0$. Luminance L and signal E are normalized to the maximum values, so that $L = 1$ for $E = 1$. By this normalization the influence of the contrast control is eliminated, which is in fact a gain control. The exponent γ is called the Gamma value of the CRT and has values between 2.2 and 2.5.

For the LCD the relation between color signal and luminance is less simple and we will characterize it by a function $L = f(E)$. This relation obeys the same conditions $L = 0$ for $E = 0$ and $L = 1$ for $E = 1$.

The influence of ambient light scattered back from the screen can be expressed by the addition of a value S, so that $L = (1 - S)f(E) + S$. In this case for video signal $E = 0$ we have $L = S$, the amount of scattered light being visible even if the display is switched off. The variation of the brightness control acts as the addition of a dc signal value offset T to the video signal, which can be expressed in the form $L = f((1 - T)E + T)$.

This results in a luminance $L = f(T)$ for zero video signal $E = 0$. If both modifications are applied, we finally get the relation

$$L = \text{TF}(E) = (1 - S)f\big((1 - T)E + T\big) + S, \tag{15.3}$$

or in the case of a CRT display we can write the transfer function

$$L = \text{TF}(E) = (1 - S)\big((1 - T)E + T\big)^{\gamma} + S. \tag{15.4}$$

If the video signal E_V is measured in millivolts or, in the case of a 8 bit digital signal, by the numbers 0–255, we have $E = E_V/E_W$, where E_W designates the maximum value. The same holds for the luminance L_V measured in Candela/m^2; the normalized luminance L is then given by $L = L_V/L_W$.

Equation (15.3) or (15.4) is a complete description of the relation between video signal and luminance of a separate channel, because it is applicable for all cases of different amounts of scattered light as well as for different settings of the brightness control. We have only to determine the shape of the function $f(E)$.

For a CRT display the determination of the function TF means the determination of the CRT Gamma. This can best be done by a series of measurements of luminance values for different video signal values. If the normalized video signals as well as the measured and normalized luminances are plotted in a diagram with logarithmic scales, the measurement points will lie approximately on a straight line (fig. 15.9). The slope of this line is the gamma value of this channel (Roberts 1991).

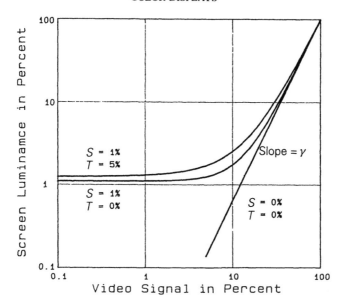

Fig. 15.9. Transfer functions of the CRT display on logarithmic scales; S is scattered light, T is a signal offset.

Two curves with different values of T and S are plotted in fig. 15.9. A certain amount of scattered light causes the curve to become flat at the lower end. There is no differentiation in the shadows on the screen, but details at low luminance are hidden by the scattered light. If the brightness is increased, the flat lower part is raised and produces a higher slope. The aim of the brightness setting is therefore to render the information in the shadows visible in the presence of scattered light.

The transfer function TF of an LC display can generally not be described in a simple analytical form. It has to be measured point by point and the results can be stored in a look-up table. The influence of scattered light S and of a signal offset T can be taken into account in the same manner.

Finally, it should be pointed out that the transfer functions of its three channels should be exactly the same in a high quality display. Differences between the transfer functions result in chromatic shifts in the grey scale. Depending on the quality of the display and on the intended accuracy of the calibration, it may be necessary to apply different transfer functions for the three channels.

15.3.3. Determination of primary tristimulus values

The tristimulus values X, Y, Z of the primaries can be determined by a spectroradiometric measurement of the emission spectra of the three primaries and subsequent calculation of tristimulus values, or by a colorimetric measurement of their tristimulus values or of their chromaticities (Cowan 1983; Brainard 1989; CIE 1990; EBU 1992).

If the method of *spectroradiometric measurement* is applied, the emission spectra of the single channels have to be evaluated throughout the visible spectrum, i.e., at least between 380 nm and 750 nm. The wavelength accuracy of the instrument is important,

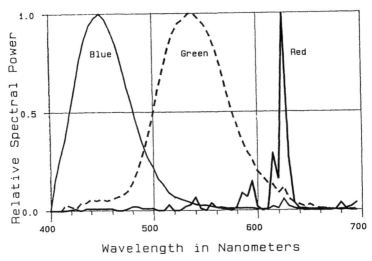

Fig. 15.10. Example for the spectral distribution of light emitted by CRT phosphors.

because even wavelength errors less than 0.5 nm cause noticeable color errors in the calculated tristimulus values (Berns et al. 1993a, 1993b). Figure 15.10 gives examples of emission spectra of the three phosphors of a shadow mask tube. Especially the red phosphor has a spectrum consisting of a number of narrow lines, being typical for the rare earth elements contained in this phosphor to enhance its luminous efficiency. The calculation of the tristimulus values X, Y, Z from the emission spectra is described in Section 2.2.6 of Chapter 2.

The tristimulus values of the primaries can be measured directly with a *colorimeter*. The colorimeter can be a spectroradiometer with built-in software for colorimetric evaluation of the measured spectrum. This kind of instrument has the advantage of delivering spectral information as well. Another class of photometers are wideband instruments including three photoelectric receivers with spectral responses modified by filters, so that they correspond to the standard response curves $\overline{x}(\lambda)$, $\overline{y}(\lambda)$, $\overline{z}(\lambda)$ of the 1931 standard observer. The outputs of the filtered photoelectric receivers are then proportional to the tristimulus values of the radiation falling on the receiver (Kane 1990; Lucassen and Walraven 1990; Laesie 1991; Grambow 1993).

It is useful not only to measure the tristimulus values of the primaries and of the display neutral or white point at the maximum value of the video signal, but also at lower values. The chromaticities x, y of the primaries and of the neutral greys, defined by three equal video signals, must be independent of the video signal level. If the chromaticity of a primary changes with video level, it is probably not really isolated and a certain amount of one or both of the other primaries is present. In this case the chromaticities of the greys change with video level, the display has a grey scale tracking error. The tristimulus values of the primaries are completely described by the chromaticity coordinates of the primaries and of the display neutral or white point if these errors are absent (Cowan and Rowell 1986).

primary/white	R	G	B	D65
chromaticity coord. x:	x_r	x_g	x_b	x_w
chromaticity coord. y:	y_r	y_g	y_b	y_w
tristimulus value X:	X_r	X_g	X_b	X_w
tristimulus value Y:	Y_r	Y_g	Y_b	Y_w
tristimulus value Z:	Z_r	Z_g	Z_b	Z_w

Of course the chromaticity coordinates can be calculated from the tristimulus values according to the relations $x = X/(x + Y + Z)$, $y = Y/(X + Y + Z)$, $z = (1 - x - y)$. From the law of additive color mixture, the following equations between the tristimulus values must hold:

$$X_r + X_g + X_b = X_w,$$
$$Y_r + Y_g + Y_b = Y_w, \tag{15.5}$$
$$Z_r + Z_g + Z_b = Z_w.$$

The X-, Y-, Z-values of the display white can be calculated from the tristimulus values of the primaries. If for any color on the display the display tristimulus values R, G, B are known, and they can be calculated from the video signals if the transfer function is known, the CIE tristimulus values of this color can be calculated according to the rules of additive mixture by the equations:

$$X = RX_r + GX_g + BX_b,$$
$$Y = RY_r + GY_g + BY_b, \tag{15.6}$$
$$Z = RZ_r + GZ_g + BZ_b.$$

It may often happen that only the chromaticity coordinates of the primaries and of the display neutral are known, for example from the display manual, or that a display is adjusted to a different neutral chromaticity. In this case the tristimulus values $X_r, \ldots, Z_b$ can be obtained from the chromaticities. We shall therefore derive the relationship which allows the calculation of the tristimulus values of the primaries starting from the chromaticity coordinates of the primaries and the display neutral.

The tristimulus values are products of the chromaticity coordinates and normalizing factors P_r, P_g, P_b, which we have to determine. We can directly derive the tristimulus values from the chromaticity coordinates for the display white, because the value Y_w, the relative luminance of the display white, $Y_w = 1$. We have therefore the following equations:

$$X_r = P_r x_r, \quad X_g = P_g x_g, \quad X_b = P_b x_b, \quad X_w = x_w/y_w,$$
$$Y_r = P_r y_r, \quad Y_g = P_g y_g, \quad Y_b = P_b y_b, \quad Y_w = 1, \tag{15.7}$$
$$Z_r = P_r z_r, \quad Z_g = P_g z_g, \quad Z_b = P_b z_b, \quad Z_w = z_w/y_w.$$

We can now use the relations (15.5) for the determination of the normalizing factors P_r, P_g, P_b:

$$P_r x_r + P_g x_g + P_b x_b = x_w/y_w,$$
$$P_r y_r + P_g y_g + P_b y_b = 1, \tag{15.8}$$
$$P_r z_r + P_g z_g + P_b z_b = z_w/y_w.$$

These relations have to be solved for the factors P_r, P_g, P_b. The solution can be written in the form

$$P_r = M_{rx} x_w/y_w + M_{ry} + M_{rz} z_w/y_w,$$
$$P_g = M_{gx} x_w/y_w + M_{gy} + M_{gz} z_w/y_w, \tag{15.9}$$
$$P_b = M_{bx} x_w/y_w + M_{by} + M_{bz} z_w/y_w.$$

The derivation of the coefficients $M_{rx}, \ldots, M_{bz}$ from the coefficients $x_r, \ldots, z_b$ is given in the Appendix 15.6, together with an example. With the normalizing factors we can calculate the tristimulus values X_r, Y_r, Z_r, X_g, Y_g, Z_g, X_b, Y_b, Z_b of the normalized primaries according to (15.7).

15.3.4. Stability, uniformity and signal-independence

The aim of the display calibration procedure is to enable the user to predict the color of a pixel from the video signals applied to this pixel. The knowledge of the transfer functions TF of the three channels and of the transformation matrix from the display tristimulus values R, G, B to the CIE tristimulus values X, Y, Z are necessary conditions but they are not sufficient. They would only be sufficient if a number of additional assumptions were true, which are in fact only approximately true. The most important of these assumptions are: temporal stability of the display, the spatial uniformity of the screen, channel independence, and the spatial independence. I will briefly discuss the deviations from these assumptions found in CRT displays. Calibration of LC displays has until now not been described extensively in the literature.

The display has a certain warm-up time after it has been switched on and constant video signals are applied to the three channels. For a CRT display this time may range between 20 minutes and some hours, depending on the display and on the required final stability (Berns et al. 1993a, 1993b). But if the displayed image is changed, the different distribution of electric charge on the mask will lead to a new stabilization process, which may last some minutes.

If three constant video signals are applied to the display, screen luminance and color should be uniform over the whole area (Cook et al. 1993). However, a decrease in luminance towards the edges is inevitable in CRT displays because of the oblique impact of the electron beams. More disturbing is a non-uniformity of chromaticity caused by a purity error (see Section 15.2.2), which can be corrected or at least reduced by degaussing the CRT. In direct-sight LC displays an angular dependence of the luminance is caused by the crossed polarizers.

If the intensity of one primary is influenced by the intensities of the other primaries at the same pixel, this violates the validity of the additivity of color mixing on the screen. In fact, such interdependence of the channels is present in CRT displays, especially in the case of high signal values and high contrast. It can be reduced considerably by a reduction of the contrast or peak luminance.

Spatial independence means that the luminance and color of a pixel is independent of the luminance and color of the pixels in the neighborhood. However, a certain fraction of light generated in the bright parts of the image will be scattered in the CRT or in the lens of a projection display and travel to other parts of the image. Another cause for spatial interdependence may be a decrease in the frequency response towards high frequencies. This will limit the maximum possible modulation between adjacent pixels in the horizontal direction.

The above-mentioned effects are present more or less in all displays and their magnitudes determine the quality of a display. It is very cumbersome to measure them. Most of them are not noticed in normal work with a display, because the human visual system is not very sensitive to slow temporal and low frequency spatial changes of luminance and color, or even compensates, for example, for spatial interdependence by simultaneous contrast. However, for an accurate calibration of a monitor they have to be kept in mind.

15.4. Colorimetry of displays

15.4.1. Device-dependent color spaces for displays

Colors on a display are specified by their video signals E_R, E_G, E_B. Figure 15.11 is a representation of color space as a coordinate system given by the axes of the video signals of the display. In this coordinate system every color is represented by a point with E_R, E_G, E_B as coordinates. It is reasonable to normalize these values to the amounts

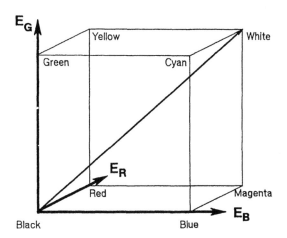

Fig. 15.11. Cube in RGB color signal space containing the colors reproducible on a display.

necessary for forming the display white. Allowed (or 'legal') video signals on the display obey the inequalities $0 \leqslant E_R, E_G, E_B \leqslant 1$. Every color complying with these inequalities is located inside a cube in this color space with side 1. A color outside this cube would have at least one video signal < 0 or > 1 and is therefore not an allowed (or an "illegal") color for this display. The corners of the cube represent the following eight colors:

Color		Primary video signals		
		E_R	E_G	E_B
Black	(S)	0	0	0
Red	(R)	1	0	0
Green	(G)	0	1	0
Blue	(B)	0	0	1
Yellow	(Y)	1	1	0
Cyan	(C)	0	1	1
Magenta	(M)	1	0	1
White	(W)	1	1	1

$$(15.10)$$

The spatial diagonal, the straight line connecting the black and the white corner in fig. 15.11, is characterized by the equality $R = G = B$. The chromaticies of the colors on this line are equal to that of the white point and this line is called the *gray* or *neutral axis*.

We use here the video signals E_R, E_G, E_B as coordinates for the colors on the display and not the tristimulus values R, G, B, because the latter can only be determined after calibration of the transfer function. Using the video signals has, at least in the case of CRT displays, an important advantage. Scaling of the colors by the video signals is more appropriate to color perception than scaling by the tristimulus values. The neutral bars formed by video signals forming a stair case with equal steps appear to be perceptually equidistant, despite the fact that the luminance steps near black are smaller than those near white.

The reason for this is that brightness perception is not proportional to luminance. This is expressed by the psychometric (perceived) brightness function L^*, which is a nonlinear function of the luminance (see Section 2.3.6 of Chapter 2). The nonlinear transfer function of the CRT is approximately the inverse of the psychometric brightness function, with the result that equidistant scaling in video signals produces nearly perceptually equidistant steps in psychometric brightness (fig. 15.12).

Defining colors by their R-G-B-components is unequivocal, but it does not correspond to perceptual color categories. Perceptually we categorize colors according to their hue, brightness and saturation or chroma. If we see a yellow color, we describe its hue by saying that it is a pure, a reddish or a greenish yellow, we describe its brightness in saying that it is more or less light or intense and we finally say that it is saturated or has high chroma or not. But we cannot describe its appearance by telling that it consists of say 70% of red and green and of 30% blue. Several proposals have been made to define *perceptual color spaces* for display colors. Examples are color spaces named HSV (Hue, Saturation, Value) (Travis 1991, p. 79), HSI (Hue, Saturation, Intensity) (Williams

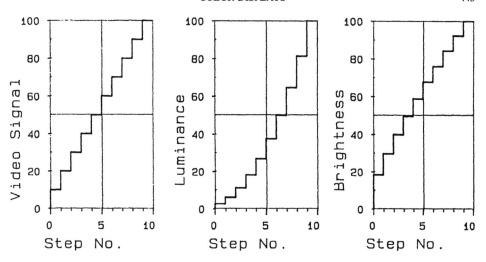

Fig. 15.12. Luminance and psychometric brightness in a grey scale with equal video signal steps.

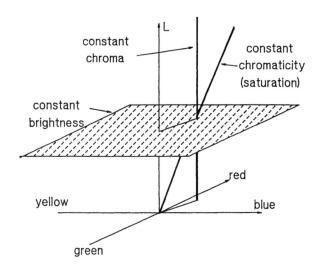

Fig. 15.13. Saturation and chroma in color space.

1988; Welch et al. 1991), HVC (Hue, Value, Chroma) (Gan et al. 1991; Levkowitz and Herman 1993).

Value is a term used in the Munsell color system (see color plates 1 and 2, Chapter 1, Section 1.5.1, and Chapter 2, Section 2.3.3) to designate the lightness of a related color, or its brightness relative to white. The intensity in HSI space is defined as the sum $I = R + G + B$. A pure red and a pure green and a pure blue of maximum intensity therefore have equal intensities, but they obviously have different values of brightnesses, because the red is lighter than the blue and darker than the green.

Saturation and chroma both depend on the distance of a given color to the neutral axis. While chroma is a measure of that distance, saturation is the ratio of chroma to the value or brightness. Figure 15.13 illustrates these relations. In all color spaces colors of the same chromaticity are situated on a line through the origin (black point), and these colors have the same saturation. If for example the illumination of a surface color is decreased, its saturation (and chromaticity) remains constant. However, its chroma is reduced, because it approaches the neutral axis. Colors of the same chroma have the same distance from the neutral axis in perceptual color spaces (Shinoda et al. 1993).

Characterizing colors in perceptual color spaces like the HSV, HSI or HVC spaces have the advantage of being perceptually meaningful. To generate a bright pink on a display is simple if one can start from a neutral of appropriate brightness and then add some chroma with a red hue, but is difficult if one has to find out the appropriate amounts of red, green and blue. As the color on the display is always generated by additive mixture of the primaries R, G, B using a perceptual color space means that the software of the computer has to transform the coordinates H, S, V into R, G, B and vice versa. Such transformations are described in the cited literature.

15.4.2. Transformation into device-independent color systems

The color spaces described in the last section refer to the primaries of the display. But different displays have more or less different primaries. Two colors on two different displays having the same R-, G-, B-video signals will therefore generally look different, unless the primaries of the two displays have the same colors and the neutral points are equal. The RGB color space and all color spaces derived from it are therefore device-dependent. If we want to describe the relation between a color on the display and a color in the world outside that display we need a device-independant color coordinate system.

Device-independent color coordinates are for example the CIE 1931 X, Y, Z tristimulus values. They refer to the *2 degree and 10 degree CIE normal observers*, defined by their spectral tristimulus values (see Section 2.2.6 of Chapter 2 and color plate 4). We restrict ourselves for the rest of this chapter to the CIE 1931 2 degree observer, because color areas on a display are normally seen under less than 10 degrees (see Section 15.2.1). If the primaries of a display can be specified by the CIE tristimulus values and the transfer function of the display is known, it is possible to relate the device-dependant video signals E_R, E_G, E_B to device-independent tristimulus values X, Y, Z (Taylor et al. 1991).

We have seen in Sections 15.3.1 and 15.4.1, that the calibration of the display is the necessary condition for proceeding from a device-dependent to a device-independent color system. The first step is the calculation of the tristimulus values R, G, B from the video signals E_R, E_G, E_B using the transfer function. For the second step we have derived the equations transforming the display tristimulus values R, G, B into the CIE tristimulus values X, Y, Z, as given in eq. (15.6).

If a color is given by its CIE tristimulus values X, Y, Z and we want to produce it on a display, we have first to derive the display tristimulus values R, G, B with the following equation:

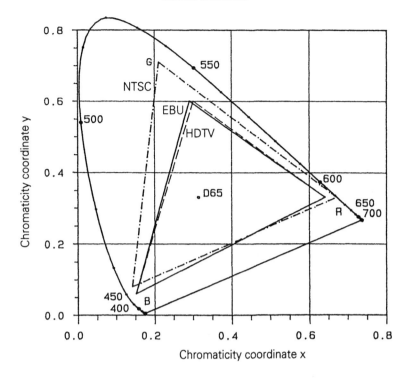

Fig. 15.14. Display primary triangles on the CIE chromaticity diagram of color plate 4; NTSC is the USA, EBU the European, and HDTV the new international TV system.

$$R = R_x X + R_y Y + R_z Z,$$
$$G = G_x X + G_y Y + G_z Z, \qquad\qquad (15.11)$$
$$B = B_x X + B_y Y + B_z Z.$$

The coefficients $R_x, \ldots, R_z$ can be derived by a matrix inversion (see the appendix, Section 15.6). As a second step, the display tristimulus values R, G, B have to be transformed into video signals using the inverse transfer function.

Figure 15.14 shows a CIE chromaticity diagram with triangles formed by three different standards of TV display primaries. They have the following chromaticity coordinates (ITU: International Telecommunication Union):

TV-System	x_r	y_r	x_g	y_g	x_b	y_b	
NTSC	0.670	0.330	0.210	0.710	0.140	0.080	(TV USA)
EBU	0.640	0.330	0.290	0.600	0.150	0.060	(TV Europe)
ITU-R BT 709	0.640	0.330	0.300	0.600	0.150	0.060	(HDTV international)

Let us suppose that two displays with different primaries are seen side by side and the same video signals are displayed on them. We further assume that their transfer functions are the same (which can be supposed in the case of two CRT displays but is rarely the case if one is an LC projection display) and that the white points of all displays are adjusted to the same chromaticity and luminance. Then the neutral colors displayed by the video signals $E_R = E_G = E_B$ look the same. Other colors generated by the same triplets of video signals would be different on the two monitors, especially for high saturation.

If the color generated on one of the displays by a triplet of video signals E_R, E_G, E_B has to be reproduced on another display, the device-independent tristimulus values X, Y, Z must be calculated using the transfer function and transformation matrix $X_r, \ldots, Z_b$ of the first display. As a second step, the tristimulus values R, G, B of the second display must be calculated using the transformation matrix $R_x, \ldots, B_z$ of this display and then the video signals E_R, E_G, E_B can be derived using the inverse transfer function of the second display (see the appendix of Section 15.6).

15.4.3. The color gamut of a display

Can a given color, be it an illuminated or self-luminous surface, be adequately reproduced on a certain display? This question can only be answered in the framework of a device-independent color space. We will use for the description of the area of reproducible colors of a display, its *color gamut*, the 1976 CIELUV color space, which is proposed by the CIE as being approximately perceptually uniform (see Section 2.3.7 of Chapter 2 and color plate 5). This color space has a chromaticity diagram (see color plate 5) with the chromaticity coordinates u', v', derived from the CIE tristimulus values according to the following equations:

$$u' = \frac{4x}{(X + 15Y + 3Z)}, \qquad v' = \frac{9Y}{(X + 15Y + 3Z)}. \qquad (15.12)$$

The CIELUV color space is formed by three coordinates, the psychometric brightness L^* and the two psychometric chroma coordinates u^* and v^*:

$$\begin{aligned}
L^* &= 116(Y/Y_n)^{1/3} - 16 && \text{for} \quad (Y/Y_n) > 0.008856, \\
L^* &= 903.3(Y/Y_n) && \text{for} \quad (Y/Y_n) \leqslant 0.008856, \\
u^* &= 13L^*(u' - u_n'), \\
v^* &= 13L^*(v' - v_n').
\end{aligned} \qquad (15.13)$$

The index 'n' denotes a neutral white reference. The psychometric or perceived lightness L^* of reference white has therefore the value $L_n^* = 100$ and the chroma coordinates vanish for white and all neutral colors: $u_n^* = v_n^* = 0$.

The chromaticities of the primaries of a display form a triangle in the chromaticity diagram. Figure 15.15 shows the $u'-v'$-chromaticity diagram of the CIELUV system. The chromaticities of colors that can be generated on this display lie inside the triangle of the display's primaries. This results from the fact that all colors of the display are generated by additive mixture and that the chromaticities of all colors additively mixed of two primaries lie on a straight line connecting the chromaticities of these mixed primaries.

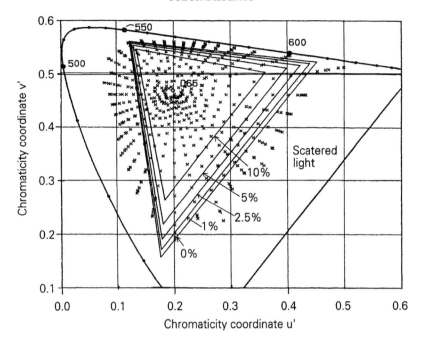

Fig. 15.15. Color gamut of display on the u'–v'-chromaticity diagram of color plate 5. The different triangles represent the color ranges in the presence of different amounts of scattered light; x: pointer colors.

As all physically realizable colors have chromaticities lying inside the spectrum locus in the chromaticity diagram, colors with chromaticities outside the primary triangle and inside the spectrum locus cannot be represented on this display. If the display is expected to have a large gamut, the primaries must be as saturated as possible and be near to the spectrum locus.

The largest primary triangle in fig. 15.15 presupposes a display being able to reproduce an ideal black with tristimulus values $R = G = B = 0$. This would allow an infinite contrast on the display. In practice the influence of ambient light falling on the screen will limit the darkest black to values well above zero. In fig. 15.15 therefore a number of smaller triangles are shown representing reduced color gamuts for different amounts of scattered light reflected from the screen.

The two-dimensional chromaticity diagram is however not sufficient for a complete description of the color gamut of the display. This can only be given in three-dimensional color space. Figure 15.16 therefore shows four different planes of constant L^* in CIELUV color space with the limit of the display gamut. A perspective view of the limits of a display color gamut is shown on color plate 39.

The adequacy of a certain display gamut depends, of course, on the application of the display. For example, in the case of the displays used in television it is desirable to be able to reproduce all real exiting surface colors. Pointer (1980) has published a collection of 16×36 color coordinates, describing real surface colors with the highest chroma values in 16 different L^*-planes. These Pointer colors are shown as crosses in

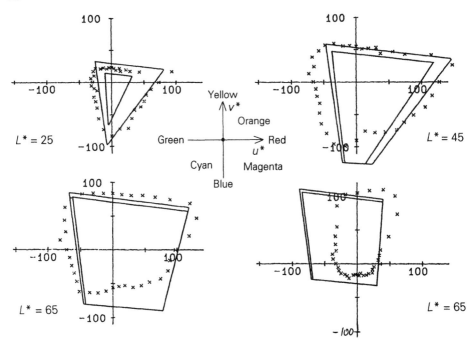

Fig. 15.16. Color gamut of a display in planes of constant brightness L^* in CIELUV color space. The limits of the display color solid are given for two different amounts of scattered light on the screen (0%, outer and 2.5%, inner limit); x: Pointer colors.

figs 15.15 and 15.16. Whereas in fig. 15.16 the green and cyan for low and medium brightness the Pointer colors lie outside the color gamut of the display, in the blue and magenta region the limit of the color gamut lies outside the limits of real surface colors.

15.4.4. Color discrimination on a display

It is often stated that a computer display with 8 bit quantized color video signals can display $(2^8)^3 = 2^{24} = 16\,777\,216$ different colors. This formulation is not correct, because these numbers refer to the possible color signal triplets and not to different colors. Many of the colors generated by different video signals cannot be discriminated by the eye and colors that the eye cannot distinguish are by definition the same color.

In the case of 6 bit quantization the neutral axis from black to white can be divided into 63 steps from 0 to 63. They are, if the eye is adapted to the display brightness and the amount of scattered light is low enough, all discriminable by the eye. The color differences dE between the steps of this grey scale can be calculated according to the CIELUV formula (see Section 2.4.3 of Chapter 2). They fall between $dE = 3.3$ in the shadows and $dE = 1.2$ near white. A color difference of $dE = 1$ is noticeable and above the discrimination threshold. But if a green color with digital video signals $E_R = E_B = 1$, $E_G = 62$ is displayed and the red signal E_R is changed by $+1$ or -1, then the calculated color difference is $dE = 0.5$, which is at or below threshold. On the other hand, the color difference between two colors produced by the two video signal triplets $E_R = E_B = E_G = 1$ and $E_R = E_B = 0$, $E_G = 2$ is $dE = 8$!

In the case of an 8 bit quantization with 255 possible steps per channel, the differences between colors of adjacent video signal triplets is one quarter of those for 6 bit quantization. Most color differences calculated for changes of $+1$ or -1 in an 8 bit system are below threshold, but even then in some parts of color space the differentiation is not high enough to make all quantization steps invisible. From the given numbers it follows that for this end at least a 10 bit quantization is necessary. A consequence that can be drawn from this is that the color space defined by the primary video signals is not perceptually uniform, despite the fact mentioned in Section 15.4.1 that the grey scale of this space has better uniformity than that of the space of tristimulus values.

The question of how many colors a display can produce makes sense only if we ask how many discriminable colors can be produced. This can be answered roughly if we calculate the volume of the three-dimensional color gamut in a uniform color space. If the CIELUV color space were perfectly uniform, we could think of the discriminable colors forming a cubic three-dimensional grid with a distance $dE = 1$ between adjacent colors. The calculated volume would then be equal to the number of discriminable colors inside the display's color gamut (Roger et al. 1992).

For the standard high-definition TV display (HDTV, see Section 15.2.1) the volume of the color gamut in CIELUV space is about 1 400 000. This number can be calculated by determining the areas inside the color gamut in different planes of constant L^* (fig. 15.16) and by adding these slices for all values of L^* between 0 and 100. As the CIELUV color space is only approximately perceptually uniform, and because of the influence of stray light, the calculated number can only be a rough estimate.

15.5. Applications

15.5.1. Television displays

Television displays in the form of home receivers are the most widely distributed color displays. The chromaticities for primaries as well as for the neutral point and the display transfer function are standardized for television displays. While the deviations from these standardized specifications may be considerable in home receivers, television studio monitors used for the estimation of the image quality in studios are quite close to the specifications.

Studio monitors have controls for the adjustment of the black and white balance. This is necessary, because in a control room different monitors must reproduce greys and whites with the same chromaticity. Otherwise it is impossible to decide if the images of two different cameras used in the same production fit together. The standardized neutral chromaticity in television is the chromaticity of the standard light source D65, which simulates daylight with a color temperature of 6500 K (see Section 2.2.3 of Chapter 2).

The input signal for a television home receiver is not a triplet of E_R-, E_B-, E_G-signals, but a coded composite video signal, containing the information of the E_R-, E_B-, E_G-signals. The composite video signal is in the case of analog television systems, modulated on a carrier frequency. The home receiver therefore needs a demodulator and a decoder to obtain the video signals E_R, E_B, E_G necessary to drive the three electron guns of the CRT.

15.5.2. Displays in design and in reproduction media

Typical applications in these fields are soft proofing in printing technology and the drafting of designs for textiles or for architectural structures on a graphic station. A common characteristic of these applications is that on the display screen a color pattern or image is created that has to simulate a pattern or image which is later produced as a print on paper, on textile or on a building. The textile design can, for example, be approved on the monitor without expensive test runs of textile dyeing, or a printing process can be optimized by soft-proofing on a screen.

Simulating different color reproduction processes on a display is difficult in more than one respect. The first problem is that a color on a self-luminous screen has an appearance differing from that of a surface color with the same color coordinates, for example a color printed on paper. If the color on the display simulates a surface color, it has, for example, to be surrounded by an area simulating the white paper, or, in the case of textile design, the texture of the textile surface has to be simulated. A comparison is only possible if the light illuminating the surface color has the same chromaticity as the neutral point of the display.

A second problem is that the color gamut of a display is different from the color gamut, for example, of the printing process. There may be colors on the display which cannot be printed, and print colors that cannot be reproduced on the display. For these cases a mapping of one color gamut into the other has to be defined, which gives the optimal appearance. Often the problem is complicated by the need for a hard copy of the design which can be handed out to the client for approval. The printer has a color gamut which is itself different from the display gamut and from the color gamut of the simulated process.

Gamut mapping is therefore an important issue for this kind of application. Each specific color reproduction process has its own color gamut. If a photographic image has to be displayed on a monitor and is then printed, a relation has to be established between the colors on the photographic image, on the display and in the print. The relationship should preserve the topological relations between neighbouring colors and different colors in one process should not collapse into one color in the other process. This is not a trivial task, especially in the areas of color space where the three color gamuts do not overlap (Lang 1995).

We cannot go into further details of gamut mapping here; the interested reader will find more information in the cited literature.

15.5.3. Displays in physiological optics and ophthalmology

The color display gives the unique possibility to generate a color pattern and to vary its colors along predetermined lines in color space. This makes it a very valuable tool for the investigation of our visual system. In most cases the generation of a desired pattern, and especially its variation, is far easier to realize on a display than, for example, by an arrangement of projection devices.

In some cases however the spatio-temporal structure of the display image (see Section 15.2.1) interferes with the required characteristic of the lest pattern. The generation of grids of high spatial frequency, for example, may lead to a disturbing interference with the pixel structure, known as Moiré pattern or aliasing. These interferences change drastically with the frequency and with the orientation of the grid. Temporal modulation of a stimulus on a display is not possible at frequencies higher than about 20 Hz without interference with the display's frame rate. If the color discrimination is investigated on a display, a 10 bit quantization is needed as a minimum for quantization levels to be invisible in all parts of color space (see Section 15.4.4).

Another possible application is the testing for color vision deficiencies. The method of the pseudoisochromatic tables, also known as Ishihara tables, can be applied very effectively on a display screen. A number of the test plates used in this method consist of colored test figures appearing on a background of a different color, where the test and background colors lie on a dichromatic confusion line (see Chapter 3).

Dichromacy and anomalous trichromacy are the most widespread deficiencies and are explained as the absence or malfunction of one type of cones in the retina. As all colors which can be discriminated by the normal eye must belong to three different cone signals, a person with only two cone types will confuse all color stimuli differing only in the missing cone signal. These color stimuli are located on the socalled confusion lines in three-dimensional color space. Three confusion lines can be drawn through every color, corresponding to the three cone types. A person with one missing cone type will confuse all colors lying on the corresponding confusion line.

On a display it is easy to create test and background colors lying on one of the confusion lines. If the different test-background combinations are presented to the person to be tested, it appears rather quickly if he or she has a color vision deficiency and of what type it is: protanopy, deuteranopy or tritanopy, depending on the kind of missing cone signal, or protanomaly, deuteranomaly or tritanomaly in the case of reduced function of the corresponding cone type. In contrast to the printed tables, the color contrast between test and background can be varied on the display and the severeness of the deficiency can be quantified.

15.5.4. Displays in color teaching

The flexibility of the color display as a tool for the generation of color test patterns is also very useful for color education. Not only can fundamental effects of color vision and relations between colors be demonstrated comfortably on the screen, but the student can interactively vary the colors and the parameters of the test and thereby actively train his experience and gain a deeper understanding of the phenomena. A number of phenomena are suited for demonstration on a color display[1].

---

[1] A software package COLDEMO (English version) or FARBDEMO (German version) containing 30 interactive training programs for PC covering different fields of color science is available from the author. Contact Dr. H. Lang, Magdeburger Str. 12a, D-64372 Ober-Ramstadt, Germany.

As the colors on the display screen are generated by additive mixture, the screen is an ideal tool to study the rules of *additive mixtures*. Let four colors F_1, F_2, F_3 and F be created on the screen with the following video signals:

$$F_1: \quad (E_R, 0, 0),$$
$$F_2: \quad (0, E_G, 0),$$
$$F_3: \quad (0, 0, E_B),$$
$$F: \quad (E_R, E_G, E_B),$$

then color F is the additive mixture of the colors F_1, F_2, F_3. By variation of the signals E_R, E_G, E_B it can be demonstrated, that all the colors of the screen are indeed mixtures of the three primaries (F_1, F_2, F_3) with varying intensities. However, metameric matches are not possible on the display screen, only between a color on the screen and another external color. Equal colors on the screen always have equal physical stimuli!

It may be worthwhile to point out that if we have three colors F_1, F_2, F_3 on the screen with the video signals

$$F_1: \quad (E_{R1}, E_{G1}, E_{B1}),$$
$$F_2: \quad (E_{R2}, E_{G2}, E_{B2}),$$
$$F_3: \quad (E_{R1} + E_{R2}, E_{G1} + E_{G2}, E_{B1} + E_{B2}),$$

then F_3 is not the additive mixture of the two colors F_1 and F_2! The reason lies in the nonlinear transfer function (see Section 15.3.2), from which it follows that the addition of video signals is not equivalent to addition of tristimulus values.

Simultaneous contrast can be demonstrated in different ways. If, for example, the display screen is divided in two or four fields of different color, and in each of these fields there is a small field with identical tristimulus values $R = G = B$ (neutral), the appearance of the four small fields will be different (as in color plate 8). The colors of the small fields can then be modified until they have the same appearance. If the colors of the surrounding fields are then switched to neutral, the amounts of the induced color shifts can be determined. A very impressive demonstration is the apparent nonuniformity of a bar of constant tristimulus values between broader bars with a continuous change of luminance and/or color.

Further visual effects that can be studied on a display are after-images (successive contrast) and chromatic adaptation, Mach bands, the Bezold-effect as an example of color assimilation, and the flicker fusion frequency for luminance and/or color contrast.

Visualization of color coordinate systems or color order systems is another attractive possibility of the color display. The borders of the color gamut of the display may be studied in different planes of a given color space, for example planes of constant luminance or of constant hue (see color plate 39).

If such planes are visualized in so-called perceptually uniform color spaces like the CIELUV or CIELAB spaces, the perceptual *uniformity* can easily be tested visually. In every plane the spatial distances between colors should be proportional to their perceptual differences. Equally spaced colors must appear to have equal color differences.

Color thresholds can be investigated if it is possible to generate colors with very fine quantization (10 bit or more). Thresholds can be demonstrated even with relative coarse quantization with the help of dithering. Dithering means the generation of a fine spatial raster of two different colors differing by only one minimum step per channel. As the colors are very similar, the eye cannot perceive the raster and the area appears in the average color.

15.6. Appendix
Derivation of transformation matrices

We want to carry out here the calculations mentioned in Sections 15.3 and 15.2 together with a numeric example, because these calculations are necessary for each display calibration.

We begin with the equations (15.8) of Section 15.3.3. It is convenient to write these equations as a matrix equation:

$$\begin{pmatrix} x_r & x_g & x_b \\ y_r & y_g & y_b \\ z_r & z_g & z_b \end{pmatrix} \begin{pmatrix} P_r \\ P_g \\ P_b \end{pmatrix} = \begin{pmatrix} x_w/y_w \\ 1 \\ z_w/y_w \end{pmatrix}. \tag{15.8'}$$

For the solution, the matrix of the chromaticities has to be inverted and then the normalizing factors P_r, P_g, P_b can be calculated from the matrix equation:

$$\begin{pmatrix} P_r \\ P_g \\ P_b \end{pmatrix} = \begin{pmatrix} M_{rx} & M_{ry} & M_{rz} \\ M_{gx} & M_{gy} & M_{gz} \\ M_{bx} & M_{by} & M_{bz} \end{pmatrix} \begin{pmatrix} x_w/y_w \\ 1 \\ z_w/y_w \end{pmatrix}, \tag{15.9'}$$

where the matrix of the coefficients $M_{rx}, \ldots, M_{bz}$ is the inverse of the matrix $x_r, \ldots, z_b$

$$\begin{pmatrix} M_{rx} & M_{ry} & M_{rz} \\ M_{gx} & M_{gy} & M_{gz} \\ M_{bx} & M_{by} & M_{bz} \end{pmatrix}^{-1} = \begin{pmatrix} x_r & x_g & x_b \\ y_r & y_g & y_b \\ z_r & z_g & z_b \end{pmatrix}. \tag{15.14}$$

A matrix is inverted according to the following procedure. We assume a matrix A with the coefficients $a_{11}, \ldots, a_{33}$ to be inverted, the inverse matrix being B with coefficients $b_{11}, \ldots, b_{33}$:

$$A = \begin{pmatrix} a_{11} & a_{12} & a_{13} \\ a_{21} & a_{22} & a_{23} \\ a_{31} & a_{32} & a_{33} \end{pmatrix}, \quad B = A^{-1} = \begin{pmatrix} b_{11} & b_{12} & b_{13} \\ b_{21} & b_{22} & b_{23} \\ b_{31} & b_{32} & b_{33} \end{pmatrix}. \tag{15.15}$$

The coefficients $b_{11}, \ldots, b_{33}$ of the inverted matrix are:

$$b_{11} = (a_{22}a_{33} - a_{23}a_{32})/D,$$
$$b_{21} = (a_{23}a_{31} - a_{21}a_{33})/D, \quad\quad\quad\quad (15.16)$$
$$b_{31} = (a_{21}a_{32} - a_{22}a_{31})/D,$$

$$b_{12} = (a_{32}a_{13} - a_{33}a_{12})/D,$$
$$b_{22} = (a_{33}a_{11} - a_{31}a_{13})/D,$$
$$b_{32} = (a_{31}a_{12} - a_{32}a_{11})/D,$$

$$b_{13} = (a_{12}a_{23} - a_{13}a_{22})/D,$$
$$b_{23} = (a_{13}a_{21} - a_{11}a_{23})/D,$$
$$b_{33} = (a_{11}a_{22} - a_{12}a_{21})/D.$$

D is the determinant of the matrix A and is

$$D = a_{11}a_{22}a_{33} + a_{12}a_{23}a_{31} + a_{13}a_{21}a_{32}$$
$$\quad - a_{13}a_{22}a_{31} - a_{11}a_{23}a_{32} - a_{12}a_{21}a_{33}. \quad\quad (15.17)$$

If $D = 0$, then the inverse of matrix A is not defined and the system of equations cannot be solved.

We will now put numbers into the formula and choose the chromaticities of the standard HDTV primaries given in Section 15.4.2 and a display neutral chromaticity of standard daylight D65 ($x_w = 0.3127$, $y_w = 0.3291$). Matrix equation (15.8′) is then:

$$\begin{pmatrix} 0.640 & 0.300 & 0.150 \\ 0.330 & 0.600 & 0.060 \\ 0.030 & 0.100 & 0.790 \end{pmatrix} \begin{pmatrix} P_r \\ P_g \\ P_b \end{pmatrix} = \begin{pmatrix} 0.9501 \\ 1.0000 \\ 1.0885 \end{pmatrix}. \quad\quad (15.18)$$

If the matrix is inverted we can calculate the values P_r, P_g, P_b and we find

$$P_r = 0.644, \quad P_g = 1.193, \quad P_b = 1.200. \quad\quad\quad (15.19)$$

With these numbers we obtain the tristimulus values of the display primaries according to the relations (15.7) of Section 15.3.3:

$$X_r = P_r x_r = 0.412, \quad X_g = P_g x_g = 0.358, \quad X_b = P_b x_b = 0.180,$$
$$Y_r = P_r y_r = 0.213, \quad Y_g = P_g y_g = 0.716, \quad Y_b = P_b y_b = 0.072, \quad (15.20)$$
$$Z_r = P_r z_r = 0.019, \quad Z_g = P_g z_g = 0.119, \quad Z_b = P_b z_b = 0.948.$$

With the help of the CIE tristimulus values of the primaries we can derive the CIE tristimulus values of any color given by its tristimulus values R, G, B on the display:

$$X = 0.412\,R + 0.358\,G + 0.180\,B,$$
$$Y = 0.213\,R + 0.716\,G + 0.072\,B, \qquad (15.21)$$
$$X = 0.019\,R + 0.119\,G + 0.948\,B.$$

If we want to calculate the display tristimulus values R, G, B from the CIE tristimulus values X, Y, Z, we have to invert the matrix of the coefficients:

$$
\begin{pmatrix} R_x & R_y & R_z \\ G_x & G_y & G_z \\ B_x & B_y & B_z \end{pmatrix}
= \begin{pmatrix} X_r & X_g & X_b \\ Y_r & Y_g & Y_b \\ Z_r & Z_g & Z_b \end{pmatrix}^{-1}
$$

$$
= \begin{pmatrix} 0.412 & 0.358 & 0.180 \\ 0.213 & 0.715 & 0.072 \\ 0.019 & 0.119 & 0.951 \end{pmatrix}^{-1}
$$

$$
= \begin{pmatrix} 3.247 & -1.540 & -0.499 \\ -0.972 & 1.875 & 0.042 \\ 0.057 & -0.205 & 1.060 \end{pmatrix}. \qquad (15.22)
$$

With this matrix we can write the relations we were looking for:

$$R = 3.247\,X - 1.540\,Y - 0.499\,Z,$$
$$G = -0.972\,X + 1.875\,Y + 0.042\,Z, \qquad (15.23)$$
$$B = 0.057\,X - 0.205\,Y + 1.060\,Z.$$

With these equations we have all the relations necessary to transform between device tristimulus values R, G, B and the device-independent CIE tristimulus values X, Y, Z of any color on the display – if the assumptions mentioned in Section 15.3.4 can be made.

References

Becker, M.E. and J. Neumeier, 1992, Measuring LCD electro-optical performance. Soc. for Inf. Display SID'92 **23**, 50–53.

Berns, R.S., R.J. Motta and M.E. Gorzynski, 1993a, CRT colorimetry. Part I: Theory and practice. Color Research and Application **18**, 299–314.

Berns, R.S., M.E. Gorzynski and R.J. Motta, 1993b, CRT colorimetry. Part II: Metrology. Color Research and Application **18**, 315–325.

Brainard, D.H., 1989, Calibration of a computer-controlled color monitor. Color Research and Application **14**, 23–34.

Chen, L.M. and S. Hasegawa, 1992, Visual resolution limits for colour matrix displays. Displays: Technology and Application **13**, 179–186.

CIE, 1990, TC 2-26, Measurement of color self-luminous displays (Draft).

Compton, K., 1989, Selecting phosphors for displays. Information Display **1**, 20–25.

Cook, J.N., P.A. Sample and R.N. Weinreb, 1993, Solution to spatial inhomogenity on video monitors. Color Research and Application **18**, 334–340.

Cowan, W.B., 1983, An inexpensive scheme for calibration of a colour monitor in terms of the CIE standard coordinates. Computer Graphics **17**, 315–321.

Cowan, W.B. and N. Rowell, 1986, On the gun independence and phosphor constancy of colour video monitors. Color Research and Application **11** Suppl., 34–38.

DeMarsh, L.E., 1974, Colorimetric standards in US color television. A report by the Subcommittee on system colorimetry of the SMPTE Television Committee. SMPTE Journal **83**, 1–5.

DeMarsh, L., 1993, Display phosphors/primaries – some history. SMPTE Journal, 1095–1098.

EBU (European Broadcasting Union) DT/004528/II-A: 1992, The methods of measurement of the colorimetric performance of studio monitors (Draft) G4/Col (1992) (Copenhagen).

Gan, Q., M. Miyahara and K. Kotani, 1991, Characteristic analysis of color information based on (R, G, B)–(H, V, C) color space transformation, in: Visual Communications and Image Processing '91, SPIE, Vol. 1605, pp. 374–381.

Grambow, L., 1993, Farbmessung an Monitoren mit Schattenmaskenröhren. Fernseh- und Kino-Technik **47**, 97–102.

Kane, J.J., 1990, Instrumentation for Monitor Calibration. SMPTE Journal, 744–752.

Laesie, B., 1991, A color analyzer for monitor adjustment. World Broadcast News, 48–50.

Lang, H., 1995, Farbwiedergabe in den Medien Fernsehen Film Druck (Muster-Schmidt Verlag, Göttingen).

Levkowitz, H. and G.T. Herman, 1993, A generalized lightness, hue, and saturation color model. CVGIP: Graphical Models and Image Processing **55**, 271–285.

Lucassen, M.P. and J. Walraven, 1990, Evaluation of a simple method for color monitor recalibration. Color Research and Application **15**, 321–326.

Munsell Color Science Laboratory, 1988, Colorimetric Calibration of Cathode Ray Tubes.

Pointer, M.R., 1980, The gamut of real surface colours. Color Research and Application **5**, 145–155.

Raue, R., A.T. Vink and T. Welker, 1989, Phosphor screens in cathode-ray tubes for projection television. Philips Tech. Rev. **44**(11/12), 335–347.

Reuber, C., 1993, Flachbildschirme – ... der weite Weg zu Nipkows Vision. Fernseh- und Kino-Technik **47**, 231–242.

Roberts, A., 1991, Methods of measuring and calculating display transfer characteristics (Gamma). Research Department Report No. BBC RD 1991/6.

Roger, T., N. Jung and F.W. Vorhagen, 1992, Nutzungstiefe eines Videosystems. Wieviele Farben sind visuell tatsächlich unterscheidbar? Elektronik **7**, 84–88.

Shinoda, H., K. Uchikawa and M. Ikeda, 1993, Categorized color space on CRT in the aperture and the surface color mode. Color Research and Application, **18**, 326–333.

SMPTE Television Video Technology Committee, 1982, Report on standardization of monitor colorimetry by SMPTE. SMPTE Journal, 1201–1202.

Sproson, N.W., 1983, Colour Science in Television and Display Systems (Adam Hilger, Bristol).

Taylor, J., A. Tabayoyon and J. Rowell, 1991, Device-independent color matching you can buy now. Information Display **4–5**, 20–49.

Travis, D., 1991, Effective Color Displays Theory and Practice (Academic Press, London, New York).

Welch, E., R. Moorhead and J.K. Owens, 1991, Image processing using the HSI color space. IEEE Proceedings of the SOUTHEASTCON'91, 722–725.

Williams, T., 1988, HSI conversion brings true color image processing to life. Computer Design, 28–29.

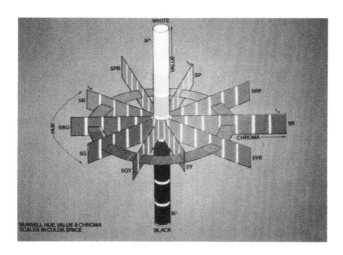

Color plate 1. The three-dimensional color space of Munsell, described by *hue*, *value*, and *chroma*. Courtesy Munsell Color.

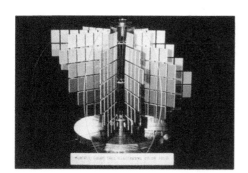 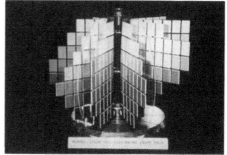

Color plate 2. Two views of the Munsell "Color Tree", illustrating the Munsell color space. Courtesy Munsell Color.

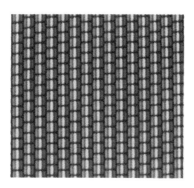

Color plate 3. Close-up of one type of television screen showing red, blue, and green phosphors excited by cathodoluminescence; stripes are about 0.25 mm wide. Photograph K. Nassau.

458

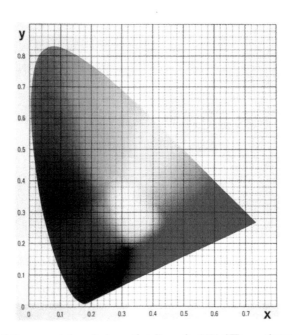

Color plate 4. An artist's representation of where colors lie on the 1931 CIE x, y chromaticity diagram. The CIE does not associate specific colors with regions on this diagram. Courtesy of Minolta Corporation.

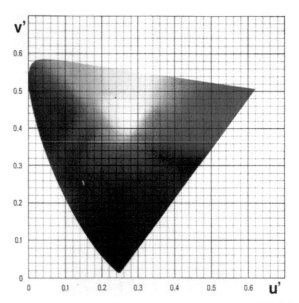

Color plate 5. An artist's representation of where colors lie on CIELUV, the 1976 CIE u', v' chromaticity diagram. The CIE does not associate specific colors with regions on this diagram. Courtesy of Minolta Corporation.

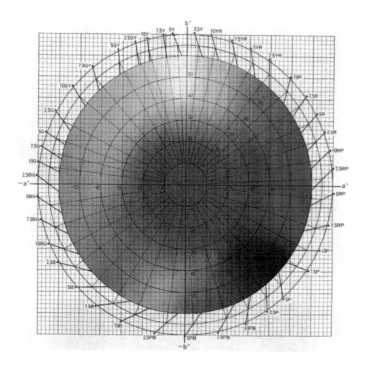

Color plate 6. An artist's representation of where colors lie on CIELAB, the 1976 CIE a^*, b^* diagram. The CIE does not associate specific colors with regions on this diagram. Surrounding the hue circle are approximations of the Munsell hues for several lightness ranges: the outermost circle represents the Munsell hue at L^* about 60 to 90 (Munsell values about 6 to 9); the next circle similarly represents 40 to 60 (4 to 6); and the innermost circle 20 to 40 (2 to 4). Courtesy of Minolta Corporation.

Color plate 7. Examples of collections of colors consisting of ink on paper, paint on paper, dyed textiles, and plastic chips that are used for the specification of color. Courtesy of Pantone, Inc.

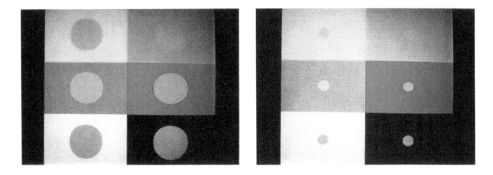

Color plate 8. Simultaneous contrast: all the central regions are neutral gray, but their apparent color depends on the colored surround as well as on size. Photographs J. Krauskopf.

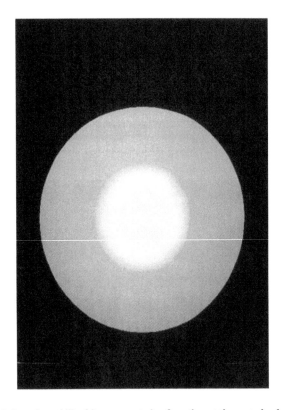

Color plate 9. Simulation of a stabilized image: on staring for a time at the central colored spot, its color may disappear. Photograph J. Krauskopf.

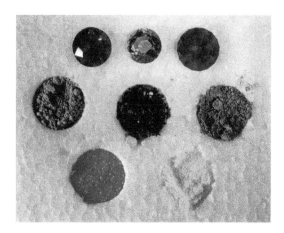

Color plate 10. Some colors produced by chromium. Above: alexandrite, emerald, and ruby (impurity ligand field colors of Cr^{3+}); center: chromium carbonate, chloride, and oxide (idiochromatic ligand field colors of Cr^{3+}); below: ammonium dichromate and potassium chromate (charge transfer colors of Cr^{6+}). Photograph. K. Nassau.

Color plate 11. A synthetic alexandrite gemstone, 5 mm across, changing from a reddish color in the light from an incandescent lamp to a greenish color in the light from a fluorescent tube lamp. Photograph K. Nassau.

Color plate 12. Six blue gemstones with different causes of color. Left to right: above: Maxixe-type beryl (radiation-induced color center), blue spinel (ligand field color from a Co impurity), spinel doublet (colorless spinel containing a layer of organic dye); below: shattuckite (ligand field in an idiochromatic Co compound), blue sapphire (Fe–Ti intervalence charge transfer); lapis lazuli (S_3^- anion–anion charge transfer); the largest stone is 2 cm across. Photograph K. Nassau.

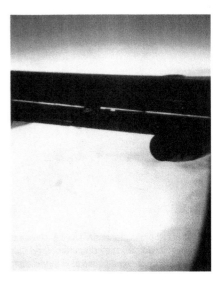

Color plate 13. Four chemoluminescent "Cyalume" light sticks, 15 cm long, made by the American Cyanamid Company. Photograph K. Nassau.

Color plate 14. The "glory", caused by interference of light scattered from small particles in a cloud, surrounding the shadow of a plane. Photograph courtesy of E.A. Wood.

Color plate 15. Mixed crystals of yellow cadmium sulfide CdS and black cadmium selenide CdSe, the intermediate band-gap colors as in fig. 4.20. Photograph K. Nassau.

Color plate 16. Multiple thin film interference produces a green metalliclike reflection of a photographic flash from the eyes of a cat. Photograph K. Nassau.

Color plate 17. Irradiation-produced color centers. Left to right: above: century-old glass bottle irradiated to form desert amethyst glass, a colorless synthetic quartz crystal as grown, and one that has been irradiated to form smoky quartz; below: a man-made citrine quartz colored yellow by Fe, and one that has been additionally irradiated to form the amethyst color center. Photograph K. Nassau.

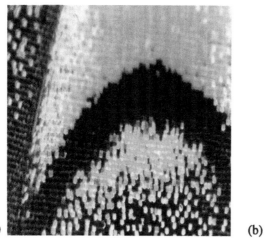
(a)

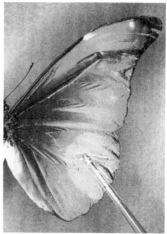
(b)

Color plate 18. (a) Scales, 0.1 mm in size, on the wing of the butterfly *Papilio polamedes*, with red and brown colors derived from organic pigments and blue from scattering. (b) Iridescent metalliclike colors produced by multiple interference on the 4-cm long wing of the butterfly *Chrysyridia madagascarensis*. Photographs courtesy of F. Mijhout.

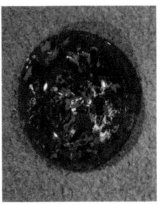

Color plate 19. North American fall foliage (organic colorants). Photograph K. Nassau.

Color plate 20. Color produced by diffraction in a 12 mm diameter synthetic black opal, grown by Ets. Ceramique Pierre Gilson. Photograph K. Nassau.

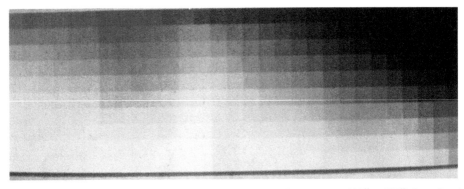

Color plate 21. Painting by Robert Swain, *UNTITLED*, 1973, acrylic on canvas, 120″ × 360″. Reproduced with permission of the artist.

Color plate 22. Painting by Sanford Wurmfeld, *II-9 (Y-white)*, 1987, acrylic on canvas, 57″ × 57″.

Color plate 23. Painting by Sanford Wurmfeld, *II-21+Black*, 1978, acrylic on canvas, 72″ × 168″.

Color plate 24. Painting by Doug Ohlson, *Swan Song*, 1995, acrylic on canvas, 108″ × 60″. Reproduced with permission of the artist.

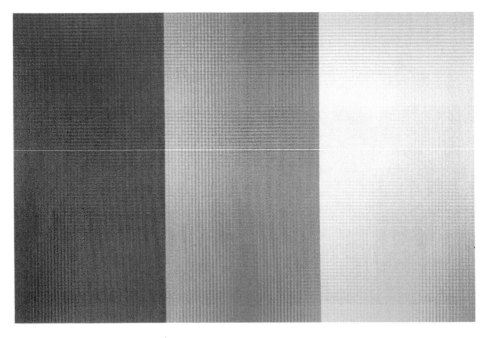

Color plate 25. Painting by Sanford Wurmfeld, *II-45 (DN/N/NL)*, 1991, acrylic on canvas, 90″ × 136½″.

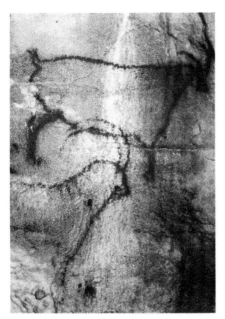

Color plate 26. Wall painting of goats from the Cougnac cave in France, about 20 000 years old. Courtesy of Het Studiecentrum voor Prehistorische Kunst of Bert Schaap, Maastricht, The Netherlands.

Color plate 27. Facial decoration designating the status of a New Zeeland native chief. Courtesy of the Amsterdam Tattoo Museum, Amsterdam, The Netherlands.

Sun Chemical Code and Description	Typical Shades at Both Masstone and Tint Levels		
	Masstone	50:50 Tint	5:95 Tint
228-0013 Pigment Red 122			
228-0641 Pigment Red 122			
228-7410 Pigment Red 122			
228-1215 Pigment Red 202			
228-0645 Pigment Red 209			

Color plate 28. Five quinacridone magenta pigments, shown at masstone as well as at two dilutions. Courtesy of Sun Chemical Corp.

Color plate 29. Road testing of traffic-marking paints using the monoarilide pigment Yellow 65. Note that the stripes are oriented across the road so as to receive maximum wear. Courtesy of Sun Chemical Corp.

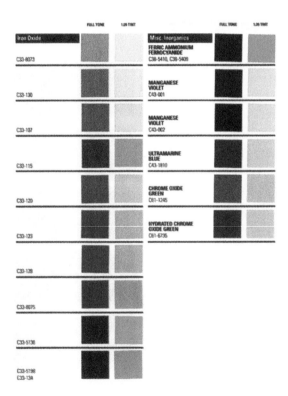

Color plate 30. Ten iron oxide pigments and six other inorganic pigments, shown at masstone and at one dilution. Courtesy of Sun Chemical Corp.

Colour Index Name & Number	Description	Masstone	1:5 Tint
Pigment Blue 15 Type	Phthalocyanine Blue R/S		
Pigment Blue 15 74160	Phthalocyanine Blue CRS		
Pigment Blue 15:1 74160	Phthalocyanine Blue NC/RS		
Pigment Blue 15:3 74160	Phthalocyanine Blue GS		
Pigment Blue 62 N-A	Victoria Blue		
Pigment Green 36 74265	Phthalocyanine Green YS		
Pigment Green 7 74260	Phthalocyanine Green BS		

Color plate 31. Seven phthalocyanine blue and green pigments, shown at masstone and at one dilution. Courtesy of Sun Chemical Corp.

Color plate 32. Ten samples of polyester fabric colored with the disperse dyes of fig. 6.11(b). Samples courtesy of Ciba Speciality Chemical, Textile Products Division, Photograph by Kurt Nassau.

Color plate 33. The white, yellow, magenta, cyan, red, green, and blue (black has been omitted for clarity) micro areas that are fused or blended by the eye when evaluating a photomechanical reproduction. (This should also be judged from far enough away so that fusion occurs.) Courtesy Graphic Arts Technical Foundation.

Color plate 34. Left: reflection image made using standard colorimetric values measured from a photographic transparency film (note that the image is very dark, high in luminance contrast, and cyan-blue in overall color balance); right: normal reflection color image. Courtesy Eastman Kodak Co.

Color plate 35. Left: color gamut limited to video standards (note the loss of saturation in the cyan balloons); right: full color gamut photographic image. Courtesy Eastman Kodak Co.

Color plate 36. Left: luminance dynamic range limited to video standards (note the loss of highlight details in the clouds); right: full luminance dynamic range photographic image. Courtesy Eastman Kodak Co.

Color plate 37. Left: quantization of linear tristimulus values, using 5-bits per color channel (note the quantization artifacts, especially in the darker skin-tone areas); right: quantization of nonlinear tristimulus values, using 5-bits per color channel; (note the reduction in quantization artifacts, especially in the darker skin-tone areas). Courtesy Eastman Kodak Co.

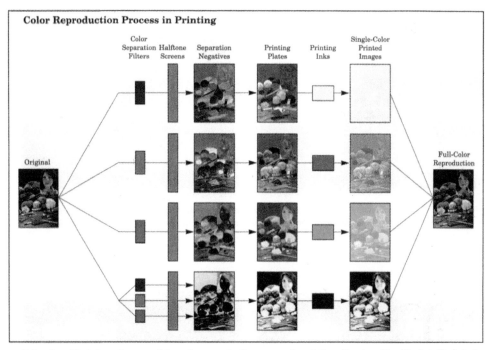

Color Reproduction Process in Printing

Color plate 38. A flow diagram of the color reproduction process in printing. Courtesy Graphic Arts Technical Foundation.

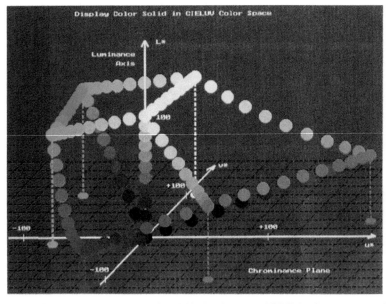

Color plate 39. Limits of the color gamut of an EBU TV display in CIELUV color space, reproduced from a computer display. The edges of the display gamut are indicated by colored circles in the u^*, v^* plane. Photograph H. Lang.

INDEX

AATCC (American Association of Textile Chemists and Colorists) 63, 72, 76, 77, 81, 91, 311, 316, 328, 343
absorbtion 10
absorption basking 231
absorptivity 369
acceptor 148, 149
acceptor groups 138, 139
accuracy of color-measuring instruments 82
actinides 134
action spectra 224–229, 251
acutance 372
adaptation 114–117
additive mixing 12–15, 452
– simultaneous 15
– spatial 15
– temporal 15
additive mixtures 452
additive primary colors 15, 358
additivity failure 369, 370, 383
Adebayo, O. 64, 91
aesthetic color expression 171
afterimage 116, 174, 183–185, 187
Agoston, A.A. 21, 30
Akita, M. 207, 208
Albers, Josef 182, 183, 187, 188, 190–193
albino animals 233, 241
albino flowers 234
albinos 158
Aldhous, M.E. 258, 276
Alessi, P.J. 82, 93
alexandrite 136
Alhazen 174
all-field 181, 183, 184, 186
all-over field 178
all-over pattern 183
Allan, J.S. 254, 278
Allen, E. 67, 91
Allen, R. 264, 267, 280
allochromatic 137, 138
allowed transition 128, 141
alloys 142

aluminum flake 309
amacrine cells 101, 102, 106
American Association of Textile Chemists and Colorists, *see* AATCC
"American Luminist" painters 190
American National Standard 76, 77, 91
American Psychiatric Association 262, 276
American Society for Testing and Materials, *see* ASTM
amethyst 150
Anderson, J.L. 264, 265, 267, 281
Anderson, R.R. 252, 279
animal coloration 221–246
– adaptation to background 232, 233
– seasonal adaptation 233
animal disguise 236
animal display 237–239
animal folklore 203
animal whiteness 233, 234
anomalous dispersion 153, 154
anomalous trichromacy 451
Anon 229, 231, 235–238, 242–244
anthocyanins 230
anthropology 195–208
antireflection coatings 160
Anuszkiewicz, R. 188, 190, 191
aperture mode 9
APS advanced photosystem 382
Aquila, F.J. 231, 244
Aquilonius, F. 173
Araus, J.L. 231, 244
arc 131
arc lamp 130
Arendt, J. 258, 260, 276, 280
Argamaso, S.M. 256, 267, 280
Aristotle 4, 174, 211
Armstrong, S.M. 258, 279
Arnall, D. 264, 278
Arnheim, R. 184, 193
art, science vs. 25–28
art, color in 169–194
arousal 260–262
Aschoff, J. 252, 254, 276

aspect ratio 425
Aspland, J.R. xv, 316, 320, 332, 343
assimilate 187
assimilated colors 181
assimilation 172, 175, 181, 184, 185, 188, 190–193
ASTM (American Society for Testing and Materials) 40, 46–48, 51, 52, 54, 55, 61, 64–68, 70, 72, 73, 75–79, 81, 83–87, 91–93
atmosphere, color in 166
atmosphere, interaction with colorants 348–351
atomic orbitals 142, 143
Attenborough, D. 243, 244
aurora 131
Avery, D.H. 263, 265, 267, 276
azo based oranges 301
azo reds 288, 289

bacterial photosynthesis 224
bacteriochlorophyll pigments 224
bacteriorhodopsin 225–227
Badia, P. 264, 276
Ballare, C. 229, 244
Bamford, C.R. 167
band 142
– gap 145–149
– theory 142–152, 167
Barker, F.M. 256–259, 267, 277
Barr, W.M. 225, 244
Bartleson, C.J. 374, 387, 401, 419, 422
Batesian mimicry 236
bathochromic 138
Baum, M.J. 252, 281
Bauman, L. 256, 281
Baylor, D.A. 107, 121
Beacham, S. 267, 277, 280
Beattie, K.A. 225, 244
Beck, J. 187, 193
Becker, M.E. 432, 455
Beckmann, W. 319, 326, 343
Beer–Lambert law 88
Bell, J.S. 27–30
Benham effect 188
Benshoff, H.M. 256, 257, 267, 276
Benson, D.M. 263, 265, 267, 281
benzimidazolone based reds 294
Benzimidazolone Orange 301
benzimidazolone yellows 300
Berga, S.L. 264, 279, 280
Berkow, R. 271, 276
Berlewi, H. 178
Berlin, B. 205, 207, 208
Berman, K. 264, 276
Berns, R.S. 81, 94, 438, 440, 455
betalains 233

Bezold, see von Bezold
bible of the folk beliefs 204
bidirectional illuminating, viewing geometry 69, 70
Bill, Max 182, 187
Billmeyer Jr., F.W. 46, 54, 70, 75, 82, 90, 93, 394, 422
binary color systems 201
Binkley, S. 252, 254, 276
biological clock 254
biological colorations 164
biological effects of color 247–281
biological rhythms 252–255
bioluminescence 167, 237
bipolar cells 101, 102, 106
Birren, F. 21, 22, 30, 171, 193, 267, 276
Bishop's ring 163, 166
Bismuth Vanadate/Molybdate Yellow 307
Bjorn, L.O. 224, 227, 229, 234, 243, 244
black body colors 129–131
black dogs, stories 204
black image 356
black printer 364, 368, 376
Blair, S.M. 231, 245
Blane 22
bleaching 150
blind 259
Blind Mary 214
blind spot 99–102
blind testing 270–276
blood pressure 260–262, 270
blue moon 166
blue topaz 150
Blue Wool Scale 286
blue-greens 224, 225
Bly, K.S. 268, 280
body decoration 199–201, 206
body decoration motives 199
Boecker, M. 264, 276
Bohm, David 29, 30
Bohr, Niels 28
Bojkowski, C.J. 258, 276
Bolte, M.A. 263, 265, 267, 276
Bolton, R. 205, 206, 208
BON reds 291
border, image 119, 120
Boring, E.G. 172, 174, 193
Born, M. 167
Boulos, Z. 264, 277, 281
boundary shadow concealment 236
Bowman, M. 204, 208
Boynton, R.M. 10, 19, 30, 116, 120, 121
Brain, R. 199–201, 207, 208
Brainard, D.H. 437, 455
Brainard, G.C. xv, 247, 256–267, 269, 270, 276–281

Brandt, W. 234, 244
brass 144
Bray, M. 264, 265, 267, 281
Breneman, E.J. 401, 419, 422
Brewer, W.L. 383, 387
Brewster 22
Briggs, W.R. 231, 245
brightness 8, 369
Brill, T.B. 352
brilliance in gemstone 154
British Standard 63, 72, 93
bromine 134
Bronstein, D.M. 256, 277
Brotman, D.J. 259, 278
brown 9
Brown, E.N. 264, 278
Brown, N. 231, 244
Bruce, A.J. 238, 244
Brugh, L. 264, 278
Bruno, M.H. 359, 387
Bube, R.H. 167
Budde, W. 72, 93
buffers, dye 341
bulimia 264
Bunney, W.E. 256, 278, 279
Bunsen burner 134
burials 198, 199
Burns, G.R. 167
Burrows, G.D. 258, 279
Burt Jr., E.J. 243, 244
Buyer's Guide (annual) 343
Byrne, B. 263, 265, 267, 281

Cadmium Mercury Orange 307
Cadmium Orange 307
cadmium reds 305
Cadmium Sulfide Yellow 307
cadmium yellow, orange, red 146, 147
Cadmium Zinc Yellow 306
calendar customs 203
calibration 400
camera 358
camouflage 236, 241
Campbell, S.S. 264, 277, 278, 281
Carbazole Violet 298
Cardinali, D.P. 256, 277
Carnot, S. 217
carotenoids 224, 226, 227, 230, 232, 233, 237
carriers, dye 341
Carter, E.C. 75, 93
Cassirer, E. 192
Cassone, V. 256–259, 277
cathode ray tube, see CRT
cathodoluminescence 166, 167

cave painting 198, 204, 206
CCDs 359
CCIR 408, 410, 415, 422
CCTR-709 primaries 421
CCIR Recommendation 601-1, ITU-R BT.601-1 415
CCIR Recommendation 709, ITU-R BT.709 408, 410
Cerulean Blue 306
charge coupled device (CCD) 358, 366
charge transfer 140–142, 167
chemiluminescence 167
chemistry of color 129–152
chemistry of colorant interactions 348, 349
chemistry of organic colorants 287–304
Chen, L.M. 427, 455
Chevreul, M.E. 22, 173, 174
Chickering, K.D. 64, 93
Chinese Blue 305
chlorine 134
chlorophyll 224–231, 232, 233, 237
choroid 100, 101
chroma 8, 410
chroma subsampling 410
chromatic adaptation 402
chromatic inversion 214
chromaticity coordinates 49, 50
chromaticity diagram 12–16, 49, 51
chromaticity gamut boundaries 407
Chrome Green 137, 308
Chrome Orange 307
Chrome Yellow 306
chromium 134–137, 141, 144, 164, 351
Chromium Oxide Green 308
chromo-lithography 174
chromophore 139, 288
chromotherapy 253, 267–270
CIE (Commission Internationale de l'Eclairage) 35–64, 69, 72, 74, 75, 83, 90, 93, 94, 250, 277, 399, 401, 422, 437, 455
CIE Chromaticity Diagram 12–16
CIE94 color difference equation 64
CIELAB, 1976 CIE $L^*a^*b^*$ color space 56–59, 61, 62
CIELUV, 1976 CIE $L^*u^*v^*$ color space 56, 59, 60, 62, 447
cinnabar 147 146
circadian 252, 259, 264
circadian disruptions 253
circadian physiology 255
circadian regulation 253, 256
circadian rhytms 254, 267
circadian systems 269
citrine 152
Clapper, F.R. 362, 387

Clarke, F.J.J. 63, 82, 94
CMC($l : c$) color difference equation 63
Cobalt Blue 306
Codd, G.A. 225, 244
coevolution 235
– antagonistic 235
– cooperative 232, 234, 235
Cohen, S. 263, 265, 267, 276
Cole, C. 257, 277
color achromatic 8
color adaptation 114–117
color appearance 9–10
color-appearance colorimetry, encoding 419
color balance 376
color, biological and therapeutic effects 247–281
color blindness 112
color casts 377
color centers 149–152, 167
color changes 345–352
color chart, for printing 363, 367
color chromatic 8
color circle 6, 20–22
color communication 367
color constancy 114, 118, 190
color contrast 174, 175, 181, 182, 206
color correction 377
color cures 203, 207
color deficiency 112
color, definition of 3–4
color difference equations 55, 56, 60
color display discrimination 448, 449
color displays 423–456
color, early views 4
color encoding in the *Photo CD* system 389–422
color fastness 286, 318, 345–352
color, fifteen causes of 123–128
color, fundamentals of science 1–30
color gamut 368, 377, 406, 415, 446–450
color image constraints 368
color imaging: printing and photography 353–387
color-imaging systems 394
color in abstract painting 169–194
color in anthropology and folklore 195–208
color in art 169–194
color in plants, animals and man 221–246
color-management 419
color matching 87, 110–112
color matching functions 39, 40, 41, 45, 46, 49
color "meanings" 206
color, measurement of 31–96
color mixing 11–17, 87–89, 172, 173, 175
– additive 12–15
– subtractive 15–17
color models 171

color monitors 358, 376, 378
color negative film 377
color originals 376
color perception 24, 25
color, philosophy of 209–219
color photography 356, 364–366
color preferences 366
color preservation 345–352
color printing 353–387
color reconstruction 395
color rendering index 367
color reproduction 374, 378, 383
color response differences 366
color scanners 359, 382
color science, fundamentals of 1–30
color separation 364, 394
color, subtractive mixing 15–17
color surround 120
color symbolism 200
color teaching displays 451
color temperature 367
color theories 17–22, 106, 115
color therapy 267–270
– double blind testing 270–276
color thresholds 453
color tolerances 65, 66
– instrumental 67
color transparencies 368
color triad 200, 201, 203, 205, 206
color units, conversions 6
color viewing 367
color vision 17–19, 97–121, 232
– deficiencies 260, 451
– in animals 106
color wheel, *see* color circle 6
colorant interactions 348
colorants: dyes
– textile applications 318
– other applications 318
colorants: organic and inorganic pigments 283–312
coloration by coincidence 232, 237
COLORCURVE 55
color symbolism 199
colorimeters 76
colorimetric calibration 435
colorimetric response 366
colorimetric scanners 401
colorimetric transformation 414
Coloroid, color order system 55
colors, subtractive primary 15
Colour Index 287, 317, 320, 322, 343, 344
Commission Internationale de l'Eclairage, *see* CIE
Committee on Colorimetry 3, 30
communication 199, 206

compatibilizers, dye 341
complementary colors 13, 116, 117
complementary dominant wavelength 13
Compton, J.A. 82, 94
Compton, K. 431, 456
computer color matching 89
Comte de Buffon 174
concealment 236
conduction band 145
cone cells 101
cone pigments evolution 235
cones 101–103, 107–109, 115–117
conjugated system 138, 139, 350
Connelly Sr., R.L. 81, 94
contralateral 104
contrast 175, 181, 184, 185, 192, 193
contrasted colors 181
Coohill, T.P. 251, 267, 277
Cook, J.N. 440, 456
Cook, W.L. 312
cooperative charge transfer 140
copper 143–145
Copper Phthalocyanine Blue 297
Copper Phthalocyanine Green 302
cornea 99–101, 253
Cornsweet, T.N. 102, 121
corona 163 162
cortex 99–100, 103–105, 109
Cott, H.B. 235, 236, 244
Cotton, F.A. 167, 168
Cowan, W.B. 437, 438, 456
Cox, G.B. 263, 265, 267, 276
Crane, N. v
Cregory, R.L. 236, 244
Crisp, D. 205, 208
CRT displays 423–456
CRT Gamma 436
CRT phosphors 438
Cruz-Coke, R. 232, 244
crystal field 135, 167
CTS 83, 94
Culpepper, J. 264, 276
cyan 356, 368
cyanobacteria 224, 225
Czeisler, C.A. 254, 259, 264, 278, 279
Czygan, F.-C. 230, 244

da Vinci, Leonardo 156
Daan, S. 264, 278
Dager, S.R. 263, 265, 267, 276
Dalal, E.N. 380, 387
Dalton, John 18
Daltonism 18, 113
Darwin, Robert Waring 174

data metrics for color encoding 405
Davenport, Y. 262, 263, 265–267, 269, 280
Dawson, D. 264, 277, 278
de Kooning, Willem 184
de Sausmarez, M. 193
DeCoursey, P.J. 256, 281
Deegan, J.F. 267, 278
degaussing 431
DeMarsh, L.E. 394, 422, 431, 456
Demichel equations 362
Democritus 211
Deng, M.H. 256, 257, 279
Dennett, D.C. 218, 219
density of states 143
depression non-seasonal 253
depression seasonal 253
DeRoshia, C.W. 264, 281
Descartes, R. 215, 219
desert "amethyst" glass 149
deuteranopes 113
device-independent color 418, 441, 444
device-independent color, myths of 418
Deyo, R.A. 269, 281
diamond 146, 147, 154
diarylide yellows 299
dichromacy 451
dichromats 113 112
diffraction 161–165, 167, 234
diffraction grating 163
diffuse highlights 409
diffuse illuminating, viewing geometry 70–72
diffuse reflectance 68–73, 81, 82
diffuse reflection 9
diffuse transmittance 83, 84
digital halftone images 361
digital quantization 410
Dijk, D.-J. 264, 277, 281
DIN System, German 22, 54
Dinitroaniline Orange 301
direct imaging printing 360
disazo condensation reds 295
dispersion 153–155, 167
dispersion, anomalous 153, 154
dispersion curve 128, 129
display 425
– white 439
– color 423–456
– distractive 238
– measuring of 86, 87
– projection 433
disruptive coloration 236
disruptive marginal pattern 236
distractive display 238
Ditchburn, R.W. 167, 168

dithering 453
Doghramji, K. 261–263, 265, 267, 270, 281
donor 147–149
donor groups 138, 139
dopant 147
Dorazio, Piero 182
dose-response 257, 259
double beam spectrophotometers 74
double blind testing 270–276
double refraction 155, 161
Dry Color Manufacturers Association 311
dualism, mind–body 214
Duffy, J.F. 264, 278
Dunner, D.L. 263, 265, 267, 276
duration 183, 193
Durie, B. 242, 244
dye lasers 139
dye layer absorptions 364
dyeing machinery, see also textiles
– A-frame 337
– beam 336
– beck 336
– dyer and pre-dryer 338
– jet 337
– jig 337
– package 333–335
– steamer 339, 340
– thermosol oven 338
– wash-boxes 339, 340
dyeing terms, see also textiles
– auxiliary chemicals 341
– batch dyeing 329, 333
– carriers 341
– chemical potential 329
– continuous dyeing 330, 337
– diffusion 329
– equilibrium 330
– exhaustion 330
– fixation 342
– isotherms (Freundlich, hybrids, Langmuir, Nernst) 330–332
– leveling 342
– liquor-ratio 330, 336, 337
– migration 338
– padding 338, 339
– partition coefficient 330
– substantivity 332
– sorption 329, 330
– stain blocking 342
– thermofixation 338
dyes 164, 313–344, 350, 351
dyes for textile applications, see also textiles
– azoic combinations 317, 327
– acid 317, 324, 325, 342
– anionic for non-ionic fibers 320
– basic (cationic) 317, 325, 326, 342
– direct 317, 320, 321, 342
– dye solution properties 329
– disperse 318, 326, 327, 342
– indigo 317, 320, 322
– ionic for ionic fibers 324
– mordant 317, 324, 325
– natural 317
– non-ionic for manmade fibers 326
– premetallized 317, 325, 326
– reactive 317, 323, 324
– sulfur 317, 321, 322
– synthetic 317
– vat 317, 321, 322

earth temperature 225, 230
Eastlake, C.L. 20–22, 29, 30
Eastman, C.I. 264, 269, 277, 278, 281
Eastman Kodak Company, also see Kodak 392, 404, 410, 422
eating disorder 264
Ebbinghaus, Hermann 175
Ebin, V. 199, 207, 208
Ebling, F.J.G. 199, 207, 208
EBU (European Broadcasting Union) DT/004 437, 456
Edgerton, S.Y. 192, 193
Edwards, C. 225, 244
effect pigments 79
Ehr, S.M. 67, 94
Ehrich, F.F. 311
Einstein 27–30
electroluminescence 148, 149
electroluminescent display (ELD) 435
electromagnetic spectrum 5–6
electromechanical engraving 367
electron donor 147–149
electron energy band 142–152, 167
electron hole 144
electron hopping 140
electron Volt (eV) 7
electronic color scanning 366
electronic scanning 359
electrons, paired and unpaired 134–137
electrons, pi-bonded 138
electrophotographic 361
Emens, J.S. 259, 278
emerald 132, 133, 136
Emmeche, C. v
emulsion sensitivities 364
enamels 166
energy 7
energy absorption 223

– animal 231
– greenery 223–231
– reinforcement 231
energy band, *see* band 142
energy band gap 145–149
energy diagram 127
energy gap 145–149
energy units, conversions 7
Engeldrum, P.G. 379, 387
English, J. 258, 260, 276, 280
environment construction 238, 239
epiphenomenalism, mind–body 216
Erb, W. 81, 94
erythema, *see* sunburn
Evans, Ralph M. 187, 193, 383, 387, 422
evergreens 230
excitations 129–134
exposure 286
eye 99–103
eye color 233, 234, 241
eye movements 118–120
Eynard, R.A. 383, 387

F-center 150
Fabian, J. 167, 168
fading 150
Fairchild, M.D. 41, 81, 94, 95
Fairman, H.S. 46, 65, 93, 94
far red/red ratio 229
Farge, Y. 167, 168
fastness 286
Faulhaber, G. 316, 344
feathers 234
Fechner's colors 188
Feinstein, A.A. 271, 278
Fell, N. 225, 244
Felleman, D.J. 105, 121
Feller, R. 173, 194
Fenchel, T. 227, 244
Fermi surface 144
fibers for textile applications, *see also* textiles
– acetate 318, 319, 332
– acrylic 318, 319, 332
– blends 320, 332, 341, 342
– cellulosic 317, 323, 332
– cotton 317, 323, 332
– manmade 318
– modacrylic, *see* acrylic
– natural 318
– nylons 318, 319, 332, 341, 342
– polyesters 318, 319, 332, 341
– regenerated 318, 332
– silk 317, 319, 332, 341, 342
– synthetic 318

– viscose rayon 317, 332
– wool 317, 319, 332, 341, 342
fibrous polymer substrates, *see* textiles
Field, G.G. xv, 90, 94, 353, 359, 362, 374, 377, 383, 387
Figgis, B.N. 167, 168
figure and field 178, 181–186, 192
figure–field organization 177
film color 185–188, 190, 191, 193
fire in gemstone 154
first surface light scattering 370
fixation 184, 185, 187
fixation time 183, 184
fixing agent, dye 342
flame test 131
flamingo 225
flash bulb 130
flavins 233
flavonoids 233
Fleming, S. 264, 278
flexography 360
flicker fusion frequency 428
flower folklore 203
flowers, color of 231–235
flowers, shaped 234, 235
Fludd, Robert 172
fluorescence 10, 22, 78, 136, 139, 148–150, 164, 166, 167, 232, 349, 409
fluorescent brighteners 139
fluorescent colors 409
fluorescent lamp 22
fluorescent lighting 349
fluorescent samples 78
fluting effect 183, 184, 187
focus 187, 188
Fogg, L.F. 269, 278
folklore 195–208
Fontana, M.P. 167, 168
forbidden transition 128, 142
Fordyce, W. 269, 281
Forsius, A.S. 172
Foster Jr., W.H. 46, 94
Foster, R.G. 256, 267, 280
fovea centralis 99–102
frame rate 427
Francis, W. 312
Franey, C. 258, 276
Frank, A. 263, 265, 267, 280
Fraunhofer, *see* von Fraunhofer
Fredrickson, R.H. 256–259, 277
Freitag, W.O. 254, 278
French, C.S. 227, 245
French, J. 264, 277, 278
frequency 7

Fresnel, Augustin 158, 159, 161
Freundlich isotherm 330–331
Friel, D.D. 234, 244
Fritsch, E. 167, 168
Fuller, C.A. 264, 279
fundamentals of color science 1–30
Fytelman, M. 311

Gabo, N. 26
Gaddy, J.R. 258, 259, 261–263, 265, 267, 270, 277, 278, 281
Gage, J. 171, 193
Galileo, G. 211, 219
Gallin, P.F. 263, 265, 267, 281
gamma value 436
gamut of color, see color gamut
Gan, Q. 443, 456
ganglion cells 101, 102, 106, 115
Gans, R. 46, 94
Gartside, M. 173
gas excitation 130–132, 166
gas lasers 132
Gates, D.M. 229, 231, 243, 244
Gebel, J. 230, 244
Geccelli, J. 192
Geissler, G. 311
gems 154, 165, 167
gemstone, brilliance and fire 154
general brightness adaptation 402
geometric metamerism 49
Geritsen, F. 21, 22, 30
Gerrard, R.M. 261, 262, 270, 278
gestalt 178, 181, 182, 185
Gibson, W. 264, 278
Gifford, Sanford 190
Gilbert, L.E. 235, 244
Giles, C.H. 329, 344
Gillin, J.C. 262–267, 269, 279, 280
Gimbal, T. 267, 278
Giorgianni, E.J. xv–xvi, 389, 392, 394, 396, 403, 404, 409, 422
glaciers 134
glass 166, 167
glazes 166
glory 163, 166
gloss 370, 380
glossy 10
Goethe, J.W. von 19, 20, 21, 25–27, 174
gold 144
Goldner, E.M. 264, 276, 279
Goldstein, E.B. 184, 193
Goldsworthy, A. 227, 244
Gombrich, E.H. 236, 244
goniochromatism 49, 79

Goodwin, F.K. 254, 257, 262, 263, 265–267, 269, 279, 280
Gordon, P.F. 167, 168
Gorzynski, M.E. 438, 440, 455
Gouveia, A.D. 225, 245
graininess 374, 379, 380
Grambow, L. 438, 456
granularity 374
gravure 359, 360, 370
gray balance 377
gray component replacement (GCR) 364
green flash 154–156, 166
Green Gold 304
green toxic scums 225
greenery under stress 230, 231
greenness in animals 233
greenness in plants 223–231
Greenewalt, C.H. 234, 244
Greenler, R. 167, 168
Gregory, P. 167, 168
Grimaldi, F. 161
Grota, L.J. 264, 265, 267, 281
Grum, F. 81, 94
Gundlach, D. 78, 95
Gunn, G.J. 225, 244

Hård, A. 54, 94
Haak, K.A. 256, 277
Häder, D.-P. 252, 278
haemocyanin 232
haemoglobin 232
Haidinger's brushes 100–102
Halbertsma, K. 172, 193
halftone
– process 361
– screen frequency 372
– screen moiré 361
– screening 370
Hall, M. 171, 193
Hallada, D.P. 319, 344
Hamilton III, W.J. 231, 241, 242, 244
Hamilton, M. 266, 278
Hanifin, J.P. 256–260, 264, 267, 277–280
Hannon, P.R. 264, 267, 277, 278, 280
Hansa Yellow G 298
Hanson Jr., W.T. 383, 387
hard image 358
Hardin, C.L. xvi, 209, 212, 218, 219
Harold, R.W. 9, 10, 30, 56, 81, 90, 94
Harris, M. 21, 173, 174, 193
Hartmann, H. 167, 168
Harvey, P. 237, 244
Hasegawa, S. 427, 455
Hayter 22

Hearle, J.W.S. 319, 344
Heathcock, C.H. 167, 168
Heisenberg, Werner von 28
heliotropism 229
Hellekson, C.J. 263, 265, 267, 280
Helmholtz, H., *see* von Helmholtz
Hemmendinger, H. 75, 82, 95
hemoglobin, *see* haemoglobin
hemocyanin, *see* haemocyanin
Hendry, G. 230, 234, 237, 244
Henning, W. 268, 278
Hering, K.E.K. 18, 22, 56, 94, 174, 175, 213, 219
Hering's theory 106
Herman, G.T. 443, 456
Herring, P.J. 237, 244
Hershel 22
Hertz 7
heterocyclic yellows 300
heteronuclear charge transfer 141
Heuberger, K.J. 362, 363, 387
high fidelity color 383
Hilbert, D. 218, 219
Hiley, B.J. 29, 30
Hill, A.R. 232, 244
Hills, P. 171, 193
Hoban, T.M. 262, 263, 265, 267, 269, 279
Hoffman, R.J. 256, 257, 267, 277
hole, electron 144
Holfeld, W.T. 319, 344
Holick, M.F. 252, 278, 279
Holley, D.C. 264, 281
Holtz, M.M. 256, 278
Holzel 22
Homer, W.I. 171, 193
homonuclear charge transfer 141
Hooke, Robert 28
Hope diamond 125, 148
Hopmeir, A.P. 311
horizontal cells 101, 102, 106
hormones 256–259, 262
Horsfall, J. 231, 244
Hotz, M.M. 256, 279
Howe-Grant 167, 168
Howell, B. 264, 278
HSI space 443
Hudson, D. 256–259, 277
hue 8
hue shifts 376
Huffman encoding 412
Hulse, F.S. 243, 244
human coloration 221–243
Hunt, R.W.G. 10, 30, 38, 54, 55, 60, 72, 86, 90, 94, 378, 384, 387, 396, 399, 419, 422
Hunter, R.S. 9, 10, 30, 56, 81, 90, 94, 380, 384, 387

Hurlbut, E.C. 256, 277
Hurvich, L.M. 175, 193, 218, 219
Hutchings, J.B. xvi, 125, 197, 202, 204, 207, 208, 221, 223, 225, 233, 235, 244, 249, 278
Huygens, Christian 28
hybrid color-imaging system 393
hybrid sorption isotherm 331, 332
hydrated Chromium Oxide Green 308
hydrogen peroxide 348, 351
hydrogen sulfide 349
hypothalamus 254, 256

ice 134
idiochromatic 137
IFT, *see* ink film thichness
Ikeda, M. 444, 456
Illnerova, H. 256, 281
illuminant 37, 38
illuminant metamerism 48
illuminant mode 9
illuminating, viewing geometry 69, 83
image
– border 119, 120
– capture 394
– compression 410
– definition 383
– definition requirements 379
– display 395
– sharpness 370
– structure 372
image, stabilized 120, 119
immune stimulation 253
immune suppression 252, 253
incandescence 129, 130, 166, 167, 348
incandescent lamp 130
incandescent lighting 349
incidental coloring 232, 237
Indanthrone Blue 298
indigo 139
indoles 233
Ingamells, W.C. 316, 343, 344
injection luminescence 149
ink film thickness (IFT) 379
ink films, printed 368
ink jet printer 361
inorganic colorant degradation 286, 347–351
inorganic colorants 283–287, 304–311
input compatibility 396, 398
intaglio, relief printing 359
integral tripack system 364
interactionism, mind–body 216
interference 158, 165, 167, 380
interference in difraction, *see* diffraction 159
interference patterns 380

interimage reflections 369
interlaced scanning mode 428
intrasubstrate reflections 369
iodine 134
ipsilateral 104
iridescence 160, 231, 234
iris, colors 158
iron 144 143
Iron Blue 305
iron oxide reds 304
iron oxide yellows 307
Ishihara Pseudoisochromatic Test 113
isotherms 330–332
Itten 22

Jackson, F. 214, 219
Jacobs, G.H. 256, 267, 277, 278
Jacobs, M. 21, 30
Jacobsen, F.M. 262, 263, 265–267, 269, 280
Jacobson-Widding, A. 207, 208
Jagger, J. 252, 278
Jameson, D. 175, 193
Jenkins, F.A. 167, 168
jet lag 253, 264, 265
Jimmerson, D.C. 256, 278, 279
Jing, Z.M. 362, 363, 387
Johnson, M.P. 264, 278
Johnson, T. 378, 387
Johnston, S.H. 264–267, 269, 280
Jorgensen, G.W. 374, 379, 387
Joseph-Vanderpool, J.R. 264–267, 269, 280
Judd, D.B. 52, 54, 94, 95, 422
Julesz, B. 120, 121
Jung, N. 449, 456

Kaiser, P.K. 10, 19, 30, 116, 120, 121, 261, 267,
   270, 278, 279
Kandinsky, Wassily 182
Kane, J.J. 438, 456
Kaplan, R. 268, 279
Katz, David 185–188, 190, 191, 193
Kay, P. 205, 207, 208
Keegan, H.J. 229, 244
Kelly, D.H. 120, 121, 183, 184, 187, 265, 267, 277
Kelly, K.A. 264, 267, 280
Kemp, M. 171, 193
Kern, H.A. 262, 263, 265, 267, 269, 279
Kettlewell, B. 232, 244
King, T.S. 256, 257, 277
Kirkpatrick, M. 237, 244
Kittel, C. 167, 168
Klee 22
Klein, D.C. 254, 256, 257, 279

Klein, T. 259, 278
Klerman, E.B. 259, 278
Kline, F. 179, 184
Kliun, I. 182, 191
Kobsa, H. 319, 344
Kodak *Photo CD* System 389–422
Kodak *PhotoYCC* Color Interchange Space 392, 406
Koffka, K. 120, 121
Kortmulder, K. 239, 244
Koslow, S.H. 256, 281
Köstner, B. 230, 244
Kotani, K. 443, 456
Krauskopf, J. xvi, 97, 252, 254, 257, 279
Kripke, D.F. 264, 279, 280
Kronauer, R.E. 254, 264, 278
Kubelka–Munk equations 17, 88
Kubelka, P. 17, 88, 94
Kuehni, R.G. 89, 94
Kuhn, Thomas 192, 193

Lacouture, Charles 173
Laesie, B. 438, 456
lake pigments 348
lamps, measuring of 86, 87
Lam, R.W. 264, 276, 279
Land, E.H. 19, 30
Lang, H. xvi, 423, 425, 450, 456
Lange, O.L. 230, 244
Langford, M.J. 358, 387
Langmuir isotherm 331
lanthanides 134
lapis lazuli 141, 306
Larin, F. 256, 277
laser, color center 150
laser, semiconductor 149
lateral brightness adaptation 402
lateral geniculate nucleus (LGN) 99–100, 103, 104,
   107, 108, 115
lattice vibrations 128
Le Blon, J.C. 15, 18, 21, 30, 173, 193
Le Grand, Y. 120, 121
leaf color 225–231
– adaptation to stress 229–231
– variegation 231
leaf geometry 230, 231
leaf orientation 229
Lederman, A.B. 256, 267, 280
Lee, D.W. 231, 245
Leibniz, G.W. 214, 215, 219
Leibniz's Mill 214
Leivy, S.W. 257, 277
lemon chrome yellows 306
lens 253
Leonardo da Vinci 18
letterpress 360

leukemism 233
levelers, dye 342
Levendosky, A.A. 264, 267, 280
Levine, J. 217, 219
Levkowitz, H. 443, 456
Lewis, P.A. xvi, 283, 311, 316, 318, 320, 328, 344
Lewy, A.J. 254, 256–259, 262–267, 269, 277–281
LGN, see lateral geniculate nucleus
Liebeknecht, A.L. 64, 94
Lieberman, J. 267, 268, 279
ligand field 134–138, 142, 167
light, biological and therapeutic effects 247–281
light, color 6
light, "colored" 6
light devices 263
light-emitting diodes 149
light mixing 12–15
light, nature of 28, 29
light sources 22, 23, 36–38, 48, 86, 87, 166
light sources for double blind testing 271–276
light therapy 263–270
light therapy devices 263
light treatment 262, 269
lightfastness 286
lightness 8
lightness constancy 114
lightning 131
lightsticks 139
limelight 130
Lin, M.C. 264, 279
Link, M.J. 263, 265, 267, 281
liquid crystal 163, 431
liquid crystal displays (LCD) 431
Liss, P. 225, 244
lithography 359, 372, 382, 383
Lithol Red 290
Lithol Rubine Red 291
Littleman-Crank, C. 264, 278
Littler, D.S. 231, 245
Littler, M.M. 231, 245
Livingstone, W. 167, 168
Locke, J. 216, 219
Loeser, J.D. 269, 281
Lohse, R. 182
Longo, V. 188, 189
look up table (LUT) 362, 363, 377
Loughlin, L.L. 256, 281
Louis, M. 182
Lowry, J.B. 231, 245
Lubs, H.A. 312
Lucassen, M.P. 438, 456
luma 410
luma and chroma channels 412
luminance 8

– contrast 402
– dynamic range 409
luminescence 166
Luminist painters, American 190
luminosity 193
luminous film color 190–192
luminous film mode 190
LUT, see look up table
Lynch, D.K. 167, 168
Lynch, G.R. 256, 257, 267, 276
Lynch, H.J. 256, 257, 278, 279
Lythgoe, J.N. 236, 238, 245
Lytle, L.D. 256, 277

Mabon, T.J. 72, 73, 94
MacAdam, D.L. 4, 29, 30, 54, 95, 120, 121
Macdowall, F.D. 227, 245
Mach, Ernst 174
MacLaughlin, J.A. 252, 279
macula lutea 99
Madden, T.E. xvi, 389, 394, 403, 409, 422
magenta 356, 368
magno-cellular 104, 105, 109
Malevich, K. 176, 177, 191
Malinowski, T. 198, 206, 208
man, color in 221–246
Mandoli, D.F. 231, 245
Manganese Violet 305
Marchetti, K. 238, 245
Marcus, R.T. xvi, 31, 70, 82, 89, 93, 96
Marden, B. 186
Marfunin, A.S. 167, 168
Markey, S.F. 254, 257, 279
Markley, C.L. 264, 281
Marshack, A. 198, 208
Martens, H. 259, 278
masking equations 362
Mason, C.W. 167, 168
materialism, mind–body 214
matte 10
Matthews, S.A. 256, 257, 277
Maxixe beryl 152
Maxwell, James Clerk 18, 99, 100, 102, 110, 111
Maxwellian view 102
Maxwell's spot 99, 100, 102
McCormack, C.E. 256, 279
McDonald, R. 63, 89, 90, 94, 95
McGrath, P.J. 262, 263, 265–267, 269, 281
McIntyre, I.M. 258, 279
McKinnon, R.A. 78, 95
McTavish, H. 225, 245
measurement of color 31–96
measuring lamps, light sources, displays 86
media displays 450

medical color effects, double blind testing 270–276
medium chrome 306
Meissl, H. 256, 281
melanins 232–234, 240–243
melatonin 256–260, 267
Menaker, M. 256–259, 277, 281
menstrual cycle disturbances 253
menstrual disorders 264
Mercury Cadmium Red 305
mercury vapor lamp 131, 132
Mesolithic sites 198
Messenger, J.B. 233, 245
metallic and pearlescent samples 79
metals 142–145
metals as pigments 309
metamerism 22–24, 48, 49, 212, 367, 401
Metelli, F. 182, 193
Mie scattering 145, 157, 158
Milette, J.J. 256, 278, 279
Miller, L.S. 256–259, 262, 263, 265, 267, 269, 277, 279
Miller, S. 265, 267, 281
Millet, M.S. 263, 265, 267, 276
Mills, W.G.B. 312
Milori Blue 305
mimicry 232, 324–236
– Batesian 236
– Müllerian 236
minerals 165, 167
Minnaert, M. 167, 168
mixed metal oxide browns 308
mixed metal oxide greens 308
mixed metal oxide yellows 307
mixed metal oxides 304
Miyahara, M. 443, 456
mode, object 9
Moholy-Nagy, L. 182, 191
moiré 361, 374, 380, 382
molecular orbitals 138, 139, 142, 143
Moles, A. 171, 193
Molinari, G. 185, 192
Mollon, J.D. 232, 235, 245
Molybdate Orange 305
Momiroff, B. 75, 82, 95
Mondrian, Piet 177
monoarylide yellows 298
monochromats 112, 113
moon, blue 158
Moore, P.D. 230, 245
Moore, R.Y. 254, 256, 279
Moore-Ede, M.C. 264, 279
Moorhead, R. 443, 456
Morgane, P.J. 254, 279
Morris, D. 243, 245

Morton, W.E. 319, 344
Moser, F.H. 312
Mostofi, N. 264, 280
Motherwell, R. 179, 184
motion picture photography 356
Motta, R.J. 438, 440, 455
mottle 374, 380
Moul, D.E. 263, 265, 267, 280
Mousterian sites 198
Mueller, P.S. 262, 263, 265–267, 269, 280
Mullaney, D.J. 264, 279
Müllerian mimicry 236
multiple internal reflection 370
Munk 17, 88
Munsell Color Science Laboratory 456
Munsell Color Space 53
Munsell System 8, 22
Munsell's Color Order System 52
Munsell, A.H. 8, 20, 22, 25, 30, 52–54, 96, 171, 172, 175, 184, 193, 367, 456
Muntz, W.R.A. 232, 245
Murdoch, M. 240, 245
Murray, M.G. 263, 265, 267, 280
Muzall, J.M. 312
Myers, B. 264, 276

Nacreous 304
Naphthol Reds 291
Nassau, K. xv, 1, 123, 125, 133, 167, 168, 234, 243, 245, 270, 345, 352
Natural Color System, see NCS
natural iron oxides 308
Nayatani, Y. 399, 419, 422
NCS System, Swedish 22, 54
Neanderthal sites 198
nectar guides 234
negative film 402
Neitz, J. 256, 277
Nelson, D.E. 256, 257, 279
Nelson, W.H. 232, 245
Nemcsics, A. 55, 96
neon tube 131
Nernst isotherm 331, 332
Neugebauer equations 362, 363
Neumeier, J. 432, 455
neuroendocrine system 252, 253, 259, 269
neurophysiology 105–109
Newhall, S.M. 52, 54, 95
Newman, B. 179
Newsome, D.A. 254, 257, 262, 263, 265–267, 269, 279, 280
Newton, Isaac 4–6, 8, 17, 18, 25–30, 125, 128, 153, 160, 163, 173, 212, 219
Newton's color sequence 160, 162

Nguyen, D.C. 256, 279
nickel powders 309
Nickerson, D. 52, 54, 95
Noland, K. 182, 187
non-seasonal depression 253, 264
Norbert, Ph. 224, 245
Norman, T.R. 258, 279
Norris, J.N. 231, 245
North, A.D. 41, 95
Northmore, D.P.N. 238, 245
Novros, D. 180, 181
NPCA Raw Materials Index 312
NPIRI Raw Material Data Handbook 312

object mode 9
obliterative shading 236
observer metamerism 48
ochre 197–199
ocular tissues 251
offset lithography 359
Ohlson, D. 185
Olson, J.M. 269, 280
opal 164 163
opaque 88, 89
ophthalmology 450
opponent color theory 18, 115
opponent mechanisms 115
opponent signals 115
optic chiasm 100, 103
optic disk 102
optic nerve 99–101, 103
optical mixing 15, 173, 188
opthalmology displays 450, 451
Optical Society of America, see OSA
opto-electronic transfer characteristic 410
Oren, D.A. 263, 265–267, 269, 280
organic colorant degradation 286, 328, 348–351
organic colorants 283–304, 309–311
organic compounds 138, 139, 167, 283–304
Orthonitroaniline Orange 301
OSA Uniform Color Scales 22
OSA-UCS color order system 54
Ostwald, W. 22, 171, 172, 175, 184, 194
Ott, J.N. 267, 280
Otto, J.L. 268, 280
output calibration 417
output characterization 417
output signal processing 414
Owens, J.K. 443, 456
Ozaki, Y. 256, 279
ozone 348

Pagel, M. 237, 245
Paige, K.N. 235, 245

paint 347, 348
painting, body 199, 201
painting, color in 169–194
Palaeolithic sites 198
Pang, S.F. 256, 281
Panksepp, J. 254, 279
Panofsky, E. 192, 194
paradigms, universal xi
paramerism 49
Parfitt, G.D. 312
Park, J. 89, 95
Parkhurst, Ch. 173, 194
Parry, B.L. 264, 279, 280
Partridge, B.L. 239, 245
parvo-cellular 104, 105
pattern 175, 193
Patterson, D. 312
Patton, T.C. 312
pearlescent pigment 304
Pearson, C.A. 240, 245
Pearson, M. 384, 387
Pelham, R.W. 256, 281
pentimenti 348
perceptual effects 402
perfect white 409
perfect white diffuser 44
Permanent Red 2B 291
Persiev, S. 362, 363, 387
perspective 192
Perylene Red 294
Petterborg, L.J. 256, 257, 277
philosophy of color 209–219
phosphorescence 10, 164
phosphors 148, 149, 167, 430
Photo CD System 389–422
photodegradation 349–351
photo-oxidation 350
photobiology 249–281
photochemical deterioration 349
photodegradation 350, 351
photographic grain 374
photographic imaging systems 358
photographic reflection prints 368
photographs 348, 364–366
photomechanical color separations 366
photomechanical reproduction 356
photomechanical reproduction process 362, 380
photomechanical reproductions 368
photomultiplier tube (PMT) 359, 366
photon 8, 127
photoperiodism 226, 229
photopic 259, 260, 271
photopigments 111–113, 115, 232, 251, 256, 267
photoreceptors 251, 256, 260, 265, 267

photoreduction 351 350
photosensitive material 358
photosensory 267
photosurgery 253
photosynthesis 223–231, 233
phototendering 350
phototherapy 253
phototropism 226, 229
PhotoYCC Space data metric 406
physics of color 129–163
physiological color effects, double blind testing 270–276
physiological displays 450, 451
physiological optics 450
phytochrome pigments 229
phytoplankton 225
Pickard, G.E. 254, 256, 280
Pierce, P.E. 89, 95
piezochromism 137
pigment applications 309–311
pigment dispersion 309–311
pigment Green 304
pigment Yellow 298
pigments 164, 316, 347, 348, 356
pigments for textiles applications, *see also* textiles
– binders 316
– dyes vs. pigments 316
– exhaust pigmenting 343
– padding 338, 343
– printing 343
pigments, inorganic 283–287, 304–311
pigments, organic 283–304, 309–311
pineal gland 255–260
pinnaglobin 232
pitch 430
pixels 425
placebo 268–271, 276
Planck, Max 25, 26, 28, 129, 130
plants, color in 221–246
plasma display 435
Platt, T. 225, 245
Plochere 22
PMT, *see* photomultiplier tube
Podolin, P.L. 256, 257, 267, 277, 280
Podolsky, B. 27, 29, 30
pointer 407
Pointer colors 447
Pointer, M.R. 378, 387, 407, 422, 447, 456
pollination 233–235
Pollock, Jackson 178, 184, 186
polymers, fibrous 318–320
Poons, Larry 183, 184, 187, 191
Pope, Alexander 27, 28, 30
Popova, L. 182

porphyrins 227, 232, 233
Post, R.M. 256, 278, 279
Potts, J.T. 252, 281
Poulton, A.L. 258, 276
Prescott, W.B. 81, 96
preservation of color 345–352
Press, M. 231, 244
Price, M.V. 234, 246
primaries, CCIR-709 421
primary colorants 356
primary colors 173, 356
– additive 15
– subtractive 15
primary tristimulus values 437
Primrose Chrome 306
principle of univariance 111, 112
printing, color 353–387
printing machine 359
process color printing 356
product standards 65
progressive scanning mode 428
projection displays 433
properties of textiles fibers, *see also* textiles
– amorphous areas 319
– barré 319
– chemical 319
– crystalline areas 319
– end groups 319
– hydrophilicity 319
– hydrophobicity 319
– ionicity 319
– moisture regain 319
– morphology 319
– physical 319
proportionality failure 370, 371, 382
Prota, G. 232, 245
protanopes 113
Provencio, I. 256, 267, 280
Prussian Blue 141, 348
psychophysics 109–120
purity 8, 13
purity errors 431
pyrometer 130
pyrotechnic devices 130

quality 368
quantization artifacts 410
quantum 8, 127
quartz, smoky 151, 152
Quinacridone Reds 292
quinones 233
Quitkin, F.M. 262, 263, 265–267, 269, 281

Rabbani, M. 410, 412, 422
radiation, visual 100, 105

radiometers 86
radiometric colorimeters 86, 87
Rafferty, B. 262, 263, 265–267, 269, 281
rainbows 166
rainbows, primary and secondary 154–156
random dot screens 372, 374
random screening 361
Ratliff, F. 171, 187, 194
Raue, R. 431, 456
Ravindran, P. 225, 245
Rayleigh scattering 10, 165–168, 234
Rayleigh, Lord 156–158
receptive fields 107–109
red ochre 197–199
reduction screen 187
reference image-capturing device 404
reference image-capturing device spectral responsivities 408
reflectance 68–73, 81, 82
reflection 10
– diffuse 9
– metal 144
– specular 9
reflection basking 231
reflection print 402
refraction, double 155, 161
refractive index 128, 129, 153–155
regular transmittance 84
Reinen, D. 137, 167, 168
Reinhardt, A. 183, 191
Reiter, R.J. 256, 257, 267, 277, 280, 281
relative luminance levels 409
relief printing processes 359, 360
Remick, R.A. 264, 279
Remington, J.S. 312
repeatability of color-meaning instruments 82
Reppert, S.M. 254, 256, 279
reproduced colors 362
reproducibility of color-measuring instruments 82
resolution 370, 372, 379, 383
retina 99–103, 106, 251, 253–255
retinex color theory 19
retinohypothalamic tract 254, 255
retinotopic 105
retroreflection 87
Reuber, C. 435, 456
Rey, M.S. 252, 257, 258, 280
Rhodes, W.L. 378, 387
rhodopsin 115, 256
rhythms 252, 256, 258, 262
Rich, D.C. 72, 75, 82, 83, 93, 95
Richardson, B.A. 256, 257, 277
Richardson, G.S. 254, 264, 278, 279
Richter, J.P. 174, 194

Richter, M. 54, 95
Rigg, B. 63, 64, 91, 94
Riley, B. 182, 188, 190
Rios, C.D. 254, 278
rites of passage 201–204, 207
Rizzo III, J.F. 259, 278
Roberts, A. 436, 456
Roberts, Sh.L. 239, 243, 245
Robins, A.H. 241–243, 245
rod cells 101
rods 101–103, 107, 115
Roger, T. 449, 456
Rollag, M.D. 256, 257, 258, 259, 260, 264, 267, 276, 277, 278, 279, 280
Ronda, J.M. 254, 264, 278
Rood 21
Roper, T. 237, 238, 245
Rosen, N. 27–29, 30
Rosenthal, N.E. 262–267, 269, 277, 279, 280, 281
Rosofsky, M. 263, 265, 267, 281
Ross, M. 269, 280
Rossman, G.R. 167, 168
rotations 132–134
Rothko, Mark 179, 187, 191
Rowell, J. 444, 456
Rowell, N. 438, 456
Rowland, W.J. 238, 245
Rubens, Peter Paul 173
Ruberg, F.L. 258–260, 277, 278, 280
Rubin, E. 177, 178
ruby 135–137
ruby glass 144
Runge 22, 172
Ruskin, John 188
Ryan, M.J. 237, 244
Rys, P. 324, 344

Sabido, J. 231, 244
saccades 119
Sack, D.A. 262, 263, 265–267, 269, 280, 281
Sack, R.L. 262, 263, 265, 267, 269, 279
SAD (seasonal affective disorder) 253, 262–267, 269, 270
SAE (Society of Automobile Engineers) 65, 76, 81, 83, 95
Saltzman, M. 90, 93, 394, 422
sample preparation for reflectance measurements 81
Sample, P.A. 440, 456
Sanchez, R. 254, 278
Sanchez, R.A. 229, 244
Sandock, K.L. 256, 281
Sanford, B. 267, 280
sapphire 134, 135, 140, 141
Sathyendranath, Sh. 225, 245

saturation 8
saturation adjustment 377
saturation compression 377
Savides, T.J. 264, 279
scanning 184, 185
scanning mode, progressive 428
scanning systems 366
scarification 199
scattering 10, 155–158, 165, 167
SCE (specular component of reflection excluded) 71
Schivelbusch, W. 251, 280
Schlager, D. 263, 265, 267, 281
Schleter, J.C. 229, 244
Schmelzer, H. 73, 95
Schrammel, P. 230, 244
Schultz, E.T. 238, 246
Schultz, P.M. 264, 267, 280
Schwindt, W. 316, 344
SCI (specular component of reflection included) 71
science vs. art 25–28
Scientific American 223, 245
sclerotin 232
Scopel, A.L. 229, 244
scotopic 259, 271
screen frequency 361
screen printing process 360
seasonal affective disorder, see SAD
seasonal depression, see SAD
seasonal reproduction 267
seaweed color 225–229, 231
secondary color 356
selection rules 128
Sellmeier dispersion formula 153
semantic color expression 171
semiconductor laser 149
semiconductors 145–149, 164
– doped 147–149
– pure 145–147
Sependa, P.A. 264, 280
Seurat 188
sexual dichromatism 239
sexual dimorphism 238, 239
sexual interest releasers 242, 243
shadow, geometrical and wave 161, 162
shadow mask tube 429
Shanahan, T.L. 259, 278
sharpness 372, 379
Shepp, A. 384, 387
Sherry, D. 265, 267, 277
Shetye, S.R. 225, 245
shiftwork 253, 264, 265
Shinoda, H. 444, 456
Shopenhauer, Arthur 18
Shore, J. 320, 344

short run color printing 382
signal processing 395
silver 144, 145
Silverman, A.J. 254, 256, 280
Simon, H. 234, 245
Simons, P. 230, 245
simultaneous color contrast 174, 175, 183, 187, 192, 452
single beam spectrophotometers 74
Sivik, L. 54, 94
Skene, D.J. 258, 260, 276, 280
skin 251, 253, 265, 269
skin cancer 241, 252
skin color adaptation 240
skin color assessment 240
skin color, natural selection for 243
Skwerer, R.G. 262, 263, 265–267, 269, 277, 280
sky, blue 157
Slater, J. 143, 168
sleep disorders 253, 264
Sliney, D. 267, 277, 280
Sloane, P. 20, 26, 29, 30
Smith, A.G. 232, 245
Smith, G. 167, 168
Smith, H. 229, 245
Smith, K.C. 249, 251, 252, 267, 280
smoky quartz 150, 151, 152
smoothness 369
SMPTE Television Video Technology Committee 431, 456
soap bubble 160
social signals 238, 239, 241
Society of Automobile Engineers, see SAE
Society for Light Treatment, Biological 263–265, 280
sodium doublet 131
sodium vapor lamp 131, 132
soft images 358
soft proofing 450
solar energy at earth's surface 224, 229–231
solar energy under water 227, 228
Soll, M. 328, 344
Solyom, L. 264, 279
Sontag, C.R. 256, 279
Sorek, E. 265–267, 269, 280
Souetre, E. 264, 267, 280
Sowerby, J. 173
space 193
spark 131
spatial 117–120, 175, 184, 185
spatial experience 192
spatial organization 181, 183
spatial perception 183
spatial relationships 181

spectral light units, conversions 6
spectrophotometers 73, 74
spectrum 5, 6, 127
specular component of reflection 71
specular highlights 409
specular reflectance 68–73, 82, 83
specular reflection 9, 68–73
Spitler, H.R. 268, 280
Spooner, D.L. 78, 95
spreading effects 117
Sproson, N.W. 425, 456
St. Elmo's Fire 166
Staarup, B.J. 227, 244
stabilized image 119, 120
stage illumination 12–15
stainblockers, dye 342
Stanczak, J. 190, 191
standard illuminant 48
standard observer 38, 39
standard viewing conditions 367
Stanziola, R.H. 55, 75, 82, 95
Starz, K.E. 264, 267, 280
status patterns 232, 237, 238
Stearns, E.I. 46, 94, 95
Stearns, R.E. 46, 81, 94
Steinlecher, S. 256, 277
Stella, Frank 179, 181
stencil printing process 360
Sterns, E.I. 81, 95, 96
Stetson, M.H. 256, 257, 267, 277, 279
Stevenson, R.W.W. 232, 244
Stewart, J.W. 262, 263, 265–267, 269, 281
Stewart, K.T. 261–263, 265, 267, 270, 281
Stiles, W.S. 56, 90, 96, 116, 120, 121, 422
stochastic screening 361, 374, 380, 383
stomata 226, 229
Storm, W. 264, 277
Streitweiser, A. 167, 168
Streletz, L. 267, 277, 280
Strens, R.G.J. 167, 168
Strocka, D. 47, 73, 95
Strogatz, S.H. 254, 278
Strontium Yellow 306
structural color 231, 234, 239, 241
Strzeminski, W. 178, 186, 188
Sturge, J.M. 384, 387
subject moiré 361
substrates 359
subtractive color reproduction 356
subtractive primary colors 15
Sulzman, F.M. 264, 279
sun glasses, self-darkening 150
sunburn (erythema) 224, 240, 241, 252
superior colliculus 99, 100, 103

supplementary colors 383
suprachiasmatic nucleus 254, 255
surface characteristics 370, 380
surface color 185–188, 190, 191, 356
surface reflection 369
surround color 120
Swain, R. 184, 187
Swanton, P.C. 380, 387
Swedish Standard 54, 95, 96
symbolism 206
symbols 200
Symmons, M. 227, 245
synthetic brown oxide 308
syntonic optometry 253, 268
syntonic therapy 268
Szabo, M. 229, 245

Tabayoyon, A. 444, 456
Takahashi, J.S. 256, 257, 278, 279, 281
Tamarkin, L. 264, 280
tanning 240, 242, 252
tattooing 199
Taylor, J. 444, 456
Technical Manual (annual) 344
Technicolor 364
television displays 449
temporal 175
temporal organization 183, 184
term diagram 135, 136
Terman, J.S. 262, 263, 265–267, 269, 281
Terman, M. 262–267, 269, 277, 281
territorial defense 239
Terstiege, H. 78, 94
testing, double blind 270–276
Tevini, M. 252, 278
textiles
– color fastness, see color fastness
– dyeing, see dyeing of textiles
– dyeing machinery, see dyeing machinery
– dyeing terms, see dyeing terms
– dyes, see dyes for textile applications
– fiber properties, see properties of textiles fibers
– fibers, see fibers for textile applications
– pigmentation, see pigments for textile applications
texture 380
textured substrates 370
therapeutic applications 253
therapeutic color effects 247–281
therapeutic color effects, double blind testing 270–276
thermal dye transfer images 361
thermochromism 137
thermoluminescence 167
Thiele, G. 256, 281

Thimann, K.V. 230, 245
Thomas, A.L. 312
Thompson, E. 218, 219
Thomson, G. 352
Thomson, P.H. 232, 245
Thorington, L. 258, 281
Thornton, W.A. 41, 96, 261, 262, 270, 281
time 183–185, 187, 192
Timson, J. 231, 245
Tinbergen, N. 238, 245
tint strength 285
Toluidine Red 291
tonal compression 374
tone 8
tone reproduction 374–376
topaz 152
total appearance concept 204, 205
total internal reflections 154
transfer function 435–437
transform 396
transformation matrices 453–455
transition elements 134–138 134
transition metal compounds 137
transition metal impurities 137, 138
transition metals 134–142, 164
transitions, allowed and forbidden 128, 141, 142
translucent 77, 88
transmission 10
transmittance measurements 83–86
transparency 181–184, 186–188, 191, 402
transparent 88, 89
trapping level 148, 149
Travis, D. 425, 442, 456
triboluminescence 131
trichromacy color theory 18
trichromats 112, 113
tristimulus values 39, 43–49, 404, 437–440
tritanopes 113
Trost, C. 240, 245
Turek, F.W. 256, 278, 279
Turner, J.A. 269, 281
Turner, V. 200, 201, 208
Turrell, J. 188
Twidell, J. 230, 245
Twilley, G. 207, 208
Tyndall effect 234
Tyndall, J. 156, 157

Uchikawa, K. 444, 456
UCR, see under color removal
ultramarine 141
Ultramarine Blue 306
Ultramarine Violet 305
ultraviolet, deterioration by 349

ultraviotel, control of 349
under color removal (UCR) 364
uniform color space 55, 56
uninvariance principle 111, 112
unlucky green 203, 204
Urbach, F. 252, 281
US Congress 264, 281

valence band 145
value 8
Van Doesberg, T. 177
Van Essen, D.C. 105, 121
Van Gogh 22
Van Norman, R.W. 227, 245
Vance, L.C. 67, 96
Vanecek, J. 256, 281
Vat Red 294
Vaughan, G.M. 256, 281
Vaughan, M.K. 256, 267, 277, 281
Venkataraman, K. 312
vermillion 146, 147, 351
vibrations 132–134, 136
vibrations, lattice 128
video signal 427
viewing conditions 358, 367
viewing distance 426
Vink, A.T. 431, 456
Vision Research 267, 281
vision, color 97–121
visual field 103
visualization 452
vitamin A 233
vitamin B_2 233
vitamin D 252
vitamin D biosynthesis 241, 242
vitamin D synthesis 253
vitamin synthesis 232
vitamins K_1, K_2 233
Volf, M.B. 167, 168
volume color 191
von Bezold, Wilhelm 21, 174
von Fraunhofer, Joseph 161
von Goethe, Johann Wolfgang 19, 20, 21, 25, 27, 174
von Helmholtz, Herman Ludwig Ferdinand 18, 20, 106, 110, 114, 174
von Korff, M. 269, 281
Vorhagen, F.W. 449, 456

Wagner, R.H. 242, 245
Walraven, J. 438, 456
Walworth, V.K. 384, 387
Wandell, B.A. 121
Ward, P. 236, 245

Warner, R.R. 238, 245, 246
warning colors 237 236
Waser, N.M. 234, 246
Wasserman, G.S. 120, 121
water 132–134
wavelength dominant 8
wavelength units, conversions 7
wavelengths 5–6
wavenumber 7
Waxler, M. 265, 267, 277
weatherability, *also see* color fastness 286
Wehr, T.A. 254, 257, 262–267, 269, 279–281
Weidner, V.R. 229, 244
weighting factors 46, 47, 48
Weinreb, R.N. 440, 456
Welch, E. 443, 456
Weleber, R.G. 256–259, 277
Welker, T. 431, 456
Weller, J.L. 256, 257, 279
Wells, M. 233, 246
West, B. 64, 96
Westphal, J. 218, 219
Wetterberg, L. 254, 256, 281
Wever, R.A. 264, 281
Weyer, M. 263, 265, 267, 276
white balance 431
white standard 75
white substrate 356
whites, perfect 409
Whitehead, A.N. 212, 219
whiteness 233, 369
– in animals 233, 234
– in flowers 234
Whitham, T.G. 235, 245
Wilkinson, G. 167, 168
Williams, T. 443, 456
Wilson, Ch. 81, 96

Wilson, K.M. 256, 281
Wilson, L.G. 263, 265, 267, 276
Winget, C.M. 264, 281
winter depression, *see* SAD
Witt 54, 95
Wolf, E. 167
Wong, S. 256, 267, 280
Wood, D.L. 133, 168
Wood, J. 207, 208
Wood, R.W. 167, 168
Wreschner, E.E. 198, 199, 206, 208
Wright, W.D. 234, 246, 422
Wurmfeld, S. xvi, 169, 174, 186, 188, 191, 194
Wurtman, R.J. 252, 256, 257, 277–279, 281
Wyszecki, G. 56, 90, 96, 120, 121, 422

yellow 356, 368
Yellowstone Park 224
Yerevanian, B.I. 264, 265, 267, 281
Yglesias, M. 261, 262, 270, 281
Yoshida, N. 207, 208
Young, M.A. 269, 278
Young, Thomas 28, 106, 110, 161, 173, 174, 194
Young's theory 111
Young–Helmholtz color theory 18, 106
Yuhas, B. 67, 91
Yule, J.A.C. 368, 384, 387

Zajonc, A. 27, 29, 30
Zellner, H. 230, 244
zinc chromate 306
zinc pigment 309
Zivin, G. 261, 262, 270, 281
Zollinger, H. 324, 344
Zolman, J.F. 271, 281
zone color theories 19
Zwinkels, J.C. 74, 96

Printed and bound by CPI Group (UK) Ltd, Croydon, CR0 4YY

08/05/2025

01864828-0001